DNA Viruses

METHODS IN MOLECULAR BIOLOGY™

John M. Walker, SERIES EDITOR

METHODS IN MOLECULAR BIOLOGY™

DNA Viruses

Methods and Protocols

Edited by

Paul M. Lieberman

The Wistar Institute, Philadelphia, PA

HUMANA PRESS ✳ TOTOWA, NEW JERSEY

Library of Congress Cataloging-in-Publication Data

DNA viruses : methods and protocols / edited by Paul M. Lieberman.
     p. ; cm. -- (Methods in molecular biology ; 292)
  Includes bibliographical references and index.
  ISBN 1-58829-353-X (hardcover : alk. paper)
  1. DNA viruses--Laboratory manuals.
  [DNLM: 1. DNA Viruses. 2. Genetic Techniques. 3. Virology--methods. QW165 D6288 2005] I. Lieberman, Paul M. II. Series: Methods in molecular biology (Clifton, N.J.) ; v. 292.
  QR394.5.D634 2005
  579.2'4--dc22
                                                                                      2004012488

Preface

The application of modern methods in molecular biology and biotechnology to the study of human, animal, and plant viruses continues to revitalize the age-old discipline of virology. Modern virology remains at the vanguard of contemporary biomedical research largely owing to the impact of viruses in human disease and pathogenesis, but also because of the utility of viruses as model systems for investigation of basic biological processes. *DNA Viruses: Methods and Protocols* describes innovative approaches to solving important problems in modern virology and also provides methodologies that can equally be applied to numerous other biological systems. Since virology, like cell biology, covers a vast expanse of methodological approaches, it is virtually impossible to cover all aspects of this dynamic field. The scope of the book is limited to DNA viruses, and it includes only a small sample of the many exciting methodological innovations of the last few years. This book does not include any specific applications to RNA viruses, but some of the methods describe techniques that have general applications to RNA viruses, as well as to cell biology.

In *DNA Viruses: Methods and Protocols* I have tried to include a sample of exciting advances in what I see as the major areas of DNA virology today. The methods presented here are representative of, but do not exhaust, the many important contributions to this field. I have divided the book into nine parts that include: viral detection, structure, entry, gene expression, replication, pathogenesis, complex cellular models, and recombinant genetics, with the addition of computational/systems approaches toward virology. Some of these divisions are arbitrary and have obvious overlaps. Nevertheless, I thought it useful to divide this volume into sections to emphasize the various methodological approaches as they are applied to important questions in virology.

Although *DNA Viruses: Methods and Protocols* attempts to cover numerous aspects of modern virology, it is apparent that many significant methodological advances have not been included. I ask those readers who would have preferred either a more focused or a more comprehensive volume to understand the book's constraints, and those authors who should have been asked to contribute to accept my apology for the oversight. Regardless of these obvious limits, I hope you find this book of interest and value in your experimental molecular biology pursuits.

Paul M. Lieberman

Contents

Contributors

J. S. AGUILAR • *Department of Molecular Biology and Biochemistry and Center for Virus Research, University of California, Irvine, Irvine, CA*

CONSTANDACHE ATANASIU • *The Wistar Institute, Philadelphia, PA*

ANTOINETTE A. T. P. BRINK • *Department of Pathology, VU Medical Centre, Amsterdam, The Netherlands*

THOMAS R. BROKER • *Department of Biochemistry & Molecular Genetics, University of Alabama at Birmingham, Birmingham, AL*

XULIN CHEN • *Department of Microbiology, School of Dental Medicine, University of Pennsylvania, Philadelphia, PA*

TOH HEAN CH'NG • *Department of Molecular Biology, Princeton University, Princeton, NJ*

LOUISE T. CHOW • *Department of Biochemistry & Molecular Genetics, University of Alabama at Birmingham, Birmingham, AL*

HENRI-JACQUES DELECLUSE • *CR-UK Institute for Cancer Studies, Department of Pathology, University of Birmingham, Birmingham, UK*

DIRK P. DITTMER • *Department of Microbiology and Immunology, The University of Oklahoma Health Sciences Center, Oklahoma City, OK*

DORJBAL DORJSUREN • *Laboratory of Antiviral Drug Mechanisms, SAIC-Frederick, National Cancer Institute, Frederick, MD*

LYNN WILLIAM ENQUIST • *Department of Molecular Biology, Princeton University, Princeton, NJ*

VITALY ERUKHIMOVITCH • *The Institute for Applied Biosciences, Ben-Gurion University of the Negev, Beer-Sheva, Israel*

FRAUKE FEHRMANN • *Department of Microbiology-Immunology, Feinberg School of Medicine, Northwestern University, Chicago, IL*

E. ALEXANDER FLOOD • *Department of Molecular Biology, Princeton University, Princeton, NJ*

LORI FRAPPIER • *Department of Medical Genetics and Microbiology, University of Toronto, Toronto, Ontario Canada*

PETER GHAZAL • *Scottish Centre for Genomic Technology and Informatics, University of Edinburgh Medical School, The Chancellor's Building, Little France Crescent, Edinburgh, UK*

JENNIFER L. GRAVEL • *Department of Primary Industries, Agency for Food and Fibre Sciences, St. Lucia, Queensland, Australia*

DONNA R. HOCHBERG • *Department of Pathology, Tufts University School of Medicine, Boston, MA*

REBECCA HINES-BOYKIN • *Department of Microbiology and Immunology, The University of Oklahoma Health Sciences Center, Oklahoma City, OK*

MAHMOUD HULEIHEL • *The Institute for Applied Biosciences, Ben-Gurion University of the Negev, Beer-Sheva, Israel*

MASATO IKEDA • *Department of Microbiology-Immunology, Northwestern University Feinberg School of Medicine, Chicago, IL*

HELLE LONE JENSEN • *Department of Pathology, Rigshospitalet, University Hospital of Copenhagen, Copenhagen, Denmark*

MING JIANG • *Yorkshire Cancer Research P53 Laboratory, Department of Biology, University of York, York, UK*

GARY L. JOHANNING • *Department of Veterinary Sciences, University of Texas M. D. Anderson Cancer Center, Bastrop, TX*

PRIYA KAPOOR • *Department of Medical Genetics and Microbiology, University of Toronto, Toronto, Ontario Canada*

JASON S. KNIGHT • *Department of Microbiology and the Abramson Comprehensive Cancer Center, University of Pennsylvania Medical School, Philadelphia, PA*

RIKA KOMAGOME • *Laboratory of Molecular and Cellular Pathology, Graduate School of Medicine, Hokkaido University; CREST, JST, Sapporo, Japan*

SZU-HAO KUNG • *Faculty of Medical Technology and Institute of Biotechnology in Medicine, National Yang-Ming University, Taipei, Taiwan, Republic of China*

LAIMONIS A. LAIMINS • *Department of Microbiology-Immunology, Feinberg School of Medicine, Northwestern University, Chicago, IL*

KE LAN • *Department of Microbiology and the Abramson Comprehensive Cancer Center, University of Pennsylvania Medical School, Philadelphia, Pennsylvania*

MANFRED LEE • *Division of Infectious Diseases, School of Public Health, University of California, Berkeley, CA*

DAVID A. LEIB • *Department of Ophthalmology and Visual Sciences; Department of Molecular Microbiology Washington University School of Medicine, St. Louis, MO*

PAUL M. LIEBERMAN • *The Wistar Institute, Philadelphia, PA*

LARISSA LEZINA • *The Wistar Institute, Philadelphia, PA*

KAI LIN • *Department of Biochemistry and Biophysics, School of Medicine University of Pennsylvania, Philadelphia, PA*

FENYONG LIU • *Division of Infectious Diseases, School of Public Health, University of California, Berkeley, CA*

RICHARD LONGNECKER • *Department of Microbiology-Immunology, Northwestern University Feinberg School of Medicine, Chicago, IL*

GARY D. LUKER • *Department of Radiology, University of Michigan Medical School, Ann Arbor, MI*

TIMOTHY J. MAHONY • *Department of Primary Industries, Agency for Food and Fibre Sciences, St. Lucia, Queensland, Australia*

ALEXANDER J. MALKIN • *BioSecurity and NanoSciences Laboratory, Department of Chemistry and Materials Science, Lawrence Livermore National Laboratory, Livermore CA; and Department of Molecular Biology and Biochemistry, University of California, Irvine, CA*

FIONA M. MCCARTHY • *Department of Primary Industries, Agency for Food and Fibre Sciences, St. Lucia, Queensland, Australia*

ALEXANDER MCPHERSON • *Department of Molecular Biology and Biochemistry, University of California, Irvine, CA*

MARISA P. MCSHANE • *Department of Microbiology and Immunology, Northwestern University Feinberg School of Medicine, Chicago, IL*

JAAP M. MIDDELDORP • *Department of Pathology, VU Medical Centre, Amsterdam, The Netherlands*

JO MILNER • *Yorkshire Cancer Research P53 Laboratory, Department of Biology, University of York, York, UK*

BERNHARD NEUHIERL • *GSF-National Research Centre for Environment and Health, Department of Gene Vectors, Munich, Germany*

BODIL NORRILD • *Institute of Molecular Pathology, University of Copenhagen, Copenhagen, Denmark*

JAMES PAPIN • *Department of Microbiology and Immunology, The University of Oklahoma Health Sciences Center, Oklahoma City, OK*

MARCO PLOMP • *Biosecurity and Nanosciences Laboratory, Department of Chemistry and Materials Science, Lawrence Livermore National Laboratory, Livermore, CA*

ROBERT P. RICCIARDI • *Departments of Microbiology and Biochemistry and Biophysics, University of Pennsylvania, Philadelphia, PA*

ERLE S. ROBERTSON • *Department of Microbiology and the Abramson Comprehensive Cancer Center, University of Pennsylvania School of Medicine, Philadelphia, PA*

CARMEN SAN MARTÍN • *The Wistar Institute, Philadelphia, PA; Biocomputing Department, Centro Nacional de Biotecnología (CSIC), Madrid, Spain*

MARTIN SAPP • *Institute for Medical Microbiology and Hygiene, University of Mainz, Mainz, Germany*

HIROFUMI SAWA • *Laboratory of Molecular and Cellular Pathology, Graduate School of Medicine, Hokkaido University; 21st Century COE Program for Zoonosis, CREST, JST, Sapporo, Japan*

NANCY M. SAWTELL • *Division of Infectious Diseases, Cincinnati Children's Hospital Medical Center, Cincinnati, OH*

SHIZUKO SEI • *Laboratory of Antiviral Drug Mechanisms, SAIC-Frederick, National Cancer Institute, Frederick, MD*

HANS-CHRISTOPH SELINKA • *Institute for Medical Microbiology and Hygiene, University of Mainz, Mainz, Germany*

ROBERT SHOEMAKER • *Screening Technology Branch, Developmental Therapeutics Program, National Cancer Institute, Frederick, MD*

YELENA SOUPRUN • *The Institute for Applied Biosciences, Ben-Gurion University of the Negev, Beer-Sheva, Israel*

SERVI J. C. STEVENS • *Department of Pathology, VU Medical Centre, Amsterdam, The Netherlands*

MARINA TALYSHINSKY • *The Institute for Applied Biosciences, Ben-Gurion University of the Negev, Beer-Sheva, Israel*

DAVID A. THORLEY-LAWSON • *Department of Pathology, Tufts University School of Medicine, Boston, MA*

BRIAN A. VAN TINE • *Department of Pathology, University of Alabama at Birmingham, Birmingham, AL,*

WOLFGANG VAHRSON • *Department of Microbiology and Immunology, The University of Oklahoma Health Sciences Center, Oklahoma City, OK*

SANDRA A. W. M. VERKUIJLEN • *Department of Pathology, VU Medical Centre, Amsterdam, The Netherlands*

SUBHASH C. VERMA • *Department of Microbiology and the Abramson Comprehensive Cancer Center, University of Pennsylvania School of Medicine, Philadelphia, PA*

EDWARD K. WAGNER • *Department of Molecular Biology and Biochemistry, Institute for Genomics and Bioinformatics, and Center for Virus Research, University of California, Irvine, CA*

FENG WANG-JOHANNING • *Department of Veterinary Sciences, University of Texas MD Anderson Cancer Center, Houston, TX*

PETER L. YOUNG • *Department of Primary Industries, Agency for Food and Fibre Sciences, St. Lucia, Queensland, Australia*

YAN YUAN • *Department of Microbiology, University of Pennsylvania School of Dental Medicine, Philadelphia, PA*

I

VIRAL DETECTION

1

Viral Detection

Feng Wang-Johanning and Gary L. Johanning

Summary

The focus of this chapter is the detection of DNA viruses. The emphasis is on amplification reactions that include reverse transcription-polymerase chain reaction (RT-PCR), PCR, real-time RT-PCR, and real-time PCR methods. Amplification of the E6 and E7 oncoproteins of HPV16 is described in detail, and primers and probes that can be used to amplify these oncogenes are described. Techniques to quantify these oncogenes in infected human tissue specimens are presented, and analysis of data resulting from real-time PCR detection of the E6 and E7 oncogenes is discussed. Other methods for viral nucleic acid detection, including nested PCR amplification, ligase chain reaction, and enzyme-linked immunosorbent assay (ELISA), are also briefly discussed.

Key Words: Viral detection; DNA viruses; cervical cancer; E6; E7; real-time PCR; RT-PCR.

1. Introduction

There are several methods for detecting DNA viruses. One widespread method that has been used in several studies is nested polymerase chain reaction (PCR) analysis of viral DNA. Nested PCR amplification has been used to detect the Epstein-Barr virus latent membrane protein-1 *(LMP-1)* gene in clinical infections *(1)*. Similarly, human cytomegalovirus (CMV) DNA was detected in human plasma by nested PCR amplification *(2)*. Other techniques that have been used for viral detection include ligase chain reaction and enzyme-linked immunosorbent assay *(3)*.

Another common method for detection of DNA viruses is signal amplification-based tests such as the hybrid capture assay (HC; Digene, Gaithersburg, MD). HC assays have been used to detect human papillomavirus (HPV), cytomegalovirus (CMV), and hepatitis B virus (HBV) DNA. The HPV HC 2 assay using probe B is capable of detecting 13 carcinogenic types of HPV that include oncogenic HPV16 and HPV18. The potential predictive value of this

From: *Methods in Molecular Biology, vol. 292: DNA Viruses: Methods and Protocols*
Edited by: P. M. Lieberman © Humana Press Inc., Totowa, NJ

test was suggested by the observation that about one of every six women with a negative papanicolaou (Pap) test and a positive HPV HC 2 test will develop an abnormal Pap smear within approx 5 yr *(4)*. A comparison of PCR and HC assays for the detection of CMV DNA in different blood compartments showed that there was a significant correlation between the results of HC with whole blood and PCR with peripheral blood lymphocytes *(5)*. The sensitivity of the PCR and HC assays was comparable when whole blood was assayed.

One method of detection that has assumed greater importance in recent years is real-time PCR detection of viral DNA or RNA. The power of the real-time PCR method lies in the ability to quantitate viral gene load and expression, using relatively straightforward manipulations of biological samples. The major factor limiting more extensive use of real-time PCR methodology is probably the prohibitive cost of the instrumentation. However, as prices fall owing to technological advances and improved economies of scale, this limitation will probably be less significant over time. In recent studies, real-time PCR has been used to assess Epstein Barr viral DNA load *(6)* and to detect viral DNA in both high- and low-titered mock specimens of vaccinia, herpes simplex, and varicella-zoster viruses *(7)*. The HC assay described above was compared with real-time PCR for detection of HBV DNA in serum samples from individuals positive for markers of HBV infection. They found that the prevalence rates for HBV DNA in the HBV-positive serum samples were 95 and 56% for real-time PCR and HC, respectively, and that there was a more than 500-fold increase in sensitivity with the PCR assay compared with the standard HC test *(8)*. We describe here the real-time PCR method for detection of DNA viruses, using as an example the E6 and E7 oncoproteins of HPV16 *(9)*.

2. Materials

1. DNeasy Tissue and RNeasy kits (Qiagen, Valencia, CA).
2. Oligonucleotide primers and probes (Primer Express software package; Perkin Elmer, Wellesley, MA) specific for HPV16 E6 or E7 cDNA sequences (GenBank accession no. NC001526.1).
3. Conventional RT-PCR or PCR primers specific for HPV16 E6 or E7 open reading frames (GenBank accession no. NC001526.1):
   a. E6 forward primer (nt 83–103): CTCT<u>GAATTC</u>GCCACCATGCACCAAAA-GAGAACTGCA* (*Eco*RI site underlined).
   b. E6 reverse primer (nt 575–555): CC<u>CTCGAG</u>GTATCTCCATGCATGATTA-CA (*Xho*I site underlined).
   c. E7 forward primer (nt 562–582): CTCT<u>GAATTC</u>GCCACCATGCATGGA-GATACACCTACA (*Eco*RI site underlined).
   d. E7 reverse primer (nt 874–853): CC<u>CTCGAG</u>GATCAGCCATGGTACAT-TATGG (*Xho*I site underlined).

*All primer sequences are written 5′-3′ and were synthesized by Invitrogen Life Technologies (Carlsbad, CA).

4. TaqMan probe and primers specific for HPV16 E6 and E7 open reading frames (GenBank accession no. NC001526.1):
   a. HPV16 E6 (nt 99–178):
      Forward primer: CTGCAATGTTTCAGGACCCA.
      Reverse primer: TCATGTATAGTTGTTTGCAGCTCTGT.
      Probe: FAM-AGGAGCGACCCGGAAAGTTACCACAGTT-BHQ.
   b. HPV16 E7 (nt 739–816):
      Forward primer: AAGTGTGACTCTACGCTTCGGTT.
      Reverse primer: GCCCATTAACAGGTCTTCCAAA.
      Probe: FAM-TGCGTACAAAGCACACACGTAGACATTCGTA-BHQ.
5. Oligonucleotide primers and probes specific for *Homo sapiens* ribosomal protein S9 (GenBank accession no. BC000802.1-BC000802) and *Homo sapiens* β-actin (GenBank accession no. BC004251.1- BC004251) as endogenous controls for quantitation of RNA and DNA.
   a. S9 (nt 419–504):
      Forward primer: ATCCGCCAGCGCCATA.
      Reverse primer: TCAATGTGCTTCTGGGAATCC.
      Probe: FAM-AGCAGGTGGTGAACATCCCGTCCTT-TAMRA.
   b. β-actin (nt 537–831):
      Forward primer: TCACCCACACTGTGCCCATCTACGA.
      Reverse primer: CAGCGGAACCGCTCATTGCCAATGG.
      Probe: FAM-ATGCCCTCCCCCATGCCATC-TAMRA.
6. TA cloning vector (Invitrogen).
7. *E. coli* strains DH5α, BL21, and M15.
8. Restriction enzymes, AmpliTaq DNA polymerase (Applied Biosystems, Foster City, CA), DNase I (RNase-free, Ambion, Austin, TX), and T4 DNA ligase and 10X ligase reaction buffer (New England Biolabs, Beverly, MA).
9. Agarose and polyacrylamide gel electrophoresis equipment.
10. LB medium and ampicillin.
11. Ready-To-Go You-Prime First-Strand cDNA synthesis beads (Amersham Biosciences, Piscataway, NJ).
12. pd(N)$_6$ random primer (Amersham Biosciences).
13. QIAquick gel purification kit (Qiagen).
14. SOC medium.
15. QIAquick gel purification kit (Qiagen).
16. MAXIscript kit for in vitro transcription (Ambion).
17. ABI PRISM 7700 sequence detector (Applied Biosystems).
18. 1X TaqMan buffer (Perkin Elmer), dNTPs (0.3 m*M* each), AmpliTaq Gold (Applied Biosystems), RNase inhibitor, 2% glycerol, murine leukemia virus (MuLV) reverse transcriptase.
19. Genomic DNA from a healthy human woman (Promega, Madison, WI).

3. Methods

3.1. Sample Collection

Exfoliated cervical cells from human subjects were collected with a cervical brush and placed in a ThinPrep vial containing PreservCyt preservative solution (Cytyc, Boxborough, MA) (*see* **Note 1**). Human cervical cancer cell lines including HPV-positive (CaSki, SiHa, and HeLa) and HPV-negative (C33A) cell lines were also used in these studies.

3.2. Conventional RT-PCR and PCR Amplification

Both conventional and real-time RT-PCR and PCR were used for E6 and E7 detection. For conventional detection by PCR, DNA or RNA was isolated from the cells and then reverse-transcribed prior to PCR for RNA analysis or directly amplified for DNA analysis.

3.2.1. RNA and DNA Isolation

Extract DNA (DNeasy Tissue Kit) or RNA (RNeasy Kit) from cervical cancer cells by following the kit manufacturer's directions.

3.2.2. RT-PCR

1. Determine the concentration of DNA or RNA by spectrometry.
2. Add DNase I (RNase-free; 1–2 U of DNase I per μg of DNA sample; Ambion.) to RNA samples to remove contaminating DNA (1 μL/10 μg of RNA) and incubate at 37°C for 30 min, followed by 75°C for 5 min to destroy residual DNase activity.
3. For cDNA synthesis.
   a. Incubate isolated RNA at 65°C for 10 min, and then incubate on ice for 2 min prior to reverse transcription.
   b. Reverse-transcribe total RNA using cDNA synthesis beads per the manufacturer's directions.
   c. RNA is reverse-transcribed at 37°C for 1 h, in a volume of 33 μL containing 10 μg of RNA and 1 μL pd(N)$_6$ random primer.
4. Amplify the reverse-transcribed samples in a volume of 50 μL containing 3.3 μL of the first-strand cDNA synthesis mixture (corresponding to 1 μg of input RNA), 5 μL of 10 X PCR buffer (Qiagen), 0.5 μL of AmpliTaq Gold (2.5 U), and various sense and antisense oligonucleotide primer pairs at 50 pmol each (*see* **Note 2**).
5. Amplify 1 μg of RNA from each sample without reverse transcriptase, to control for genomic DNA contamination.
6. Use SiHa or CaSki cell RNA (HPV16-positive cervical cell lines) as a positive control, and an equal amount of C33A or HeLa cell RNA (HPV16-negative cervical cell lines) as a negative control.
7. Analyze each sample in parallel with β-actin sense and antisense primer pairs to control for sample-to-sample variations in PCR amplification efficiency.

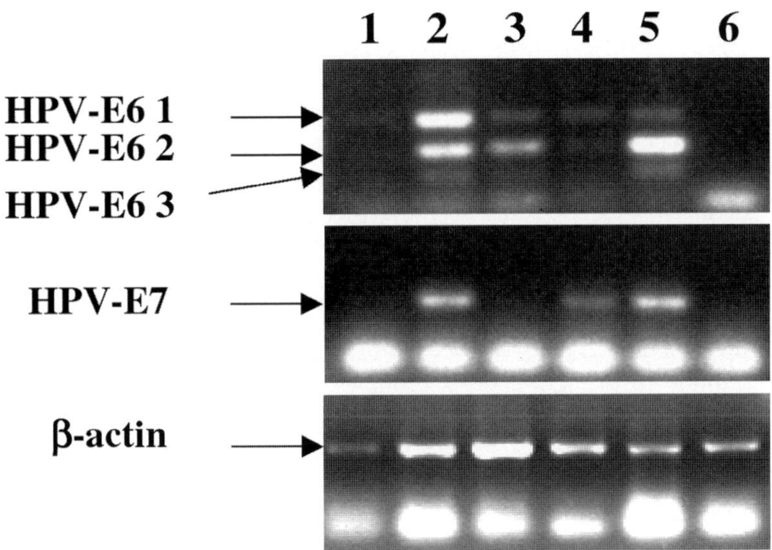

Fig. 1. Detection of human papillomavirus (HPV)16 E6/E7 mRNAs by RT-PCR. Atypical squamous cells of undetermined significance (lane 1), cervical cancers (lanes 2 and 5), high-grade squamous intraepithelial lesion (lane 3), low-grade squamous intraepithelial lesion (lane 4), and normal cervical tissue (lane 6).

8. Denature PCR reactions at 94°C for 3 min, followed by 30 cycles of denaturation (94°C for 1 min), annealing (55°C for 1 min), and extension (72°C for 1 min).
9. Analyze amplified products on a 2% agarose gel with a 100-bp marker in one lane (*see* **Note 3**).
10. Gel-purify the PCR products with a QIAquick gel purification kit (Qiagen). A sample gel of products obtained after electrophoresis of cervical tissues subjected to RT-PCR is presented in **Fig. 1** (*see* **Note 4**).

3.2.3. PCR Amplification

1. Use approx 200 ng of DNA to amplify E6 or E7 DNA directly.
2. Amplify E6 or E7 open reading frames with the same E6 or E7 forward and reverse primers that were used for RT-PCR amplification. Follow the same protocol that was used for RT-PCR, except do not treat samples with DNase or reverse transcriptase.
3. Carry out amplification reactions in a volume of 50 µL containing 200 ng of DNA, 5 µL of 10X PCR buffer (Qiagen), 0.5 µL of AmpliTaq DNA polymerase (2.5 U), and various sense and antisense oligonucleotide primer pairs at 50 pmol each, plus 0.3 m*M* each of dNTPs.
4. Analyze amplified products on a 2% agarose gel with a 100-bp marker in one lane.

3.3. Cloning and Sequencing of HPV16 E6 and E7 cDNAs

3.3.1. Ligation of HPV16 E6 and E7 Into TA Cloning Vector

1. Clone PCR products into a pCR-II vector (Invitrogen): In a total of 10 μL reaction mixture, include 5 μL of E6 or E7 DNA obtained after gel purification of RT-PCR products, 2 μL of pCR-II vector, 1 U of T4 ligase, and 1 μL of 10X ligase buffer.
2. Ligate the mixture overnight at 14°C.
3. Transform DNA (2 μL of ligation mixture) into *E. coli* competent cells (50 μL) by adding DNA to the cells and incubating the cells on ice for 10 min. Place the cells in a water bath (42°C) for 45 s to heat-shock the cells, and then put tubes back on ice for 2 min. Add 500 μL of SOC medium to the mixture, and incubate with shaking for 1 h at 37°C.
4. Plate the mixture (100 μL) onto a Petri dish containing LB medium plus ampicillin (100 μg/mL) with isopropyl thiogalactose and X-galactose (40 μg/mL each) for color selection, and incubate overnight at 37°C.

3.3.2. Preparation of HPV16 E6 and E7 Plasmids

1. Pick several white colonies and grow a single colony in 2 mL LB medium plus 100 μg/mL ampicillin in an incubator with shaking overnight at 37°C.
2. Extract plasmid DNA using a miniprep DNA isolation kit (Qiagen).
3. Excise the insert by digestion with *Eco*RI. Digest a total of 20 μL of mixture that includes 1 μg plasmid obtained from the miniprep, 2 μL buffer (10X buffer for *Eco*RI enzyme) and 0.5 μL *Eco*RI by incubation for 1 h at 37°C.
4. Check 10 μL of the *Eco*RI digest by gel electrophoresis on a 2% agarose to verify that the E6 or E7 inserts are the correct size. The E6 insert will contain 492 bp, and the E7 insert will contain 312 bp.
5. Make frozen stocks of the bacteria containing correctly sized inserts of E6 or E7 oncogene by suspending in glycerol (final glycerol concentration of 15%) and storing at –70°C.
6. Grow the stock culture with the correctly sized insert in LB medium (1 L) plus ampicillin overnight in an incubator at 37°C, with shaking.
7. Extract DNA from the culture using a maxiprep DNA plasmid isolation kit (Qiagen).
8. Digest the DNA again to confirm that plasmids with the expected sizes of HPV16 E6 or E7 oncogene inserts are present. Verify the gene sequences by DNA sequencing using universal primer or M13 reverse primer.

3.4. Quantitation of HPV16 E6 and E7 mRNA Using Real-Time RT-PCR

Housekeeping gene (ribosomal protein S9 for RT-PCR and β-actin for PCR) probes and primers are used to amplify known amounts of RNA and DNA, to construct standard curves. The same housekeeping gene probe and primers are used to determine the RNA or DNA concentrations of unknown samples. An analogous process is used to determine HPV16 E6 or E7 copy numbers, using an E6 or E7 standard curve. Samples are analyzed in triplicate.

3.4.1. In Vitro Transcription

1. Digest E6 or E7 cDNA in the TA cloning vector with a restriction enzyme to linearize the plasmid.
2. Synthesize HPV16 E6 or E7 cRNA templates using a MAXIscript in vitro transcription kit (Ambion), starting with 1 μg of linearized E6 or E7 plasmid.
3. Quantitate the resulting E6 or E7 cRNA by measuring the absorbance at 260 nm (*see* **Note 5**).
4. Calculate the E6 or E7 cRNA copy number according to the following formula: copy number = (mass)($6.023 \times 10_{17}$/molecular weight)(Nuc)(length), where the mass is in μg, the molecular weight = 339.8 for RNA, Nuc = 1 for single-strand cRNA and 2 for double- strand cRNA, and length = the length of the target in bp.

3.4.2. Real-Time RT-PCR

1. Carry out one-step real-time RT-PCR or PCR using an ABI PRISM 7700 sequence detector.
2. Perform the reverse transcription and amplification reaction in 25 μL final volume containing RNA plus 1X TaqMan buffer (Perkin Elmer), dNTPs (0.3 m*M* each), 0.625 U of AmpliTaq Gold, RNase inhibitor, 2% glycerol, and 0.625 U of MuLV reverse transcriptase (*see* **Note 6**).
3. Reverse transcription and thermal cycling conditions are 30 min at 48°C followed by 10 min at 95°C and 40 cycles of 15 s at 95°C and 1 min at 60°C.
4. Prior to amplification of samples, first optimize the concentrations of E6 and E7 forward and reverse primers, and then optimize the concentrations of E6 and E7 probes (*see* **Note 7**).
5. Use PCR premixes containing all reagents except for total RNA as a no-template control, and use ribosomal protein S9 primers and probe as an internal control.
6. Generate a standard curve with known amounts of cDNA from reverse-transcribed CaSki cell RNA (10.0, 5.0, 1.0, and 0.5 ng total RNA) and optimized concentrations of S9 primers and probe (*see* **Note 8**).
7. Determine the RNA concentration in the unknown sample by extrapolating the cycle threshold (C_t) value of the unknown to the copy number value obtained from the standard curve (*see* **Note 9**).
8. Generate an absolute standard curve with known amounts of E6 or E7 cRNA molecules (1×10^8, 1×10^6, 1×10^4, 1×10^2, and 0) and E6 or E7 probe and primers. A narrower range of E6 or E7 cRNA molecule concentrations can be used if the sample copy number can be expected to fall within this narrower range.
9. Amplify the RNA extracted from unknown samples in separate reactions using the same E6 or E7 primers and probe.
10. Determine the copy number of E6 or E7 mRNA of the samples by linear extrapolation of the C_t values using the equation of the line obtained from the absolute E6 or E7 standard curve (*see* **Note 10**).
11. Divide the copy numbers of E6 or E7 by the relative amounts of S9 (*see* **Notes 11–13**).

3.4.3. Real-Time PCR

1. For real time PCR, β-actin primers and probe, rather than S9, are used to amplify known cellular DNA concentrations to construct a standard curve. The same β-actin primers and probe are used to determine the DNA concentration of unknown samples, and the C_t value of the unknown will be compared with the C_t value obtained from the standard curve to calculate the concentration of unknown samples.
2. The amplification reaction is performed in 25 μL final vol containing 1X Universal PCR Master Mix (Applied Biosystems) and the optimized amounts of E6 or E7 primers and probes.
3. Thermal cycling conditions are 1 min at 94°C followed by 40 cycles of 30 s at 94°C, 30 s at 55°C, and 30 s at 72°C.
4. Amplify β-actin to generate a standard curve using known amounts of genomic DNA from a healthy human woman (200, 20, 2, 0.2, and 0 ng).
5. Use E6 or E7 probes and primers to amplify known copy numbers of CaSki cell DNA (1×10^8, 1×10^6, 1×10^4, 1×10^2, and 0; or another more appropriate range of concentrations).
6. Amplify the DNA extracted from unknown samples in separate reactions using the same E6 or E7 primers and probe.
7. Determine the copy number of E6 or E7 DNA of the samples by linear extrapolation of the C_t values using the equation of the line obtained from the absolute E6 or E7 standard curve. These values are then divided by the relative amounts of β-actin (*see* **Notes 11–13**).

4. Notes

1. The cells in PreservCyt were stored at room temperature (approx 22°C). Since only a small amount of cells are used for ThinPrep cytology, the residual cells stored in PreservCyt can be a valuable resource for additional laboratory investigations, including viral DNA detection. A recent study reported that there was some time-dependent DNA degradation (as assessed by PCR amplification of β-globin DNA) during a 5-yr period of temperature-controlled storage of PreservCyt specimens *(10)*. We find that HPV DNA and transcripts can be quantitated within 3–4 mo of collection, using real-time PCR and RT-PCR *(9)*. It is possible that the short primers employed in real-time PCR or RT-PCR enable partially degraded DNA or RNA to be more readily amplified.
2. No dNTP is required because cDNA synthesis beads contain enough dNTP for PCR amplification.
3. If a sample has a β-actin band (600 bp) and an E6 or E7 band, but none of these bands is apparent from the same RNA sample without reverse transcription, the PreservCyt sample is HPV16-positive. If a sample has a β-actin band, but no E6 or E7 band, and no band from the same RNA sample without reverse transcription, the PreservCyt sample is HPV16-negative. If a sample has no β-actin band, the sample does not have enough cDNA for PCR amplification.

Table 1
Calculation of HPV16 E6 and E7 Copy Numbers in Cervical Cancer Cell Lines

E6 C_t	No. of E6 molecules	E7 C_t	No. of E7 molecules	S9 C_t	No. of S9 molecules
A. Generation of standard curve					
18	1.00E + 06	17.27	1.00E + 06	18.21	1.00E + 01
21.55	1.00E + 05	20.67	1.00E + 05	19.07	5.00E + 00
25.46	1.00E + 04	20.76	1.00E + 04	20.09	1.00E + 00
28.01	1.00E + 03	27.05	1.00E + 03	21.49	5.00E-01
31.52	NTC[a]	30.94	NTC	40	NTC
31.09	NTC	31.1	NTC	40	NTC
31.11	NTC	31.49	NTC	40	NTC

Cell line	E6 C_t	Quantity[b]	E7 C_t	Quantity	S9 C_t	Quantity	E6 copy no.[c]	E7 copy no.[c]
B. Analysis of samples								
C33A	31.12	1.50E + 02	31.34	1.40E + 01	22.13	3.70E + 01	4.05E + 00	3.78E-01
CaSki	21.02	1.40E + 05	21.07	4.20E + 04	26.15	3.40E-03	4.12E + 07	1.24E + 07
HeLa	30.72	2.00E + 02	31.15	1.60E + 01	22.08	2.10E-01	9.52E + 02	7.62E + 01
SiHa	20.48	2.10E + 05	20.67	5.80E + 04	23.72	4.00E-02	5.25E + 06	1.45E + 06

[a] No-template control.
[b] Value obtained after extrapolating C_t to "quantity" axis in absolute standard curve.
[c] Value obtained after dividing E6 quantity or E7 quantity by S9 quantity.

4. E6 mRNA is able to undergo splicing to full-length transcripts and two alternative smaller spliced transcripts in cervical cancer cells and tissues, whereas E7 is present as a single band upon polyacrylamide gel electrophoresis.

5. RNA concentration is measured by spectrophotometry at 260 nm, and the purity is determined by the A_{260} to A_{280} ratio. However, the spectrometric reading is not always accurate. To verify the absorbance readings, serial dilutions of commercially available RNA of known concentration are loaded on an agarose gel side by side with RNA extracted from cells, to estimate the concentration.

6. The reverse transcription and amplification reactions are performed in the same tube. All RT-PCR reactions were performed in optical reaction tubes (Applied Biosystems) designed for the ABI PRISM 7700 sequence detector system.

7. To optimize the primer and probe concentrations, arbitrarily select a probe concentration, and then amplify a sample with different concentrations of primer pairs. The concentration of primer pairs giving the lowest C_t value is then used to determine the optimal concentration of probe, by varying probe concentrations. The primer and probe concentrations that give the lowest C_t values are used for the remainder of the experiments. The optimized concentrations of E6 and E7 forward

and reverse primers and probe are typically 100 nM for the forward primers, 100 nM for the reverse primers, and 150 nM for the probe.

8. Typical optimized concentrations of S9 forward, reverse, and probe concentrations are 200, 200, and 100 nM, respectively.
9. S9 amounts are quantitated by linear extrapolation from the C_t values of the unknown samples using the equation of the line obtained from the S9 standard curve.
10. The C_t values obtained from the unknown samples are compared with C_t values obtained from the known copy numbers of E6 or E7 cRNA, and copy numbers for the unknowns are determined by extrapolation.
11. Because the concentration of nucleic acid in each sample will be slightly different, even though the initial concentration is approximately the same (as assessed by spectrophotometry), it is necessary to normalize the concentration of each sample by dividing the E6 or E7 copy number by the S9 concentration.
12. In each experiment, SiHa cell RNA is used as a positive control, and an equal amount of C33A or HeLa cell RNA is used as a negative control.
13. **Table 1** depicts the calculation of E6 and E7 copy numbers in cervical cancer cell lines. In part A, the C_t values for various known numbers of E6 or E7 molecules are shown. These data are used to generate a standard curve. In part B, the C_t values for various cervical cancer cell lines are extrapolated on the standard curve to give a numerical quantity of molecules. These quantities are normalized to the quantity of S9 molecules by dividing the E6 or E7 quantity by the S9 quantity. The resulting value is the E6 or E7 copy number.

References

1. Walling, D. M., Brown, A. L., Etienne, W., Keitel, W. A., and Ling, P. D. (2003) Multiple Epstein-Barr virus infections in healthy individuals. *J. Virol.* **77,** 6546–6550.
2. Preiser, W., Brink, N. S., Ayliffe, U., et al. (2003) Development and clinical application of a fully controlled quantitative PCR assay for cell-free cytomegalovirus in human plasma. *J. Clin. Virol.* **26,** 49–59.
3. Castle, P. E., Escoffery, C., Schachter, J., et al. (2003) *Chlamydia trachomatis,* herpes simplex virus 2, and human T-cell lymphotrophic virus type 1 are not associated with grade of cervical neoplasia in Jamaican colposcopy patients. *Sex. Transm. Dis.* **30,** 575–580.
4. Castle, P. E., Wacholder, S., Sherman, M. E., et al. (2002) Absolute risk of a subsequent abnormal Pap among oncogenic human papillomavirus DNA-positive, cytologically negative women. *Cancer* **95,** 2145–2151.
5. Weinberg, A., Schissel, D., and Giller, R. (2002) Molecular methods for cytomegalovirus surveillance in bone marrow transplant recipients. *J. Clin. Microbiol.* **40,** 4203–4206.
6. Merlino, C., Cavallo, R., Bergallo, M., et al. (2003) Epstein Barr viral load monitoring by quantitative PCR in renal transplant patients. *New Microbiol.* **26,** 141–149.
7. Espy, M. J., Uhl, J. R., Sloan, L. M., Rosenblatt, J. E., Cockerill, F. R. 3rd, and Smith, T. F. (2002) Detection of vaccinia virus, herpes simplex virus, varicella-

zoster virus, and *Bacillus anthracis* DNA by LightCycler polymerase chain reaction after autoclaving: implications for biosafety of bioterrorism agents. *Mayo Clin. Proc.* **77,** 624–628.

8. Ho, S. K., Yam, W. C., Leung, E. T., et al. (2003) Rapid quantification of hepatitis B virus DNA by real-time PCR using fluorescent hybridization probes. *J. Med. Microbiol.* **52,** 397–402.

9. Wang-Johanning, F., Lu, D. W., Wang, Y., Johnson, M. R., and Johanning, G. L. (2002) Quantitation of human papillomavirus 16 E6 and E7 DNA and RNA in residual material from ThinPrep Papanicolaou tests using real-time polymerase chain reaction analysis. *Cancer* **94,** 2199–2210.

10. Castle, P. E., Solomon, D., Hildesheim, A., et al. (2003) Stability of archived liquid-based cervical cytologic specimens. *Cancer* **99,** 89–96.

2

Quantitative Detection of Epstein-Barr Virus DNA in Clinical Specimens by Rapid Real-Time PCR Targeting a Highly Conserved Region of EBNA-1

Servi J. C. Stevens, Sandra A. W. M. Verkuijlen, and Jaap M. Middeldorp

Summary

Here we describe a LightCycler-based real-time PCR for quantitative detection of EBV DNA in clinical samples such as unfractionated whole blood, serum, or plasma. This assay is based on amplification of a highly conserved 213-bp region of the *EBNA-1* gene, a single-copy gene of EBV required for maintenance of the EBV genome within the infected host cell. For real-time detection of amplicons, two internal hybridization probes are added, labeled with the fluoregenic dyes fluorescein and LCRed640, respectively. Simultaneous hybridization of these probes to the amplification products brings them in close proximity. Subsequent excitation of the fluorescein label by filtered excitation light from a light source in the LightCycler device will lead to fluorescence energy transfer (FRET) from the fluorescein label to the LCRed640 label. The light emitted from the LCRed640 label is then measured and correlates to the amount of product generated. The cycle at which the fluorescence exceeds the background is designated the *threshold cycle*. By comparing the threshold cycle of a clinical specimen with those of standard curve samples, the amount of EBV DNA in clinical samples can be determined. This real-time PCR approach is extremely rapid owing to efficient heat conduction by using glass capillaries, small reaction volumes, and air as heating medium. The "closed-tube" system eliminates the risk of PCR contamination by product carryover and also the need for post-PCR detection.

Key Words: Epstein-Barr virus; herpesviruses; DNA quantification; real-time PCR; viral load; EBNA-1; LightCycler.

1. Introduction

Epstein-Barr virus (EBV) is a human γ-herpesvirus infecting more than 90% of the population worldwide. In adolescents, EBV is the causative agent of infectious mononucleosis. Furthermore, EBV is associated with a still increas-

From: *Methods in Molecular Biology, vol. 292: DNA Viruses: Methods and Protocols*
Edited by: P. M. Lieberman © Humana Press Inc., Totowa, NJ

ing number of malignant proliferative disorders of both lymphoid and epithelial origin, including Burkitt's and Hodgkin's lymphoma, B- and T-cell non-Hodgkin's lymphoma, and nasopharyngeal and gastric carcinoma *(1,2)*. In immunocompromised individuals such as transplant recipients and AIDS patients, EBV is the driving force behind lymphoproliferative disorders, originating from infected B cells, which initially are benign and polyclonal but may progress to malignant lymphoma if left untreated *(1)*.

In healthy carriers of EBV, the number of EBV-infected B cells remains stable throughout life, reflecting the tightly controlled balance between EBV-driven B-cell proliferation and the host immune response. This is also characterized by persistent high levels of anti-EBV T cells and serum antibodies to latent and lytic EBV antigens *(3,4)*. Disturbance of this balance, e.g., by iatrogenic or natural immunosuppression, leads to increased numbers of EBV-infected B cells and EBV DNA loads in blood (reviewed in **ref. 5**). Consequently, in defined high-risk populations such as solid organ and hematopoetic stem cell transplant recipients, monitoring of EBV DNA load in peripheral blood is a suitable and widely used diagnostic tool for predicting posttransplant lymphoproliferative disease (PTLD) and guidance of pre-emptive therapeutic intervention *(5)*. Introduction of EBV DNA load monitoring in the routine diagnostic patient care of most transplant centers in the late 1990s has greatly increased transplant success and decreased EBV-related morbidity and mortality in transplant settings.

In immunocompetent patients with EBV-linked malignancies, such as nasopharyngeal carcinoma (NPC), monitoring of EBV DNA loads in peripheral blood may be useful in predicting the efficacy of therapeutic interventions *(6–8)*.

Although at present EBV DNA load monitoring is widely applied, basic interlaboratory standardization of clinical specimen type, polymerase chain reaction (PCR) technique, unit of measurement and clinically relevant cutoff values has not yet been established. Thus, a variety of clinical specimens are being used for determination of EBV DNA loads in the circulation, including whole blood, plasma, serum, and isolated leukocyte cell fractions (reviewed in **ref. 5;** *see* also **Note 5**). Concerning PCR target sequences, the *Bam*HI W-repeat region, which is often used for sensitive qualitative detection of the virus, is unsuited for quantification purposes because the number of *Bam*HI W repeats differs between clinical isolates of EBV *(9)*. Consequently most, but not all, studies use a single-copy EBV gene target in PCR. However, for some of these genes virtually nothing is known about nucleotide polymorphism in clinical EBV isolates. Furthermore, variation in amplicon size may influence quantification owing to differences in amplification efficiency. Real-time PCR is currently the most widely used technique for EBV load determination, a technique that depends on the use of external standard dilution series. It is proposed that the Namalwa cell line, which stably contains two integrated copies of the EBV genome, is most suited for standardization *(9)*.

This chapter describes a reproducible LightCycler-based real-time PCR assay for quantitative detection of EBV DNA with an analytical sensitivity of 10 copies of viral DNA or <1 EBV-positive lymphoblastoid cell line in whole-blood DNA background *(10)*. This method allows rapid quantification of the EBV DNA load in clinical specimens owing to use of small reaction volumes in thin borosilicate glass capillaries for efficient heat conduction and air for heating and cooling *(11)*. Using this assay, EBV DNA load can be determined in less than 2.5 h including DNA isolation, LightCycler reaction setup, EBV DNA amplification, and data analysis. The closed-tube format of real-time PCR strongly reduces the risk of PCR contamination by product carryover. Real-time quantification of PCR products by double-probe hybridization provides high specificity, and fluorescence detection eliminates the need for post-PCR detection procedures and allows quantification in the log-linear phase of amplification over a wide dynamic range. These features strongly favor the use of real-time PCR-based methods in routine molecular diagnostic settings over conventional semiquantitative, limiting dilution or quantitative competitive PCR methods.

The EBV real-time PCR described in this chapter is based on amplification of a 213-bp region of Epstein-Barr nuclear antigen-1 *(EBNA-1)*, a single-copy gene of EBV. *EBNA-1* protein is expressed in all EBV-carrying cells and is essential for maintenance of the viral genome and establishment of latency by anchoring the viral episome to the host chromosome *(12)*. Mutational hot spots within *EBNA-1* have been mapped extensively in clinical isolates of EBV *(13–15)*, permitting primer selection in a highly conserved region *(16)*. This *EBNA-1* region of the EBV genome allowed reliable EBV DNA load quantification in samples from patients with various EBV-associated diseases worldwide *(16,17)*. To increase assay specificity and allow quantification, two internal oligonucleotide hybridization probes, labeled with fluorescein and LCRed640, respectively *(10)*, are added to the PCR. During PCR amplicon formation, simultaneous annealing of the two probes to the EBNA-1 PCR product brings them into close proximity. Excitation of the fluorescein label by filtered excitation light from a light source in the LightCycler device will lead to fluorescence energy transfer from the fluorescein label to the LCRed640 label. The light emitted from this LCRed640 label correlates with the amount of PCR product generated. The cycle at which the fluorescence exceeds the background is designated the *threshold cycle* (C_t). By relating the C_t of a clinical specimen to a defined series of C_t values in a standard curve, the initial amount of EBV DNA in the specimen can be determined.

2. Materials

1. LightCycler device and computer with LightCycler software (Roche Diagnostics, Basel, Switzerland).
2. LightCycler accessories sample carrousel, centrifuge, capillary adapters and capillary cooling block and Capillaries and stoppers.
3. LightCycler-FastStart DNA Master Hybridization Probes kit containing 10X stock LC Fast Start Reaction Mix Hybridization Probes, LC Fast Start Enzyme, $MgCl_2$, and water (Roche Diagnostics).
4. Forward primer: EBV F QP1L (5′-gccggtgtgttcgtatatgg-3′) (*see* **Note 1**).
5. Reverse primer: EBV R QP2 L (5′-caaaacctcagcaaatatatgag-3′) (*see* **Note 1**).
6. Hybridization probe 1: *EBNA-1* FLN (5′-tctcccctttggaatggccccctg-fluorescein) (*see* **Note 2**).
7. Hybridization probe 2: *EBNA-1* LCN (5′-LCRed640-acccggcccacaacctg-3′) (*see* **Note 2**).
8. For real-time PCR standard curve:
   EBV-positive Burkitt's lymphoma cell line Namalwa (American Type Culture Collection [ATCC] CRL-1432; *see* **Note 3**) or purified and spectrophotometrically quantified plasmid DNA containing the *EBNA-1* gene, e.g., pBR322 containing the *Bam*HI-K fragment of the EBV genome *(18)*; (*see* **Note 4**).
9. 0.1 *M* HCl.

3. Methods

1. Isolate DNA from the clinical specimen of choice (*see* **Notes 5** and **6**).
2. Prepare the FastStart DNA Master Hybridization Probes mix stock solution (containing enzyme and buffer) by adding 60 µL of the LC Fast Start Reaction Mix Hybridization Probes (tube 1b from the kit mentioned in **Subheading 2., item 3**) to the LC Fast Start Enzyme (tube 1a; *see* **Note 7**). Mix gently by pipeting up and down. Do not vortex.
3. Prepare PCR master mix (*see* **Notes 8** and **9**) by combing the following reagents in a sterile precooled plastic Eppendorf tube (volumes given per reaction; multiply volumes by the amount of samples to be amplified): 7.6 µL sterile water, PCR grade (colorless cap); 2.4 µL $MgCl_2$ (25 m*M* stock solution, blue cap; *see* **Note 10**); 0.5 µL EBV F QP1L forward primer (20 pmol/µL); 0.5 µL EBV R QP2L reverse primer (20 pmol/µL); 1 µL hybridization probe 1 EBNA-1 FLN (4 pmol/µL); 1 µL hybridization probe 2 EBNA-1 LCN (4 pmol/µL); and 2 µL FastStart DNA Master Hybridization Probes mix.
4. Pipet 15 µL of PCR master mix to a precooled capillary placed in an adapter in the cooling block (*see* **Note 11**)
5. Add 5 µL of isolated DNA to the 15 µL PCR mix in the capillary.
6. For the standard curve, add, for example, 10, 10^2, 10^3, 10^4, and 10^5 copies of plasmid DNA or the DNA equivalent of 5, 50, 500, 5000, and 50,000 Namalwa cells in a volume of 5 µL to five separate reactions of 15 µL PCR mix (*see* **Note 12**).
7. Seal each capillary with a stopper and place capillaries in the carrousel in LightCycler centrifuge. Close centrifuge, press *start*, and open the lid when the centrifuge is finished.

8. Remove the carrousel, and place it in LightCycler device.
9. The PCR consists of three programs (preincubation, amplification, and cooling). Use the following amplification conditions.
    a. Program 1: preincubation ("activation" of *Taq* polymerase and denaturation of template DNA).

Cycle program data	**Value**
No. of cycles	1
Analysis mode	None
Temperature targets	**Segment 1**
Target temperature (°C)	95
Incubation time (min:s)	10:00
Temperature transition rate (°C/s)	20.0
Secondary target temperature (°C)	0
Step size (°C)	0.0
Step delay (cycles)	0
Acquisition mode	None

    b. Program 2: amplification.

Cycle program data	**Value**		
Cycles	45		
Analysis mode	Quantification		
Temperature targets	**Segment 1**	**Segment 2**	**Segment 3**
Target temperature (°C)	95	55	72
Incubation time (s)	10	10	10
Temperature transition rate (°C/s)	20.0	20.0	20.0
Secondary target temperature (°C)	0	0	0
Step size (°C)	0.0	0.0	0.0
Step delay (cycles)	0	0	0
Acquisition mode	None	Single	None

    c. Program 3: cooling.

Cycle program data	**Value**
Cycles	1
Analysis mode	None
Temperature targets	**Segment 1**
Target temperature (°C)	40
Incubation time (s)	30
Temperature transition rate (°C/s)	20.0
Secondary target temperature (°C)	0
Step size (°C)	0.0
Step delay (cycles)	0
Acquisition mode	None

10. Set the fluorescence gains as follows: channel F1 gain = 1; channel F2 gain = 15; channel F3 gain = 30 (*see* **Note 13**).
11. Define the name of the samples and sample type (*positive, negative, standard,* or *unknown*) in the software field *Run* under the heading *edit samples.* For standards, give known concentration (for example in *copies/reaction*). When completed, select *Done* and then *Save* and provide the *settings file* with an appropriate name. Then select *Run* and then again *Save* and provide *data file* with an appropriate name. Select *Done.* The LightCycler run is initiated.
12. After completion of the run, perform data analysis using the channel setting F2/F1, baseline adjustment to *arithmetic,* and Fit Points setting at 2. First determine the baseline in *Step 1: Baseline* in the *LightCycler Data Analysis* mode of the software. Set noise band in the log-linear phase of the amplification in *Step 2: Noise Band* in the *LightCycler Data Analysis* mode of the software. Then, in *Step 3: Analysis,* the software generates a standard curve ($y = ax + b$) by plotting *cycle number* against *log concentration (copies per reaction)* for standard samples (with known copy numbers; *see* **Note 14**). The log amount of EBV *(x)* in an "unknown sample" can calculated from its threshold cycle *(y).* The threshold cycle is determined by the software as the cycle in which the fluorescence signal exceeds the mean background fluorescence at baseline by 3 SD.
13. The software automatically calculates the number of EBV DNA copies present in each reaction. This can be seen in the LightCycler Quantification Report. Recalculate this to copies/mL for whole blood, plasma, or serum samples (*see* **Note 15** and **16**).
14. (Optional). In addition to the quantification using the standard curve approach, an optional melting curve analysis can be included after the last cycle of amplification to determine the specificity of the amplified region hybridizing with the probes. For this, before starting a run include the following program directly after after program 2 **(step 9)**, and omit program 3.

    a. Melting curve program.

Cycle program data		**Value**	
Cycles		1	
Analysis mode		Melting curve	
Temperature targets	**Segment 1**	**Segment 2**	**Segment 3**
Target temperature (°C)	95	40	95
Incubation time (s)	0	60	0
Temperature transition rate (°C/s)	20.0	20.0	0.1
Secondary target temperature (°C)	0	0	0
Step size (°C)	0.0	0.0	0.0
Step delay (cycles)	0	0	0
Acquisition mode	None	None	Continous

15. When the run is completed, analyze as described above (*see* **steps 12** and **13**). For melting curve analysis, select the melting curve region of the PCR by placing vertical cursors at the beginning (40°C) and the end (95°C) of the temperature profile in the analysis front screen.

16. Use the fluorescence settings F2/F1 and select the *temperature* option for the *x*-axis.
17. Analyze data in the *melting curve* field, in which –d(F2/F1)/dT is plotted against the temperature increase. In step 1 *(Melting peaks)*, set method to *linear with background correction*. Place "end cursors" (blue) into the flat part of the melting curve after the end of the melting process. Place "beginning cursors" (green) into a region before the product melting begins for all samples. Do not include regions of the curve that still rise.
18. Determine T_m (for each sample individually or collectively for all samples in case of a homogeneous pattern) in the menu *Extra: manual Tm* by placing vertical cursors at the points where –d(F2/F1)/dT reaches its maximum (and the decrease in fluorescence per temperature unit is at the highest). The T_m of a perfectly matching product–probe combination (e.g., for the EBV prototype B95-8 strain) is approx 58.1°C. Mismatches will yield lower T_m.

4. Notes

1. Primers should be high-performance liquid chromatography (HPLC)-purified, and for each batch of primers, PCR products should always first be tested on standard agarose gel electrophoresis for a discrete 213-bp band after PCR.
2. Hybridization probes with the fluorogenic labels fluorescein and LCRed640 can be ordered, for example, from TIB Molbiol (Berlin, Germany) at www.tib-molbiol.de and should be dissolved and stored according to the manufacturer's instructions.
3. The EBV-positive Burkitt's lymphoma cell line Namalwa contains two copies of EBV per cell, integrated into chromosome 1 *(19)*. Cells can be counted microscopically using a Burker-T rk chamber or by an automated cell counting device. The Namalwa cell line can be obtained from the ATCC under number CRL 1432.
4. The concentration *(C)* of purified plasmid DNA can be quantified spectrophotometrically by measuring absorption at 260 nm (A_{260}) and using **Eq. 1:**

$$C \text{ (ng/µL)} = A_{260} \times \text{dilution factor} \times 50 \tag{1}$$

where 1 A_{260} U of double-stranded DNA corresponds to a concentration of 50 ng/µL.
The number of plasmid copies (no. plasmid) can be calculated using **Eq. 2:**

$$\text{no. plasmid (copies/µL)} = (C \times N_{\text{avogadro}})/(660 \times \text{plasmid size} \times 10^9) \tag{2}$$

where C is given in ng/µL, $N_{\text{avogadro}} = 6.022 \times 10^{23}$/mol, 660 Daltons is the average molecular mass of a DNA bp, plasmid size is given in bp, and 10^9 is the conversion factor from nanograms to grams.
5. For quantification of circulating EBV DNA loads in clinical settings, we recommend the use of unfractionated whole blood because it combines all blood compartments, including both cell-associated and cell-free EBV DNA, and it gives a standardized and absolute value for the EBV DNA load (EBV DNA copies/mL total circulatory compartment) *(20)*. Some studies report plasma and serum as suitable clinical specimens in hematopoetic stem cell transplant recipients *(21)*, but we were unable to

detect elevated EBV DNA loads in serum of solid organ transplant recipients or AIDS patients, despite extremely elevated (cell-associated) EBV DNA loads in unfractionated whole blood *(17,22)*. In addition, nonstandardized blood clotting and plasma isolation conditions may introduce incontrollable variables, such as DNA release from apoptosed or fragile cells. At present little is known about the physical characteristics and origin of elevated EBV DNA loads in most populations, and the load may represent virion-derived EBV DNA, cell-associated EBV DNA, or cell-free EBV DNA released from in vivo or in vitro lysed cells. Detection of a certain form of EBV DNA may have additional clinical value in a given population. Because this value is currently undefined, it must be determined for each population independently. Isolated peripheral blood mononuclear cells or B cells as clinical specimens, and calculation of the EBV DNA load per 10^6 cells or per μg cellular DNA is not recommended because cell counts may vary considerably in immuno-suppressed patients, which would influence the relative amount of EBV DNA obtained from these clinical materials, whereas absolute EBV load may not vary.

6. DNA purity is of vital importance for quantitative real-time PCR analysis, as remaining PCR inhibiting substances such as heparin, ethylene adraminetetraacetic acid (EDTA), lipids, and hemoglobin may lead to lower quantification or negative results. Extraction-based DNA isolation methods, e.g., by the MagNA Pure LC Instrument (Roche Diagnostics) or NucliSens silica-based DNA extraction (BioMerieux) are preferable, as they remove interfering substances more effective-ly in particular from peripheral whole blood samples *(23,24)*.

7. Avoid repeated freezing and thawing of LC Fast Start Reaction Mix Hybridization Probes (tube 1b and LC Fast Start Enzyme (tube 1a). The Fast Start DNA Master Hybridization Probes mix stock solution should be stored at 2–8°C and used with-in 1 wk. Prepare the PCR master mix in tubes placed in the precooled LC cooling block, and keep all reagents in this block after thawing. The cooling block should be stored at 2–8°C and can be used for up to 4 h outside the refrigerator.

8. To decrease the risk of DNA contamination and false positivity of PCR, clean the laboratory bench and pipets with 0.1 *M* HCl or 10% bleach and subsequently water before starting the experiment. During the experiment, open tubes only when nec-essary. Wear gloves when pipeting. Use filter tips for all pipeting. Use separate lab-oratories for preparation of PCR mixes, isolating DNA and amplification. Aliquot all PCR reagents *(25)*.

9. To decrease further the risk of contamination owing to PCR product carryover, heat-labile uracil-DNA-glycosylase (UNG; Roche Diagnostics) can be optionally added to LightCycler reactions. This technique relies on incorporation of deoxyuri-dine triphosphate during the LightCycler reaction instead of dTTP. UNG cleaves DNA at any site where a deoxyuridylate residue has been incorporated. Subsequently, at high temperatures the abasic sites are hydrolyzed, and the UNG enzyme is inactivated. Native (template) DNA does not contain uracil and is there-fore not degraded.

10. The final $MgCl_2$ concentration in the PCR is 4 m*M,* as the Fast Start DNA Master Hybridization Probes mix already contains 1 m*M,* and separate $MgCl_2$ (2.4 μL/reaction) is added to achieve the final 4 m*M* concentration.

11. Do not touch the surface of the glass capillaries, and always wear gloves when handling them. Handle capillaries very careful, as they are extremely fragile. When needed, samples can be analyzed by standard agarose gel electrophoresis after real-time PCR. For this, capillaries (without stoppers) can be placed upside down in an Eppendorf reagent tube and centrifuged in a benchtop centrifuge for 10 s at low speed, to collect the PCR product.

12. Once the number of plasmid DNA copies per µL standard curve DNA stock is calculated (*see* **Note 4**), dilute in steps of 10-fold dilutions to the required concentrations. To decrease pipeting errors, pipet volumes of at least 5 µL in all dilution steps. Always keep plasmid and Namalwa DNA dilutions on ice because the diluted standard curve samples contain extremely small amounts of DNA, which may easily be degraded. Avoid repeated freezing and thawing of DNA stocks, and aliquot quantified DNA standards at relatively high concentrations or add carrier DNA such as herring sperm DNA.

13. With the LightCycler software version 3.5 or higher, no fluorescence gain setting is required.

14. The efficiency *(E)* of the LightCycler PCR can be calculated from the slope of the standard curve: $E = 10^{-1/slope}$. Thus, a theoretical efficiency of 2 (= 100%) will yield a slope of –3.3. For accurate quantification, it is crucial for clinical samples to have the same amplification efficiency as the standard curve samples. This can be judged from the graph in which log fluorescence (F2/F1) is plotted against the cycle number. The slope of the amplification curves of clinical samples and standard samples should be parallel. If not, it may be necessary to perform additional DNA purifications or dilute the sample.

15. The clinically relevant cutoff value of EBV DNA load monitoring should be determined for each population individually. It should be based, for example, on EBV DNA loads in healthy carriers or matched patient controls without EBV-associated diseases.

16. For LightCycler-negative clinical samples, the presence of amplifiable DNA in the sample can be checked by performing PCR for a cellular target, e.g., β-globin. The presence of substances interfering with PCR can be checked by spiking the sample with a low amount (e.g., 100 copies) of EBV plasmid or cell line DNA and subsequent reamplification. Spiked samples should be accurately quantified within a predefined range of assay variation *(10)*.

References

1. International Agency for Research on Cancer (1997) *IARC Monographs on the Evaluation of Carcinogenic Risks to Humans,* vol. 70: *Epstein-Barr Virus and Kaposi's Sarcoma Herpesvirus/Human Herpesvirus 8.* WHO, Lyon, France.

2. Middeldorp, J. M., Brink, A. A. T. P., van den Brule, A. J. C., and Meijer, C. J. L. M. (2003) Pathogenic roles for Epstein-Barr virus (EBV) gene products in EBV-associated proliferative disorders. *Crit. Rev. Oncol. Hematol.* **45,** 1–36.

3. Thorley-Lawson, D. (2001) Epstein-Barr virus: exploiting the immune system. *Nat. Rev. Immunol.* **1,** 75–82.

4. Yao, Q. Y., Rickinson, A. B., and Epstein, M. A. (1985) A re-examination of the Epstein-Barr virus carrier state in healthy seropositive individuals. *Int. J. Cancer.* **35,** 35–42.
5. Stevens, S. J. C., Verschuuren, E. A. M., Verkuijlen, S. A. W. M., van den Brule, A. J. C., Meijer, C. J. L. M., and Middeldorp, J. M. (2002) Role of Epstein-Barr virus DNA load monitoring in prevention and early detection of post-transplant lymphoproliferative disease. *Leuk. Lymphoma* **43,** 831–840.
6. Chan, A. T., Lo, Y. M., Zee, B., et al. (2002) Plasma Epstein-Barr virus DNA and residual disease after radiotherapy for undifferentiated nasopharyngeal carcinoma. *J. Natl. Cancer Inst.* **94,** 1614–1619.
7. Lo, Y. M. (2001) Quantitative analysis of Epstein-Barr virus DNA in plasma and serum: applications to tumor detection and monitoring. *Ann. NY Acad. Sci.* **945,** 68–72.
8. Lo, Y. M., Chan, A. T., Chan, L. Y., et al. (2000) Molecular prognostication of nasopharyngeal carcinoma by quantitative analysis of circulating Epstein-Barr virus DNA. *Cancer Res.* **60,** 6878–6881.
9. Rowe, D. T., Webber, S., Schauer, E. M., Reyes, J., and Green, M. (2001) Epstein-Barr virus load monitoring: its role in the prevention and management of post-transplant lymphoproliferative disease. *Transplant. Infect. Dis.* **3,** 79–87.
10. Stevens, S. J. C., Verkuijlen, S. A. W. M., van den Brule, A. J. C., and Middeldorp, J. M. (2002) Comparison of quantitative competitive PCR with LightCycler-based PCR for measuring Epstein-Barr virus DNA load in clinical specimens. *J. Clin. Microbiol.* **40,** 3986–3992.
11. Wittwer, C. T., Ririe, K. M., Andrew, R. V., David, D. A., Gundry, R. A., and Balis, U. J. (1997) The LightCycler: a microvolume multisample fluorimeter with rapid temperature control. *Biotechniques* **22,** 176–181.
12. Leight, E. R. and Sugden, B. (2000) EBNA-1: a protein pivotal to latent infection by Epstein-Barr virus. *Rev. Med. Virol.* **10,** 83–100.
13. Bhatia, K., Raj, A., Guitierrez, M. I., et al. (1996) Variation in the sequence of Epstein Barr virus nuclear antigen 1 in normal peripheral blood lymphocytes and in Burkitt's lymphomas. *Oncogene* **13,** 177–181.
14. Snudden, D. K., Smith, P. R., Lai, D., Ng, M. H., and Griffin, B. E. (1995) Alterations in the structure of the EBV nuclear antigen, EBNA1, in epithelial cell tumours. *Oncogene* **10,** 1545–1552.
15. Wrightham, M. N., Stewart, J. P., Janjua, N. J., et al. (1995) Antigenic and sequence variation in the C-terminal unique domain of the Epstein-Barr virus nuclear antigen EBNA-1. *Virology* **208,** 521–530.
16. Stevens, S. J. C. Vervoort, M. B. H. J., van den Brule, A. J. C., Meenhorst, P. L., Meijer, C. J., and Middeldorp, J. M. (1999) Monitoring of Epstein-Barr virus DNA load in peripheral blood by quantitative competitive PCR. *J. Clin. Microbiol.* **37,** 2852–2857.
17. Stevens, S. J. C., Verschuuren, E. A. M., Pronk, I., et al. (2001) Frequent monitoring of Epstein-Barr virus DNA load in unfractionated whole blood is essential for early detection of posttransplant lymphoproliferative disease in high-risk patients. *Blood* **97,** 1165–1171.

18. Arrand, J. R., Rymo, L., Walsh, J. E., Bjorck, E., Lindahl, T., and Griffin, B. E. (1981) Molecular cloning of the complete Epstein-Barr virus genome as a set of overlapping restriction endonuclease fragments. *Nucleic Acids Res.* **9,** 2999–3014.
19. Lawrence, J. B., Villnave, C. A., and Singer, R. H. (1988) Sensitive high resolution chromatin and chromosome mapping in situ: presence and orientation of two closely integrated copies of EBV in a lymphoma line. *Cell* **42,** 51–61.
20. Stevens, S. J. C., Pronk, I., and Middeldorp, J. M. (2001) Toward standardization of Epstein-Barr virus DNA load monitoring: unfractionated whole blood as preferred clinical specimen. *J. Clin. Microbiol.* **39,** 1211–1216.
21. van Esser, J. W., Niesters, H. G., Thijsen, S. F., et al. (2001) Molecular quantification of viral load in plasma allows for fast and accurate prediction of response to therapy of Epstein-Barr virus-associated lymphoproliferative disease after allogeneic stem cell transplantation. *Br. J. Haematol.* **113,** 814–821.
22. Stevens, S. J. C, Blank, B. S., Smits, P. H., Meenhorst, P. L., and Middeldorp, J. M. (2002) High Epstein-Barr virus (EBV) DNA loads in HIV-infected patients: correlation with antiretroviral therapy and quantitative EBV serology. *AIDS* **16,** 993–1001.
23. Boom R, Sol CJ, Salimans MM, Jansen CL, Wertheim-van Dillen PM, van der Noordaa, J. (1990) Rapid and simple method for purification of nucleic acids. *J. Clin. Microbiol.* **28,** 495–503.
24. Witt, D. J. and Kemper, M. (1999) Techniques for the evaluation of nucleic acid amplification technology performance with specimens containing interfering substances: efficacy of Boom methodology for extraction of HIV-1 RNA. *J. Virol. Methods* **79,** 97–111.
25. Kwok, S. and Higuchi, R. (1989) Avoiding false positives with PCR. *Nature* **339,** 237–238.

3

Profiling of Epstein-Barr Virus Latent RNA Expression in Clinical Specimens by Gene-Specific Multiprimed cDNA Synthesis and PCR

Servi J. C. Stevens, Antoinette A. T. P. Brink, and Jaap M. Middeldorp

Summary

We describe a two-step RT-PCR method for simultaneous detection of *EBNA-1* (QK and Y3K splice variants), *EBNA-2*, *LMP-1*, *LMP-2a* and *-2b*, *ZEBRA*, and *BARTs* RNA encoded by Epstein-Barr virus. As a control for RNA integrity, the low-copy-number transcript derived from U1A snRNP, a cellular housekeeping gene, is coamplified. Copy DNA (cDNA) for these nine targets is simultaneously synthesized in a gene-specific, multiprimed cDNA reaction, which strongly reduces the amount of required clinical specimen and allows more sensitive detection than random hexamer or oligo-dT priming. For amplification, cDNA synthesis is followed by nine separate PCRs for the mentioned targets. Primers were designed either as intron-flanking, to avoid background DNA amplification, or in different exons, allowing identification of differentiallly spliced RNA molecules. To increase specificity, PCR products are detected by autoradiography after hybridization with radiolabeled internal oligonucleotide probes. The method described is highly suitable for profiling EBV latent RNA expression in tissue biopsies, cultured or isolated cells, and unfractionated whole blood and for definition of EBV latency type I, II, or III gene expression in these samples.

Key Words: Epstein-Barr virus; herpesviruses; RNA profiling; reverse transcriptase polymerase chain reaction; latency type; *EBNA-1; EBNA-2; LMP-1; LMP-2; ZEBRA; BARTs; U1A snRNP.*

1. Introduction

Epstein-Barr virus (EBV) is a ubiquitous human γ-herpesvirus associated with a still increasing number of lymphoid and epithelial proliferative disorders, including infectious mononucleosis, Burkitt's and Hodgkin's lymphoma, B- and T-cell non-Hodgkin's lymphoma, lymphoproliferative disorders in the immunosuppressed, and nasopharyngeal (NPC) and gastric carcinoma *(1,2)*. Although >80 open reading flames (ORFs) have been predicted on the EBV genome thus

From: *Methods in Molecular Biology, vol. 292: DNA Viruses: Methods and Protocols*
Edited by: P. M. Lieberman © Humana Press Inc., Totowa, NJ

far *(1)*, only very restricted numbers of genes are expressed at the RNA and protein level in EBV-associated tumors. The so-called EBV latency RNAs include those encoding Epstein-Barr nuclear antigens (EBNA-1, EBNA-2, EBNA-3a, -b, and -c, and EBNA-4; also referred to as EBNA-1–6), the latent membrane proteins (LMP-1 and LMP-2a and -b), the noncoding Epstein-Barr-encoded small RNAs (EBER-1 and -2) and the *Bam*HI rightward transcripts (BARTs). BARTs contain the predicted putative ORFs *Bam*HI-A rightward frame (BARF)-0, RK-BARF-0, A73, and Royal Postgraduate Medical School (RPMS)-1, of which the in vivo protein coding function remains to be proved. In some EBV-associated diseases such as AIDS-related lymphoma and posttransplant lymphoproliferative disease and in lymphoblastoid cell lines, expression of *Bam*HI leftward frame-1 (BZLF-1; or ZEBRA; responsible for the switch between latency and lytic infection) is observed in a minority of cells *(3)*. Potential functions of the EBV latent genes mentioned are discussed in detail elsewhere *(1,2)*. Classically, three types of latent EBV infection are distinguished. EBER-1 and -2 and BARTs are expressed in all three latency types *(4)*. In latency type 1 EBV infection, as is found in Burkitt's lymphoma, expression of "EBNA-1 only" (derived from the EBV Q promotor) is observed. In latency type 2, which is characteristic of EBV infection in Hodgkin's lymphoma and NPC, additional expression of LMP-1 and LMP-2a and -b is seen, besides Qp-driven EBNA-1. Finally, in latency type 3, expression of all EBNAs (mainly originating from the C/W promotor) is observed in combination with LMP-1 and LMP-2a and -b expression. Latency type 3 infection is observed in EBV-transformed lymphoblastoid cell lines and in lymphomas arising in immunocompromised hosts. The distinct patterns of viral gene expression observed in different EBV-associated diseases may reflect differing pathogenesis and/or cell-specific regulation of viral gene expression. A yet undefined latency type with expression of EBER-1/2, EBNA-1, LMP-2a,b, BARTs, and BARF-1 is found in EBV-positive gastric carcinoma *(5)*. BARF-1 mRNA expression is also detected in NPC and appears to be a carcinoma-specific marker not expressed in EBV-positive lymphomas *(6,7)*. However, because BARF-1 mRNA is nonspliced, specific precautions are needed for its detection, as described in **refs. 5–7.**

This chapter describes a sensitive reverse transcription polymerase chain reaction (RT-PCR) assay for determination of EBV latent RNA expression profiles in tumor biopsies or smears, isolated or cultured cells, and unfractionated whole blood. It targets transcripts for EBNA-1 (QK and Y3K spliced variants), EBNA-2, LMP-1, LMP-2a and -2b, BARTs, ZEBRA, and the low-copy transcript derived from U1A small nuclear ribonucleoprotein (snRNP), a human housekeeping gene used as control for RNA integrity. The multiprimed RT strategy is based on simultaneous cDNA synthesis for the eight different EBV transcripts and host-encoded U1A snRNP mRNA by priming with gene-specific antisense primers.

Subsequently, neosynthesized cDNA is amplified using two gene-specific primers for each of the nine targets in separate PCRs. Primers were designed either as intron-flanking, to avoid background DNA amplification, or in different exons, allowing differentiation between cDNA amplification of differentially spliced mRNA molecules. To increase specificity, PCR products are detected by autoradiography after hybridization with radiolabeled internal oligonucleotide probes. Multiprimed EBV gene-specific copy (c)DNA synthesis is highly suitable for profiling of RNA expression patterns because it requires minimal amounts of clinical specimen in comparison with single-primed cDNA reactions. Furthermore, the use of specific antisense priming allows more sensitive detection of the EBV latent RNAs than random hexamer or oligo-dT priming, as shown in a previous study by us *(4)* and provides an internal control simultaneously. Suitability of the described method for analyzing EBV RNA expression profiles has been shown in Hodgkin's disease, T- and B-non-Hodgkin's lymphoma *(4)*, gastric carcinoma *(5)*, and peripheral blood of HIV-infected individuals *(8)* and transplant recipients *(9)*.

2. Materials (*see* Note 1)

2.1. RNA Isolation From Clinical Specimens

1. RNA-Bee RNA Isolation Solvent (Tel-Test, Friendswood, TX; *see* **Note 2**).
2. Positive control (EBV-positive cell line with latency type 3 infection, e.g., JY, Raji, or JC5).
3. Chloroform.
4. Isopropanol.
5. 75% Ethanol.
6. Benchtop centrifuge.

2.2. EBV Gene-Specific Multiprimed cDNA Synthesis

1. The primer sequences for the EBV targets and the control housekeeping gene and the expected amplicon sizes are given in **Table 1.**
2. Multi-RT-primer mix (containing 2 pmol/μL each of antisense primers EBNA-1 K, EBNA-2 H1, LMP-1.2, LMP2ab2, A4, Z2, and U1A2 (*see* **Note 3**).
3. 10X RT buffer: 500 mM Tris-HCl, 600 mM KCl, 30 mM MgCl$_2$, pH 8.3.
4. 100 mM dithiothreitol (DTT).
5. dNTP stock solution containing 2 mM each of dATP, dTTP, dCTP, and dGTP.
6. 40 U/μL; RNasin RNase inhibitor (Promega, Madison, WI).
7. 9 U/μL; Avian myeloblastosis virus reverse transcriptase (AMV-RT; Promega).
8. Two water baths or heating blocks.

2.3. cDNA Amplification by PCR

1. RT reaction supplement buffer: 1 mM of each dNTP, 50 mM Tris-HCl, 60 mM KCl, 3 mM MgCl$_2$, pH 8.3. This can be prepared from the dNTP and 10X RT buffer stock solutions.

Table 1
Sequences (5′ to 3′) of Forward and Reverse Primers and Internal Oligonucleotide Probes and Expected Amplified Fragment Sizes for Multiprimed EBV RT-PCR Analysis[a]

Transcript	Oligonucleotide	Sequence (5′-3′)	Expected amplicon size (bp)
	EBNA-1 Q	GTGCGCTACCGGATGGCG	
EBNA-1 QK	EBNA-1 K	CATTTCCAGGTCCTGTACCT	236
	pU	AGAGAGTAGTCTCAGGGCAT	
	EBNA-1 Y3	TGGCGTGTGACGTGGTGTAA	
EBNA-1 Y3K	EBNA-1 K	CATTTCCAGGTCCTGTACCT	265
	pU	AGAGAGTAGTCTCAGGGCAT	
	EBNA-2 Y2	TACGCATTAGAGACCACTTTGAGCC	RNA: 195
EBNA-2	EBNA-2 H1	AAGCGCGGGTGCTTAGAAGG	DNA: 400
	probe = EBNA-1 Y3	TGGCGTGTGACGTGGTGTAA	
	LMP-1.1	TGTACATCGTTATGAGTGAC	
LMP-1	LMP-1.2	ATACCTAAGACAAGTAAGCA	247
	pLMP-1.3	ACAATGCCTGTCCGTGCAAA	
	LMP-2a1	ATGACTCATCTCAACACATA	
LMP-2a	LMP-2ab2	CATGTTAGGCAAATTGCAAA	280
	pLMP-2ab	ATCCAGTATGCCTGCCTGTA	
	LMP-2b1	CAGTGTAATCTGCACAAAGA	
LMP-2b	LMP-2ab2	CATGTTAGGCAAATTGCAAA	280
	pLMP-2ab	ATCCAGTATGCCTGCCTGTA	
	A3	AGAGACCAGGCTGCTAAACA	
BARTs	A4	AACCAGCTTTCCTTTCCGAG	232
	pA	AAGACGTTGGAGGCACGCTG	
	Z1	CGCACACGGAAACCACAACAGC	RNA: 227
ZEBRA (BZLF-1)	Z2	CGGGGGGATAATGGAGTCAACATCC	DNA: 400
	pZ	GCTTGGGCACATCTGCTTCAACAGG	
	U1A1	CAGTATGCCAAGACCGACTCAGA	
U1A snRNP	U1A2	GGCCCGGCATGTGGTGCATAA	215
	pU1A3	AGAAGAGGAAGCCCAAGAGCCA	

BART, *Bam*HI rightward transcript; EBNA, Epstein-Barr nuclear antigen; LMP, latent membrane protein; snRNP, small nuclear ribonucleoprotein; ZEBRA, BZLF-1 EBV replication activator.

[a] Forward primers are given first, reverse (antisense) primers second, and internal oligonucleotide probes (oligonucleotide name starts with "p") as third for each target. Most primers are designed to be intron-flanking to prevent background genomic DNA amplification. For some targets, background DNA amplification will yield higher bands (indicated in column 4). The sequence of the probe for EBNA-2 transcripts is the same as for the EBNA-1 Y3 forward primer.

2. 10X PCR buffer: 50 m*M* Tris-HCl, 440 m*M* KCl, 12 m*M* $MgCl_2$, pH 8.3.
3. dNTP stock solution containing 2 m*M* each of dATP, dTTP, dCTP, and dGTP.
4. Forward and reverse primers (*see* **Table 1**) in a concentration of 100 pmol/µL.
5. 5 U/µL, AmpliTaq DNA polymerase (Applied Biosystems, Foster City, CA).
6. PCR thermocycler (e.g., Applied Biosystems PE 9600).

2.4. Detection of PCR Products by Hybridization With Radiolabeled Internal Oligonucleotide Probes

2.4.1. Electrophoretic Separation of PCR Products and Blotting to Nylon

1. Agarose
2. TBE: 90 m*M* Tris, 80 m*M* boric acid, 2 m*M* EDTA, pH 8.0.
3. Electropheresis unit.
4. Nylon membrane, e.g., GeneScreen Plus (Nen Life Science, Boston, MA).
5. Blotting buffer: 0.4 *M* NaOH.
6. Capillary blotting setup.
7. Hybridization buffer: 0.5 *M* Na_2HPO_4, 7% sodium dodecyl sulfate [SDS], pH corrected to 7.2 with H_3PO_4.
8. Plastic tray with hermetic seal.

2.4.2. Radioactive 5′-End-Labeling of Oligonucleotides by T4 Polynucleotide Kinase

1. T4 polynucleotide kinase (New England Biolabs, Beverley, MA).
2. 10X kinase buffer (New England Biolabs).
3. [γ-^{32}P]ATP (370 MBq/mL; 10 mCi/ml; Amersham Bioscience, Piscataway, NJ).
4. Glass Pasteur pipet (diameter 5 mm).
5. Glass wool.
6. Sepharose G50 (Pharmacia Biotech, Uppsala, Sweden).
7. TE: 10 m*M* Tris-HCl, 1 m*M* EDTA, pH 8.0.
8. Water bath with agitation.
9. Wash buffer: 3X SSC/0.1% SDS (0.45 *M* NaCl, 0.045 *M* Na-citrate, 1% SDS, pH 7.0).
10. Geiger-Mülller counter, e.g., Mini Monitor Series 900 G-M tube (Mini-Instruments, Burnham-on-Crouch, UK) or Liquid Scintillation Counter, e.g., TriCarb 1900 TR (Packard Bioscience, Boston, MA)
11. Film caset.
12. Photographic film, e.g., Kodak X-OMAT AR

3. Methods

3.1. RNA Isolation From Clinical Specimens

1. Cut 5–10 cryosections of 5 µ*M* (depending on biopsy size) and immediately add 1 mL of RNA-Bee RNA Isolation Solvent. For isolated or cultured cells, add 0.2 mL RNA-Bee per million cells. Vortex thoroughly, and store on ice for 5 min (*see* **Notes 4** and **5**). For each experiment include a positive control, such as an EBV-

positive cell line with latency type 3 (e.g. JY, JC5, or Raji) or clinical specimens collectively expressing the EBV genes to be amplified.

2. Add 1/10th vol of chloroform. Vortex for 15 s, and store on ice for 5 min.
3. Centrifuge 1 mL of the suspension for 15 min at maximum speed in a standard high-speed benchtop centrifuge (12,000g).
4. Transfer the aqueous (colorless) phase to a new reaction tube, add an equal volume of isopropanol, and incubate on ice for 15 min (*see* **Notes 6** and **7**).
5. Centrifuge 250 μL of the isopropanol-RNA solution for 30 min at maximum speed in standard benchtop centrifuge (*see* **Note 8**)
6. Remove supernatant (*see* **Note 9**) and add 500 μL of cold (−20°C) 75% ethanol. Vortex briefly, and centrifuge for 5 min at maximum speed.
7. Remove supernatant and dry RNA pellet at room temperature (*see* **Note 10**).

3.2. EBV Gene-Specific Multiprimed cDNA Synthesis

1. Dissolve the RNA pellet in 5 μL of multi-RT-primer mix.
2. Incubate at 65°C for 5 min to denature secondary RNA structures. After incubation, immediately place the sample on ice. Centrifuge briefly after 3 min.
3. Add 15 μL of RT reaction mix to the RNA sample. The RT reaction mix contains per reaction (for preparation of RT master mix multiply volumes by the amount of samples to be analyzed): − 2 μL 10X RT buffer, 2 μL 100 m*M* DTT, 10.6 μL 2 m*M* dNTPs, 0.2 μL RNasin, and 0.2 μL AMV-RT (*see* **Note 11**).
4. Incubate at 42°C for 60 min for cDNA synthesis (*see* **Note 12**).

3.3. cDNA Amplification by PCR

1. After cDNA synthesis, supplement the RT reaction with 25 μL of RT reaction supplement buffer to obtain a volume large enough for the subsequent nine different PCR reactions.
2. Divide 5 μL aliquots of of supplemented RT reaction in nine PCR tubes.
3. Add 45 μL PCR mix, respectively, with forward and antisense primers for either of the following EBV transcripts:

    a. EBNA-1 QK splice variants (primers EBNA-1 Q and EBNA- 1 K).
    b. EBNA-1 Y3K splice variants (primers EBNA-1 Y3 and EBNA-1 K).
    c. EBNA-2 (primers EBNA-2 Y2 and EBNA-2 H1).
    d. LMP-1 (primers LMP-1.1 and LMP-1.2).
    e. LMP-2a (primers LMP-2a1 and LMP-2ab2).
    f. LMP-2b (primers LMP-2b1 and LMP-2ab2).
    g. BARTs (primers A3 and A4).
    h. ZEBRA (primers Z1 and Z2).
    i. The cellular housekeeping gene U1A snRNP (primers U1A1 and U1A2).

    The PCR mix contains per reaction (*see* **Notes 13** and **14**): 36.8 μL water, 5 μL 10X PCR buffer, 0.25 μL forward primer (100 pmol/μL), 0.25 μL reverse primer (100 pmol/μL), 2.5 μL 2 m*M* dNTPs, and 0.2 μL AmpliTaq DNA polymerase.

4. Cycle the samples in a PCR device using the following PCR program: 4 min at 95°C; 40 cycles of 1 min at 95°C, 1 min at 55°C, and 1 min at 72°C; 7 min at 72°C, and finally a hold at 4°C (*see* **Note 15**).

3.4. Detection of PCR Products by Hybridization With Radiolabeled Internal Oligonucleotide Probes

3.4.1. Electrophoretic Separation of PCR Products and Blotting to Nylon

1. Separate PCR products and a molecular size marker (e.g., a 100-bp ladder or 2-log ladder available from New England Biolabs) by standard 1.5% agarose gel electrophoresis in TBE buffer for 1–2 h at approx 100 mA.
2. Transfer the PCR products from agarose gel to nylon by standard alkaline capillary blotting in blotting buffer.
3. Mark orientation of the samples on the nylon membrane.
4. Neutralize the nylon membrane by washing three times 5 min in 2X SSC.
5. Air-dry the nylon membrane.
6. Incubate the nylon membrane with 50 mL hybridization buffer in a fully sealed plastic tray placed in a water bath at 55°C for at least 15 min (*see* **Note 16**).

*3.4.2. Radioactive 5′-End-Labeling of Oligonucleotides by T4 Polynucleotide Kinase and Detection of PCR Products by Autoradiography (see **Note 17**)*

1. Label the internal oligonucleotides for detection of PCR products using the following reaction): 14 µL water, 1 µL olignucleotide probe (20 pmol/µL), 2 µL 10X kinase buffer (New England Bio Labs), 1 µL T4 polynucleotide kinase (New England Bio Labs), and 2 µL [γ-^{32}P]ATP.
2. Incubate for 60 min at 37°C.
3. Pipet labeling reaction onto Sepharose G50 column (*see* **Note 18**) to separate the labeled oligonucleotide probe from free (nonincorporated) label by gel filtration.
4. Repeatedly pipet 200 µL fractions of TE onto the column and collect fractions of 200 µL.
5. Measure activity of the collected fractions by β-radiation monitor or scintillation counter. Gel filtration will yield two peaks of activity, the first peak containing the labeled oligonucleotide probes (activity >500 counts/s, typically fractions three to five) and the second peak containing the free label. Pool the fractions containing the oligonucleotide probe, and add to the nylon membrane in hybridization buffer (*see* **Subheading 3.4., step 4**).
6. Incubate for 16–20 h in a fully sealed plastic tray (max. 7 × 20 cm) in a water bath at 55°C with agitation (*see* **Note 19**).
7. Collect hybridization buffer containing the labeled probe for reuse (*see* **Note 20**). Wash three times by adding a small volume of wash buffer (covering the nylon membrane completely) and incubating for >15 min at 55°C using agitation. Measure activity of wash buffer after the last washing step. If this is still active, additional washing steps may be performed until no activity is detected.

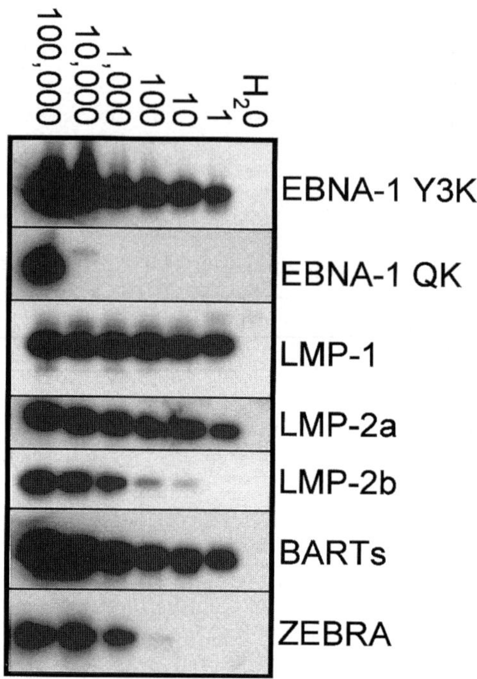

Fig. 1. Autoradiograph showing relative sensitivity of multiprimed EBV RT-PCR analysis. A dilution series of 1–100,000 EBV-positive JY cells was made in a whole blood DNA background, originating from an EBV-negative donor. JY is a lymphoblastoid cell line that expresses EBNA-1 mainly from the C/W promotor and shows limited ZEBRA expression in <1% of the cells. For most transcripts, RNA from at least one cell equivalent could be detected, except for EBNA-1 QK and ZEBRA RNA, which are expressed in only a minority of JY cells. Analytical sensitivity is at least 10 RNA, copies for each target (*4*).

8. Seal nylon membrane in plastic. First, carefully remove remaining liquid by rubbing the partially sealed membrane with force and using tissues to absorb excess liquid. Then seal the membrane completely and expose overnight at –80°C in a cassete to a X-ray film (e.g., Kodak safety Film AR O).
9. Develop the film in an X-ray developing machine or manually using photo developing solutions (e.g., Agfa). An example can be seen in **Fig. 1.**

4. Notes

1. Because RNases are omnipresent, it is essential to use RNase-free disposables such as reaction tubes and pipet tips. The use of filter tips is strongly recommended. Glassware should be sealed with aluminium foil and baked for >2 h at 180°C to degrade RNAses, which are resistant to autoclaving. Solutions and reagents should

be ordered either RNase free or treated with diethyl pyrocarbonate (DEPC; add 0.1% [(v/v)] of DEPC and autoclave). Note that some chemicals, e.g., Tris-HCl are reactive with DEPC and should not be DEPC-treated. Aliquoting of all reagents including primers is recommended to avoid contamination with RNases or PCR products by carryover.

2. RNA-Bee should be stored in the dark at 2–8°C. It contains guanidium thiocyanate, which is an irritant, and phenol, which is toxic by ingestion, skin contact, and inhalation of vapor. Discard waste as appropriate, according to environmental safety guidelines.

3. For cDNA synthesis from the EBNA-1 QK and EBNA-1 Y3K splice variants, the same antisense primer (EBNA-1 K) is used. The same holds true for the LMP-2a and -2b (primer LMP-2ab2). Thus, a total of seven antisense primers is used for multiprimed cDNA synthesis of nine targets.

4. Because RNA is extremely sensitive to omnipresent RNases, precautions should be taken to protect RNA from enzymatic degradation. These include wearing gloves when handling RNA, opening tubes only if necessay for short periods, and cleaning lab benches with bleach, ethanol, or 0.1% SDS before starting the experiment.

5. For isolation of RNA direct from unfractionated whole blood, the RNA-Bee method is unsuited. For this we prefer the NucliSens® Nucleic Acid Isolation Method (BioMerieux, Boxtel, The Netherlands), which is a silica-based RNA extraction method efficiently removing substances putatively interfering with amplification in RT-PCR *(10,11)*.

6. Make sure that no organic phase is removed. It is advised to leave a small amount of aqueous phase on top to avoid this.

7. The RNA/isopropanol solution can be stored long term at –80°C. Always keep on ice when pipeting the required volume from the RNA/isopropanol stock solution and return solution to –80°C immediately after use.

8. The amount of RNA/isopropanol to be used for RNA precipitation may be varied when preferred or according to availability of clinical material. For EBV RNA profiling in tissue biopsies, we routinely use the RNA equivalent of 2.5 cryosections (5 μm).

9. Place tubes in centrifuge with marked orientation, as the RNA pellet may not always be visible after precipitation. Be careful not to touch the pellet when discarding the supernatant.

10. When visible, the pellet may have a white appearance but should be glassy and transparent after drying. White pellets after drying indicate too much salt present. To avoid this, an additional washing step with 75% ethanol may be required. It is important not to let the RNA pellet dry completely, as this greatly decreases its solubility. Do not dry RNA by centrifugation under vacuum. Dissolve the RNA by passing the solution through a pipet tip and/or incubating for 10–15 min at 55–60°C. The final preparation of RNA has a A_{260}/A_{280} ratio 1.6–1.9.

11. Store all RT reagents at –20°C. After thawing, keep reagents on ice. Make sure all reagents are completely dissolved before use, especially DTT, which easily precipitates at low temperatures.

12. cDNA may be stored for the long term at –20°C until used for amplification by PCR.

13. Prepare PCR mix on ice. Keep all reagents for PCR on ice after thawing to avoid nonspecific *Taq* DNA polymerase activity and primer–dimer formation. Store PCR reagents at –20°C.
14. To decrease the risk of DNA contamination of the PCR reactions and false positivity, clean laboratory bench and pipets with 0.1 *M* HCl or 10% bleach and subsequently with water before starting the experiment. During the experiment open tubes only if necessary. Wear gloves when pipeting. Use filter tips for all pipeting. Use separate laboratories for preparation of PCR mixes, isolating DNA, and amplification. Aliquot all PCR reagents *(12)*.
15. PCR products can be stored at 4°C for 1–2 d. For long-term storage, it is advised to use –20°C to avoid enzymatic PCR product degradation by 5′-exonuclease activity of *Taq* polymerase.
16. Nylon membranes can be placed on top of each other to a maximum of three, as long as they are completely covered in hybridization buffer and are able to move independently of each other.
17. **Caution:** For working with ^{32}P, a β-emitter, precautions should be taken to protect the environment and the technician. Wear protective glasses, use protective plexiglass screens and tube holders, and monitor benches for putative contamination using a Geiger-Müller tube (*see* **Subheading 2.**). Discard as short-lived radioactive waste.
18. Prepare a Sephadex G50 suspension by adding 30 g Sephadex to 500 mL of water. Autoclave and store at 4°C. Prepare a column by applying a small dot of glass wool on the narrow part inside of a Pasteur glass pipet (5-mm diameter). Pipet the Sephadex suspension in the pipet until the Sephadex level is approx 5 mm below the top of the Pasteur pipet. Equilibrate the column with 1 mL of TE before adding the end-labeling reaction mixture. Do not leave the column standing dry at any moment but continually keep adding 200-μL TE fractions. Collect fractions of 200 μL in different tubes for radioactivity measurements.
19. Hybridize and wash at exactly 55°C. Higher temperatures may lead to negative results, as the probe may be unable to bind and lower temperatures may increase nonspecific background hybridization.
20. Radiolabeled probes in hybridization buffer can be stored at –20°C for reuse. Probes can be used several times, but keep in mind that ^{32}P has a half-life of 14.3 d.

References

1. International Agency for Research on Cancer (1997) *IARC monographs on the Evaluation of Carcinogenic Risks to Humans,* vol. 70: *Epstein-Barr Virus and Kaposi's Sarcoma Herpesvirus/Human Herpesvirus 8.* WHO, Lyon, France.
2. Middeldorp, J. M., Brink, A. A. T. P., van den Brule, A. J. C., and Meijer, C. J. L. M. (2003) Pathogenic roles for Epstein-Barr virus (EBV) gene products in EBV-associated proliferative disorders. *Crit. Rev. Oncol. Hematol.* **45,** 1–36.
3. Oudejans, J. J., Jiwa, M., van den Brule, A. J., et al. (1995) Detection of heterogeneous Epstein-Barr virus gene expression patterns within individual post-transplantation lymphoproliferative disorders. *Am. J. Pathol.* **147,** 923–933.

4. Brink, A. A., Oudejans, J. J., Jiwa, M., Walboomers, J. M., Meijer, C. J., and van den Brule, A. J. (1997) Multiprimed cDNA synthesis followed by PCR is the most suitable method for Epstein-Barr virus transcript analysis in small lymphoma biopsies. *Mol. Cell. Probes* **11,** 39–47.

5. zur Hausen, A., Brink, A. A., Craanen, M. E., Middeldorp, J. M., Meijer, C. J., and van den Brule, A. J. (2000) Unique transcription pattern of Epstein-Barr virus (EBV) in EBV-carrying gastric adenocarcinomas: expression of the transforming BARF1 gene. *Cancer Res.* **60,** 2745–2748.

6. Brink, A. A. T. P., Vervoort, M. B. H. J., Middeldorp, J. M., Meijer, C. J. L. M., and van den Brule, A. J. C. (1998) Nucleic acid sequence based amplification (NASBA), a new method for analysis of spliced and unspliced Epstein-Barr virus latent transcripts and its comparison with RT-PCR. *J. Clin. Microbiol.* **36,** 3164–3169.

7. Hayes, D. P., Brink, A. A. T. P., Vervoort, M. B. H. J., Middeldorp, J. M., Meijer, C. J. L. M., and van den Brule, A. J. C. (1999) Expression of Epstein-Barr virus (EBV) transcripts encoding homologues to important human proteins in diverse EBV-associated disease. *Mol. Pathol.* **52,** 97–103.

8. Stevens, S. J. C, Blank, B. S., Smits, P. H., Meenhorst, P. L., and Middeldorp, J. M. (2002) High Epstein-Barr virus (EBV) DNA loads in HIV-infected patients: correlation with antiretroviral therapy and quantitative EBV serology. *AIDS* **16,** 993–1001.

9. Stevens, S. J. C, Verschuuren, E. A., Pronk, I., et al. (2001) Frequent monitoring of Epstein-Barr virus DNA load in unfractionated whole blood is essential for early detection of posttransplant lymphoproliferative disease in high-risk patients. *Blood* **97,** 1165–1171.

10. Boom, R., Sol, C. J., Salimans, M. M., Jansen, C. L., Wertheim-van Dillen, and P. M., van der Noordaa, J. (1990) Rapid and simple method for purification of nucleic acids. *J. Clin. Microbiol.* **28,** 495–503.

11. Witt, D. J. and Kemper, M. (1999) Techniques for the evaluation of nucleic acid amplification technology performance with specimens containing interfering substances: efficacy of boom methodology for extraction of HIV-1 RNA. *J. Virol. Methods* **79,** 97–111.

12. Kwok, S. and Higuchi, R. (1989) Avoiding false positives with PCR. *Nature* **339,** 237–238.

4

Quantitative Detection of Viral Gene Expression in Populations of Epstein-Barr Virus-Infected Cells In Vivo

Donna R. Hochberg and David A. Thorley-Lawson

Summary

The method described in this chapter uses limiting dilution analysis in conjunction with RT-PCR to determine quantitatively what percentage of EBV-infected cells within a given population are expressing the viral genes *EBNA-1 Q-K, EBNA-2, LMP-1, LMP-2, BZLF-1,* and the *EBERs.* Because this technique involves limiting dilution analysis, it is possible to define which viral transcription programs are being used at the single-cell level. This assay takes 3–4 d to complete and involves the following steps: (1) sample preparation and isolation of the cell population of interest; (2) DNA-PCR limiting dilution analysis to determine the frequency of infected cells within the cell population; (3) RNA isolation; (4) cDNA synthesis; (5) PCR; (6) visualization of PCR products by Southern blotting; and (7) calculations. As an example, we have used PBMCs from the blood of an acute infectious mononucleosis patient. However, this technique can be applied to other cell populations, such as B cells, and other patient groups, such as healthy long-term virus carriers and immunosuppressed organ transplant recipients.

Key Words: EBV; Epstein-Barr virus; RT-PCR; DNA-PCR; infectious mononucleosis (IM); quantitative PCR; EBNA1; EBNA2; LMP1; LMP2; BZLF1; EBERs.

1. Introduction

The Epstein-Barr virus (EBV) is a ubiquitous, persistent virus implicated in a number of neoplasias including Burkitt's lymphoma, Hodgkin's disease, and nasopharyngeal carcinoma *(1)*. EBV is also the causative agent of infectious mononucleosis (IM) *(2,3)*. In an effort to understand the etiology of these diseases, EBV has been the focus of much study both in clinically affected patients and in healthy long-term virus carriers. Early studies on tumor cells revealed that different transcription programs are expressed by the cells of different tumors (**Table 1**) *(1)*. Virus-infected cells are relatively rare *(4,5)* in healthy carriers, making studies on this population more difficult. Continuing advances in polymerase chain reac-

From: *Methods in Molecular Biology, vol. 292: DNA Viruses: Methods and Protocols*
Edited by: P. M. Lieberman © Humana Press Inc., Totowa, NJ

Table 1
EBV Gene Expression Programs

Program	Genes expressed	Found in:
Growth program (or latency III)	All latent genes	Cells infected in vitro Naive B cells of the tonsil
Default program (or latency II)	*EBNA-1 QK, LMP-1, LMP-2, EBERs*	Hodgkin's disease, NPC, germinal center and memory cells of the tonsil
EBNA-1 only	*EBNA-1 QK, EBERs*	Burkitt's lymphoma, dividing memory B cells of the peripheral blood
Latency program (or Latency 0)	*EBERs*	Memory B cells of the peripheral blood
Lytic program	All lytic genes	Plasma cells of the tonsil

Abbreviations: EBER, Epstein-Barr-encoded small RNA; EBNA, Epstein-Barr nuclear antigen; LMP, latent membrane protein; NPC, nasopharyngeal carcinoma.

tion (PCR) and reverse-transcription (RT)-PCR technologies have aided studies on the healthy population *(5–7)*. By these approaches it is possible to detect mRNA and DNA from only a few copies of the target sequence or from a single infected cell. Such studies have shown that different populations of B cells in the healthy carrier express different transcription programs (**Table 1**) *(8,9)*. These programs are the same as those originally described for different EBV-associated tumors. This information has allowed for the development of a comprehensive model of how EBV establishes and maintains a persistent infection while continuing to produce infectious virus. In this model EBV uses these transcription programs to mimic and induce normal B-cell differentiation in infected cells *(10)*.

RT-PCR studies for EBV genes were and are generally performed on bulk cell populations *(11,12)*. One shortfall of this approach is the inability to distinguish whether the amplification products are derived from transcripts present in a single cell or from many cells *(13)*. This is because a typical sample will contain 10^6–10^7 B cells. Without knowing the number of infected cells present in the sample, it is impossible to determine whether a positive result means that all the infected cells are expressing a given gene or whether only a very small fraction of the infected cells are expressing the gene. Erroneous conclusions can be made from data collected in this way (*see* **Note 30**) *(11–13)*. To determine the percentage of infected cells expressing a given gene, quantitative approaches are necessary. The first step in the quantitative RT-PCR method described below is to determine the number of infected cells present by limiting dilution DNA PCR for EBV. The next step is to perform limiting dilution RT-PCR for each gene. Poisson statistics are used to calculate the absolute number of EBV-infected cells and the percentage of these cells expressing the relevant viral

Table 2
Diagnostic Genes for EBV Expression Programs[a]

Gene expression program	Gene					
	EBNA-1 QK	*EBNA-2*	*LMP-1*	*LMP-2*	*BZLF-1*	*EBER*
Growth	–	+	+	+	–	+
Default	+	–	+	+	–	+
Latency	–	–	–	–	–	+
EBNA-1	+	–	–	–	–	+
Lytic	–	–	–	–	+	–

[a] Although there are many other EBV genes, we have chosen the six listed here to distinguish the various transcription programs. BZLF, BamH1 Z fragment, left reading frame; for other abbreviations, see **Table 1** footnote.

genes. One important aspect of this method is that by examining results at the limit dilution (defined as the dilution at which one-third or less of the samples expressed any one gene), it is possible to determine which viral transcription programs are being used by individual infected cells (Table 2).

The method described below allows for quantitative assessment of what percentage of EBV infected cells are expressing the Epstein-Barr nuclear antigen *(EBNA)-1 Q-K, EBNA-2,* latent membrane protein *(LMP)-1, LMP-2, BZLF-1,* and Epstein-Barr-encoded small RNA *(EBER)* genes. These genes have been chosen because they are diagnostic of the common transcription programs found in EBV-infected cells **(Table 2)**. Statistically accurate results cannot be obtained unless mRNA for each gene is detectable from a single cell. Therefore, before this method is applied to clinical samples, it is vital to demonstrate the sensitivity of each RT-PCR. The technique described will be valuable in studying other EBV genes as well as studying other virus populations, particularly when infected cells are rare, such as is the case for other herpesviruses. However, successful application of this procedure to other EBV genes (or to other viral or cellular genes) will rely on the development of PCR assays sensitive enough to detect transcription from a single cell. We have presented results in this chapter from the peripheral blood mononuclear cells (PBMCs) of an IM patient; however, this procedure can be applied to the study of both other patient populations and to other cell populations.

2. Materials

2.1. Sample Preparation

1. EBV$^+$ control cell lines: IB4, AKATA, and B958.
2. EBV$^-$ filler cells.

3. Heparinized blood sample.
4. Ficoll-Hypaque (Pharmacia).
5. PBS/BSA: 0.5% bovine serum albumin/1X phosphate-buffered saline.
6. 0.5 *M* Acetic acid.

2.2. Quantitative Determination of the Frequency of Virus-Infected Cells: DNA-PCR

1. 10 mg/mL Proteinase K.
2. 10X Buffer: 20 mM MgCl$_2$, 500 mM KCl, 100 mM Tris-HCl, pH 8.3.
3. 4.5% Tween-20.
4. 4.5% NP-40.
5. 96-Well V-bottom plate (Falcon).
6. 55°C Incubator.
7. dNTP mix (10 mM each dNTP; Invitrogen).
8. Diethyl pyrocarbonate (DEPC) or high-performance liquid chromatography (HPLC) treated water.
9. 20 pM of each primer (*see* **Table 3**).
10. *Taq* (Applied Biosystems).
11. *Taq* Supplied Buffer (Applied Biosystems).
12. 95°C Heat block.

2.3. Quantitative Determination of the Percentage of Gene-Expressing Cells: RT-PCR

2.3.1. RNA Isolation

1. 55°C Heat block.
2. 200-µL MicroAmp reaction tubes with caps (Applied Biosystems).
3. Trizol reagent.
4. Chloroform.
5. Isopropanol.
6. Ethanol.

2.3.2. cDNA Synthesis

1. 42°C and 68°C Heat blocks.
2. Random primers.
3. dNTPs.
4. Superscript enzyme including dithiothreitol (DTT) and First Strand Buffer (Invitrogen).
5. 25 mM MgCl$_2$.

2.3.3. PCR

1. *Taq* Polymerase (Applied Biosystems).
2. 10X *Taq* Supplied PCR Buffer (Applied Biosystems).
3. 25 mM MgCl$_2$.
4. GeneAmp 9600/2400 reaction thermocycler (Perkin Elmer).

Table 3
Primer Sequences and PCR Conditions

Target gene	Primer	Annealing temp. (°C)	Vol. MgCl$_2$ (A) (μL)	Vol. of H$_2$O (B) (μL)	Product size (bp)	Ref.
RT-PCR						
EBNA-1 QK	RT3: 5′-TGG CCC CTC GTC AGA CAT GAT T-3′	62	0	17	220	From Sam Speck
	QB: 5′-AGC GTG CGC TAC CGG AT-3′					
EBNA-2	E2F: 5′-CAT AGA AGA AGA AGA GGA TAG AGA-3′	62	2	15	271	*14*
	E2R: 5′-GTA GGG ATT CGA GGG AAT TAC TGA-3′					
LMP-1	L1F: 5′-TTG GTG TAC TCC TAC TGA TGA TCA CC-3′	65	1	16	129	*14*
	L1R: 5′-AGT AGA TCC AGA TAC CTA AGA CAA GT-3′					
LMP-2	L2F: 5′-ATG ACT CAT CTC AAC ACA TA-3′	55	0	17	280	*12*
	L2R: 5′-CAT GTT AGG CAA ATT GCA AA-3′					
BZLF-1	ZF: 5′-TTC CAC AGC CTG CAC CAG TG-3′	59	0	17	182	*12*
	ZR: 5′-GGCAGC AGC CAC CTC ACG GT-3′					
EBERs	EBF: 5′-AAA ACA TGC GGA CCA CCA GC-3′	65	0	17	167	*12*
	EBR: 5′-AGG ACC TAC GCT GCC CTA GA-3′					
DNA-PCR						
Wrepeat	W1: 5′-CTT TAG AGG CGA ATG GGC GCC A-3′	66	1	31	°480	*5*
	EM2: 5′-TCC AGG GCC TTC ACT TCG GTC T-3′					

BZLF-1, BamH1 Z fragment, left reading frame; for other abbreviations, *see* **Table 1** footnote.

2.3.4. Visualization and Detection of PCR Products by Southern Blotting

1. 50X TAE: 968 g Tris, 228.4 mL glacial acetic acid, 148.8 g EDTA, up to 4 L with water.
2. 6X Sample buffer: 0.25% bromophenol blue, 0.25% Zylene cyanol, 30% glycerol.
3. One-phor-all buffer (Pharmacia).
4. 100-bp Ladder (Invitrogen).
5. Alkalization solution: 1 M NaCl (232 g/4L), 0.5 M NaOH (80 g/L) in water.
6. Neutralization solution: 1 M NaCl (232 g/4L), 1 M Tris-HCl (484 g/L), pH to 7.4, up to 4 L with water.
7. Supercharged Nytran (Schleicher and Schuell).
8. QIAquick Gel Extraction Kit (Qiagen).
9. Nuseive GTG agarose (BMA Bioproducts).
10. Seakem LE agarose (BMA Bioproducts).
11. [α-^{32}p]dATP at 3000 Ci/mM.
12. [α-^{32}p]dCTP at 3000 Ci/mM.
13. Random Primed labeling kit (Roche).
14. Hybridization oven.
15. 100X Denhardt's solution: 10 g Ficoll 400, 10 g polyvinyl prolidone, 10 g BSA up to 400 mL with water.
16. 20X SSC: 701 g NaCl, 352.8 g sodium citrate up to 4 L with water.
17. 1X TE: 10 mM Tris-HCl, 1 mM EDTA pH 8.0.
18. Wash A: 150 mL 20X SSC, 6.25 mL 20% sodium dedecyl sulfate (SDS) up to 500 mL with water.
19. Trichloracetic acid at 5 and 10%.
20. 95% EtOH.
21. Whatman filters (934-AH).
22. Kodak X-OMAT AR film.
23. Prehybridization solution: 6X SSC, 1X Denhardt's solution, 1% SDS.
24. Hybridization solution: 6X SSC, 50% formamide, 1% SDS, 2% dextran sulfate.
25. Semilog paper.

3. Methods

3.1. Sample Preparation

3.1.1. Cell Line Controls

Several control experiments demonstrating the ability to detect expression of each gene from a single cell should be performed before proceeding to the analysis of clinical samples. Cell line controls should also be included in each experiment thereafter. Prepare separate controls for each of the necessary cell lines: IB4, AKATA, and B958 (*see* **Note 1**).

1. Split cells the day before use.
2. Count and resuspend cells to 1×10^3 cells/mL in PBS/BSA.

3. Prepare 10-fold serial dilutions to give concentrations of 1×10^2 and 1×10^1 cell/mL.
4. Aliquot appropriate volumes of these dilutions into microcentrifuge tubes to give six to eight tubes each of 1, 5, 10, and 100 cells per tube (*see* **Notes 2** and **3**).
5. Count and resuspend EBV carrier cells to 5×10^6 cells/mL in PBS/BSA. Add 1 mL to each of the tubes prepared in **step 4** (*see* **Note 4**).
6. Proceed to **Subheading 3.3.**

3.1.2. Purification of Mononuclear Cells Via Ficoll (see **Note 5**)

1. Turn centrifuge to 25°C, and *turn off the brake;* leave it off for entire purification.
2. Obtain heparinized blood through routine venipuncture (*see* **Note 6**).
3. Dilute blood 1:1 in 1X PBS.
4. *Slowly* layer 30 mL of diluted blood onto 20 mL of room-temperature Ficoll in a 50-mL conical tube. Be careful to maintain an interface between the two layers.
5. Place tubes in centrifuge, and spin for 30 min at 400*g*, 25°C.
6. Aspirate off plasma layer (top layer; yellow in color); be careful not to disturb the buffy coat layer (white interface between the plasma and Ficoll layers).
7. Carefully remove buffy coat cells with a sterile transfer pipet to a fresh 50-mL conical tube. Buffy coat cells from two tubes may be combined to a total volume of 50 mL in PBS/BSA.
8. Invert tubes to mix, and centrifuge at 4°C for 15 min at 350 *g*.
9. Aspirate supernatant.
10. Resuspend all pellets together to a total volume of 50 mL in PBS/BSA.
11. Count cells (*see* **Note 7**).
12. Centrifuge cells for 10 min at 4°C and 1300*g*.
13. Resuspend cells to 5×10^6 cells/mL in PBS/BSA.
14. From the total number of cells (*see* **Note 8**) available, determine a series of dilutions to be used for RT-PCR analysis (*see* **Note 9**) and a separate set of dilutions to be used for DNA PCR (*see* **Note 10**).
15. Aliquot the RT-PCR dilutions into microcentrifuge tubes, and add EBV– filler cells to each tube to give a total of 5×10^6 cells per tube (*see* **Note 4**). Proceed to **Subheading 3.3.**
16. Aliquot the DNA-PCR dilutions into a 96-well plate (*see* **Note 11**). Proceed to **Subheading 3.2.**

3.2. Quantitation of EBV-Infected Cells: DNA-PCR

1. Centrifuge microtiter plate for 15 min at 4°C, 400*g*.
2. Aspirate supernatant from each well, being careful to not dislodge the cell pellet
3. Prepare digestion buffer as follows: 100 µL 4.5% Tween-20, 100 µL 4.5% NP-40, 100 µL PCR buffer, 50 µL proteinase K (10 mg/ml; Invitrogen; *see* **Note 12**), and 650 µL of H_2O.
4. Resuspend cell pellets in 10 µL of digestion buffer.
5. Incubate the microtiter plate at 55°C overnight (*see* **Note 13**).
6. Add 5 µL of water to each well; mix well. The samples are now ready for PCR.

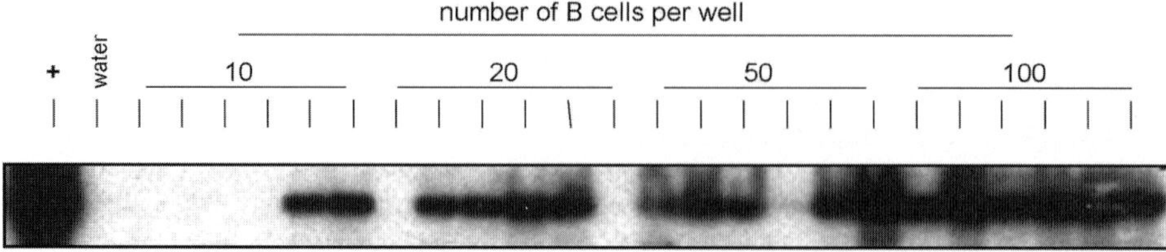

Fig. 1. Limiting dilution DNA PCR for the determination of the frequency of EBV-infected cells. Example of a Southern blot for a limiting dilution DNA PCR for EBV in a population of B cells from the peripheral blood of an infectious mononucleosis (IM) patient. Note that the numbers of cells per well are low. This is because IM patients have very high frequencies of infected cells. If this were a frequency analysis for a healthy carrier, the cell dilutions would need to be much higher (10^3–10^6 per well).

7. Preheat the PCR machine to 95°C.
8. Count the number of sample wells, add 10 to this number, and call it x (*see* **Note 14**). Carefully label 0.2-mL thin-walled PCR tubes.
9. Prepare PCR master mix for x tubes as follows: $1x$ μL 20 pM EM2 primer, $1x$ μL 20 pM W1 primer, $1x$ μL 10 mM dNTP mix, $1x$ μL 25 mM MgCl$_2$, $5x$ μL *Taq* Supplied Buffer, and $31x$ μL HPLC H$_2$O.
10. Aliquot 40 μL of master mix into each PCR tube.
11. Add 5 μL of sample DNA.
12. Heat samples for 10 min at 95°C.
13. Prepare *Taq* solution for X tubes as follows: $4.3x$ μL of H$_2$O, $0.5x$ μL of *Taq* supplied buffer, and $0.2x$ μL of *Taq*.
14. Add 5 μL of this mix to each tube while the tubes are still in the heat block.
15. Transfer samples directly from the heat block to a preheated PCR machine (*see* **Note 15**).
16. Perform amplification with the following program: 35 cycles at 95°C for 30 s, 65°C for 1 min and 1 cycle at 72°C for 5 min.
17. Visualize the PCR products on agarose gel according to the Southern blotting procedure detailed below in **Subheading 3.4.** Results are tabulated as described in **Subheading 3.5.** **Figure 1** shows example results from this procedure.

3.3. Quantitation of EBV Gene-Expressing Cells: RT-PCR

3.3.1. RNA isolation

1. Turn on all heat blocks (55°C, 42°C, 68°C, 95°C).
2. Centrifuge samples, including controls, prepared in **Subheading 3.1.** at 300g for 5 min.
3. Pour off supernatant. Dab on a paper towel to remove as much liquid as possible.
4. Resuspend each sample in 1 mL of Trizol reagent (*see* **Notes 16** and **17**).
5. Let stand at room temperature for 5 min.

6. Add 200 µL of chloroform, and shake vigorously for 15 s.
7. Let stand for 2 min at room temperature; you should see a clear layer on top and a pink layer on bottom. If the layers do not separate, shake again and let stand for 2 min.
8. Centrifuge at 12,000*g* for 15 min.
9. During the spin of **step 5,** prepare a new set of tubes (one for each sample), and add 500 µL of room-temperature isopropanol to each tube.
10. When the spin of **step 5** is complete, carefully remove the clear aqueous layer from the top of each sample, and add this to the isopropanol (*see* **Note 18**).
11. Vortex each sample thoroughly and let stand at room temperature for 10 min.
12. Centrifuge at 12,000*g* for 10 min at 4°C.
13. Carefully pour off the isopropanol (*see* **Note 19**).
14. Add 1 mL of ice-cold freshly prepared 70% EtOH to each tube; vortex briefly.
15. Centrifuge samples at 7500*g* for 5 min.
16. Pour off EtOH at 4°C (*see* **Note 20**).
14. Air-dry for 8 min (*see* **Note 21**).
16. Add 7 µL of DEPC or HPLC H$_2$O to each tube; do not pipet up and down.
17. Heat for 10 min at 55°C, then vortex gently, and return to 55°C for a further 5 min.
18. Quick-spin to gather liquid to the bottom of the tube.
19. Proceed *immediately* to cDNA synthesis to prevent RNA degradation.

3.3.2. cDNA Synthesis

1. Add 5 µL of random primers (50 ng/µL) to the bottom of one 0.2-mL PCR tube for each sample.
2. Add 7 µL of RNA, and mix gently by pipeting.
3. Incubate at 68°C for 8 min.
4. Condense at –20°C for 2 min.
5. Quick-spin tubes to bring the total volume to the bottom of the tube.
6. Prepare First Strand Master Mix as follows. For one reaction: 4 µL of First Strand Buffer (Invitrogen), 2 µL of DTT (Invitrogen), and 1 µL of 10 m*M* dNTPs (Invitrogen).
7. Add 7 µL of First Strand Master Mix to each sample; mix gently with pipet tip, but do not pipet up and down.
8. Incubate at room temperature for 10 min.
9. Add 1 µL of Superscript, and mix with pipet tip; do not pipet up and down.
10. Incubate for 10 min at room temperature.
11. Incubate at 42°C for 50 min.
12. Stop reaction by incubation for 15 min at 68°C.
13. Add 180 µL of HPLC H$_2$O, and vortex thoroughly; cDNA is now ready for PCR (*see* **Note 22**).

3.3.3. PCR (see **Note 23**)

1. Preheat PCR machine(s) to 95°C.
2. Carefully label one set of PCR tubes for each target gene. Include two water controls for each gene.

Fig. 2. Limiting dilution RT-PCR. **(A)** Results of a control experiment using EBV$^+$ cell lines. This experiment demonstrates that each RT-PCR assay can detect RNA from a single infected cell. **(B)** and **(C)** Results of limiting dilution analysis performed on cells from infectious mononucleosis (IM) patients. For patient 1, the experiment was performed exactly as described in this chapter. For patient 2, B cells were first isolated by negative selection using the Stem Sep System from Stem Cell Technologies. For both patients it is clear that only a small fraction of cells is expressing each gene, as there are 100s of infected cells in each lane, yet many samples were negative for gene expression. The exact percentage of cells expressing each gene is indicated below the blots. For patient 1 the frequency of infected cells was determined to be 1 infected cell per 6250 peripheral blood mononuclear cells (PBMCs; data not shown), and for patient to the frequency was determined to be 1 infected cell per 45 B cells. BZLF, BamH1 Z fragment, left reading frame; EBER, Epstein-Barr virus-encoded small RNA; EBNA, Epstein-Barr nuclear antigen; LMP, latent membrane protein.

3. Prepare a separate master mix for each target gene. Make enough mix for the number of RT samples plus two H_2O controls. For one PCR reaction: 1 µL 20 pM forward primer, 1 µL 20 pM reverse primer, 1 µL 10 mM dNTPs (Invitrogen), (A)* µL MgCl$_2$, 5 µL 10X *Taq* Supplied PCR Buffer, and (B)* µL of H_2O. *, The volumes of MgCl$_2$ and H_2O are different for each target gene. The values for each are given in **Table 3.**
4. Aliquot 25 µL of the appropriate master mix into each PCR tube.
5. Add 20 µL of sample cDNA (*see* **Note 24**).
6. Incubate tubes for 5 min at 95°C.
7. Prepare *Taq* master mix for 10 tubes: 43 µL of HPLC H_2O, 5 µL of *Taq* Supplied Buffer, and 2 µL of *Taq*.
8. Add 5 µL of *Taq* mix to each tube while it is still in the heat block.
9. Transfer tubes directly to a preheated PCR machine (*see* **Note 15**).
10. All PCR programs are as follows: 1 cycle at 95°C for 5 min and 40 cycles at 95°C for 15 s, annealing temperature for 30 s, 72°C for 30 s, and 1 cycle at 72°C for 5 min. Annealing temperatures for each gene can be found in **Table 3.**
11. PCR products are then visualized via the Southern Blotting protocol detailed below in **Subheading 3.4.** An example of results for a clinical IM sample and controls are shown in **Fig. 2.**

3.4. Visualization and Detection of PCR Products by Southern Blotting

3.4.1. Electrophoresis and Transfer of DNA

1. Prepare a 2% Nuseive/1% Seakem agarose gel in 1X TAE containing ethidium bromide.
2. Mix 12 µL of PCR product with 2 µL of sample buffer and incubate at 65°C for 5 min; prepare a 100-bp ladder by mixing 50 µL of the ladder with 25 µL of sample buffer and 25 µL of One-Phor-All buffer; do not heat the bp ladder.

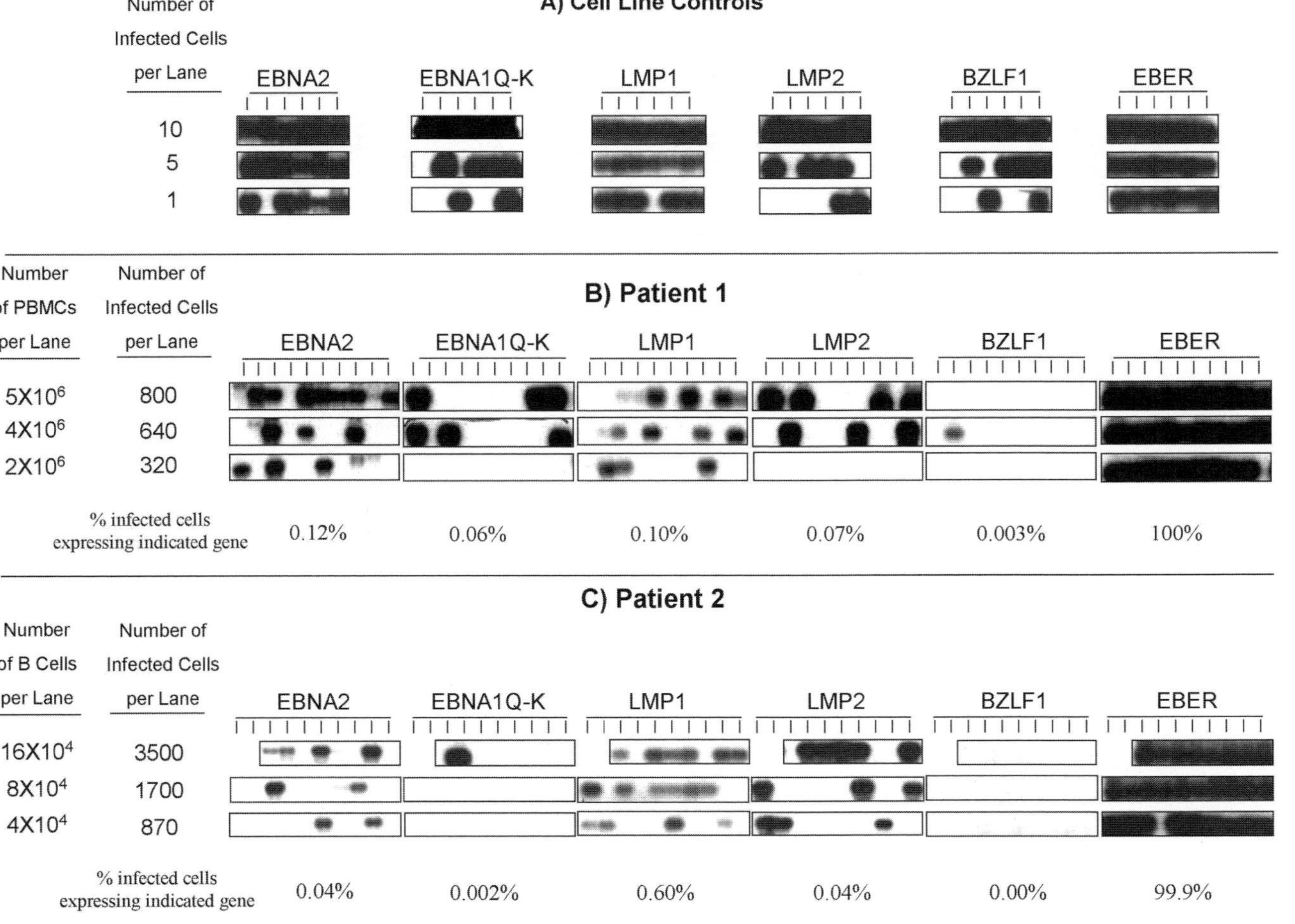

A) Cell Line Controls
Number of Infected Cells per Lane
EBNA2 EBNA1Q-K LMP1 LMP2 BZLF1 EBER
10
5
1
B) Patient 1
Number of PBMCs per Lane
Number of Infected Cells per Lane
EBNA2 EBNA1Q-K LMP1 LMP2 BZLF1 EBER
5X10^6 800
4X10^6 640
2X10^6 320
% infected cells expressing indicated gene
0.12% 0.06% 0.10% 0.07% 0.003% 100%
C) Patient 2
Number of B Cells per Lane
Number of Infected Cells per Lane
EBNA2 EBNA1Q-K LMP1 LMP2 BZLF1 EBER
16X10^4 3500
8X10^4 1700
4X10^4 870
% infected cells expressing indicated gene
0.04% 0.002% 0.60% 0.04% 0.00% 99.9%

3. Quick-spin to gather liquid to the bottom of the tube.
4. *Immediately* load 10 µL of the sample onto the gel
5. Run gel at 100 V until upper dye front is approx 5 cm from wells.
6. Wash gel twice in alkalization solution for 15 min each.
7. Wash gel twice in neutralization solution for 15 min each.
8. Wash gel for 15 min in 5X SSC.
9. Wash gel for 15 min in 2X SSC.
10. Transfer DNA to Supercharge Nytran via capillary action according to the manufacturer's instructions.
11. Remove Nytran from transfer apparatus and allow to air-dry for 2 min.
12. Crosslink DNA to Nytran using a UV crosslinker set for optimal crosslinking (*see* **Note 25**).

3.4.2. Prehybridization

1. Turn on hybridization oven to 42°C.
2. Place Nytran in a hybridization bottle, making sure it lies flat against the side.
3. Rinse blot with 2X SSC.
4. Add 10 mL of prehybridization buffer.
5. Boil (10 mg/mL) single-stranded (ss)DNA for 10 min.
6. Add ssDNA to the hybridization bottle to a final concentration of 50 µg/mL.
7. Incubate for 2 h to overnight at 42°C in the hybridization oven. Make sure the hybridization oven rotator is turned on.

3.4.3. Preparation of Probe DNA

1. Add 5 µL of probe DNA (10 ng/µL; *see* **Note 26**) to a microcentrifuge tube.
2. Boil for 10 min.
3. Add 1 µL of reagent #4 of the Random Primed Labeling Kit.
4. Add 1 µL of reagent #5 of the Random Primed Labeling Kit.
5. Add 2 µL of reagent #6 of the Random Primed Labeling Kit.
6. Add 5 µL of ^{32}P-labeled α-dATP and 5 µL of ^{32}P-labeled α-dCTP.
7. Add 1 µL of reagent of the Random Primed Labeling Kit #7 (*see* **Note 27**).
8. Incubate at 37°C for 30 min.
9. Incubate at 65°C for 15 min.
10. Add 80 µL of 1X TE.
11. Quick-spin to gather liquid to the bottom of the tube.
12. Determine labeling efficiency as follows:
    a. Add 1 µL of probe to a filter, and dry by vacuum.
    b. Wash with 10% TCA and dry, 5% TCA and dry, 95% ethanol and dry.
    c. Place filter in scintillation vial, and add scintillation fluid.
    d. Determine CPM/µL.

3.4.4. Hybridization

1. Boil labeled probe and ssDNA for 10 min.
2. Replace prehybridization solution with hybridization solution.

3. Add 25 µL of labeled probe.
4. Incubate overnight at 42°C in the hybridization oven with rotation.
5. Wash blots twice in wash A for 15 min each at room temperature.
6. Wrap Nytran in plastic wrap and expose to XOMAT-AR Film at −70°C (*see* **Note 28**).

3.5. Calculations

This process can be broken into two parts. First, it is necessary to determine the frequency of EBV-infected cells in the sample from data collected in **Subheading 3.2.** This information is then used, in conjunction with data from **Subheading 3.3.** to calculate the percentage of infected cells expressing each gene. Both of these calculations employ Poisson statistics (*see* **Note 29**).

3.5.1. Calculation of the Frequency of EBV-Infected Cells

1. Begin by examining the Southern blot film for the DNA-PCR.
2. For each cell dilution, calculate the fraction of negative samples.
3. Plot the fraction negative on the *y*-axis of semilog graph paper vs the corresponding cell number for that dilution on the *x*-axis.
4. Draw a line through the data points. Start the line at the point of 0 cells (*x*-axis) and fraction negative 1 (*y*-axis).
5. Where the data line intersects 0.37 fraction negative, the corresponding *x*-axis value is the number of cells required to have one EBV-infected cell.

3.5.2. Calculation of the Percentage of EBV-Infected Cells Expressing Each Viral Gene

The calculations for each gene must be performed separately.

1. Determine the number of infected cells present in each RT reaction (*see* **Note 30**).
2. Plot the fraction negative vs the number of infected cells per sample on semilog paper.
3. Draw a line through the data points. Start the line at the point of 0 cells (*x*-axis) and fraction negative 1 (*y*-axis).
4. Where the data line intersects 0.37 fraction negative, the corresponding *x*-axis value is the number of cells required to have one EBV-infected cell expressing the gene of interest

3.5.3. Definition of Viral Transcription Programs

To determine whether an individual cell is expressing a specific transcription program, you must first identify the limit dilution. This is the dilution at which only one gene-expressing cell is present in a given sample. The limit dilution is therefore defined as the dilution at which one-third or less of the samples (at that dilution) is expressing any one of the genes of interest.

4. Notes

1. The IB4 cell line is used as a control for expression of *LMP-1, LMP-2, EBNA-2,* and *EBERs.* AKATA is used for EBNA-1 QK and *EBERs* expression. B958 is used for *BZLF-1* expression. Other possible cell line controls include Rael as a substitute for AKATAs and Raji in place of IB4.

2. As an example, to get 100 cells/tube, aliquot 100 μL of the 10^3 dilution; to get 5 cells per tube, aliquot 50 μL of the 10^2 cell dilution. An alternative to preparing cell dilutions by hand is to use a fluorescence-activated all sorting (FACS) machine to sort the desired number of control cells directly into microcentrifuge tubes containing the appropriate amount of filler cells. If dilutions are done by hand, Poisson statistics should hold, and at the single-cell dilution two-thirds of the samples will contain a cell. The other third will be negative. If the dilutions are done with a FACS sorter, then all samples should be positive for a cell minus the error of the sorter.

3. *BZLF-1* is estimated to be expressed from only 5% of B958 cells. This can be confirmed by immunofluorescence staining for BZLF-1 protein (Z). Adjust the number of cells per tube accordingly.

4. EBV-negative filler cells may be obtained from the blood or tonsils of serologically EBV individuals. These primary cells should be isolated via the mononuclear cell purification strategy detailed in **Subheading 3.1.2.** If primary cells are unavailable, EBV cell lines such as BJAB cells may be substituted, although this is a less desirable alternative. In this case, only 1×10^6 filler cells should be used.

5. PBMCs as well as mononuclear cells from the tonsil or spleen may be used for this procedure. To create a single cell suspension from a solid organ, cover the sample with PBS/BSA, and mince with forceps. Resuspend the cell solution in 400 mL of PBS/BSA, and filter through a cell strainer to remove any debris. Proceed to **step 4** of the mononuclear cell purification procedure (**Subheading 3.1.2.**).

6. Only licensed phlebotomists should draw blood. Treat all blood samples as hazardous biomaterials. To ensure the highest quality RNA, we recommend that blood samples be processed immediately after the sample is drawn and that the isolated PBMCs not be frozen for future use.

7. Dilute a small sample of cells 1:10 with 0.5 *M* acetic acid when counting PBMCs to lyse any remaining red blood cells

8. In the example given in this chapter, PBMCs were used as an unfractionated population. However, it is possible to perform this assay on purified populations of B cells or memory B cells. To isolate these cell populations we recommend the use of the Stem Cell Technologies Stem Sep System. This is a negative selection method. Perform all procedures according to the manufacturer's instructions, choosing to perform steps at 4°C when the option is given. Briefly, cells are resuspended to a concentration of 5×10^7 cell/mL and stained with 100 μL of antibody cocktail (Stem Cell Technologies, cat. no. 14054) for 30 min on ice. Antibody cocktails contain antibodies directed against all types of PBMCs except the population of interest. Cells are then stained with 60 μL of magnetic colloid (Stem Cell Technologies, cat. no. 10051) for 30 min on ice. The sample is passed over a col-

umn, in the presence of a magnet. The population of interest is collected as the flow-through fraction. Alternatively, FACS sorting can be used. Again, we recommend using a negative selection scheme.

9. The concentration of cells per dilution will vary depending on the number of cells available. Use the entire sample minus the cells needed for DNA-PCR. Twofold serial dilutions beginning from 5×10^6 cells per dilution are recommended

10. The concentration of cells and the dilutions made will vary depending on the number of cells available and the nature of the sample. Use as few cells as possible for the DNA-PCR, as the RT-PCR requires large numbers of cells. For IM patients we recommend preparation of 10 dilutions covering a wide range from 10 cells per well to 1×10^5 cells per well. Make six to eight replicates per dilution. For healthy individuals, use twofold serial dilutions ranging from 1×10^6 per well to 1×10^4. Please note that a very detailed protocol for the DNA-PCR limiting dilution assay has been published previously *(6)*.

11. Do not use the well on the outside edge of the plate, as evaporation from these areas is more dramatic.

12. Proteinase K is not stable over long periods. Follow the manufacturer's instructions for resuspension and storage very carefully. If the overnight digestion yields viscous suspensions, this is an indicator of poor digestion quality. New proteinase K should be obtained immediately.

13. To prevent evaporation, the digestion plate must be carefully sealed with an adhesive plate sealer and taped around the edges.

14. The extra 10 tubes of master mix prepared in this step should be used for water controls and for EBV DNA-positive controls. The DNA for these controls is prepared from 1×10^6 EBV[+] IB4 cells by standard phenol/chloroform methods. Resuspend the DNA preparation in 1 mL of water. Use 5 µL for each control PCR reaction.

15. The hot start procedure is very important to the success of the PCR reaction. Samples must therefore be transferred to the PCR machine rapidly from the heat block while the PCR machine is running at 95°C If the temperature of the PCR machine drops below 63°C during the transfer of tubes, inconsistent results may occur. Any thermocycler can be used here. However, it is important to remember that for each PCR machine it is necessary to optimize each PCR reaction carefully for all parameters including $MgCl_2$ concentration, annealing temperature, and other cycling parameters.

16. Carefully and slowly resuspend the pellet by pipeting up and down. Continue pipeting until the entire pellet is dissolved, and no white wisps are present. Incomplete resuspension of the pellet can lead to poor RNA yields.

17. Samples may also be stored at –70°C for up to 6 mo before proceeding. Because of the length of the procedure, preparing samples on one day and freezing the Trizol is often convenient. The Trizols must be completely thawed before proceeding.

18. Be careful not to remove the white interface between the layers. Remove as much of the top clear layer as possible.

19. Pellets may or may not be visible at this stage. This step should be done in the cold room.

20. This should be done one tube at a time, keeping a close eye on the RNA pellet. Remove as much EtOH as possible, and then place tubes on their side
21. This step is critical. Overdrying results in poor RNA resuspension, and underdrying allows the remaining EtOH to interfere with the cDNA synthesis
22. cDNA can be stored at –20°C. However, it has been our experience that best results are obtained if all PCR reactions are run on a single day.
23. An alternative to the conventional PCR and Southern blotting procedure presented below is the use of real-time PCR. Substitution of real-time PCR for conventional PCR will save a great deal of time. However, it is important that these PCR assays be optimized and shown to detect each gene from a single cell.
24. By using only 1/10th of the cDNA for each PCR reaction, 10 separate PCRs can be run from a single RT, allowing direct analysis of each target gene from the same cells.
25. Crosslinking can also be achieved by baking at 80°C for 30 min.
26. Probe DNA is prepared by amplifying DNA from control cell lines via the PCR assays described in this chapter. The PCR products are then run on agarose gels and purified using Qiagen's QIAquick gel purification kit. A small aliquot of this product is then sent for sequencing to verify the nature of the product. Probe labeling is performed with Roche's Random Primed Labeling Kit.
27. This reagent is unstable. Keep on ice at all times.
28. An exposure time of approx 15 min is usually sufficient. However, this will depend on the efficiency of probe labeling and could take as little as 5 min or as much as overnight.
29. The Poisson distribution is based on the formula $s = e^{-u}$, where s = the fraction of negative samples and u = the number of events per sample at that dilution. If an average of one event is occurring in the samples tested at a given dilution, the fraction of negative samples will be $e^{-1} = 0.37$. For example, in **Fig. 2,** for patient 1, at 4×10^6 PBMCs (640 infected cells) per sample, 7 of 10 samples were negative for *EBNA-2*. This gives a fraction negative of 0.7. From $s = e^{-u}$, we can calculate that there is an average of 0.36 *EBNA-2*-positive cells per 4×10^6 PBMC, or 640 infected cells. Therefore, 0.056% infected cells are *EBNA-2*-positive. When multiple dilutions are tested, calculation of an accurate frequency is performed by plotting the log of the fraction negative vs the cell number tested at the dilution. This is done for each dilution and should give a straight line through the point (1 fraction negative; 0 cells tested). A straight line is indicative of an assay that is sufficiently sensitive to detect single events. At fraction negative 0.37 on the line, the corresponding *x*-axis value indicates the number of cells required to observe one positive event. In our example, this equates to one infected cell in 1.1×10^7 PBMCs, or 1780 infected cells. The error in the measurement can be estimated from the upper and lower boundaries of the line. For samples with too few signals to calculate an accurate frequency, the result can be expressed as the number of positive samples for the number of cells tested along with a 95% confidence limit based on the Poisson distribution. Consult a statistics textbook for a complete description of Poisson statistics and its application.

30. For example, if the frequency of infected cells was determined to be 1 infected cell per 2000 total cells, and there were 200,000 total cells in the RT reaction, then there were 200,000/2000 = 100 infected cells in the RT reaction.

31. A good example of how incorrect conclusions can be drawn from bulk RT-PCR assays is shown in **Fig. 2.** Here, if one were simply to score results as positive and negative for each gene, the conclusion would be that patient 1 is positive for all the genes tested. However, by limiting dilution analysis, we have shown that in fact less than 1% of cells are expressing each gene.

References

1. Keiff, E. (2001) Epstein Barr virus and its replication, in *Field's Virology,* vol. 2. Lippincott, Williams & Wilkins, Philadelphia, pp. 2511–2574.

2. Henle, G., Henle, W., and Diehl, V. (1968) Relation of Burkitt's tumor-associated herpes-type virus to infectious mononucleosis. *Proc. Natl. Acad. Sci. USA* **59,** 94–101.

3. Evans, A. S., Niederman, J. C., and McCollum, R. W. (1968) Seroepidemiologic studies of infectious mononucleosis with EB virus. *N. Engl. J. Med.* **279,** 1121–1127.

4. Khan, G., Miyashita, E. M., Yang, B., Babcock, G. J., and Thorley-Lawson, D. A. (1996) Is EBV persistence in vivo a model for B cell homeostasis? *Immunity* **5,** 173–179.

5. Miyashita, E. M., Yang, B., Lam, K. M., Crawford, D. H., and Thorley-Lawson, D. A. (1995) A novel form of Epstein-Barr virus latency in normal B cells in vivo. *Cell* **80,** 593–601.

6. Babcock, G. J., Miyashita-Lin, E. M., and Thorley-Lawson, D. A. (2001) Detection of EBV infection at the single-cell level. Precise quantitation of virus-infected cells in vivo. *Methods Mol. Biol.* **174,** 103–110.

7. Babcock, G. J., Decker, L. L., Volk, M., and Thorley-Lawson, D. A. (1998) EBV persistence in memory B cells in vivo. *Immunity* **9,** 395–404.

8. Babcock, G. J., Hochberg, D., and Thorley-Lawson, A. D. (2000) The expression pattern of Epstein-Barr virus latent genes in vivo is dependent upon the differentiation stage of the infected B cell. *Immunity* **13,** 497–506.

9. Joseph, A. M., Babcock, G. J., and Thorley-Lawson, D. A. (2000) Cells expressing the Epstein-Barr virus growth program are present in and restricted to the naive B-cell subset of healthy tonsils. *J. Virol.* **74,** 9964–9971.

10. Thorley-Lawson, D. A. (2001) Epstein-Barr virus: exploiting the immune system. *Nat. Rev. Immunol.* **1,** 75–82.

11. Prang, N. S., Hornef, M. W., Jager, M., Wagner, H. J., Wolf, H., and Schwarzmann, F. M. (1997) Lytic replication of Epstein-Barr virus in the peripheral blood: analysis of viral gene expression in B lymphocytes during infectious mononucleosis and in the normal carrier state. *Blood* **89,** 1665–1677.

12. Tierney, R. J., Steven, N., Young, L. S., and Rickinson, A. B. (1994) Epstein-Barr virus latency in blood mononuclear cells: analysis of viral gene transcription during primary infection and in the carrier state. *J. Virol.* **68,** 7374–7385.

13. Hochberg, D., Middeldorp, J. M., Catalina, M., Sullivan, J. L., Luzuriaga, K., and Thorley-Lawson, D. A. (2004) Demonstration of the Burkitt's lymphoma Epstein-Barr virus phenotype in dividing latently infected memory cells in vivo. *Proc. Natl. Acad. Sci. USA* **101,** 239–244.
14. Chen, F., Zou, J. Z., di Renzo, L., et al. (1995) A subpopulation of normal B cells latently infected with Epstein-Barr virus resembles Burkitt lymphoma cells in expressing EBNA-1 but not EBNA-2 or LMP1. *J. Virol.* **69,** 3752–3758.

5

Detection and Quantification of the Rare Latently Infected Cell Undergoing Herpes Simplex Virus Transcriptional Activation in the Nervous System In Vivo

Nancy M. Sawtell

Summary

Herpes simplex virus (HSV), in contrast to most other members of the herpes virus family, has the ability to infect, enter latency, and reactivate from latency in a number of nonhuman species, including mice. This provides a unique opportunity to study the complex lytic-latent cycle of a human neurotropic virus in a mouse model. This chapter details basic methods for inducing and quantifying reactivation, with emphasis on the first strategy for detecting and quantifying the initiation of HSV reactivation in vivo.

Key Words: Herpes simplex virus; latent infection; persistent infection; viral reactivation; hyperthermic stress; initiation of reactivation; whole tissue *in situ* detection; viral latency; immunohistochemistry.

1. Introduction

Herpes simplex virus (HSV) invades the host nervous system during infection at the body surface *(1)* (reviewed in **ref. 2**). In the nervous system, the virus proceeds through the lytic replicative cycle in some neurons, whereas in others, lytic phase transcription is either not initiated or aborted, and the viral genome enters a transcriptionally repressed or latent stage (for recent review, *see* **ref. 3**). These latently infected neurons serve as a life-long reservoir of viral genetic information within the host. Periodically, in response to stressful stimuli, one to a few of the latently infected neurons that comprise this reservoir support the re-entry of the latent viral genome into lytic phase transcription, and infectious virus is produced. Undoubtedly both viral and host factors contribute to this exquisitely controlled process. However, the molecular mechanisms by which the generalized latent transcriptional repression is relaxed to yield a cellular

From: *Methods in Molecular Biology, vol. 292: DNA Viruses: Methods and Protocols*
Edited by: P. M. Lieberman © Humana Press Inc., Totowa, NJ

environment now compatible with lytic viral gene expression and infectious virus production remain largely unknown. Progress toward addressing this question in vivo has been made (reviewed in **ref. 3**), but further advances will require the ability to probe, at the molecular level, the events occurring in those few cells that are entering and progressing through the reactivation process.

A central issue is demonstrating that the viral gene transcription detected in analyses of reactivation represents the primary changes occurring in those rare reactivating neurons and not the subsequent lytic events from the secondary spread of virus in the ganglia. This is critical in explant reactivation studies, in which spread of the reactivated virus through the ganglia occurs at times beyond 22 h. Because only a few neurons in the latent pool undergo reactivation *(4–7)*, most assays are not sensitive enough to detect the initial events. Reverse transcription polymerase chain reaction (RT-PCR) seemed to hold great promise for mapping early changes in the transition from latent into lytic viral transcription. The sensitivity of this approach has the potential to detect the primary transcriptional changes linked to reactivation even in a few cells among hundreds of thousands. However, it is now clear that a low level of lytic gene-related transcription is detected in latent ganglia by RT-PCR *(8,9)* (Sawtell, unpublished data). Whether this transcriptional activity represents *bona fide* lytic gene expression or simply noise from the million or so copies of the latent viral genome has not yet been determined. Regardless of the biological significance of this RNA, on a practical level it severely complicates the analysis of reactivation by RT-PCR.

Thus analysis of the primary events of reactivation in vivo will require examination at the individual cell level. Although detection of protein expression in the neuron undergoing reactivation in vivo has been accomplished, during the past 22 yr since the first report, fewer cells undergoing reactivation in vivo have been reported *(4–7)*.

The detection of what is apparently a very rare event using standard immunohistochemistry (IHC) on sectioned tissue is simply too labor-intensive to be useful for the detailed quantitative studies necessary to address the questions. Using a method for the analysis of gene expression in fetal nervous tissue *(10)*, we have developed an approach for the analysis in whole ganglia of viral protein expression during latency and following a reactivation stimulus *(11)*. This approach was utilized to link temporally and quantitatively the infectious virus in the ganglia during latency and following a reactivation stimulus with the viral translational activity in individual cells in the ganglia. These data provide evidence that those cells expressing detectable viral proteins are indeed those undergoing reactivation. Importantly, the approach is practical for large studies in which the detection, characterization, and quantification of the rare cells undergoing reactivation in vivo are required. Combined with engineered viral mutants and specific antibodies and probes, whole ganglia *in situ* analyses should prove pivotal for advancing our understanding of the molecular events that couple the induction stimulus to virus production.

2. Materials

2.1. Inoculation

1. Viral stock; store aliquoted at –70°C, thaw just before use, and keep on ice.
2. Mice: male Swiss Webster, 18–20 g upon arrival (Harlan). All procedures involving animals must be detailed in the post inoculation (PI) protocol and approved by the PI Institutional Animal Care and Use Committee (*see* **Notes 1–3**).
3. Anesthesia (according to your animal protocol).
4. Razor blades.
5. Variable-speed, hand-held rotary tool (Dremel or similar).

2.2. Hyperthermic Stress Procedure

1. Heated water bath with built-in stirrer (Neslab model gp100 or similar; *see* **Note 4**).
2. Accurately calibrated thermometer.
3. Wire rack for holding mouse restrainers.
4. Mouse restrainers (50-mL conical tubes with holes).
5. Towels.
6. Warm oven (33°C).

2.3. Detection of In Vivo Reactivation in Ganglia (see Note 5)

2.3.1. Dissection

1. Latently infected mice post hyperthermic stress and control mice (*see* **Notes 6** and **15**).
2. CO_2 or other method of sacrificing animals.
3. Sterile dissection tools.
4. 70% Ethanol (for wetting mice).
5. Media in 1.5-mL tubes on ice.

2.3.2. Homogenization

1. Sterile 2-mL straight-wall, ground-glass tissue homogenizers (Radnoti; the number of homogenizers required = the number of samples).
2. Stirring motor (IKA RW15) fitted with flexible chuck adapter (Daigger).

2.3.3. Plating Homogenates

1. Rabbit skin cells (RSC) or other HSV-permissive cell line.
2. Media: minimum essential medium (MEM) + 5% newborn bovine serum (NBS; for rabbit skin cell [RSC]).
3. 6-Well or 60-mm tissue culture plates.
4. Overlay: 1% methyl cellulose in media.
5. Crystal violet.
6. Dissecting microscope (to count plaques).

2.4. Detecting the Initiation of Reactivation in Whole Ganglia (see *Note 5*)

1. Mouse trigeminal ganglia (TG) harvested from uninfected and latently infected mice before and after hyperthermic stress (*see* **Notes 6** and **15**).
2. CO_2 or other method of euthanizing animals.
3. 1.5-mL tubes.
4. Nutator (Adams) or similar device.
5. Phosphate-buffered saline (PBS), sterile.
6. Tris-buffered saline (TBS), sterile.
7. 0.5% Paraformaldehyde in PBS; store in the dark at room temperature. **Caution:** formaldehyde is toxic and should be handled according to your institutional biohazard guidelines.
8. Methanol.
9. Dimethyl sulfoxide (DMSO).
10. Normal horse serum (NHS).
11. 30% Hydrogen peroxide (store refrigerated).
12. 50 mg/mL Glucose oxidase (GO), 500X stock solution in PBS; store at $-20°C$.
13. $2\ M$ β-D(+) glucose 100X stock solution in double-distilled (dd)H_2O; store at $-20°C$.
14. $2\ M$, Na azide 1000X stock solution in ddH_2O; store at $-20°C$.
15. PBS containing 2% bovine serum albumin (BSA), 5% NHS, 5% DMSO.
16. Primary antibody, in this case, rabbit anti-HSV 1/2 (Accurate) (*see* **Note 16**).
17. Horseradish peroxidase (HRP)-labeled secondary antibody, in this case, HRP-labeled goat antirabbit IgG (Vector).
18. Diaminobenzidine (DAB; Aldrich; make up fresh just before use, and handle with caution.
19. Sterile glycerol.
20. Microscope and photographic apparatus.

3. Methods

3.1. Inoculation

1. Mice are anesthetized according to the recommendations of your Institutional Veterinary Services. All procedures performed on animals must be detailed in the PI's animal protocol approved by the PI's Institutional Animal Care and Use Committee.
2. Corneal surface inoculation has been described in detail elsewhere *(5)*.
3. Inoculation of whisker pad: shave the whisker pad with a sharp razor blade, and lightly abrade the shaved surface with a variable speed rototool fitted with a small, fine-grit abrasive cylinder. The result should be increased pinkness of the tissue *without* bleeding **(Fig. 1)**. Gently rub the inoculum onto the abraded surface in a 15-µL volume (*see* **Notes 1–3**).

3.2. Inducing In Vivo Reactivation by Hyperthermic Stress

Figure 2 illustrates the apparatus used for this procedure.

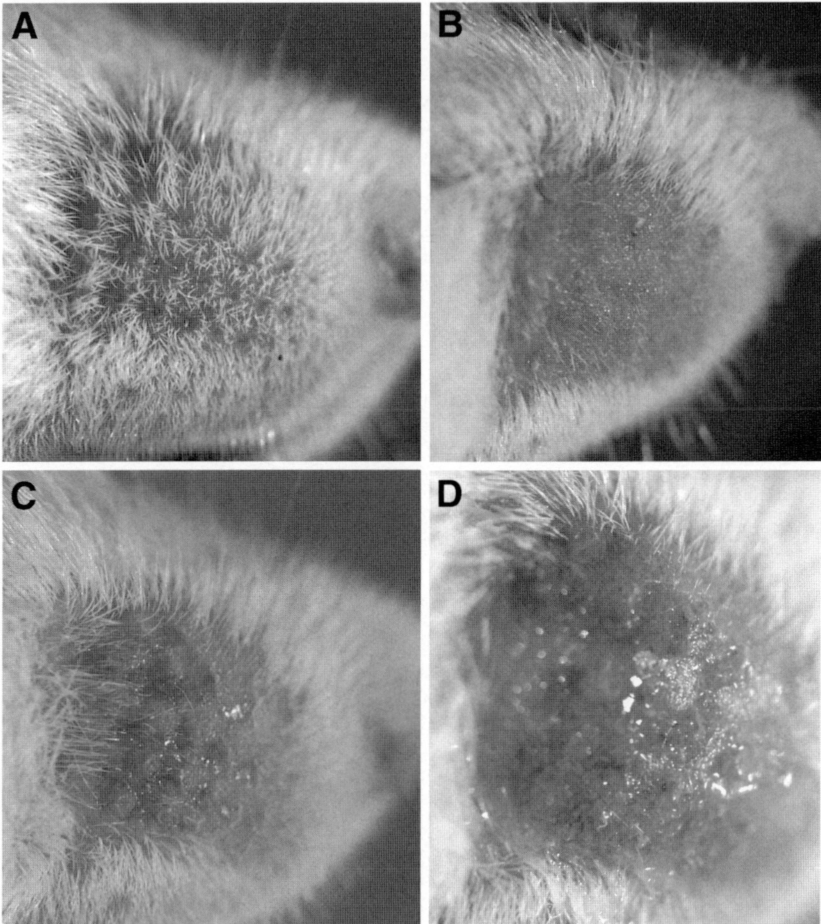

Fig. 1. Whisker pad inoculation. Whisker pad **(A)** is shaved as shown in **(B)**. Shaved area is lightly abraded with a fine-grit, abrading cylinder driven by a variable speed rotary tool **(C)**, and inoculum is gently rubbed onto the surface **(D)**. Bleeding does not occur.

1. Adjust the water bath to 42.5–43°C using an accurately calibrated thermometer. Do not assume that the digital readout is accurate, and do not allow the temperature to go above 43°C.
2. Band a wire rack with latex tubing, which serves as elastic retaining straps to secure the restrainers **(Fig. 2B)**.
3. Make a restrainer by drilling about twenty 3-mm-diameter holes, uniformly spaced over the surface of a 50-mL conical polypropylene tube **(Fig. 2A)**. Cut a slot into the tube cap for the tail of the mouse **(Fig. 2A)**.
4. Place the mice into the restrainers, and cap and strap them into the wire rack (*see* **Note 7**).

Fig. 2. Apparatus used for hyperthermic stress procedure. (**A**) 1, Restrainer: 50-mL tube with holes; 2, cap with tail slot; 3, foam restrainer plug. (**B**) Wire rack with latex tubing straps holding restrainer tubes. (**C**) Wire rack with restrainer in water bath. White lines indicate water level.

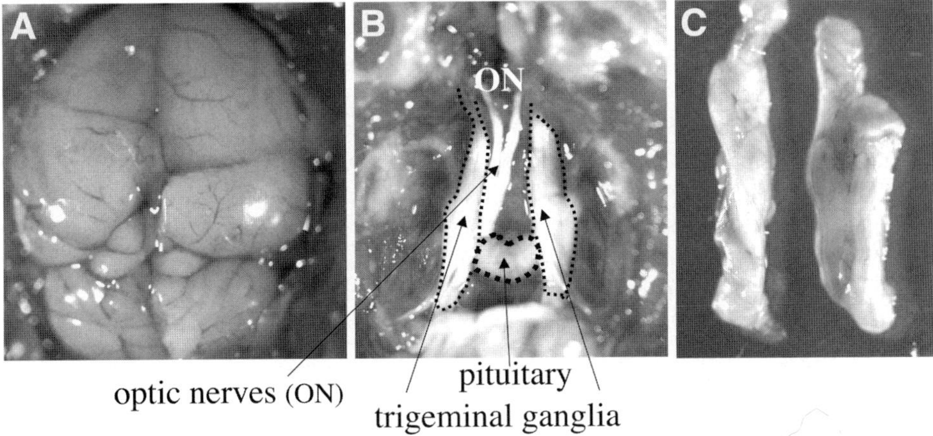

Fig. 3. Trigeminal ganglion (TG) dissection. (**A**) The skull is removed, revealing the brain. (**B**) The brain is lifted back, revealing the structures. The trigeminal ganglia lie to the right and left of the centrally positioned pituitary gland. (**C**) TG after removal.

5. Adjust the water level in the bath so that when the rack is lowered to the bottom of the bath, the water comes to the "shoulders" of the mice (**Fig. 2C**).
6. If mice are small, use a foam cushion (plug) in the bottom of the tube to keep the animal positioned properly in the tube (**Fig. 2A**).
7. Immerse the mice in the bath for 10 min.
8. Monitor the animals carefully during the procedure.
9. After the animals are removed from the bath, towel them dry and place them in an incubator warmed to 33°C for 20–30 min to prevent what can be lethal hypothermia.

3.3. Detection of Reactivation in the TG (see Note 5)

Peak infectious virus production occurs in the TG at 22–24 h after hyperthermic stress *(5)*.

3.3.1. Harvesting Tissues

Animals are euthanized according to the investigator's protocol. Sterile instruments and technique should be employed when harvesting tissues for detection of infectious virus. If surface tissues will be analyzed, these should be removed prior to wetting the animal with ethanol. Eyes are easily removed with forceps and whisker pads with small curved scissors. Wet down mouse fur with ethanol. To remove the TG, open the skull and peel back the brain (**Fig. 3**). The TG lie to the right and left of the pituitary (**Fig. 3B**).

For detection of infectious virus, ganglia are placed upon removal in ice-cold sterile 1.5-mL tubes containing 1 mL of media and maintained on ice.

3.3.2. Homogenization

1. Place on ice sterile 2.0-mL straight-wall ground-glass tissue homogenizers (Radnoti) and chill.
2. Transfer the media and the tissue to the base of the grinder, and fit the pestle into a flexible chuck adapter installed into a stirring motor (IKA-RW15).
3. Homogenize the tissue with 7–10 strokes set at 2.5 (5 g) on ice.
4. Rinse the homogenate remaining on the pestle with a small amount of media into the grinder, and use a sterile plugged 9-inch Pasteur pipet to transfer the homogenate from the grinder to a 1.5-mL tube.
5. Centrifuge homogenates (2,250 g at 4°C) for 5 min, place on ice, and immediately plate onto indicator monolayers.

3.3.3. Plating Homogenates

1. Prepare indicator cell monolayers the day before. Monolayers in 6-well or 60-mm plates should be just confluent at the time of use.
2. Pipet the entire homogenate onto the monolayer and absorb with rocking for 2 h (37°C, 5% CO_2).
3. Remove the homogenate, rinse the plates twice with media, overlay with media containing 1% methylcellulose, and incubate at 37°C in 5% CO_2 until plaques are a convenient size to count under the dissecting microscope (48–72 h).
4. Remove the overlay from plates, and rinse them three times with PBS before adding crystal violet.

3.4. Detection of the Initiation of Reactivation (Detecting Expression of Lytic Genes in Latently Infected Ganglia)

1. Day 1. Euthanize animals according to investigator's protocol. For fixation, proceed as follows:
   a. Dissect TG as described above.
   b. Immediately upon removal, place tissue in 0.5% paraformaldehyde in PBS and incubate at room temperature with gentle agitation (nutate; Adams Nutator) for 2 h. Tubes of 1.5 mL work well (*see* **Notes 8–10**).
   c. Remove paraformaldehyde. (Follow your institutional biohazard guidelines for disposal.)
   d. Rinse in PBS three times for 15 min each.
   e. Remove final PBS wash, and add methanol containing 20% DMSO.
   f. Nutate overnight at room temperature.
2. Day 2: Endogenous peroxidase pretreatment I.
   a. Remove methanol/DMSO, and nutate tissue for 1 h in methanol containing 20% DMSO and 10% H_2O_2.
   b. Remove methanol/DMSO/H_2O_2 solution, and rinse tissue in 100% methanol 3 X 15 min. on nutator.
   c. Store tissue in 100% methanol at –70°C overnight or longer (*see* **Note 11**).

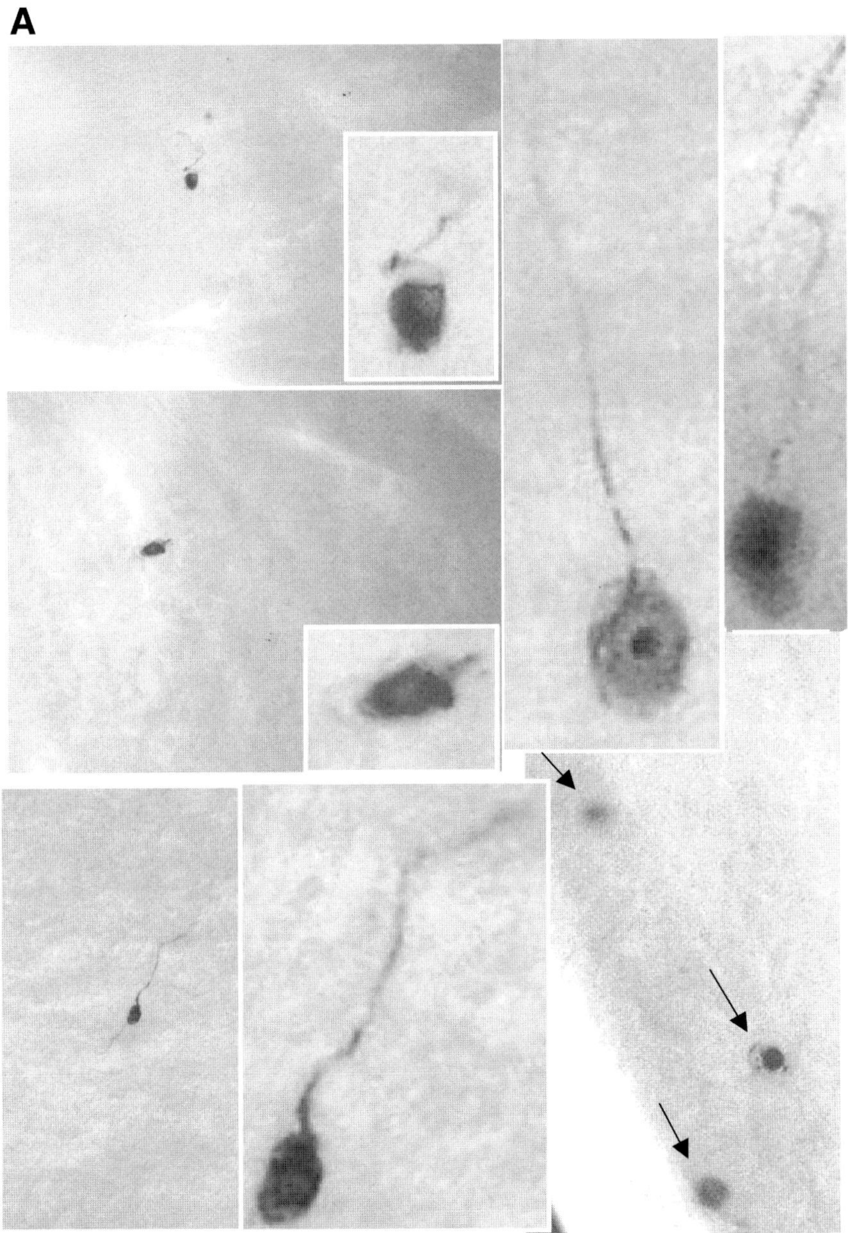

Fig. 4. Neurons in the trigeminal ganglia (TG) latently infected with HSV-expressing lytic viral proteins at 22 h post hyperthermic stress. **(A)** The rare neurons in the latently infected TG that have entered lytic viral gene expression following histochemistry. In the bottom right panel, three neurons (arrows) in this TG have initiated reactivation. **(B)** Localization of the lacrimal gland (LG) relative to the TG is shown at top. This gland contains dark brown and black structures that can easily be confused with positively stained neurons if bits of this tissue are removed with the TG during dissection, as illustrated at the bottom. *(Figure continues)*

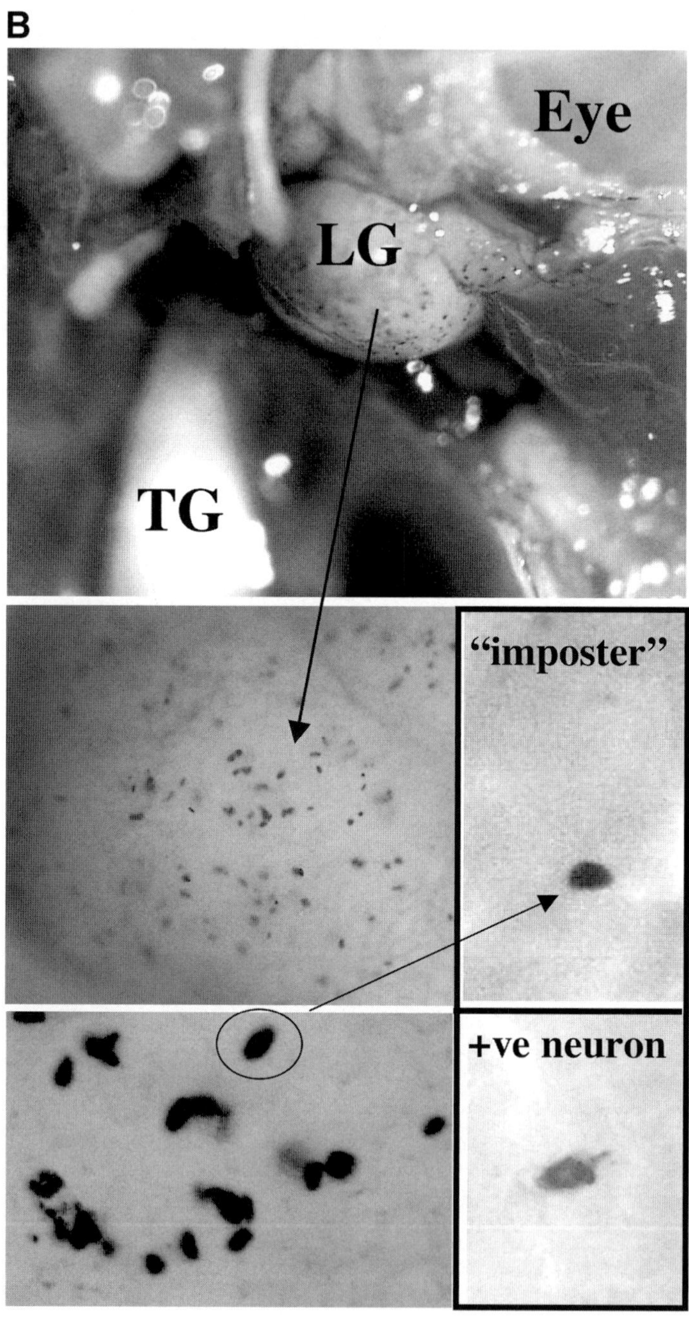

Fig. 4. *(continued)*

3. Day 3: Endogenous peroxidase pretreatment II (*see* **Note 8**).
   a. Bring tissue in methanol to room temperature. Remove methanol and rehydrate by incubating in PBS 3 X 15 min each on nutator
   b. Incubate tissue for 2 h at 37°C in PBS containing 0.002 *M* Na azide, 0.02 *M* glucose, and 100 µg/mL GO *(12)*.
   c. Remove GO solution, and rinse in PBS 3 X 15 min, on nutator.
   d. Remove PBS and add primary antibody, rabbit anti-HSV (Accurate) diluted 1:3000 in PBS containing 2% BSA, 5% NHS, 5% DMSO, and nutate at room temperature or 37°C overnight (*see* **Note 12**).
4. Day 4: remove primary antibody.
   a. Rinse in PBS 5 X 60 min each (total 5 h) on nutator (*see* **Note 13**).
   b. Remove PBS and add HRP-labeled antirabbit antibody (Vector) diluted 1:500 in PBS containing 2% BSA, 5% NHS, 5% DMSO, and nutate at room temperature overnight (*see* **Note 14**).
5. Day 5: remove HRP-labeled antibody.
   a. Rinse in PBS 5 X 1 h each (total of 5 h).
   b. Remove last PBS wash, and rinse once in TBS.
   c. Incubate in 0.1 *M* Tris-HCl, pH 8.2, containing 250 µg/mL DAB and 0.004% H_2O_2 solution. Typical development time is 5 min; however, this will need to be determined empirically.
   d. Remove DAB solution and rinse twice in ddH_2O for 15 min each.
   e. Remove ddH_2O, and nutate in 100% glycerol overnight.
   f. Press ganglia between two glass slides, tape, and examine under a microscope (**Fig. 4**); *see* **Notes 15–17**.

4. Notes

1. The mice are 18–20 g upon arrival and are housed for a minimum of 1 wk prior to inoculation. The vast majority of our studies have been performed on male Swiss Webster mice. Balb/c mice behave similarly in our model. C57 black mice are significantly more resistant. Differences in susceptibility to the virus should be considered before selecting a mouse strain.
2. Input titer: selecting the input titer will depend on several issues, including the virulence of the viral strain(s) and the susceptibility of the mouse strains utilized. Our standard input titer with strain 17syn+ is $1–2 \times 10^5$PFU. This results in an infection rate of 100%, with a 10–20% rate of mortality in male Swiss Webster mice Utilizing this inoculum strain and titer, approx 25% of the neurons in the TG become latently infected, with a mean genome copy number of around 50. Over a period of 6 yr and hundreds of groups of mice, we have found that these results are extremely reproducible. However, when analyzing mutants with mutations in genes that reduce the ability of the virus to replicate in vivo, additional issues arise. First, replication at the body surface and in the TG is directly related to the number of latent infections established in the ganglia *(13–16)*. Since the number of latent infections is directly correlated to reactivation frequency, any study with the goal of assessing the role of a particular gene or mutation in reactivation must be per-

formed on groups of mice with equivalent numbers of latent infections. By adjusting input titers of the mutant and rescue, the number of latent infections can be equalized, but this can only be done using methods to quantify latency accurately, for example *(17)*.

3. It is critically important for in vivo studies to monitor the progression of the infection and determine the variability of viral replication from animal to animal. Viral replication at the body surface and in the ganglia is correlated with the establishment of latency. In turn, the size of the latent pool is correlated with reactivation efficiency. It is imperative that the investigator know the animal-to-animal variation with respect to those parameters critical to the facets of latency and reactivation being examined. The importance of minimizing animal-to-animal variation during primary infection cannot be overemphasized. This can be done with practice, even in outbred mice. We have found that independent analysis of three animals at each time point for viral replication at the surface site of inoculation and the TG on d 2, 4, 6, and 8 provides an excellent indicator of the quality and consistency of inoculation.

4. Temperature uniformity in the water bath is important. A forced/suction circulating bath is not ideal because the shed hair and fecal matter from the animals clog the apparatus. A bath with an internal stirrer rather than a pump eliminates this problem.

5. "Reactivation" is defined operatively and by definition requires the detection of infectious virus in a tissue in which infectious virus was previously undetectable. We define the term "initiation" of reactivation as detection of lytic viral proteins in a tissue in which lytic proteins were previously undetectable.

6. Mice are considered to be latently infected at times beyond 30 d post inoculation. In light of recent data *(11,18)*, it may be prudent to delay for 40–50 d.

7. Those experienced in handling mice will have no difficulty with this. However, for those less experienced, this technique may help. Pick up the mouse as if you were going to give an intraperitoneal injection. Rotate wrist so that the mouse(abdomen down) is parallel with the benchtop. Take restrainer in free hand and put the open end of the restrainer up to the head of the mouse. Gently push the mouse forward into the restrainer while freeing your grasp. Adjust tail through slot of cap and secure cap.

8. As with all immunohistochemical procedures, optimum fixation conditions for any particular antigen/antibody combination must be determined empirically. In the case of this method, the low percentage of paraformaldehyde is important for maintaining the crosslinking of the tissue at a level consistent with penetration of primary and secondary antibody reagents.

9. Staining works equally well if the animal is fixed by perfusion rather than drop fixation of the tissue after dissection. Depending on the experimental design, one or the other method may be preferable. For example, we routinely perfusion-fix; however, if one TG is analyzed for infectious virus and the second TG is analyzed for viral protein, drop fixation is required.

10. Depending on experimental design, ganglia can be pooled and the procedure performed on all the ganglia from a given group in a single tube. Alternatively, ganglia can be placed in the wells of a 48-well plate and analyzed individually.

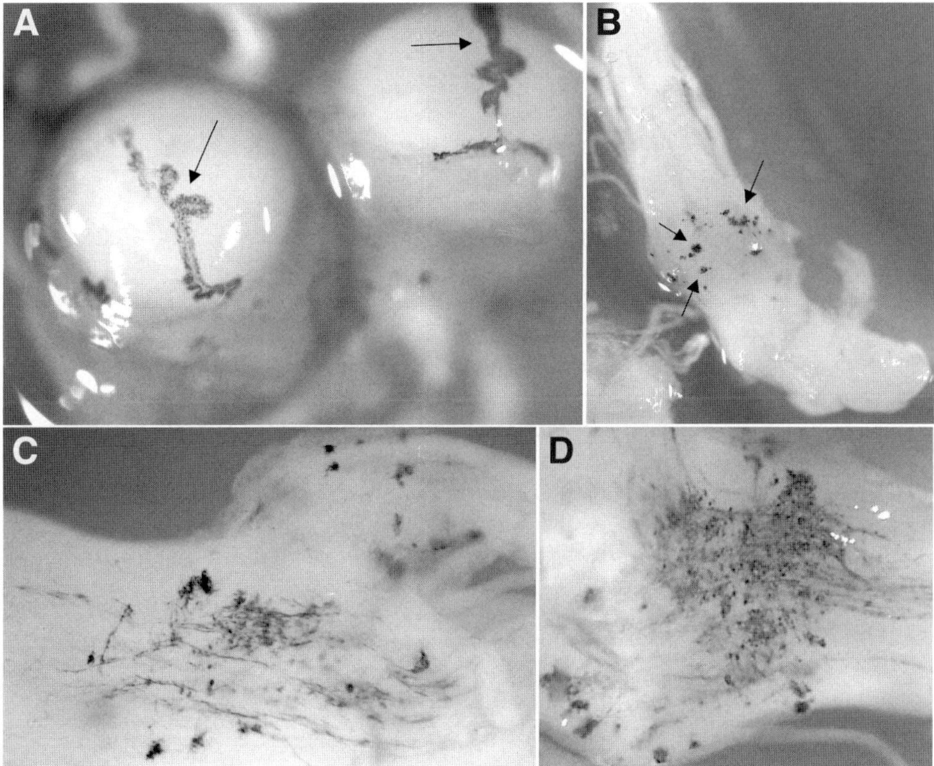

Fig. 5. Detection of HSV lytic viral protein expression in whole tissues. **(A)** Eyes d 2 post inoculation (pi) on corneas with strain 17 syn+. Arrow indicate regions of lytic protein expression on corneal surfaces. **(B)** Trigeminal ganglia 2 d pi. Neurons expressing lytic viral proteins are revealed (arrows). Increased protein expression in the TG on d 3 **(C)** and d 4 **(D)** parallels infectious virus titers.

11. It is our experience with several antigen/antibody combinations that staining is enhanced by keeping the tissue at −70°C for several days. Again, this will have to be determined for the antigen/antibody utilized.
12. Incubating the primary antibody at 37°C can increase staining intensity.
13. Five hours is the minimum wash time. Increasing washing time and/or the salt concentration in the PBS from 0.15 M to 0.5–0.75 M can help reduce nonspecific binding.
14. The choice of detection strategies is limited by the ability of the enzyme conjugates to penetrate the tissue. HRP works well. Alkaline phosphatase-labeled secondary antibody reagents and biotin/avidin systems are larger in size and do not penetrate well.
15. Interpreting the results requires training. The appropriate controls must be analyzed along with test samples. A typical experiment should include TG from uninfected

mice, TG from latently infected mice before treatment, and TG from the test group, i.e., from mice post hyperthermic stress. Do a trial run of the procedure on acutely infected ganglia **(Fig. 5)**. Infected surface tissues can be included. Eyes shown in **Fig. 5** were processed identically to the TG. Carefully compare uninfected and infected tissues under the microscope so that you will develop a sense of relevant vs irrelevant staining. The most troublesome "artifact" results from the inclusion of even very tiny pieces of lacrimal gland with the dissected ganglia. Ganglia should be checked thoroughly before proceeding with the assay to make sure that dissected ganglia are free of any contamination by this gland. As shown in **Fig. 4,** the lacrimal gland contains very dark brown to black structures, which, if freed during the staining process, can be interpreted as positive neurons. Neurons initiating reactivation will be extremely rare (*see* **Subheading 1.** for references).

16. We have utilized this method successfully with a number of different antibodies, including antibodies recognizing specific lytic viral proteins as well as host cell proteins.

17. Pressing ganglia between slides allows viewing at the higher magnification required for identifying and counting positive neurons. Other tissues, such as eyes and snouts or TGs during lytic infection, can be viewed under a dissecting microscope without pressing **(Fig. 5)**.

References

1. Goodpasture, E. W. (1929) Herpetic infections with special reference to involvement of the nervous system. *Medicine* **8,** 223–35.
2. Fields, B. N., Knipe, D. M., and Howley, P. M. (1996) *Fields Virology,* 3rd ed. Lippincott-Raven, Philadelphia.
3. Preston, C. M. (2000) Repression of viral transcription during herpes simplex virus latency. *J. Gen. Virol.* **81,** 1–19.
4. McLennan, J. L. and Darby, G. (1980) Herpes simplex virus latency: the cellular location of virus in dorsal root ganglia and the fate of the infected cell following virus activation. *J. Gen. Virol.* **51,** 233–243.
5. Sawtell, N. M. and Thompson, R. L. (1992) Rapid in vivo reactivation of herpes simplex virus in latently infected murine ganglionic neurons after transient hyperthermia. *J. Virol.* **66,** 2150–2156.
6. Ecob-Prince, M. S., Rixon, F. J., Preston, C. M., Hassan, K., and Kennedy, P. G. (1993) Reactivation in vivo and in vitro of herpes simplex virus from mouse dorsal root ganglia which contain different levels of latency-associated transcripts. *J. Gen. Virol.* **74,** 995–1002.
7. Shimeld, C., Whiteland, J. L., Williams, N. A., Easty, D. L., and Hill, T. J. (1996) Reactivation of herpes simplex virus type 1 in the mouse trigeminal ganglion: an in vivo study of virus antigen and immune cell infiltration. *J. Gen. Virol.* **77,** 2583–2590.
8. Kramer, M. F. and Coen, D. M. (1995) Quantification of transcripts from the ICP4 and thymidine kinase genes in mouse ganglia latently infected with herpes simplex virus. *J. Virol.* **69,** 1389–1399.

9. Chen, S. H., Kramer, M. F., Schaffer, P. A., and Coen, D. M. (1997) A viral function represses accumulation of transcripts from productive-cycle genes in mouse ganglia latently infected with herpes simplex virus. *J. Virol.* **71,** 5878–5884.
10. Luque, J. M., Adams, W. B., and Nicholls, J. G. (1998) Procedures for whole-mount immunohistochemistry and in situ hybridization of immature mammalian CNS. *Brain Res. Brain Res. Protoc.* **2,** 165–173.
11. Sawtell, N. M. (2003) Quantitative analysis of herpes simplex virus reactivation in vivo demonstrates that reactivation in the nervous system is not inhibited at early times postinoculation. *J. Virol.* **77,** 4127–4138.
12. Andrew, S. M. and Jasani, B. (1987) An improved method for the inhibition of endogenous peroxidase non-deleterious to lymphocyte surface markers. Application to immunoperoxidase studies on eosinophil-rich tissue preparations. *Histochem. J.* **19,** 426–430.
13. Sawtell, N. M. (1998) The probability of in vivo reactivation of herpes simplex virus type 1 increases with the number of latently infected neurons in the ganglia. *J. Virol.* **72,** 6888–6892.
14. Sawtell, N. M., Thompson, R. L., Stanberry, L. R., and Bernstein, D. I. (2001) Early intervention with high-dose acyclovir treatment during primary herpes simplex virus infection reduces latency and subsequent reactivation in the nervous system in vivo. *J. Infect. Dis.* **184,** 964–971.
15. Sawtell, N. M., Poon, D. K., Tansky, C. S., and Thompson, R. L. (1998) The latent herpes simplex virus type 1 genome copy number in individual neurons is virus strain specific and correlates with reactivation. *J. Virol.* **72,** 5343–5350.
16. Thompson, R. L. and Sawtell, N. M. (2000) Replication of herpes simplex virus type 1 within trigeminal ganglia is required for high frequency but not high viral genome copy number latency. *J. Virol.* **74,** 965–974.
17. Sawtell, N. M. (1997) Comprehensive quantification of herpes simplex virus latency at the single-cell level. *J. Virol.* **71,** 5423–5431.
18. Feldman, L. T., Ellison, A. R., Voytek, C. C., Yang, L., Krause, P., and Margolis, T. P. (2002) Spontaneous molecular reactivation of herpes simplex virus type 1 latency in mice. *Proc. Natl. Acad. Sci. USA* **99,** 978–983.

6

Reporter Cell Lines for the Detection of Herpes Simplex Viruses

Szu-Hao Kung

Summary

Virus culture has played significant roles in basic and clinical virology, with a number of advantages that cannot be attainable by modern molecular techniques. However, virus culture is generally a slower process, as it inevitably takes the period of a full replication cycle of a given virus. A genetically modified cell culture with a virus-inducible marker is described here, using a frequently isolated DNA virus (herpes simplex virus) as a model. The assay system relies on expression of the reporter gene driven by a specific viral promoter that is triggered early in the course of viral infection. The reporter gene employed was green fluorescent protein (GFP) or secreted alkaline phosphatase (SEAP), whose assays offer real-time detection or quantification, respectively. This cell-based assay is simple, rapid, sensitive, specific, and quantitative and serves as a phenotypic method for determination of antiviral susceptibilities.

Key Words: Herpes simplex virus; reporter cell line; reporter gene, green fluorescent protein; secreted alkaline phosphatase; acyclovir; antiviral susceptibility testing; ICP10 promoter.

1. Introduction

Herpes simplex viruses types 1 and 2 (HSV-1 and HSV-2; collectively HSV) cause a wide spectrum of clinical illness in immunocompromised individuals, especially newborns, transplant recipients, and patients with the acquired immunodeficiency syndrome *(1)*. Rapid and sensitive diagnostic assays based on direct detection of HSV-specific proteins and nucleic acid sequences have been developed *(1,2)*. Nevertheless, virus culture remains the only method to detect infectious virus particles and serves as the standard for diagnosis of HSV infection in clinical virology laboratories. The added value of virus culture could be to facilitate the analysis of clinically relevant viral phenotypes, such as antiviral susceptibility, given that a number of effective anti-HSV chemothera-

From: *Methods in Molecular Biology, vol. 292: DNA Viruses: Methods and Protocols*
Edited by: P. M. Lieberman © Humana Press Inc., Totowa, NJ

peutic drugs are available *(3)*. However, virus culture is often hampered by a labor-intensive procedure, a prolonged turnaround time, and a subjective way to observe the cytopathic effect (CPE). Hence, a simple and rapid cell culture-based assay for detection of HSV would be highly desirable.

With advances in the understanding of HSV replication, it is conceivable to take advantage of virus-specific events, such as transcription from a viral promoter, to identify virus-infected cells. The basic strategy is to stably deliver into a cell an expression cassette containing a reporter gene driven by a viral promoter; when a particular virus enters this cell, a virus-specific transactivation upon the promoter would take place, leading to production of an easily detected reporter protein. Lines of evidence have indicated that the promoter controlling the large subunit of ribonucleotide reductase, the ICP6 from HSV-1 or the ICP10 from HSV-2, may be suitable for this goal. Studies have shown that the background expression driven by these promoters is extremely low and that such expression is specifically induced by HSV-1 or HSV-2 infection within several hours post infection *(4–8)*. Furthermore, a cell line (VeroICP6LacZ) that was stably transfected with the *Escherichia coli LacZ* gene driven by the HSV-1 ICP6 promoter was reported *(9)* and was utilized to identify HSV-1 or -2 rapidly in clinical specimens *(10,11)*, as well as to determine antiviral susceptibility *(12,13)*. However, detection and quantitation of the *LacZ* gene expression require fixation of cells, addition of exogenous substrates, and preparation of cell lysates. Through these experimental procedures, cells of interest are killed and are not suitable for further study. Thus reporter genes, the green fluorescent protein *(GFP)* and secreted alkaline phosphatase *(SEAP)* genes, were taken advantage of to establish alternative reporter systems for continuous monitoring of HSV infections in living cells.

GFP, originally identified from jellyfish *Aequorea victoria,* has several promising features that make it ideal for the purpose. *GFP* emits bright green light after exposure to ultraviolet or blue light without extrinsic labeling or substrates. After the wild-type (wt) *GFP* gene was cloned, it was used as a reporter gene in a real-time and noninvasive fashion *(14)*. In addition, variants of *GFP* have also been designed that are better adapted to mammalian expression and signal detection. For instance, enhanced *GFP (EGFP)* has a single, strong, red-shifted excitation peak at 488 nm, which is well suited for detection by fluorescence microscopy *(15,16)*. Kung et al. *(17)* addressed a reporter cell system using *EGFP* driven by the promoter of the HSV-2 ICP10. Real-time detection of the reporter signal can be achieved under an inverted microscope as early as 6 h after infection, with progressive increase in the intensities at later time points. Moreover, each *EGFP*-positive cell reflects a plaque-forming unit *(17)*, suggesting that the sensitivity of this reporter system is equivalent to that of a plaque formation assay.

As for the SEAP reporter gene, SEAP is a highly stable, secreted enzyme whose detection only requires supernatants from the cultured cells; thus cell lysate preparation is unnecessary, and automation is amenable *(18–20)*. A reporter cell line that utilizes the *SEAP* gene was also documented *(21)*. The reporter cell system allows extremely sensitive detection, at a linear range over four orders of magnitude as measured by a chemiluminescence-based assay *(22,23)*. Moreover, evaluation of the stable line with acyclovir (ACV)-sensitive or -resistant HSV isolates demonstrated that the 50% inhibitory concentrations (IC$_{50}$s) determined by this method correlated closely with those obtained by a plaque reduction assay (PRA) *(21)*.

The reporter cell system permits detection of the reporter signal early in the very first cycle of infection, unlike the inspection of CPE, which takes more than one replication cycle to develop. This is mainly because of the expression kinetics of the viral promoter, the sensitivity of the reporter assay, and the permissiveness of the parental cell line chosen. Furthermore, the reporter cell system is considered a functional assay that measures infectious virus particles, avoiding the defective virions that are biologically less important yet detectable with nucleic acid-based tests. Taken together, both the *GFP* and *SEAP* reporter systems are useful means for simple, rapid, and sensitive detection of HSV, with the added value of *GFP* for real-time observation and *SEAP* for quantitation and automation.

2. Materials

2.1. Reagents

1. 100% Isopropanol
2. ACV (Sigma, St. Louis, MO): prepared as stock solution at 10–50 m*M* in water and stored at –20°C.
3. Dulbecco's modified Eagle's medium (DMEM) supplemented with 10% heat-inactivated fetal bovine serum and penicillin/streptomycin (referred to as DMEM-10) is used for routine cultivation of tissue culture cells.
4. DMEM supplemented with 2% heat-inactivated fetal bovine serum and penicillin/streptomycin (referred to as DMEM-2) is used for viral infection of tissue culture cells.
5. Great EscAPe™ SEAP Chemiluminescence kit (Clontech, Palo Alto, CA).
6. Lipofectamine (Gibco-BRL, Gaithersburg, MD) for transfection.
7. Neomycin (G418 sulfate, MDBio) for selection in mammalian cells.
8. Plasmid pEGFP-1 (Clontech) encoding a human codon-optimized *GFP* downstream of a multiple cloning site, and a neomycin resistance gene.
9. Plasmid pSEAP2-Basic (Clontech) encoding a human placental *SEAP* downstream of a multiple cloning site.
10. Plasmid pSV2*neo* encodes the neomycin-resistant gene for cotransfection.
11. PUREGENE™ kit (Gentra Systems, Minneapolis, MN) for viral DNA isolation.
12. *Taq* DNA polymerase (Promega, Madison, WI).

2.2. Cell Lines

1. Vero African green monkey kidney cells (ATCC deposit no. CCL-81).

2.3. Equipment

1. 37°C Incubator, gassed with 5% CO_2 in air.
2. Thermal cycler for polymerase chain reaction (PCR).
3. Inverted fluorescence microscope (Nikon TE200).
4. Luminometer (Wallac Victor[2] 1420 Multilabel Counter).

3. Methods

To establish a reporter cell line that generates the *EGFP* or *SEAP* reporter in response to HSV infection, several steps are followed. The HSV-2 ICP10 promoter is isolated by PCR amplification and cloned into the reporter gene-harboring plasmid. Stable transfection with the recombinant plasmid into Vero cells is carried out, and appropriate clones are selected following HSV infection; those with the least background expression and the highest inducible expression are chosen. These clones are evaluated for *EGFP* or *SEAP* production by fluorescence microscopy or chemoluminescent assay, respectively (*see* **Note 1**).

3.1. Extraction of Viral DNA

1. Infect a subconfluent to confluent monolayer of Vero cells with an HSV-2 (strain 186) stock in a T25 tissue culture flask at a multiplicity of infection (MOI) of 3.
2. Allow the infection to proceed for approx 24 h. All cells should be rounded up and still adherent to the flask, yet they should be just about ready to detach.
3. The following protocol follows the manufacturer's instructions for PUREGENE, with slight modifications. Add 50 µL supernatant from virus-infected cells to a sterile 1.5-mL microcentrifuge tube containing 250 µL cell lysis solution. Incubate at 65°C for 15 min to complete lysis.
4. Add 1.5 µL RNase A solution to the cell lysate, and incubate at 37°C for 30 min to cool the sample to room temperature.
5. Add 100 µL protein precipitation solution to the lysate. Vortex the sample, and place it into an ice bath for 5 min. Centrifuge at 13,000–16,000*g* for 5 min.
6. Pour off the supernatant containing the DNA (leaving behind the precipitated protein pellet) into a clean 1.5-mL microcentrifuge tube containing 300 µL 100% isopropanol.
7. Mix the sample and incubate at room temperature for at least 5 min. Centrifuge at 13,000–16,000*g* for 5 min.
8. Pour off the supernatant, add 300 µL 70% ethanol, and invert the tube several times to wash the DNA pellet.
9. Centrifuge at 13,000–16,000*g* for 1 min. Carefully pour off the ethanol.
10. Invert and drain the tube on clean absorbent paper and allow to air-dry.
11. Resuspend viral DNA in 50–100 µL TE buffer or dH_2O using wide-bore pipet tips.

3.2. PCR for the DNA Segment Encompassing the HSV-2 ICP10 Promoter

1. To obtain a DNA fragment bearing the ICP10 promoter (–535 to +113 relative to the mRNA cap site *[5]*) of HSV-2 (strain 186), use the following set of primers for a PCR reaction:
   a. Forward: 5′-CGGAGATCTGCTGCAGAACCTCTTTCCCTA-3′.
   b. Reverse: 5′-CTTAAGCTTATCAGACGACGGTGGTCCCGG-3′.
2. PCR parameters are as follows: 98°C for 10 min, then 30 cycles with a denaturation step at 94°C for 1 min, hybridization at 56°C for 2 min, and elongation at 72°C for 3 min with a final elongation at 72°C for 7 min. The 5′ and 3′ primers are engineered with the *Bgl*II and *Hind*III restriction sites, respectively.

3.3. Construction of Plasmids

The PCR fragment is subsequently restricted and directionally cloned into the corresponding sites of the multiple cloning sites (MCS) in the pEGFP-1 or pSEAP2-Basic plasmid (Clontech). The resultant recombinant plasmid is designated pICP10-EGFP or pICP10-SEAP.

3.4. Generation of a Stably Transfected Cell Line That Displays EGFP After HSV Infection

3.4.1. Transfection by Lipofectamine

1. Split Vero cells 1:6 from a fully confluent 6-well plate in DMEM-10 medium, and incubate the cells at 37°C in a CO_2 incubator for 18–24 h before transfection.
2. For transfection, prepare the following solutions in a microcentrifuge tube:
   Solution A: for each transfection, dilute 2 μg of DNA into 100 μL serum-free DMEM.
   Solution B: for each transfection, dilute 2 μL of lipofectamine reagent into 100 μL serum-free DMEM.
3. Combine the two solutions, mix gently, and incubate at room temperature for 30 min to allow DNA–liposome complexes to form.
4. While complexes form, rinse the cells once with 2 mL of serum-free DMEM.
5. For each transfection, add 0.8 mL of serum-free DMEM to the tube containing the complexes. Mix gently and overlay the diluted complex solution onto the rinsed cells.
6. Incubate the cells with the complexes at 37°C in a CO_2 incubator for 5 h.
7. Following incubation, add 1 mL of DMEM containing 20% FBS without removing the transfection mixture.
8. Replace the medium with fresh DMEM-10 medium at 18–24 h following the start of transfection.

3.4.2. Selection of Appropriate Transfected Clones

1. Forty-eight hours after the transfection, propagate the transfected cells in DMEM-10 containing neomycin at 0.8 mg/mL until the mock-transfected cells are all dead (about 2 wk).

2. After the selection period, individual colonies are isolated by limiting dilution in 96-well plates by seeding 0.5–2.0 cells per well.
3. Pick up at least 30 colonies and maintain in DMEM-10 media containing 0.2 mg/mL neomycin.
4. Observe the colonies with an inverted fluorescence microscope; subject those that show an undetectable level of *EGFP* to infection with HSV-2 stock at an MOI of 3.
5. Ten hours following the infection, select the clones that consistently exhibit approx 100% expression of *EGFP* for subsequent studies. The most appropriate clone from the procedure above is designated Vero-ICP10-EGFP (*see* **Note 2**).

3.5. Development of Stable Transfectants That Produce SEAP Following HSV Infection

The protocol is much like that given in **Subheading 3.4.2.** except that transfection is carried out with 2 μg pICP10-SEAP and 0.2 μg pSV2*neo* plasmids. The method to single out an appropriate *SEAP*-producing clone is to measure the *SEAP* activity; the clone with the lowest background level and the highest induced SEAP activity following viral infection is designated Vero-ICP10-SEAP and is used for further study.

3.6. The SEAP Assay

The protocol follows the manual instructions for the Great EscAPe™ SEAP Chemiluminescence kit (Clontech) with some modifications.

1. Remove an aliquot of supernatant from 24-well tissue culture plates and clarify by centrifugation in a microcentrifuge at 12,000*g* for 2 min.
2. Mix 90 μL of 1X dilution buffer with 10 μL of sample, and then incubate the mixture at 65°C for 30 min to eliminate the endogenous alkaline phosphatase activity.
3. Add 100 μL of assay buffer to the mixture, and incubate for 5 min at room temperature.
4. Prepare a CSPD substrate at a concentration of 1.25 m*M* by diluting with 20X chemiluminescence enhancer.
5. Add 100 μL of the diluted substrate to each diluted sample for 10 min at room temperature.
6. The chemiluminescent signal was detected by a luminometer (Wallac Victor[2] 1420 Multilabel Counter).

3.7. Reporter Cell-Based Antiviral Susceptibility Testing

1. Seed the Vero-ICP10-SEAP cells in 24-well tissue culture plates about 18–24 h before infection.
2. Inoculate the cells with virus suspension in the amount of 10–100 plaque-forming units (PFU) in 100 μL DMEM-2 medium.
3. After a 90-min adsorption, aspirate the inoculum and add 500 μL DMEM-2 medium alone or the medium with ACV to each well.

4. To estimate the IC_{50} values derived from the *SEAP* reporter system, twofold serial dilutions of ACV are applied. For the drug-sensitive strains, the concentrations tested are 2, 1, 0.5, 0.25, and 0.125 µg/mL. For the drug-resistant isolates, the concentrations 32, 16, 8, 4, and 2 µg/mL are tested. Control groups include mock-infected cells and infected cells with no ACV added.

5. Following infection for 48 h, collect supernatant from each well and measure *SEAP* activity by the method given in **Subheading 3.6.**

6. Relative *SEAP* expression is given as percentage of expression for control infections without ACV and is plotted against the ACV concentration. The IC_{50} is defined as the ACV concentration that reduced the SEAP reading by 50% from that in the untreated control wells (*see* **Note 3**).

4. Notes

1. To ensure consistency of experiments, the stable transfected cells should be assessed for their responsiveness to HSV infection from time to time (say every 3 mo). If the stable clones lose their sensitivity to viral infection, performing limiting dilution is recommended to gain an appropriate clone.

2. Serotyping of HSV is clinically essential because of their distinct modes of pathogenesis and different responses to antiviral administration. However, both Vero-ICP10-EGFP and Vero-ICP10-SEAP cells respond to HSV-1/2 infection in an indistinguishable fashion. Nevertheless, serotyping could be accomplished by conducting an immunofluorescent assay using monoclonal antibody to HSV-1 or -2 once the reporter signal is detected from these stable lines.

3. Clinically, it is important to determine in a timely manner whether the clinical syndrome of a patient is caused by a drug-resistant HSV. Clinical specimens of unknown titers could be applied for the *SEAP*-based antiviral susceptibility testing. The viral titer could be rapidly determined by simultaneous inoculation of the *GFP*-based Vero-ICP10-EGFP cell line. Concurrently, a serial 10-fold dilution of clinical specimens would be performed, followed by direct inoculation of each dilution onto the Vero-ICP10-SEAP line in the absence or presence of ACV at 2 µg/mL (the cutoff to differentiate drug susceptibility from resistance). From real-time enumeration of the *GFP*-positive cells under a fluorescent microscope, the dilution that results in a titer at the range of $10–10^3$ PFU suitable for the assay would be identified and employed; if the SEAP activity from the ACV-containing well is greater than 50% of that from the corresponding no-drug well, the virus would be considered resistant.

References

1. Whitley, R. J. and Roizman, B. (2002) Herpes simplex viruses, in *Clinical Virology,* 2nd ed. (Richman, D. D., Whitley, R. J., and Hayden F. G., eds.), ASM, Washington, DC, pp. 375–401.

2. Storch, G. A. (2001) Diagnostic virology, in *Fields Virology,* vol. 1, 4th ed. (Knipe, D. M. and Howley, P. M., eds.), Lippincott Williams & Wilkins, Philadelphia, pp. 493–531.

3. Crumpacker, C. (2001) Antiviral therapy, in *Fields Virology,* vol. 1, 4th ed. (Knipe, D. M. and Howley, P. M., eds.), Lippincott Williams & Wilkins, Philadelphia, pp. 393–433.

4. Goldstein, D. J. and Weller, S. K. (1988) Herpes simplex virus type-1 induced ribonucleotide reductase activity is dispensable for virus growth and DNA synthesis: isolation and characterization of an ICP6 *lacZ* insertion mutant. *J. Virol.* **62,** 196–205.

5. Wymer, J. P., Chung, T. D., Chang, Y. N., Hayward, G. S., and Aurelian, L. (1989) Identification of immediate-early-type *cis*-response elements in the promoter for the ribonucleotide reductase large subunit from herpes simplex type 2. *J. Virol.* **63,** 2773–2784.

6. Wymer, J. P., Aprhys, C. M., Chung, T. D., Feng, C. P., Kulka, M., and Aurelian, L. (1992) Immediate early and functional AP-1 *cis*-response elements are involved in the transcriptional regulation of the large subunit of herpes simplex virus type 2 ribonucleotide reductase (ICP10) *Virus Res.* **23,** 253–270.

7. Hanson, N., Henderson, G., and Jones, C. (1994) The herpes simplex virus type 2 gene which encodes the large subunit of ribonucleotide reductase has unusual regulatory properties. *Virus Res.* **34,** 265–280.

8. Zhu, J. and Aurelian, L. (1997) AP-1 *cis*-response elements are involved in basal expression and Vmw110 transactivation of the large subunit of herpes simplex virus type 2 ribonucleotide reductase (ICP10) *Virology* **231,** 301–312.

9. Stabell, E. C. and Olivo, P. D. (1992) Isolation of a cell line for rapid and sensitive histochemical assay for the detection of herpes simplex virus. *J. Virol. Methods* **38,** 195–204.

10. Stabell, E. C., O'Rourke, S. R., Storch, G. A., and Olivo, P. D. (1993) Evaluation of a genetically engineered cell line and a histochemical β-galactosidase assay to detect herpes simplex virus in clinical specimens. *J. Clin. Microbiol.* **31,** 2796–2798.

11. Patel, N., Kauffmann, L., Baniewicz, G., Forman, M., Evans, M., and Scholl, D. (1999) Confirmation of low-titer, herpes simplex virus-positive specimen results by the enzyme-linked virus-inducible system (ELVIS) using PCR and repeat testing. *J. Clin. Microbiol.* **37,** 3986–3989.

12. Tebas, P., Stabell, E. S., and Olivo, P. D. (1995) Antiviral susceptibility testing with a cell line that expresses β-galactosidase after infection with herpes simplex virus. *Antimicrob. Agents Chemother.* **39,** 1287–1291.

13. Tebas, P., Scholl, D., Jollick, J., McHarg, K., Arens, M., and Olivo, P. D. (1998) A rapid assay to screen for drug-resistant herpes simplex virus. *J. Infect. Dis.* **177,** 217–220.

14. Chalfie, M., Tu, Y., Euskirchen, G., Ward, W. W., and Prasher, D. C. (1994) Green fluorescent protein as a marker for gene expression. *Science* **263,** 802–805.

15. Cormack, B. P., Valdivia, R. H., and Falkow, S. (1996) FACS-optimized mutants of the green fluorescent protein (GFP) *Gene* **173,** 33–38.

16. Yang, T. T., Cheng, L., and Kain, S. R. (1996) Optimized codon usage and chromophore mutations provide enhanced sensitivity with the green fluorescent protein. *Nucleic Acids Res.* **24,** 4592–4593.

17. Kung, S. H., Wang, Y. C., Lin, C. H., Kuo, R. L., and Liu, W. T. (2000) Rapid diagnosis and quantification of herpes simplex virus with a green fluorescent protein reporter system. *J. Virol. Methods* **90,** 205–212.
18. Berger, J., Hauber, J., Hauber, R., Geiger, R., and Cullen, B. R. (1988) Secreted placental alkaline phosphatase: a powerful new quantitative indicator of gene expression in eukaryotic cells. *Gene* **66,** 1–10.
19. Cullen, B. R. and Malim, M. H. (1992) Secreted placental alkaline phosphatase as a eukaryotic reporter gene. *Methods Enzymol.* **216,** 362–368.
20. Yang, T. T., Sinai, P., Kitts, P. A., and Kain, S. R. (1997) Quantification of gene expression with a secreted alkaline phosphatase reporter system. *Biotechniques* **23,** 1110–1114.
21. Wang, Y. C., Kao, C. L., Liu, W. T., Sun, J. R., and Kung, S. H. (2002) A cell line that secretes inducibly a reporter protein for monitoring of herpes simplex virus Infection and drug susceptibility. *J. Med. Virol.* **68,** 599–605.
22. Bronstein, I., Fortin, J. J., Voyta, J. C., et al. (1994) Chemiluminescent reporter gene assays: sensitive detection of the GUS and SEAP gene products. *Biotechniques* **17,** 172–174, 176–177.
23. Bronstein, I., Martin, C. S., Fortin, J. J., Olesen, C. E., and Voyta, J. C. (1996) Chemiluminescence: sensitive detection technology for reporter gene assays. *Clin. Chem.* **42,** 1542–1546.

II

Virus Structure and Imaging

7

Unraveling the Architecture of Viruses by High-Resolution Atomic Force Microscopy

Alexander J. Malkin, Marco Plomp, and Alexander McPherson

Summary

Atomic force microscopy (AFM) has recently emerged as an effective complement to other structure determination techniques for studying virus structure and function. AFM allows the direct visualization of viruses in a hydrated state and can probe surface topography in unrivaled detail. Moreover, AFM can be used to elucidate dynamic processes associated with the life cycle of viruses in vitro. It can readily produce high-resolution, nonaveraged, single-particle images of both polymorphic and pleiomorphic viruses. Although AFM does not yield images of internal structures within an intact virion as do penetrating techniques such as electron microscopy and X-ray crystallography, nonetheless, by visualizing the surfaces of internal structures upon treatment with chemical and enzymatic agents, as we demonstrated recently with vaccinia virus, modeling of the complex architecture of a large virus is possible.

Key Words: Atomic force microscopy; structural virology; virus architecture; herpes simplex virus; vaccinia virus; plant virus; icosahedral capsid; AFM resolution; subviral structures.

1. Introduction

Detailed knowledge of virion architecture is paramount to a comprehensive understanding of the key events in a virus life cycle, and this, in turn, is of substantial value in the implementation of more efficacious preventive and therapeutic measures against emerging diseases and, potentially, various aspects of biodefense. Furthermore, the surface structures of viruses determine their physicochemical properties such as hydrophobicity, adhesion, dispersal, and response to the environment. Thus the identification and characterization of surface proteins and the internal structures of viruses is critical to delineating mechanisms of pathogenesis and host immune response. However, despite decades of study of viruses, and their pressing importance in human medicine and biodefense, many of their structural properties remain unresolved.

From: *Methods in Molecular Biology, vol. 292: DNA Viruses: Methods and Protocols*
Edited by: P. M. Lieberman © Humana Press Inc., Totowa, NJ

In the past several decades structural studies of viruses have been conducted either by X-ray crystallography *(1)*, electron microscopy (EM) *(2,3)*, or a combination of both techniques *(3)*. Whereas X-ray crystallography is typically limited to the study of virion structures with a diameter of less than about 50 nm, EM can be utilized for investigations of larger virions. In the past 5–7 yr significant advances have been made in structural studies of viruses by cryo-EM coupled with image reconstruction techniques *(2)*. However, because of their large sizes, lack of symmetry, and structural heterogeneity, the vast majority of animal and human viruses are often refractory to X-ray crystallographic analysis or reconstruction by cryo-EM.

Progress in structural virology very much depends on the development of new high-resolution techniques and tools. Atomic force microscopy (AFM), which was invented by Binnig, Quate, and Gerber in 1986 *(4)*, has become a major tool for rapid and accurate topographical characterization of various materials. It has sub-nanometer height resolution and several-nanometer lateral resolution. It spans a range of scanning dimensions from tens of nanometers up to several hundreds of microns. It is a relatively easy-to-use, rapid, and reasonably inexpensive technique. AFM has tremendous potential in structural virology and pathology since it has the capacity to provide direct high-resolution information on the structure, function, and assembly of virions in vitro. AFM can be used to image the intact structures of viruses, as well as the internal structures of viruses through chemical and enzymatic dissection. This then serves as a guide for modeling the virion architecture.

2. Materials

Turnip yellow mosaic virus (TYMV), cucumber mosaic virus (CMV), herpes simplex virus-1 (HSV-1), and intracellular mature vaccinia (IMV) virus were isolated and purified according to procedures described in detail elsewhere *(5–8)*. AFM imaging of intact virions under physiological conditions was conducted in appropriate buffer solutions, which were normally those utilized for resuspension of virus during late stages of purification. Detergents and enzymes typically used for disassembly of virions were utilized in the AFM dissection experiments.

3. Methods

The methods described below outline (1) the principle of AFM operation, (2) practical questions related to AFM operations, and (3) the visualization of intact virions and subviral structures by AFM

3.1. Principle of Operation

AFM (**Fig. 1**) utilizes an ultrasharp microfabricated tip (**Fig. 2A**) mounted on the end of a flexible cantilever 100–250 μm long. This is scanned across the sur-

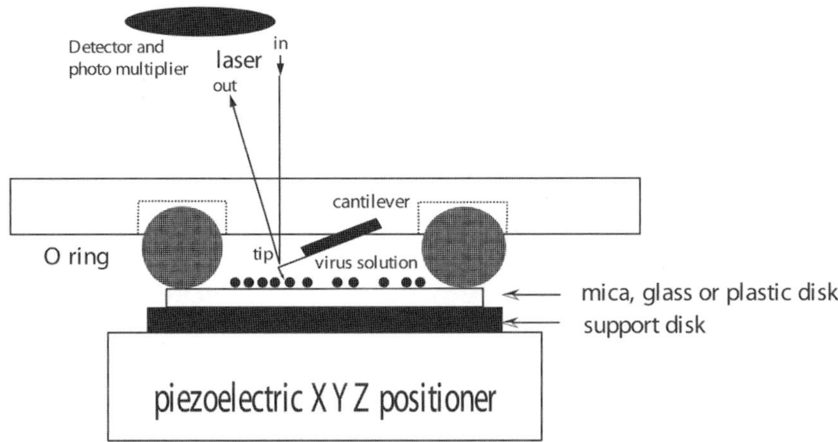

Fig. 1. Schematic diagram of an atomic force microscope. The volume of a fluid-filled cell depends on the thickness of the O-ring and is approx 30–70 µL.

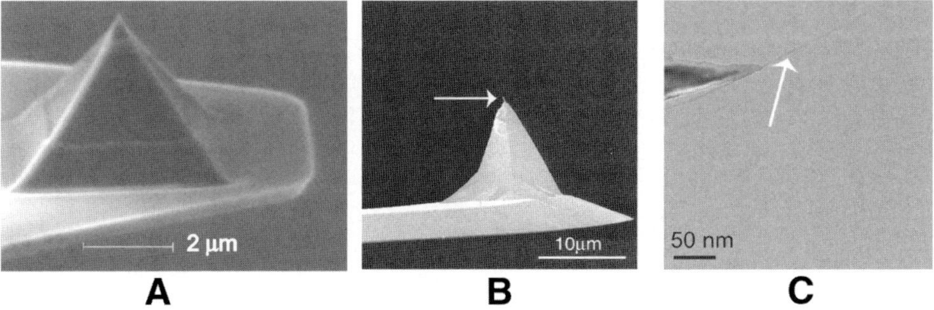

Fig. 2. (A) SEM image of the oxide sharpend Si_3N_4 AFM probe. **(B)** SEM image of a nanotube tip (arrow) attached to an Si cantelever. **(C)** TEM image of CVD grown nanotube tip (arrow) on an Si cantelever. (B and C, courtesy of A. Noy, Lawrence Livermore Laboratory.)

face of a sample in a systematic raster manner, with the lateral and vertical movements of either the tip or the sample under piezoelectric positional control. Because of interactions between atoms on the surface of the sample and those on the tip, deflection of the cantilever takes place. The deflection of the cantilever is monitored with an optical beam deflection system, whereby a laser beam focused on the top of the cantilever is reflected onto a split photodiode. Cantilever deflections as small as 1–2 Å can be readily measured and utilized to correct the tip–sample distance in order to maintain a constant imaging force.

Subsequently, the height corrections are used to construct a topographic image of the sample.

AFM can operate in either contact or tapping mode. In the former mode, the AFM probe remains in contact with the sample during the scanning, whereas in the latter mode the tip oscillates a few angstroms from the sample surface, tapping it only during a short interval in its oscillation cycle. In both tapping and contact modes, the feedback mechanism adjusts, through the piezoelectric positioner, the vertical height of the sample (or AFM probe) in order to maintain a constant amplitude of the oscillating probe (tapping mode), or deflection of the cantilever (contact mode).

In contact mode, lateral forces caused by probe–surface interactions typically complicate the imaging of soft biological samples. This often results in poor imaging or even the physical displacement of virions adsorbed on the substrate. In tapping mode, virtually no lateral forces are applied to the sample during imaging, which makes it the preferred approach to probe structures of virions and their disassembly at high resolution.

3.2. Imaging in the Fluid Environment

AFM experiments can be performed in vacuum, air, and (most importantly for imaging of biological samples) physiological fluids. Operation in a liquid environment requires utilization of a fluid cell (**Fig. 1**), commercially available from AFM manufacturers (*see* **Note 1**). Because visualization is carried out in a fluid environment, specimens suffer no dehydration, as is generally the case with electron microscopy. Additionally, no fixing or staining of the sample is required. Virus specimens can be observed over extended periods as long as they remain stationary. The great power of AFM, however, lies not only in its imaging capability but also from the nonperturbing nature of the probe interaction with the surface under study. Because the specimen ignores the presence of the probe, its natural processes continue uninhibited. This allows the investigator to record, not only a single snapshot, but a series of images that may extend over hours or even days (*see* **Note 2**). The imaging frequency depends on the scan rate of the probe, and images may be gathered rapidly, within tens of seconds, or more slowly. Since image quality generally deteriorates with increased scan speed, images are commonly collected over a period of 1–5 min.

A further property of AFM, when carried out in fluid cells, is that the liquid can be changed repeatedly during the course of an experiment without appreciably disturbing the specimen. This is of great value in the study of structure, function, and dynamic properties of viruses because it is often desirable to study these processes under different solution conditions. Temperature in the AFM fluid can be modified over a wide range using either relatively easy to build in-house temperature control devices or those commercially available from AFM manufacturers.

3.3. Sample Preparation and Image Data Type

Typically, viruses adhere well to freshly cleaved mica. Other substrates, such as graphite or glass, can also be utilized. If samples do not adhere, then it may be necessary to treat the substrate with various reagents (for example, glass substrates coated with poly-L-lysine) in order to improve adhesion. For AFM experiments, droplets of virus suspension (as small as 0.5 µL), or chemically/enzymatically treated samples thereof, are deposited directly onto freshly cleaved mica substrates and are allowed several minutes to settle (*see* **Note 3**). For imaging of fully hydrated samples in physiological buffer (*in situ* imaging), the mica substrates with the deposited sample are transferred into the AFM fluid cell, which is subsequently filled with buffer solution. For imaging in air (*ex situ* imaging) the sample droplets are allowed to settle for approx 5 min, and mica substrates are then rinsed gently with double-distilled water and quickly dried with a stream of nitrogen gas (*see* **Note 4**).

Tapping mode AFM images are typically collected from *height* data. *Height* data correspond to the change in piezo height needed to keep the oscillation amplitude of the cantilever constant. Simultaneously, images can be collected from *amplitude* or *phase* data, which correspond to the change in amplitude or phase of the cantilever oscillation respectively. *Amplitude* images (for imaging in fluid) and *phase* images (for imaging in the air) typically display the greatest amount of contrast and detail. Usually height and *amplitude/phase* data are captured simultaneously (*see* **Notes 5** and **6**).

3.4. AFM Probes and Resolution

The sharpness and apex shape of the probe tip determine the lateral apex resolution of AFM. Commercial AFM probes fabricated from Si and Si_3N_4 are typically utilized for imaging in air and under fluid, respectively. Their radii of curvature are 5–20 and 20–60 nm, respectively. As seen in **Fig. 2A,** they have square pyramidal shapes with cone angles of 20–35°. Overall tip shape and sharpness become more important in high-resolution imaging of the relatively rough structures of viruses, which have deep crevasses, and in delineating details of protein complexes on their surfaces. Tip parameters can be significantly improved by etching of Si tips and oxide sharpening of Si_3N_4 tips, which improve their aspect ratios and produce smaller tip radii of 3–5 nm (Si) and 5–10 nm (Si_3N_4). These sharpened AFM probes are available from various manufacturers (Digital Instruments, Asylum Research, Molecular Imaging, and others). Although Si probes are sharper than silicon nitride probes, because of their relative stiffness, they typically cannot be utilized for high-resolution imaging of viruses in fluid.

Silicon probes either displace virions attached to a substrate or damage the relatively soft surfaces of the virion. Thus AFM imaging *in situ* is conducted strictly by utilizing Si_3N_4 probes. For imaging in the air, Si tips with spring constants of >50 N/m can be utilized. For imaging of biological samples in fluid, Si_3N_4 probes with a spring constant of <1 N/m are typically used.

In the last several years, various techniques have been developed to fabricate carbon nanotube AFM probes *(9)*. These probes typically have an aspect ratio of more than 100, and the radius of curvature of a single walled nanotube is approx 1 nm. Carbon nanotube AFM probes are fabricated either by manual assembly, growth by chemical vapor deposition (CVD), or pick-up methods *(10)*. Examples of a manually attached carbon nanotube to an Si probe and a carbon nanotube grown by CVD on an Si AFM probe are presented in **Fig. 2B** and **C.** The high-resolution imaging capabilities of carbon nanotube probes, which are now commercially available (e.g., from Molecular Nanosystems), were demonstrated in air recently on various biomolecules, such as DNA, antibodies, proteins, and nucleosomes (thoroughly reviewed in **ref. *10***). However, various technical difficulties have so far prevented application of carbon nanotube AFM probes in fluid.

3.5. Typical AFM Image Artifacts

A major technical problem that an experimentalist performing high-resolution AFM imaging of viruses faces is variation in tip properties. Thus, the sharpness of AFM probes within the same wafer, microfabricated under the same conditions, can vary significantly (*see* **Note 7**). Additionally, probes can become dull or contaminated with debris during imaging. This results in poor-quality or erroneous images, with features that reflect the shape of the tip or the debris attached to it rather than the shape of the imaged object. Double- or multiple-tip images of the same feature can be formed with an AFM probe having two or more endpoints that contact the surface simultaneously. If the size of the probe tip has the same scale size as the virions adsorbed on a substrate, then the lateral sizes of the former appear larger in AFM images than might be expected. This happens because the image obtained is the convolution of the AFM probe shape with that of the particle. On the other hand, because of extremely high "height" resolution, which approaches 1 Å, the vertical height measurements from AFM of virions give a remarkably accurate value for their diameters.

3.6. Imaging of Icosahedral Viruses

3.6.1. T = 3 Icosahedral Plant Viruses

AFM provides a resolution sufficient not only to visualize the gross shape of virions but also to visualize the molecular details of virus surfaces. Indeed, this was demonstrated in the first direct visualization of the capsomere structure of

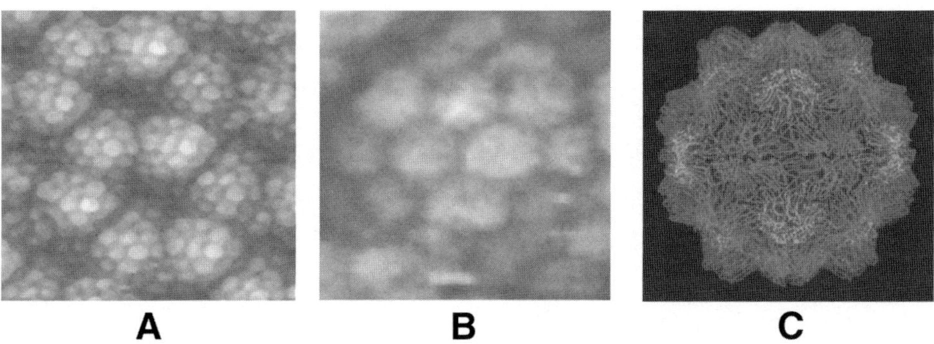

Fig. 3. (A) and (B) *In situ* AFM images of TYMV particles immobilized in the crystalline lattice clearly display capsomeres on the surface of the *T* = 3 icosahedral virions. C Structure of the capsid of TYMV based on X-ray diffraction analysis. AFM images are 140 × 140 nm (A) and 38 × 38 nm (B).

a virus by AFM *(5).* **Figure 3A** shows surface area from a TYMV crystal. The capsid of a *T* = 3 icosahedral TYMV virion is 28 nm in diameter and is composed of 180 identical protein subunits, each of about 20 kDa, organized into 12 pentameric and 20 hexameric capsomeres. The capsomeres protrude about 40 Å above the surface of the virion *(11).* In **Fig. 3A** and **B,** surfaces of individual icosahedral virions making up the plane of TYMV crystals can be clearly distinguished. Furthermore, individual capsomeres consisting of either five or six protein capsid subunits are clearly resolved. High-resolution images of a single TYMV virion, seen in **Fig. 3B,** are remarkably consistent with the 3.2-Å structural model of the virion **(Fig. 3C)** determined by X-ray diffraction analysis *(11).* From AFM images **(Fig. 3A, B)** the sizes of capsomeres and the differences between the highest and lowest points on the capsid surface were determined to be about 50 and 40 Å respectively, which correlate very well with those known from X-ray diffraction analyses.

Here it was demonstrated that the structure of the virion could be visualized in fluid under physiological conditions to high lateral resolution. This resolution is consistent with the sharpness of the best commercially available Si_3N_4 AFM probes (*see* **Subheading 3.4.**). However, occasionally, even higher resolution has been achieved by *in situ* AFM. Thus, in **Fig. 4A** not only are the approx 50-Å wide capsomeres on the surfaces of CMV virions visible, but holes/depressions at their centers could also be visualized. The structure of CMV has been solved to 3.2 Å resolution **(Fig. 4B)** by X-ray diffraction analysis *(12).* From these data, the size of the depressions is approx 20 Å, which is in good agreement with the AFM data in **Fig. 4A.** This value provides the current upper limit for AFM resolution of a virus structure, while imaging *in situ,*

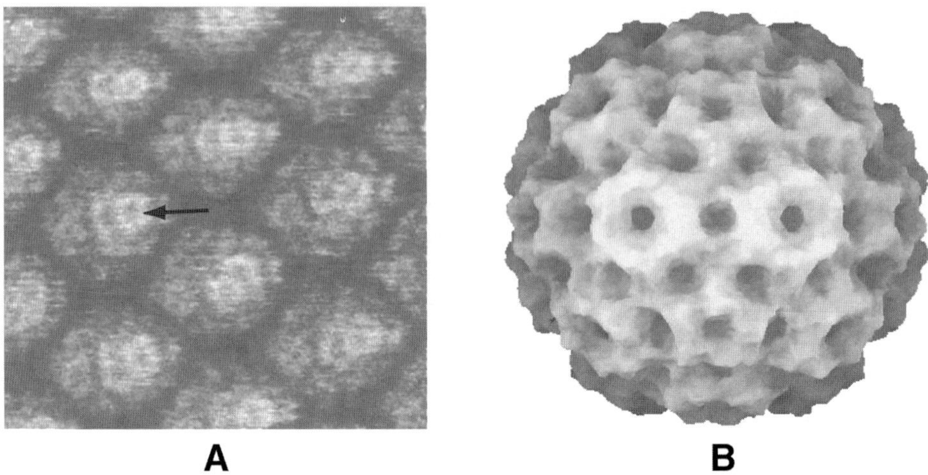

A **B**

Fig. 4. (A) A 100 × 100-nm AFM image of CMV particles immobilized in the crystalline lattice. Arrow points at a caposomere with a visible hollow channel in the center. **(B)** Structure of the capsid of CMV based on X-ray diffraction analysis.

and is similar to that achieved when carbon nanotube AFM probes were utilized for imaging in air *(10)*. This high resolution, achieved in fluid and using commercial oxide sharpened Si_3N_4 probes with end radii typically no better than 5 nm, can be explained by the existence of a nanoprotrusion at the tip end. The resolution achieved with TYMV and CMV could be obtained for the high-resolution imaging of surfaces of virus crystals as well as individual virions adsorbed to a substrate *(13)*.

3.6.2. Visualization of Icosahedral Capsid of Herpes Simplex Virus-1

Recently the structure of the icosahedral capsid of HSV-1 was deduced by cryo-EM reconstruction to 8.5 Å *(14)*. The main components of the icosahedral HSV-1 capsid, which is 125 nm in diameter and of triangulation number $T = 16$, are 12 pentameric capsomeres (pentons) and 150 hexameric capsomeres (hexons). These capsomeres are interconnected by 320 smaller protein complexes known as triplexes *(15)*. Together, these units construct a capsid consisting of 20 equivalent triangles, with pentons on the corners, a row of three hexons between them, and three more hexons in the center of each triangle **(Fig. 5)**. Pentons (hexons) consist of five (six) copies of the 150-kDa VP5 capsid protein *(15)*, and each hexon also contains six copies of the 12-kDa VP26, forming a ring on top of the VP55 *(15,16)*. Each triplex consists of two copies of 33-kDa VP23 and one 53-kDa VP19 protein *(17)*.

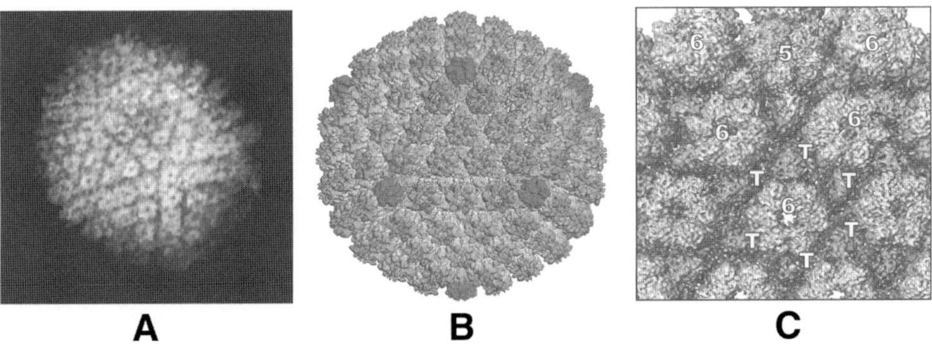

Fig. 5. HSV-1 capsid structure as seen by electron microscopy. **(A)** Single HSV-1 capsid as seen by conventional electron microscopy using negative staining. **(B)** and **(C)** HSV-1 capsid at 8.5-Å resolution, reconstructed by merging of cryo-EM images of 5860 particles *(14)*. (B) The entire capsid, with pentons in indicated by a darker color. (C) A portion of the capsid enlarged. The penton is labeled 5, hexons are denoted 6, and triplexes are labeled T. (A, reproduced with permission from http://www.uct.ac.za/depts/mmi/stannard/linda.html; B and C, reproduced with permission from **ref.** *14.* Copyright 2000 American Association for the Advancement of Science.)

The 8.5-Å cryo-EM model of the HSV structure provides a standard against which the AFM images can be compared. As illustrated in **Fig. 6A** and **B,** icosahedral HSV-1 capsids imaged in air with Si probes clearly show the capsomere structure even at relatively low resolution, with capsomere diameters of approx 15 nm being faithfully recorded by AFM *(7,18)*. Because of their different environments (each penton is surrounded by five hexons, whereas each hexon is surrounded by six other capsomeres), the pentameric and hexameric clusters in **Fig. 6A** and **B** can be discriminated from one another in AFM images.

At higher resolution, the substructure of individual capsomeres could be further visualized *(7,18)*. Thus, as seen in **Fig. 6C,** many capsomeres show a hole (indicated with arrows) with a diameter of approx 35 Å in the center, consistent with the cryo-EM reconstruction model of the HSV-1 capsid **(Fig. 5C)**. In addition, smaller protein clusters (indicated with arrows in **Fig. 6D**) with sizes in the range of 40–50 nm that link adjacent capsomeres can be visualized as well. These clusters correspond to the triplexes, which are a characteristic feature of the capsid **(Fig. 5C)**. In the AFM images some of the triplexes are situated in the cavities between three capsomeres (white arrows in **Fig. 6D**), consistent with the cryo-EM reconstruction model **(Fig. 5C)**. Other units appear positioned between two, rather than three capsomeres (black arrows). The distortion in the symmetry of capsomere packing in the capsid seen in **Fig. 6** as well as various arrangements of triplexes could be caused by drying of the capsids.

                                      Malkin, Plomp, and McPherson

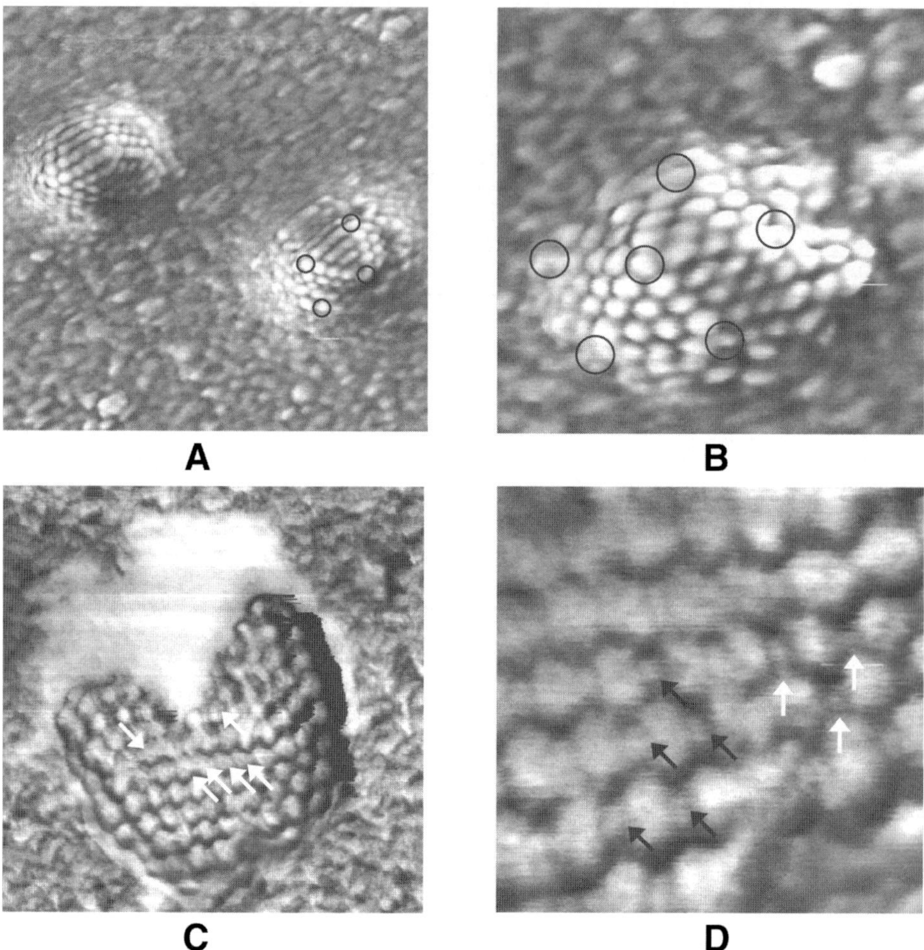

Fig. 6. (A) and **(B)** Icosahedral HSV-1 capsids with pentons denoted by black circles. **(C)** A capsid that flattened upon absorption onto mica, partly covered with (white) lipid envelope. White arrows point at capsomeres with visible hollow channels in their center. **(D)** Center region of the same capsid as in (C). In addition to the channels, units between the hexons and pentons are also visible. These are the triplexes, consistent with the cryo-EM images in **Fig. 5C.** EM shows these triplexes to be positioned in the cavities between three capsomeres. AFM images show triplexes both in these cavities (white arrows) and in the space between two capsomeres (black arrows). (Reproduced with permission from **ref. 7.**)

Although capsids seen in AFM images do not exhibit the level of perfection implied by the cryo-EM image reconstruction structure **(Fig. 4C, D)**, they are consistent, in a general sense, with that model. Not only does AFM permit us

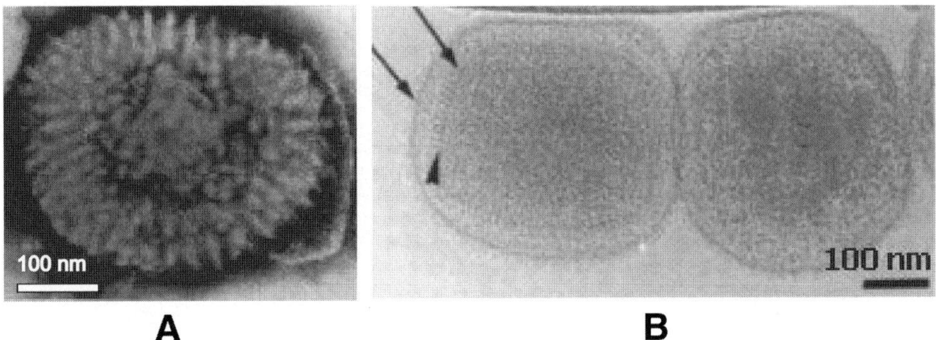

Fig. 7. Electron micrographs of vaccinia virus. **(A)** Particle negatively stained in 2% uranyl acetate. **(B)** Vaccinia viuses as observed by cryo-EM. (A, reproduced with permission from **ref. 22**; B, reproduced with permission from **ref. 21.**)

to visualize the eccentricities that characterize the individual members of the purified population, it shows us the natural variation of the mean structure. Compared with EM single images **(Fig. 5A)**, the resolution is similar or only slightly lower.

3.7. Visualization of Intact Intracellular Mature Vaccinia Virus

Because of lack of symmetry and heterogeneity, large numbers of animal and human viruses are not amenable to high-resolution EM image reconstruction analyses *(2)*. In these cases, AFM can provide the important structural information otherwise lacking. This was demonstrated in our recent AFM studies of IMV virus *(8)*. Vaccinia virus, the basis of the smallpox vaccine, is one of the largest viruses to replicate in humans *(19)*. Vaccinia has a structurally complex, markedly asymmetric virion *(19–21)*. The structure of the vaccinia virion has been extensively studied by conventional EM (thoroughly reviewed in **ref. 19**). The virion was initially described *(19)* as brick-shaped, with dimensions of approx 300 × 230 nm **(Fig. 7A)**, with inner and outer membranes sandwiching a pair of lateral bodies and enclosing a central core containing the genomic DNA.

Initially, cryo-EM studies of vaccinia virus *(22)* revealed that a number of virion features, such as membranous surface tubules, the dumbbell shape of the core, and lateral bodies seen earlier by negative staining EM, were not visible in cryo-EM images **(Fig. 7B)**. This suggested that those features may have been artifacts of virion dehydration and nonisotropic collapse during sample preparation. Nonetheless, structures reminiscent of the lateral bodies as well as the dumbbell shape have been observed even in some recent cryo-EM studies *(20,21)*. Because of these inconsistencies, the combination of conventional and cryo-EM did not provide a definitive model for the structure of the vaccinia virion.

Fig. 8. IMV virions adsorbed to mica. **(A–C)** IMV imaged *in situ* (in 50 m*M* Tris-HCl, pH 7.5). Although isolated virions were observed in all experiments, the vast majority were found, irrespective of the concentration of virions in suspension, to be aggregated into 2D arrays. AFM images are 10 × 10 μm in (A) and 1.2 × 1.2 μm in (B). (C), high-resolution, 385 × 385-nm AFM image of intact IMV virion. (Reproduced with permission from **ref. 8.**)

Figure 8 shows AFM images of the vaccinia virus *(8)* in which naturally occurring hydrated vaccinia aggregates and single particles, in physiological buffer, were adsorbed to mica. Although some virions were spherical in shape, with a diameter of approx 350 nm, most appeared more ellipsoidal, with major and minor axial dimensions in the range 320–380 and 260–340 nm, respectively, consistent with those from cryo-EM studies *(19–22)*. However, virion height, as measured by AFM, varied in the range of 240–290 nm, approximately double the virion height estimated by cryo-scanning EM (SEM) and EM (110 and 150 nm, respectively [*21,23*]). The estimation of vertical dimensions by EM techniques is, however, not straightforward. It depends on the stage tilt angle, whereas "height" information from AFM is very precise, with a resolution approaching 0.1 nm. From the AFM images, unprocessed, fully hydrated IMVs have a rounded barrel shape, which differs from the classical brick-shaped morphologies described in the literature *(19)*. Size heterogeneity of vaccinia virions observed in AFM images may arise from variability in the composition of enveloping membranes or from the alternate stages of intracellular maturation populating the virus harvest.

In contrast to the rather smooth appearance of the virion outer surface by cryo-EM *(20–22)* seen in **Fig. 7B,** *in situ* AFM images showed the virion to have an irregular surface even at relatively low resolution (e.g., **Fig. 8B**). High-resolution *in situ* AFM images **(Fig. 8C)** demonstrate a high density of protrusions on the virion surface, giving it a "knobby" appearance. The protrusions appear to be of fairly uniform size, about 25 × 30 nm, although they are irreg-

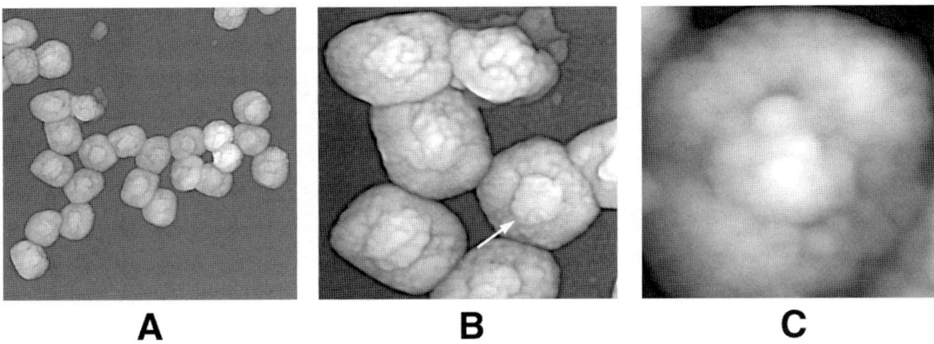

Fig. 9. (A–C) Air-dried IMV virions adsorbed to mica. In (B) the central raised area on the upper surface of one of the virions is indicated with an arrow. AFM images are: (A), 2.9 × 2.9 µm; (B), 1.0 × 1.0 µm; (C), 350 × 350 nm. (Reproduced with permission from **ref. 8.**)

ular in arrangement. The difference in surface topography of intact vaccinia virus seen in AFM and cryo-EM images is more likely owing to intrinsic differences between the two techniques. Whereas *in situ* AFM is a nonpenetrating, topographical technique, cryo-EM images provide a projection of the entire thickness of the sample. Given the virion's thickness of approx 240–290 nm and its electron-translucent properties, it would not be surprising if virion surface protrusions, which could be less than approx 20 nm in height after vitrification, were undetectable in high-resolution cryo-EM images *(20–22)* or even thin-section EM images *(24)*, leading to a smooth, featureless topographical appearance for the virion.

In some of the most detailed images **(Fig. 8C)**, subunits of approx 6 nm diameter could be discriminated within the protrusions *(8)*. This would correspond to a molecular weight of about 120–150 kDa for a monomeric globular protein. Since the molecular weights of the 16 major membrane proteins of vaccinia virus are all considerably less *(25)*, the subunits probably represent oligomers. Because of inherent softness, the virion may be deformed slightly by AFM tip pressure, making the additional fine structure of the protrusions difficult to resolve.

AFM allows direct visualization of the way overall and surface structures of virion change in response to the environment. Thus, upon dehydration **(Fig. 9)**, air-dried vaccinia virions showed relatively uniform dimensions of 300–360 × 240–280 nm (lateral) × 120–130 nm (height) and assumed more rectilinear shapes. Alternative surface features emerge, and the overall brick-shaped morphologies become consistent with that observed by EM *(19)* with a pronounced, central, raised area, which extends approx 30 nm above the virion surface.

Although dehydration of vaccinia virus does not result in loss in infectivity, a property that was integral to the development of vaccinia as a vaccine for worldwide smallpox eradication *(26)*, AFM demonstrates a remarkable 2.2–2.5-fold collapse along the vertical dimension upon drying, indicating the remarkably deformable, fluid nature of the virion structure.

3.8. AFM Imaging of Subviral Structures

An expected disadvantage of AFM would seem to be that, as a nonpenetrating technology, it would provide only topographical information of outer surfaces. To overcome this limitation, internal structural features and interior details can be probed by controlled dissection with chemical agents or enzymes, through a sequential peeling away of layers of structure that reveals those below.

3.8.1. HSV-1

The ability to utilize AFM for imaging of subviral structures was first demonstrated in work on the visualization of HSV *(7)*. Prior to treatment with any agents, the AFM imaging revealed enveloped HSV-1 virions, as presented in **Fig. 10A.** Mixing of the virus solution with detergent (0.2% Triton X-100) results in removal of the lipid envelope from most virions, which then renders the tegument and capsomere structure visible. Thus, in **Fig. 10B,** a capsid that is mostly covered with an irregular coating of particles of various sizes up to 10 nm is seen; on the upper part of the virus capsid, the underlying, highly regular packing of the proteins composing the virus capsid can be recognized as well. It was demonstrated earlier that the tegument, which lies between the lipid envelope and the capsid, has a complex nonicosahedrally ordered arrangement and contains at least 18 different viral proteins. The largest, VP1-3, has a predicted size of 336 kDa *(27)*. This corresponds well to the particles of approx 10 nm size observed in AFM images (**Fig. 10B**). More vigorous treatment of virions with detergent resulted in the complete loss of all envelope and tegument proteins and revealed the icosahedral capsid structure of the virions *(7)*, as seen in **Fig. 6.**

AFM can probe not only the structures of large macromolecular assemblies but also dynamic processes involving the assemblies. Thus, for example, we were able to observe the disassembly of viruses and the release of HSV DNA, as seen in **Fig. 10C.** Investigations of the extrusion of HSV DNA may be important for understanding DNA packing in the capsid shell. In **Fig. 10C,** virions treated with 0.5% sodium dodecyl sulfate (SDS) are seen in the presence of DNA strands, which form bundles of many intertwined DNA chains. The DNA fibers were found to have diameters of 1.5–3 nm, which corresponds well to the 2.5-nm diameter of a double-stranded DNA double helix. In the inset of **Fig.**

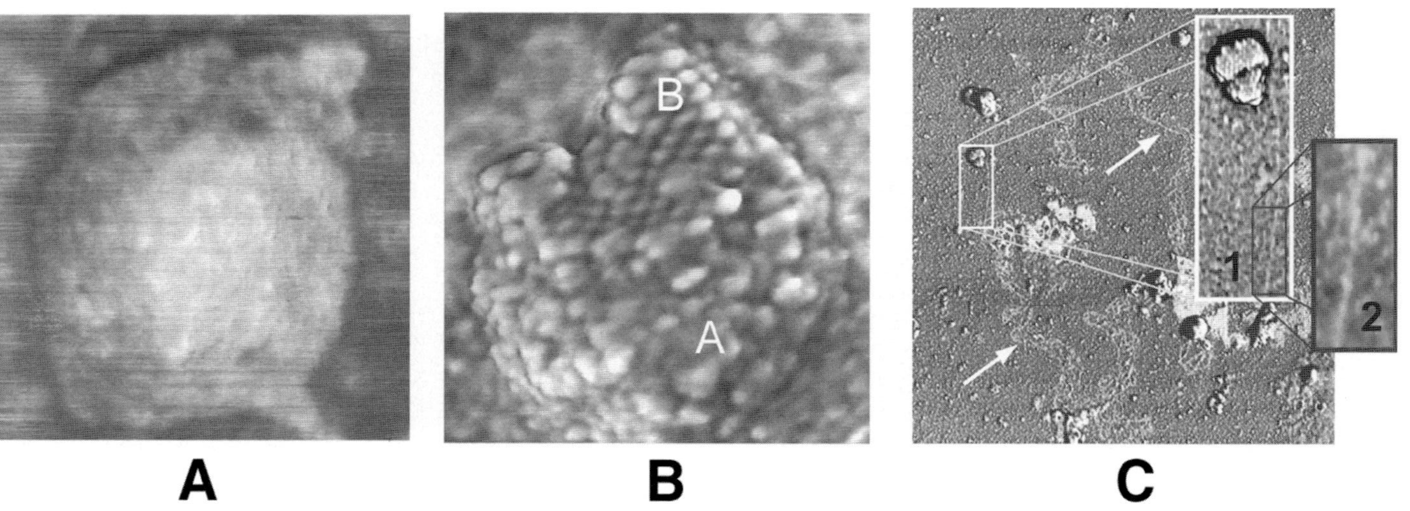

Fig. 10. (A) The intact enveloped HSV-1 virion envelope. **(B)** Addition of 0.2% Triton X-100 partially removes the lipid envelope. Most of this capsid is covered by a irregular collection of particles of approx 10 nm in size (area A), which correspond to tegument proteins. In a smaller region the underlying, highly regular capsid is exposed (area B). **(C)** DNA escapes from the capsids after treatment with 0.5% SDS. Bundles of several DNA chains are indicated with arrows. The area in the white rectangle is enlarged in (C). Fragment of double-stranded DNA escaping from a capsid is seen in insets 1 and 2. (Reproduced with permission from **ref. 7.**)

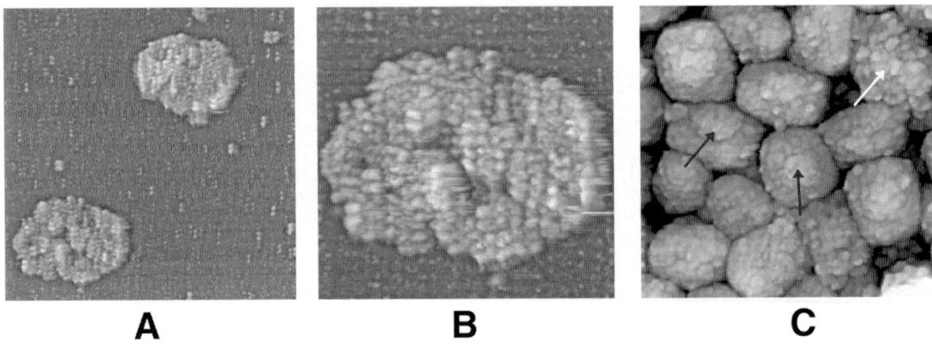

Fig. 11. Dissection of IMV virions with nonionic detergent in combination with reducing agent (1% Igepal, 2% 2-mercaptoethanol in 50 m*M* Tris-HCl, pH 7.5; ~30–45-min incubation at 37°C). **(A)** and **(B)** Intact "coats." **(C)** IMV viral cores. The "satellite domain" structures associated with intact and partially unfolded cores are indicated with black and white arrows, respectively. AFM images are: (A), 2.5 × 2.5 μm; (B), 1 × 1 μm; (C), 1.35 × 1.35 μm. (Reproduced with permission from **ref. 8.**)

10C, entire capsomeres are seen to be lost from the capsid surface, apparently as discrete units, with distinct holes seen in the capsid, and a strand of the individual double-stranded DNA molecules emerging from the HSV capsid.

3.8.2. IMV

The dissection procedures utilized in *ex situ* AFM visualization of subviral structures of HSV-1 could also be applied *in situ* under physiological conditions, as was demonstrated in our recent AFM structural studies of vaccinia virion *(8)*. In the first series of experiments, we treated intact, purified IMV with a nonionic detergent, 1% Igepal, in combination with a disulfide-reducing agent (2%, 2-mercaptoethanol). The effects of these two reagents mimic events during early infection, including stripping of viral membrane upon entry of the virus into the initially reducing environment of the cell. Among the products, seen in **Fig. 11,** were membrane patches embedded with proteins, along with viral cores. The crosslinked "studs" have the same size as the protrusions on the surfaces of untreated IMV **(Fig. 8C).** Membrane patches varied in thickness from 25 to 35 nm and were approx 700 × 900 nm in area. Surrounding the virion core, these structures, seen in **Fig. 11A** and **11B** corresponded to what has been referred to in some EM studies as the virion "outer membrane" *(19)* or "core envelope" *(22).*

The heights of the viral cores seen in **Figs. 11C** and **12A** and **B** are in the range of 170–220 nm, which is roughly the height of the intact IMV virion lacking a 30-nm-thick envelope. Lateral dimensions are in the range of 320 × 260

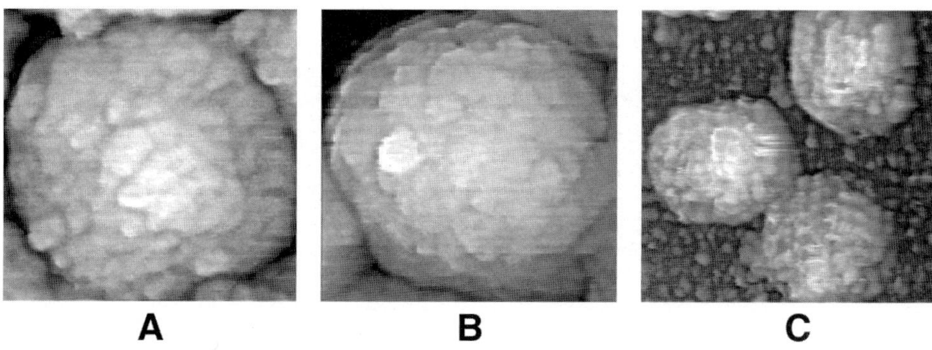

Fig. 12. High-resolution AFM images of cores after dissection of IMV virions with nonionic detergent in combination with reducing agent (1% Igepal, 2% 2-mercaptoethanol in 50 m*M* Tris-HCl, pH 7.5, treated for 30–45 min at 37°C). **(A)** and **(B)** Core surface. **(C)** 70–100-nm diameter particles. AFM images are: (A), 380 × 380 nm; (B), 365 × 365 nm; (C), 400 × 400 nm. (Reproduced with permission from **ref. 8.**)

nm. Although viral cores in **Fig. 12A** and **B** show protrusions similar in size to those seen on the surfaces of intact IMV, their shapes are not the same, consistent with the possibility of different proteins decorating the two surfaces. The surface areas of viral cores were comparable to the surface areas of the core envelopes.

Many of the intact viral cores seen in **Fig. 11C** possess satellite domains *(8)*. These protruding domains are also evident in desiccated samples **(Fig. 9)**. They consistently occupy the top surfaces of particles within aggregates. Their dimensions are similar to the subviral particles seen in **Fig. 12C,** which have dimensions of 70–100 nm, consistent with individual, isolated, satellite domains.

During viral disassembly, satellite domains can detach from the vaccinia virus. The AFM results are consistent with recent cryo-SEM studies of IMV undergoing cellular entry, which also reveal the presence of roughly spherical domains of about 50 nm diameter, independent of the virion. These particles are sometimes accompanied by cavities in the IMV of approximately the same dimensions, consistent with loss of satellite domains *(20)*. The satellite domains observed by AFM in this study may correspond to the lateral bodies observed in conventional EM virages.

More prolonged treatment of IMV with nonionic detergent plus reducing agent produces partial unfolding of IMV and the emergence of 30–40-nm-diameter tubules, as seen in **Fig. 13.** After several hours at 37°C, virions unfold more completely, as illustrated in **Fig. 13A–C,** showing them to be comprised

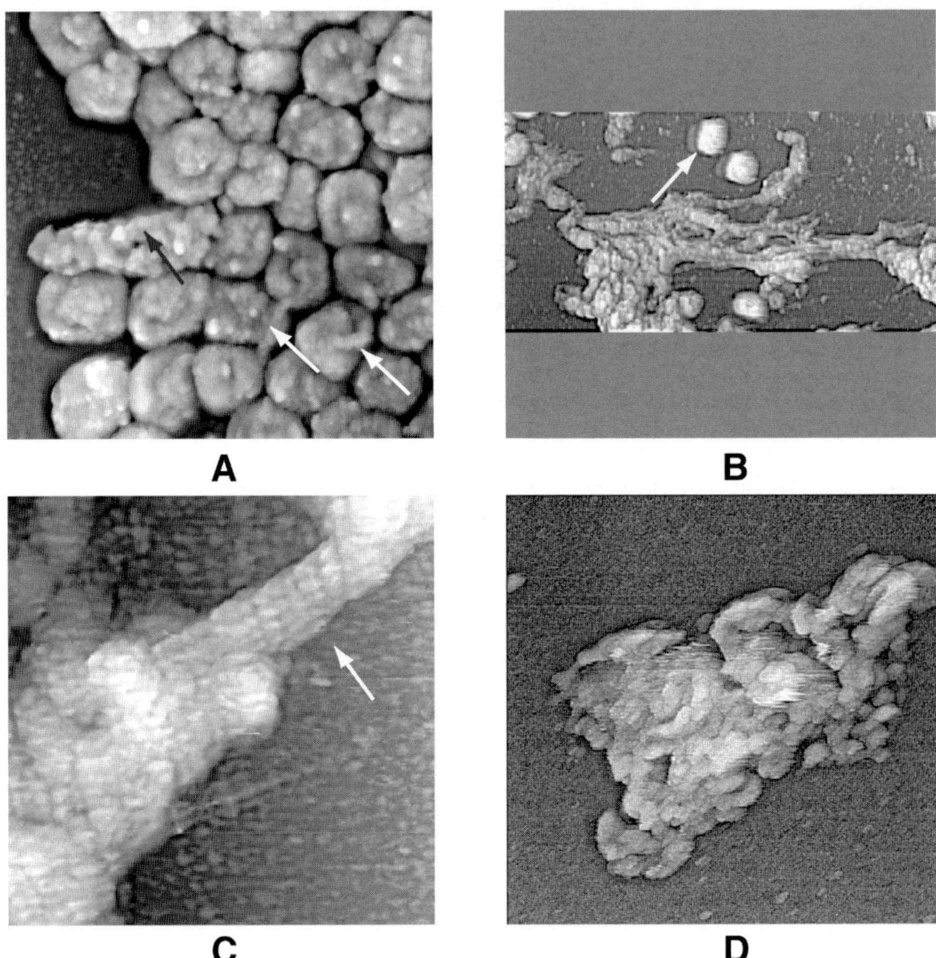

Fig. 13. Unfolding of IMV virions. **(A–C)** Upon extended treatment with nonionic detergent plus disulfide reducing agent (1% Igepal, 2% 2-mercaptoethanol in 50 m*M* Tris-HCl, pH 7.5, 120–180-min treatment at 37°C). (A) Several IMV virions (indicated with white arrows) have become partially unfolded, resulting in the appearance of 30–40-nm diameter tubules. Several virions (indicated with a black arrow) have unfolded completely. (B) and (C) Complete unfolding of IMV results in the appearance of tangled structures formed from 30–40 nm tubules. In (B), intact IMV virions (indicated with an arrow) are also seen. **(D)** 30–40-nm tubular networks are seen upon treatment of IMV with the reducing agent DTT. AFM images are: (A), 1.6 × 1.6 μm; (B), 6 × 3 μm; (C), 570 × 570 nm; (D), 1.35 × 1.35 μm. (Reproduced with permission from **ref. 8.**)

of tangled 30–40-nm tubules. A helical array of protein subunits with left-handed helicity and a pitch of approx 16-nm is evident on the surface of an extended tubule (**Fig. 13C**). The degraded viral cores in **Fig. 13** are tangled knots of 30–40-nm tubules. Similar 30–40-nm tubules were reported based on EM, leading to a novel model in which the IMV is an interconnected labyrinth of tubules and membrane cisternae *(20,21)*.

It was reported that treatment of IMV with reducing agents also affects the integrity of IMV *(20,21)* and contributes in vivo to the "activation" of vaccinia virions for early-phase transcription *(28)*. Exposure of IMV to the reducing agent dithiothreitol (DTT) leads to IMV unfolding in a manner comparable to that observed with nonionic detergent plus reducing agent. This is seen in **Fig. 13D.**

The 30–40-nm tubular structures produced under mild conditions, illustrated by **Fig. 13,** also result from more vigorous degradation using proteinase K plus SDS. Under the latter conditions, however, IMV disintegration continues beyond the 30–40-nm tubule stage, leading to the exposure and release of components within the tubules. The components initially revealed were filaments of 16-nm diameter (**Fig. 14A**). At higher resolution, the surfaces of the 16-nm filaments, like those of the 30–40 nm tubules, exhibited a helical geometry but were additionally linearly segmented (**Fig. 14B**).

With high proteinase K concentrations and longer duration, a further disintegration of the virus took place. What remained were long strands of double-helical DNA dispersed on the surface of the mica substrate (**Fig. 14C, D**). These DNA strands were associated with residual portions of the 16-nm-diameter filaments (**Fig. 14C,D**), as illustrated in **Fig. 14D,** the DNA strands were frequently observed emerging from the segments of the 16-nm tubules. That is, the DNA appears, almost certainly, to be contained within 16-nm filaments. The thickness of the DNA strands in **Fig. 14D** is in the range of 2.2–2.8 nm, again very close to the size of the DNA helix diameter of 2.5 nm. However, the thickness of large portions of DNA strands, such as the one in **Fig. 14C,** is in the range of 5–8 nm, which indicates the presence of residual filament-derived proteins.

The physical length of the vaccinia virus genome is approx 65 μm. A 65-μm linear tube with a diameter of 30–40 nm (that of the tubules), would have a volume four to six times that of the entire virion core, clearly an impossibility. Our measurements are therefore consistent with a compaction of the genome within the apparently coaxial 16-nm filaments and 30–40-nm tubules. The upper limit on the degree of compaction would depend on the proportion of the entire core occupied by the tubule. In this regard, it is unclear whether there is an additional "viroplasm" within the core, or indeed within 16-nm filament segments themselves (whose outer diameter is ~6.5 times that of the associated double-stranded DNA segment).

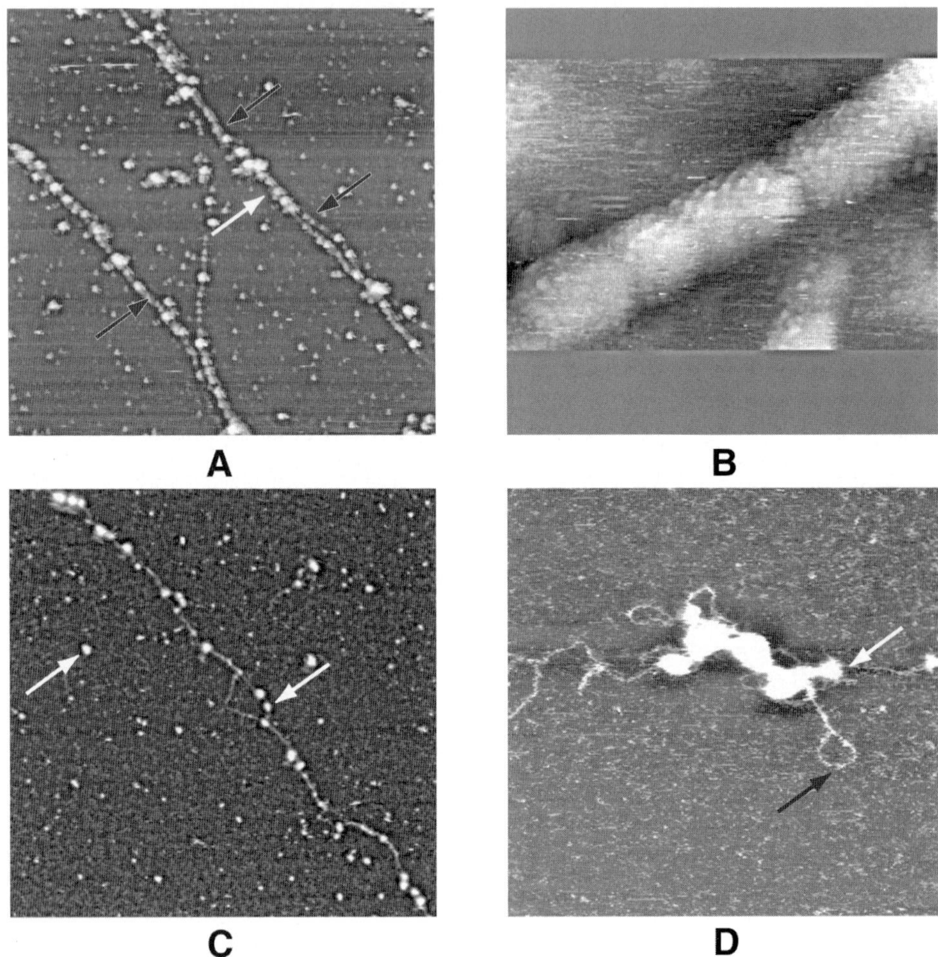

Fig. 14. Dissection of IMV virions with proteinase K in combination with ionic detergent. (0.2–1 mg/mL proteinase K in 0.1% SDS, 2 mM CaCl$_2$, 50 mM HEPES-NaOH, pH 7.5, ~45–60-min treatment at 37°C). **(A)** Partially digested portions of 30–40-nm tubules (indicated with white arrows) reveal 16-nm filaments (indicated with black arrows). **(B)** 16-nm filament. **(C–D)** DNA strands formed upon extended treatment (90–180 min at 37°C) of IMV virions with proteinase K, as in (A) and (B). In (C) and (D), isolated 16-nm filaments are seen dispersed on the mica along with 16-nm filaments and associated DNA strands. Both are indicated with white arrows. In (D), a naked DNA strand is indicated with black arrow. AFM images are: (A) 6 × 6 μm; (B) 240 × 120 nm; (C) 1.5 × 1.5 μm; (F) 580 × 580 nm. (Reproduced with permission from **ref. 8.**)

3.9. Model of the Architecture of the IMV Virion

The observations we present here are consistent with a structural model of vaccinia based on a hierarchy of substructures *(8)*. The double-stranded genomic DNA is encapsidated within 16-nm nucleoprotein filaments, formed by a helical array of protein subunits. The 16-nm filaments are constructed as contiguous linear segments, the segmentation presumably necessary to confer flexibility for folding into the more compact internal structure of the virion. The segmentation could, of course, also have physiological implications, as in transcription for example.

The 16-nm filaments in turn appear to be encased within the 30–40-nm tubules, also formed from helical sheaths of protein subunits. The 30–40-nm tubules interact with many accessory or matrix proteins that serve in conjunction with the 30–40-nm tubule sheath proteins to fold and condense the 30–40-nm tubules into a compact virion core. This process involves disulfide bond formation. The condensed tubular mass combines with a 70–100-nm diameter satellite domain, and this pairing of core and satellite domain is then enshrouded by membranes heavily studded by proteins and containing protrusions and spicules on its outer surface.

4. Notes

1. To obtain high-resolution images, prior to experiments, the fluid cell must be thoroughly cleaned using dish soap, warm water, and ethanol. Particular attention should be addressed to the cell surface area through which the AFM laser beam passes into the fluid. While cleaning, avoid any materials that can scratch the surface of the fluid cell. Blow-dry the wet fluid cell with compressed air or nitrogen until all moisture has evaporated.

2. When experiments do not require exchanging the fluid, the O-ring in the fluid cell assembly (Digital Instrument) can be omitted. This makes experiments easier, since drift during imaging can be caused by tensions on the O-ring. In this case put approx 30–40 µL of the required solution onto the tip assembly of the AFM fluid cell and lower the tip holder onto the sample. After several hours of operation, insert a pipet with a round pipet microtip into the fluid port of the cell in order to add any additional amount of solution needed to counter the evaporation losses. Avoid formation of bubbles in the fluid cell, which could interfere with the path of the laser beam and thereby prevent imaging.

3. Concentrations of virus solutions of approx 10^9–10^{10} virions/mL are typically adequate for AFM imaging. Viruses often tend to aggregate and form 2D arrays on the substrate. These arrays as well as individual virions adsorbed on the substrate are ideal for high-resolution imaging. In cases of high multilayer densities of adsorbed virions on the substrate, decrease these through dilution of the sample.

4. Unknown particle densities in virus preparations could be estimated in a fast and reliable fashion through the imaging and subsequent counting of dried virion samples of known initial volume and dilution.

5. Environmental vibrations can affect high-resolution imaging and obviously need to be eliminated as best as possible. Thus, AFM instruments are usually situated on an optics table or marble slab.
6. Unfortunately, according to our multiyear experience with high-resolution imaging of biological samples in fluid, only one of four to five AFM probes from the same wafer produces acceptable resolution. If this ratio is lower, request the manufacturer to exchange the wafer for new probes.
7. Apart from the shape and sharpness of the AFM probe, the imaging parameters also play an important role in image quality. For imaging in tapping mode in liquid, the amplitude setpoint is usually gradually decreased manually from a value corresponding to an out-of-contact level until a good tracing of the sample is reached. Another factor that influences the imaging in tapping mode is the drive amplitude of the oscillating cantilever. Too high a driving amplitude may result in damaging of the sample.

Acknowledgments

The authors thank E. K. Wagner and P. D. Gershon for providing herpes and vaccinia virus preparations and helpful discussions. We wish to thank A. Greenwood, R. Lucas, J. Zhou, and M. K. Rice for technical assistance. The National Aeronautics and Space Administration (grant NAG8-1569) supported this research. Work was performed under the auspices of the DOE by the LLNL under contract W-7405-ENG-48.

References

1. Harrison, S. C. (1990) Principles of virus structure, in *Fields Virology,* 2nd ed. (Fields, B. N. and Knipe, D. M. eds.), Plenum, New York, pp. 37–61.
2. Baker, T. S., Olson, N. H., and Fuller, S. D. (1999) Adding the third dimension to virus life cycles: three-dimensional reconstruction of icosahedral viruses from cryo-electron micrographs. *Microbiol. Mol. Biol. Rev.* **63,** 862–922.
3. Rux, J. J. and Burnett, R. M. (1998) Spherical viruses. *Curr. Opin. Struct. Biol.* **8,** 142–149.
4. Binnig, G., Quate, C. F., and Gerber, C. (1986) Atomic force microscope. *Phys. Rev. Lett.* **56,** 930–933.
5. Malkin, A. J., Kuznetsov, Y. G., Lucas, R. W., and McPherson, A. (1999) Surface processes in the crystallization of turnip yellow mosaic virus visualized by atomic force microscopy. *J. Struct. Biol.* **127,** 35–43.
6. Malkin, A. J. and McPherson, A. (2002) Novel mechanisms for defect formation and surface molecular processes in virus crystallization. *J. Phys. Chem.* **106,** 6718–6722.
7. Plomp, M., Rice, M. K., Wagner, E. K., McPherson, A., and Malkin, A. J. (2002) Rapid visualization at high resolution of pathogens by atomic force microscopy: structural studies of herpes simplex virus-1. *Am. J. Pathol.* **160,** 1959–1966.
8. Malkin, A. J., McPherson, A., and Gershon, P. D. (2003) Structure of intracellular mature vaccinia virus visualized by *in situ* atomic force microscopy. *J. Virol.* **77,** 6332–6340.

9. Dai, H., Hafner, J. H., Rinzler, A. G., Colbert, D. T., and Smalley, R. E. (1996) Nanotubes as nanoprobes in scanning probe microscopy. *Nature* **384,** 147–151.

10. Hafner, J. H., Cheung, C. L., Woolley, A. T., and Lieber, C. M. (2001) Structural and functional imaging with carbon nanotube AFM probes. *Prog. Biophys. Mol. Biol.* **77,** 73–110.

11. Canady, M. A., Larson, S. B., Day, J., and McPherson, A. (1996) Crystal structure of turnip yellow mosaic virus. *Nat. Struct. Biol.* **3,** 771–781.

12. Smith, T. J., Chase, E., Schmidt, T., and Perry, K. L. (2000) The structure of cucumber mosaic virus and comparison to cowpea chlorotic mottle virus. *J. Virol.* **74,** 7578.

13. Kuznetsov, Y. G., Malkin, A. J., Lucas, R. W., Plomp, M., and McPherson, A. (2001) Imaging of viruses by atomic force microscopy. *J. Gen. Virol.* **82,** 2025–2034.

14. Zhou, Z. H., Dougherty, M., Jakana, J., Rixon, F. J., and Chiu, W. (2000) Seeing the herpesvirus capsid at 8.5 Å. *Science* **288,** 877–880.

15. Zhou, Z. H., Prasad, B. V. V, Jakana, J, Rixon, F. J., and Chiu, W. (1994) Protein subunit structures in the herpes simplex virus A-capsid determined from 400 kV spot-scan electron cryomicroscopy. *J. Mol. Biol.* **242,** 456–469.

16. Booy, F. P., Trus, B. L., Newcomb, W. W., Brown, J. C., Conway, J. F., and Steven, A. C. (1994) Finding a needle in a haystack: detection of a small protein (the 12-kDa VP26) in a large complex (the 200 MDa capsid of herpes simplex virus). *Proc. Natl. Acad. Sci. USA* **91,** 5652–5656.

17. Newcomb, W. W., Brown, J. C., Booy, F. P., Steven, A. C., Wall, J. S., and Brown, J. C. (1993) Structure of the herpes simplex virus capsid. Molecular composition of the pentons and the triplexes. *J. Mol. Biol.* **232,** 499–511.

18. Malkin, A. J., Plomp, M., and McPherson, A. (2002) Application of atomic force microscopy to studies of surface processes in virus crystallization and structural biology. *Acta Cryst.* **58,** 1617–1621.

19. Fenner, F., Wittek, R., and Dumbell, K. R. (1989) *The Orthopoxviruses.* Academic, New York.

20. Griffiths, G., Roos, N., Schleich, S., and Locker, J. K. (2001) Structure and assembly of intracellular mature vaccinia virus: thin-section analysis. *J. Virol.* **75,** 11056–11070.

21. Griffiths, G., Wepf, R., Wendt, R., Krijnse-Locker, J., Cyrklaff, M., and Roos, N. (2001) Structure and assembly of intracellular mature vaccinia virus: isolated-particle analysis. *J. Virol.* **75,** 11034–11055.

22. Dubochet, J., Adrian, M., Richter, K., Garces, J., and Wittek, R. (1994) Structure of intracellular mature vaccinia virus observed by cryoelectron microscopy. *J. Virol.* **68,** 1935–1941.

23. Roos, N., Cyrklaff, M., Cudmore, S., Blasco, R., Krijnse-Locker, J., and Griffiths, G. (1996) A novel immunogold cryoelectron microscopic approach to investigate the structure of the intracellular and extracellular forms of vaccinia virus. *EMBO J.* **15,** 2345–2355.

24. Hollinshed, M., Vanderplassen, A., Smith, G. L., and Vaux, D. J. (1999) Vaccinia virus intracellular virions contain only one lipid membrane. *J. Virol.* **73,** 1503–1517.

25. Jensen, O. N., Houthaeve, T., Shevchenko, A., et al. (1996) Identification of the major membrane and core proteins of vaccinia virus by two-dimensional electrophoresis. *J. Virol.* **70,** 7485–7497.

26. Collier, L. (1955) The development of a stable smallpox vaccine. *J. Hyg.* **53,** 76–101.

27. Zhou, Z. H., Chen, D. H., Jakana, J., Rixon, F. J., and Chiu, W. (1999) Visualization of tegument-capsid interactions and DNA in intact herpes simplex virus type 1 virions. *J. Virol.* **73,** 13210–3218.

28. Moss, B. (2001) Poxviridae and their replication, in *Fields Virology* (Fields, B. N., Knipe, D. M., Chanock, R. M., et al., eds.), Raven, New York, pp. 661–684.

8

Studying the Structure of Large Viruses With Multiresolution Imaging

Carmen San Martín

Summary

Multiresolution imaging is an extremely useful technique for understanding in detail the structure of large DNA viruses that do not yield to the requirements of protein crystallography. The methodology consists in fitting the atomic structures of capsid components, independently solved, to medium-resolution, 3D maps of the complete virion, obtained by cryoelectron microscopy and image processing. On combining the two kinds of imaging data, one must take into account their intrinsic differences, as they have different resolution, suffer from different imaging artifacts, and are at different scales. These efforts are rewarded by "quasi-atomic" resolution models that provide valuable information about protein–protein interactions in the capsid. Difference maps calculated by subtracting the quasi-atomic model from the cryoelectron microscopy map reveal the molecular envelope of those capsid components whose atomic structure is unknown. A better understanding of the complex interactions involved in capsid assembly, stabilization, and disassembly is thus achieved.

Key Words: Cryoelectron microscopy; cryo-EM; difference imaging; DNA virus; multiresolution imaging; virus structure; X-ray crystallography.

1. Introduction

Understanding virus structure is key to understanding fundamental processes in the viral life cycle, such as capsid assembly and disassembly, receptor binding, and antibody neutralization. Two important methods used to image virus structure are X-ray crystallography and cryo-electron microscopy (cryo-EM). X-ray crystallography exploits the diffracting power of large (~0.2 mm) three-dimensional (3D) protein crystals irradiated by an X-ray beam. It requires large amounts of sample that have to be induced to form the large ordered arrays, and produces diffraction patterns that later have to be interpreted to yield the structure. Cryo-EM does not require sample crystallization, and so smaller

From: *Methods in Molecular Biology, vol. 292: DNA Viruses: Methods and Protocols*
Edited by: P. M. Lieberman © Humana Press Inc., Totowa, NJ

amounts of sample can be used. Two-dimensional projections of the biological object in solution are recorded and later combined using image processing procedures to obtain a 3D reconstruction.

X-ray crystallography gives the most detailed picture of viral capsid structure, as it can reach atomic resolution *(1)*. About 45 unique high-resolution viral structures solved by crystallographic techniques are currently available at the Protein Databank (PDB) *(2)*. Crystallographic techniques are increasingly able to deal with very large macromolecular complexes, as shown by the largest virus structures solved so far: the bluetongue virus core (700-Å diameter) *(3)*, and bacteriophage HK97 (660-Å) *(4)*. However, many viruses present particularly challenging problems for crystallography, not only because of their large size and complex capsid composition, but also because they are too difficult to obtain in the amounts required for crystallization, or because the presence of flexible capsid components hinders production of highly ordered crystals.

Individual viral components are generally more amenable to overexpression and crystallization. Their atomic structures give us a very detailed picture of each capsid component but do not provide information on their mutual relation in the virion. On the other hand, nowadays one can routinely obtain 3D virus capsid maps from cryo-EM *(5)* at medium resolution, i.e., 12–15 Å, and achieving subnanometer resolution is not that uncommon *(6–11)*. Thus, with cryo-EM, a picture of the complete virion is obtained, although the lower resolution makes it a blurry picture compared with crystallography maps.

Multiresolution imaging (**Fig. 1**) consists in cleverly combining these two sources of 3D information: atomic structure of individual components and medium resolution of whole viral particles. The known crystal structures are fitted to the cryo-EM density, resulting in a 3D model in which the atomic structures are placed in the cryo-EM map with a precision of about 4 Å. The resultant description of the virus is called a *quasi-atomic* or *pseudo-atomic* model. This model provides information at higher resolution than the cryo-EM map by itself and offers a more complete picture than the structures of the isolated components. Subsequent analysis of the quasi-atomic model reveals characteristics of the intercapsomere relationships, as well as those between protein shells and other virion components like membranes or DNA *(10,12,13)*. Quasi-atomic models also allow modeling of conformational changes related to capsid maturation *(14,15)* or assembly *(16)*. They can also be used for determining the phases of X-ray diffraction data for those viruses for which crystals are available *(3,4)*, thus saving the effort needed for isomorphous replacement or Se-methionine derivatization. Difference maps calculated between the quasi-atomic model (containing only those capsid components whose atomic structure is available) and the cryo-EM reconstruction reveal the molecular envelope of those components whose high-resolution structure is unknown *(17,18)*.

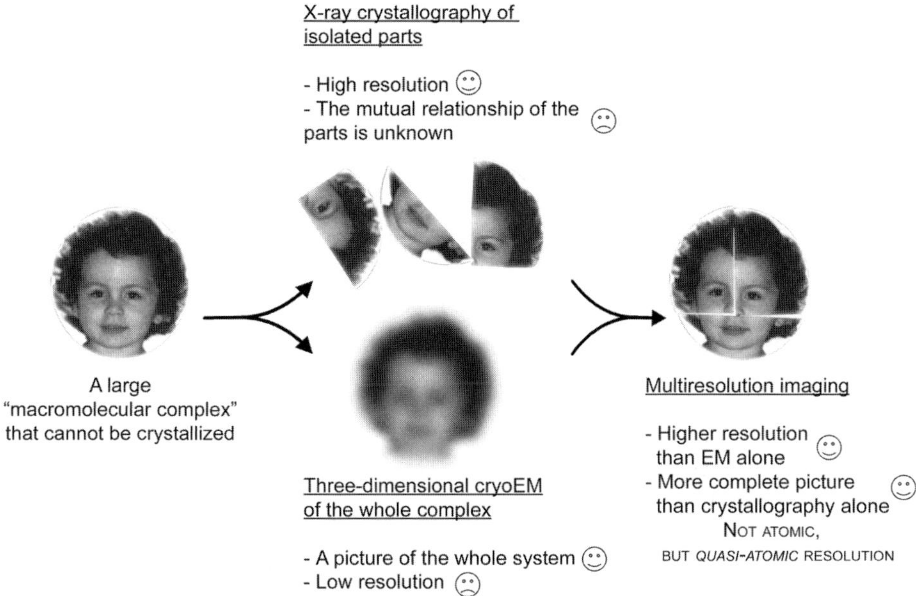

Fig. 1. Fundamentals of multiresolution imaging, illustrated with an example from the macroscopic world.

Depending on the cryo-EM map resolution and the biochemical knowledge available, these envelopes can be used to model the missing structures.

In this chapter, procedures for the calculation of a quasi-atomic model, a difference map, and their analysis are described. Both X-ray crystallography and cryo-EM are highly specialized techniques for which there is an extensive bibliography (*see,* for example, **refs.** *1,19–21*), whose description goes beyond the scope of this chapter. Fortunately for the scientific community, the 3D models obtained by both methods are publicly available in specialized databases. The PDB at http://www.pdb.org *(2)* contains atomic coordinates for about 20,000 biological macromolecules; the Macromolecular Structure Database (MSD) at http://www.ebi.ac.uk/msd/ *(22)* has recently started to store 3D electron microscopy (3DEM) maps. Thus, for the purposes of this chapter, it will be assumed that the two pieces of 3D data to be combined (cryo-EM map of a virus capsid and atomic coordinates of a capsid protein) are available for downloading from the databases. As the computational tools used in multiresolution imaging are available through crystallographic, image processing, and visualization software usually running under UNIX operating systems, some knowledge of computing, in particular of the UNIX environment *(23)*, will also be assumed.

Although the technique described here has hitherto been applied to icosahedral capsids, cryo-EM holds the promise of obtaining 3D maps of nonsymmetric viruses by way of tomographic techniques *(24)*. Thus, multiresolution imaging is likely to play a significant role in the study of large pleomorphous viruses in the near future.

2. Materials

1. A 3D cryo-EM map of the virus under study.
2. A PDB file with the atomic coordinates of a capsid protein.
3. A computer running some form of the UNIX operating system (SGI IRIX, Compaq/HP Tru64, Linux, or others).
4. Crystallography software:
   a. O *(25)*, from http://xray.bmc.uu.se/alwyn/.
   b. X-PLOR *(26)*, from http://atb.csb.yale.edu/xplor.
   c. CNS *(27)*, from http://cns.csb.yale.edu/.
   d. CCP4 *(28)*, from http://www.ccp4.ac.uk/.
   e. MAPMAN, MAMA, and MOLEMAN2, from The Uppsala Software Factory *(29)* at http://xray.bmc.uu.se/~gerard/manuals/.
   f. DEALPDB, from http://www.bmsc.washington.edu/kumar_progs/programs.html.
5. Cryo-EM image processing software:
   a. SPIDER and WEB *(30)*, from http://www.wadsworth.org/spider_doc/spider/docs/master.html.
   b. EMBL icosahedral reconstruction software *(31)*, from http://www.strubi.ox.ac.uk/strubi/cryo/STRUBI_Virus_Structure.html#I.
   c. MRC image processing package *(32)* (contact the authors at rac1@mrc-lmb.cam.ac.uk for downloading).
6. Visualization software:
   a. OpenDX, from http://www.opendx.org/.
7. Script files required to run the procedures described in this chapter are available at http://www.cnb.uam.es/~carmen/multiresolution.

3. Methods

The case of bacteriophage PRD1, a membrane-containing dsDNA bacteriophage of the Tectiviridae family *(33)*, will be used to illustrate the combination of cryo-EM and crystallographic data. Multiresolution imaging of the large (70-MDa, 700-Å diameter) PRD1 particle resulted in a model for virion structure highlighting the structural elements involved in capsid stabilization and vertex dynamics during infection *(13,17)*, as well as the structural parallels with the mammalian adenovirus *(34)*. The atomic structure of PRD1 will soon be available *(35,36)*, thus providing a direct comparison between the results obtained by multiresolution imaging and protein crystallography.

Figure 2 shows the main steps to be followed in obtaining a quasi-atomic model of a viral capsid. Essentially, the process starts with a manual fit of the

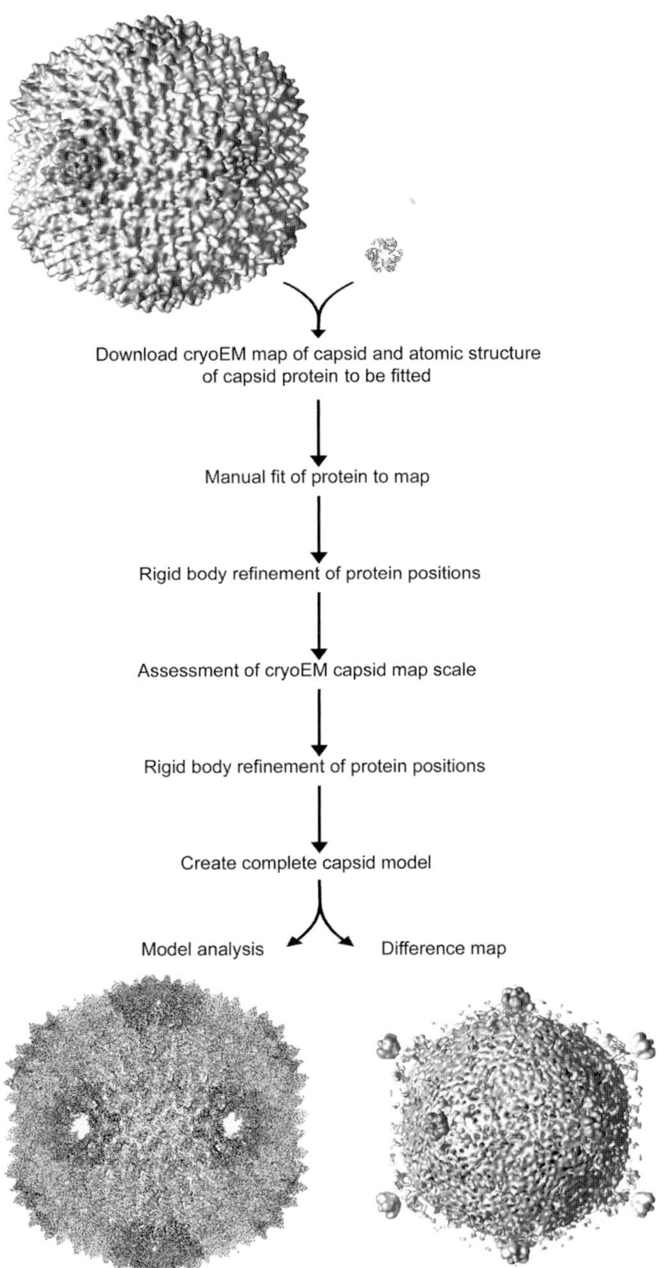

Fig. 2. Flow diagram showing the sequence of operations to be performed for obtaining a viral capsid quasi-atomic model and difference map.

atomic coordinates of the individual capsid proteins into the cryo-EM map of the complete capsid. When the virus under study is icosahedral, only those molecules included in the icosahedral asymmetric unit (AU) need to be manually fitted; the rest are generated later by symmetry operations. In a second step, the fitting is objectively optimized using the rigid body refinement approach. Each molecule in the AU is considered a rigid body, free to rotate and shift independently (as long as it does not clash with its neighbors) until the discrepancy between the atomic structures and the EM map is minimized. As the actual scale of the cryo-EM map is not known accurately, this is determined with the fitted crystallographic structure as a reference, and a second rigid body refinement is performed. At this point a complete capsid model can be created by applying icosahedral symmetry to the contents of the AU and used to analyze protein–protein contacts or to calculate a difference map.

3.1. Downloading 3D Data To Be Combined

3.1.1. Download Cryo-EM Map of Virus Particle

Use your favorite Web browser to connect to the MSD at http://www.ebi.ac.uk/msd/. Click on *Searches,* and then on *EMSearch.* Search for and download the cryo-EM map of the viral particle under study. In the example used here, we will choose accession code *1013,* corresponding to PRD1 *sus1,* a mutant that does not package DNA *(37,38)*. Download file *emd_1013.map*. Note the resolution of the map (14 Å) and the pixel size (3.44 Å) in the header file *emd_1013.xml*. Note also the map dimensions in pixels ($256 \times 256 \times 256$). This will be needed later to calculate the cell size for crystallographic programs (cell size in Å = 256×3.44).

3.1.2. Download Atomic Coordinates of Capsid Component

Connect to the PDB at http://www.pdb.org. Search for and download the atomic coordinates of the viral protein(s) to be fitted to the cryo-EM map. In our example, we will use P3, the major coat protein of bacteriophage PRD1 (PDB ID 1HX6) *(39,40)*. The PRD1 capsid contains 240 P3 trimers, arranged with $T = 25$ icosahedral geometry *(41)*.

3.2. Manual Fit of Atomic Coordinates to Cryo-EM Map

3.2.1. Prepare Cryo-EM Map

Before starting the fitting process, care must be taken that the cryo-EM map is in the right orientation and has the appropriate dimensions for the programs to be used later. For icosahedral viruses, the usual orientation convention places the center of the viral capsid at point (0, 0, 0) and the Cartesian axes *x, y, z* coincident with three mutually perpendicular twofold axes of the icosahedron (**Fig. 3**).

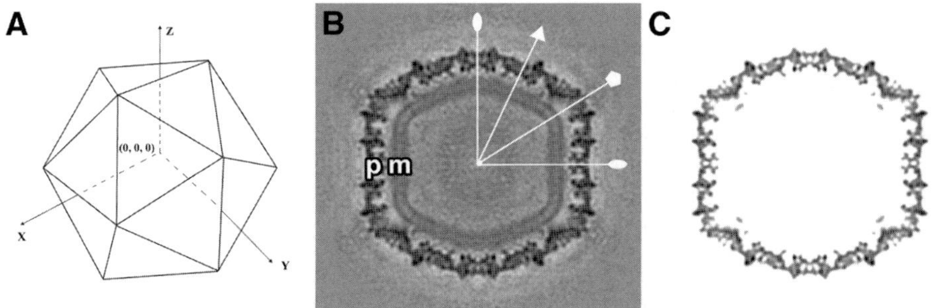

Fig. 3. Icosahedral orientation convention and masking of the cryo-EM map. (**A**) Schematic showing an icosahedral capsid centered at (0,0,0) and with coordinate axes coincident with three perpendicular twofold symmetry axes. (**B**) A section cut through the center of the PRD1 *sus1* cryo-EM map. The icosahedral twofold (ovals), threefold (triangle), and fivefold (pentagon) axes are indicated. The icosahedral protein shell (p), composed of P3 and other minor capsid proteins, and the double layer of the membrane (m) can be appreciated. Protein density is black. (**C**) The same map after preparation for fitting. Most of the density not belonging to the icosahedral protein shell has been computationally removed, as described in the text.

The dimensions of the map (in pixels) must be a power of 2. The map orientation and dimensions can be changed with the program SYMMETRIZE from the EMBL icosahedral reconstruction software. The map is then normalized so that the average density value is 0 and the standard deviation is 1; as much density as possible not corresponding to the atomic structure to be fitted is eliminated with a mask, and all negative values are set to 0, using SPIDER (**Fig. 3**). This procedure is equivalent to solvent flattening in crystallography and removes background noise and additional structures (e.g., membranes, DNA) that could induce errors during the subsequent fitting operations. It also makes graphic representations cleaner and easier to work with, an aspect that is especially relevant when dealing with large viruses. The unmasked map will be used again when analyzing the results of the fit and the difference map.

A note on the map file format is needed. MSD maps are stored in the widely used crystallographic CCP4 map format (http://www.ccp4.ac.uk/dist/html/maplib.html#description), also known as MRC format in the 3DEM field. When SPIDER is used for map preparation or other image processing operations, we need to convert the map to SPIDER file format (http://www.wadsworth.org/spider_doc/spider/docs/image_doc.html). This can be done within SPIDER itself (commands CP TO/FROM MRC) or with stand-alone programs such as EM2EM (http://www.imagescience.de/em2em/welcome.htm). In the current example, and in the script files supplied at our web page, we will be using

MRC2SPI or SPI2MRC as needed. These are stand-alone programs supplied with the EMBL icosahedral reconstruction package but are based on SPIDER routines, and so the user is required to have a SPIDER license before using them.

3.2.2. Manual Fit

We will now manually place a number of independent molecules in the icosahedral AU. This number depends on the virus triangulation number *(42)*. In the case of the pseudo $T = 25$ PRD1 capsid, the AU contains four P3 trimers. The whole capsid can be generated later by applying icosahedral symmetry to the molecules in the AU.

We first use a PDB file editing utility such as MOLEMAN2 to remove all nonprotein atoms that could be included with the protein atomic coordinates, for example, water molecules or solvent ions. Then we need to do a little more preprocessing on the cryo-EM map so we are able to read it in O, the graphics program of choice for the manual fit. The preferred map format for displaying large maps in O is called BRIX (http://xray.imsb.au.dk/~mok/brix/brix.html). Whereas in the CCP4 map format convention the map is assumed to be centered at (N/2, N/2, N/2), where N is the size of the map in pixels, in BRIX the map must be centered at (0,0,0). We take care of this discrepancy by using program IMEDIT.EXE, from the MRC image processing package, to edit the map file header and change the map starting point accordingly. Then, the map format interchange utility MAPMAN is used to write the BRIX format map file. We are now ready to start the manual fit in O. After loading the PDB file and the cryo-EM map, the O command *move_object* will be used to shift and rotate the first copy of the P3 trimer and place it in one of the four independent positions in the AU. Once an acceptable fit has been reached, we will save the new coordinates in a new PDB file and repeat the procedure for the other three independent P3 trimers. We can then use the *symmetry_object* command in O to generate a graphic representation of the whole capsid and see how our first, manually made, model looks.

3.3. Optimization of the Fit by Rigid Body Refinement

The crystallographic rigid body minimization technique (see, for example, **ref. 43**) refines the positions of rigid groups of atoms using six variables (three rotations and three translations) for each rigid group, by minimization of the crystallographic residual

$$E_x = \frac{W}{N} \sum_{\vec{h}} \left[\left| F_{obs}(\vec{h}) \right| - k \left| F_c(\vec{h}) \right| \right]^2 \tag{1}$$

where W is an overall weight, N is a normalization factor, k is a scale factor, $\vec{h} = (h,k,l)$ are the Miller indices of the selected reflections, $|F_{obs}|$ are the Fourier amplitudes of the observed (in our case, the cryo-EM map) structure factors, and $|Fc|$ are the Fourier amplitudes of the X-ray data structure factors of the molecule to be fitted. Throughout the refinement, the icosahedral symmetry will be taken into account by using noncrystallographic constraints.

As for the manual fit procedure, several preparatory operations are needed before the rigid body refinement step.

3.3.1. Calculate Cryo-EM Map Structure Factors

The structure factors of the masked cryo-EM map are calculated by Fourier transformation using the program SFALL of the CCP4 package, including data up to a resolution slightly higher than the map (e.g., 10 Å for a 14-Å cryo-EM map). Because the rigid body refinement will be run in X-PLOR, the calculated structure factors are converted to a suitable format using the CCP4 program MTZDUMP and a home-made utility called MTZ2XPLOR (available at the author's web page).

3.3.2. Prepare X-PLOR Coordinate and Structure Files

After the manual fit, we obtained four separate PDB files, one for each of the four independent P3 trimers in the AU. We now append them to create a new PDB file containing the whole AU and assign a unique chain ID and segment ID to each of the 12 P3 monomers in the file so they can be easily identified. These operations are performed with MOLEMAN2. MOLEMAN2 operation *split* is then used to create an input file for X-PLOR called *asy_generate.inp*. This script file will create coordinate files in the appropriate format for X-PLOR. It will also create the structure file (extension *.psf*) that X-PLOR requires containing information on the topology of each molecule.

3.3.3. Find the Best B Factor Value for the Fit

The rigid body refinement has to be performed including only data up to the resolution of the cryo-EM map. This implies a cutoff in frequency, which must be smooth to avoid ripple artifacts. This is accomplished by the use of a decay function of the form

$$d(s) = e^{-B/s^2} \tag{2}$$

where s is the resolution and B the so-called temperature factor. The value of B depends on the cutoff edge. To find the best B factor for a particular map, decay functions with B values ranging from 0 to 3500 Å^2 are applied to the X-ray data, and the crystallographic R factor is calculated for each case, using X-PLOR, with the resolution cutoff at the cryo-EM map resolution. (Note: all X-PLOR input files used here have been adapted from original input files included in the standard X-PLOR distribution.) The B value that gives the lowest R factor is selected and used in all subsequent fitting operations. The crystallographic R factor is defined as *(19)*

$$R = \frac{\sum_{\vec{h}} \left\| F_{obs}(\vec{h}) \right| - k \left| F_c(\vec{h}) \right\|}{\sum_{\vec{h}} \left| F_{obs}(\vec{h}) \right|}$$

(3)

where F_{obs}, F_c, $\vec{h}$, and k are defined as in **Eq. 1**. In the PRD1 study, B values in the 100–400-$Å^2$ interval were found to be appropriate for cryo-EM reconstructions at approx 15-Å resolution. Larger values (~2000 $Å^2$) were needed when the cryo-EM map was at lower (25-Å) resolution.

3.3.4. Amplitude Scaling and Rigid Body Refinement

After determining the temperature factor to be used, the amplitudes of the cryo-EM map structure factors are scaled with respect to those of the atomic model by comparing their average values in reciprocal space shells, using a homemade program called AMPLITUDE_SCALE (*see* **Subheading 2.**, **item 7**). This compensates for the variations in signal amplitude induced by the electron microscope aberrations *(20)*.

The scaled structure factors and the atomic model of all the molecules in the AU are then used as input to X-PLOR for rigid body minimization, with the appropriate 60 noncrystallographic symmetry operators, the overall temperature factor previously found, and including structure factors up to the resolution of the cryo-EM map. The value of the weight constant in **Eq. 1** is obtained by running the rigid body refinement several times for each map, with values of W ranging from 1 to 10^8 in 10-fold increments.

3.4. Assessment of the Absolute Cryo-EM Map Scale

The magnification of an electron microscope is subject to uncertainties caused by the conditions of imaging and has been shown to vary by as much as 2% in the same micrograph *(44)*, and even more between micrographs. Finding the absolute scale of the cryo-EM map is critical for the quality of the final fit, especially at resolutions better than 15 Å when one can start analyzing the quasi-atomic model in quantitative terms. For example, in the PRD1 study, it was found that at the nominal scale given by the electron microscope magnification (3.68 Å/pixel) the P3 trimers were too far apart in the capsid to establish the protein–protein contacts needed for stabilization of the structure. After scale assessment using the procedure here described, it was determined that the actual scale corresponded instead to 3.44 Å/pixel, that is, a 7% error in the nominal value. With the new scale values, the quality of the fit improved drastically (R changed from 39.9 to 33.7%), and so did the P3 trimer packing: the fitted AU presented no gaps between trimers, and they interdigitated very closely.

The availability of an X-ray model is very helpful to determine the correct scale of a reconstruction. However, in cases like the one described here, there is a circular problem: we have a cryo-EM reconstruction for the complete viral particle, but an X-ray model for only a small piece of it—the P3 trimer. We need to determine the scale for obtaining a reliable fit, but we need first to fit the X-ray model in place for estimating the absolute scale. This is why we do not address the scale problem until we have a first capsid model obtained by rigid body refinement at the nominal magnification. Now we can proceed as follows: starting from the positions found after the first fit, each one of the P3 trimers in the AU is translated toward the center of the particle in steps representing a 0.5% change in the scale, using MOLEMAN2. A suitable range of scales has to be covered, for example from 110 to 90% of the nominal pixel size. For each of the tested scales, the cryo-EM map cell size has to be changed in the header (IMEDIT.EXE) and its structure factors recalculated with SFALL. Then, the X-PLOR format coordinate *(.pdb)* and structure *(.psf)* files are generated for the new (translated) AU with *asy_trans_generate.inp,* and the *R* factor is calculated with X-PLOR. The scale giving the lowest *R* factor is chosen as the correct one.

3.5. Final Model

Once the absolute scale of the cryo-EM map has been determined, the rigid body refinement (**Subheading 3.3.**) is repeated at the newly found scale to optimize the fit (**Fig. 4**). The quasi-atomic model for the complete capsid is constructed by applying icosahedral symmetry to the AU, using MOLEMAN2.

Use the AutoDep tool to deposit the quasi-atomic model coordinates at the MSD database (http://www.ebi.ac.uk/msd/Deposition/Autodep.html) or directly at the PDB (http://pdb.rutgers.edu/adit/). Please check the recommendations at http://www.ebi.ac.uk/msd/iims/EM_PDB.html for a template to be followed when depositing coordinates based on a cryo-EM map.

3.6. Analysis of the Model

3.6.1. Quality of the Fit

The crystallographic *R* factor is used as a quality criterion throughout the fitting, as explained above. A plot showing the evolution of the *R* factor during the PRD1 *sus1* fitting is shown in **Fig. 4B.** Other estimates of the model quality can be obtained from the number of steric clashes, calculated with X-PLOR input file *check_clash.inp* or the number of atoms lying outside the cryo-EM map density, as estimated using the programs MAPMAN and MAMA. The uniqueness of the fit can be verified by applying random rotations and translations to the fitted molecules and repeating the rigid body refinement.

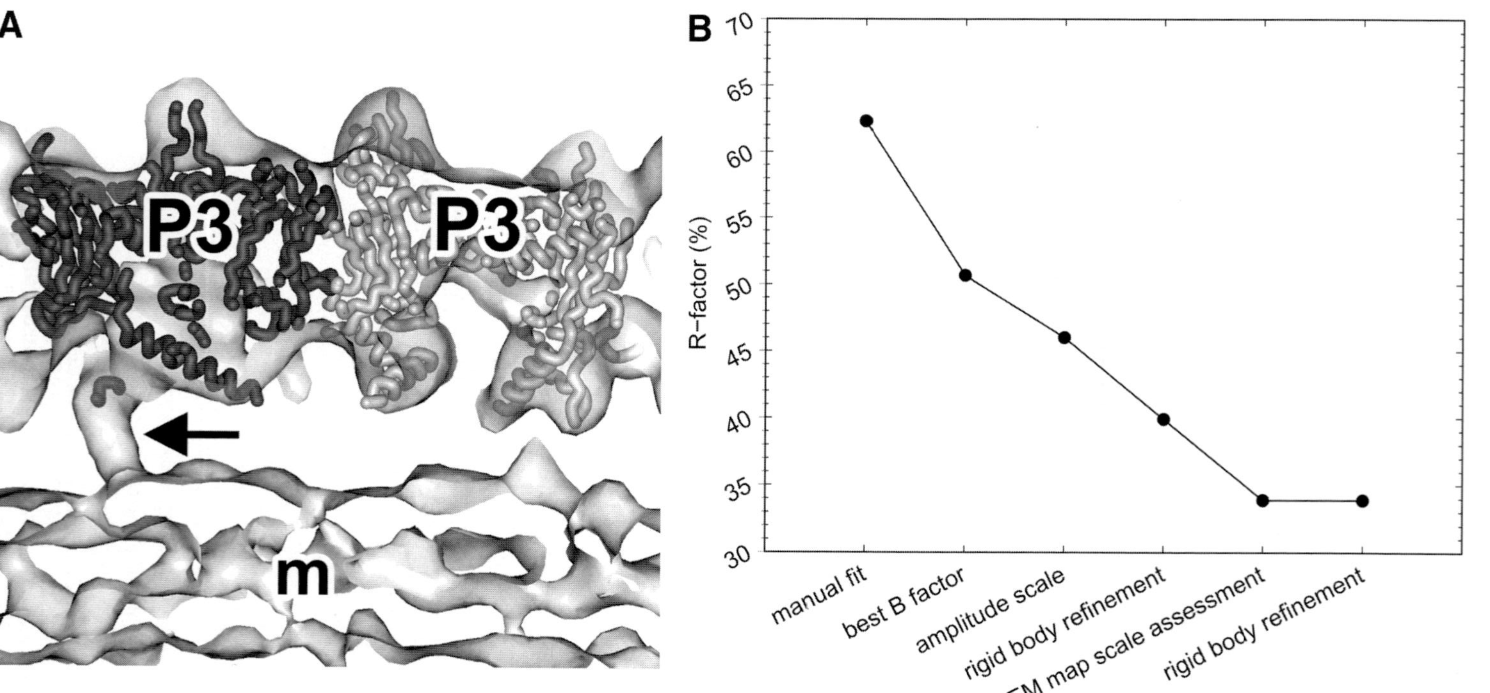

Fig. 4. Fit of the P3 trimer atomic structure to the PRD1 *sus1* cryo-EM map. (**A**) The C_α trace of two P3 trimers (P3) is shown, fitted to the PRD1 *sus1* cryo-EM density (semitransparent surface) A section across the capsid is shown. The arrow indicates tubular density (probably an α-helix) connecting the P3 layer to the viral membrane (m), a key element for viral assembly. (**B**) Progression of the *R* factor value throughout the PRD1 *sus1* fitting steps.

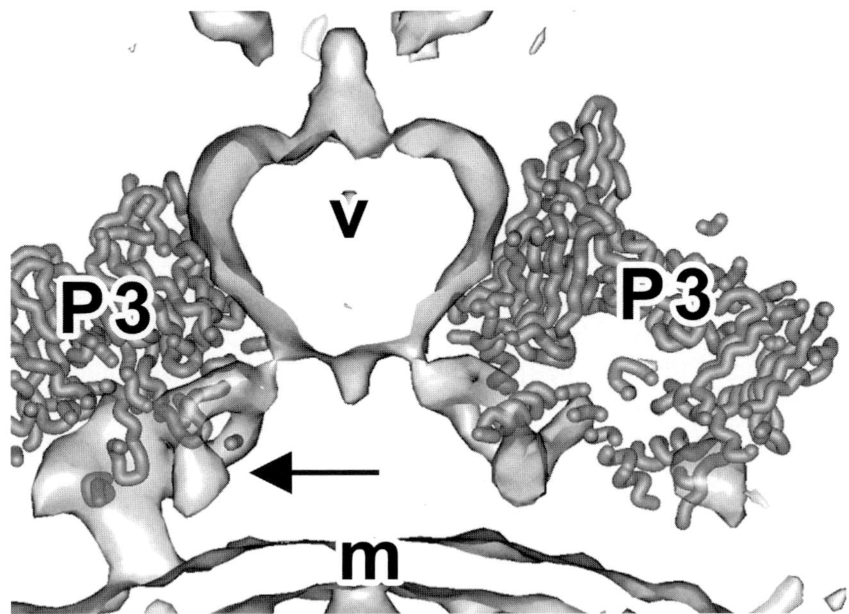

Fig. 5. The difference map (semitransparent surface) of the PRD1 wild-type virion revealed the vertex proteins (v) and their connections (arrow) to the viral membrane (m) and peripentonal P3 trimers (P3) These connections are very likely involved in vertex destabilization during DNA injection in the host.

3.6.2. Protein–Protein Contacts in the Quasi-Atomic Capsid Model

The residues involved in protein–protein contacts can be identified using the CNS input file *contacts.inp,* which should be run once for each interface in and around the AU. Residues with atoms closer than 4 Å are usually considered contacting residues. Another interesting possibility is submitting the model to the VIPER database *(45)* at http://mmtsb.scripps.edu/viper/viper. VIPER holds tools to calculate the energy of all protein–protein interfaces in the capsid, which can then be interpreted in terms of energetics of assembly and stability.

3.7. Calculation of a Difference Map

In general it is not possible to define molecule boundaries directly in cryo-EM maps, owing to their low resolution. In the case of viral particles, this means we cannot determine the location and shapes of each individual capsid component just by looking at the map. However, once a quasi-atomic model is available, difference maps can be calculated by subtracting it from the cryo-EM map. This will reveal the molecular envelope of those capsid components for which high-resolution data are not available **(Fig. 5)**. In combination with biochemical and genetic knowledge, the information thus obtained can be used to generate a model for the disposition of proteins in the virion and their role in assembly and viral life cycle. A clean difference map is vital for a meaningful and reliable interpretation, as is an objective selection of the threshold used for surface rendering of the map. These will be the main points described in the current section.

3.7.1. Create a Density Map From the Quasi-Atomic Model

The quasi-atomic model of the viral capsid is transformed into a density map by inverse Fourier transformation of its structure factors, using CCP4. The map can be calculated to the highest possible resolution given the sampling interval (that is, double the pixel size); a small B factor (e.g., 300 Å^2 for a 15-Å map) should be applied to avoid Fourier ripple artifacts. Then it is filtered to the cryo-EM map resolution using SPIDER. Unless otherwise stated, the operations involved in the difference imaging and described below will also be carried out in SPIDER.

3.7.2. Scale Map Densities and Subtract

Before subtracting the two maps, the density values in each must be set to a comparable range. (Note that for difference imaging the original, nonmasked cryo-EM map has to be used, rather than the masked map used for fitting.) The usual way to accomplish this is to calculate the average and standard deviation (σ) values for each map, and then for each pixel subtract the average value and divide by σ, thus ending up with two maps with average = 0 and σ = 1. In our experience, however, the cleanest differences were obtained when both maps were set to have the same average and σ values only in the region comprised by the molecules present in the quasi-atomic model.

First, a mask defining the region occupied by molecules in the quasi-atomic model is created by applying a threshold value to the quasi-atomic model map, so that the volume enclosed by those voxels with density values higher than the threshold accounts for 100% of the expected volume. The volume can be estimated using the standard density value for proteins (1.33 g/cm^3), or directly from the atomic coordinates, for example with MAMA. Then, the average and σ values of both the quasi-atomic model and the cryo-EM map are calculated, but taking into account only the voxels inside the mask. Finally, average subtraction and division by σ are applied to *all* voxels of each map, and the resulted density-scaled quasi-atomic model map is subtracted from the density-scaled cryo-EM map.

3.8. Analysis of the Difference Map

Most of the analysis of the difference map relies on interactive visual examination. OpenDX and O are excellent choices to perform this task. A note on how to use OpenDX to display SPIDER format volumes can be found in http://www.cnb.uam.es/bioinfo/opendx/index.html.

Cryo-EM maps characteristically do not have a sharp edge separating the specimen from the background; rather, the transition from biological material to ice is smooth and blurry, and the researcher has to decide where to place the edge based on other knowledge, such as the expected volume of the protein under study *(20)*. Therefore, regardless of which program is used, a key step for map display and analysis is choosing a threshold value that will define the surface rendered. In our case, we use the quasi-atomic model as a reference to define at which threshold the surface rendered encloses 100% of the expected mass, as described before. The threshold thus found must be applied to both cryo-EM and quasi-atomic model density-scaled maps, as well as to the difference map, for all subsequent visualization and analysis.

Each of the different peaks in the difference map can be isolated by applying a surrounding spherical mask in SPIDER. Once they are isolated, the number of voxels above the 100% mass threshold is counted and transformed to mass value by using the known protein density or volume, as explained in **Subheading 3.7.2.** The values obtained can be related to the mass of the known capsid components and so can help to assign the latter to the different areas of the difference map. However, it must be realized that the mass values obtained in this way can only be taken as rough approximations, since poor map quality, low occupancy, or disorder in a particular capsid protein can change the estimated mass value significantly.

4. Notes

1. Multiresolution imaging is an emerging and rapidly growing field for which there is still no established general protocol. Other valuable methods and software exist that can be used for the tasks described here. For example, vector quantization methods as implemented in the SITUS package *(46)* and correlation-based methods such as COAN *(47)*, DockEM *(48)*, and COLORES *(49)* can be used for general purpose fitting of atomic structures into cryo-EM maps. In particular, the real space refinement program EMfit *(50,51)* has been successfully used in the virus structure field *(10,52–54)*. Also worthy of note are the excellent software tools MOLSCRIPT *(55)*, BOBSCRIPT *(56)*, and RASTER3D *(57)* (**Figs. 4A and 5**) to produce publication-quality figures of the complicated virus structures.

Acknowledgments

This work was part of a collaborative effort among the groups of Roger M. Burnett (The Wistar Institute, Philadelphia), Dennis H. Bamford (University of Helsinki, Finland), and Stephen D. Fuller (Oxford University, UK). Invaluable support from members of all three groups is gratefully acknowledged. In particular, Sarah J. Butcher, Juha H. Huiskonen, Felix de Haas, and Stacy D.

Benson provided high-quality data and helpful suggestions and discussion. Roger M. Burnett carefully read and commented on the manuscript. Primary funding came from grants from the National Institutes of Health (AI-17270), the National Science Foundation (MCB 95-07102), the Wistar Institute Cancer Center (CA 09171), and the Human Frontiers Science Program (RGP0320/2001-M). Current funding from the Spanish Comisión Interministerial de Ciencia y Tecnología (BIO2001-1237), Comunidad Autónoma de Madrid (07B-0032), and Fondo de Investigaciones Sanitarias (G03-185) is acknowledged. C.S.M. holds a CSIC I3P Research Scientist position, supported by the European Social Fund.

References

1. Fry, E. E., Grimes, J., and Stuart, D. I. (1999) Virus crystallography. *Mol. Biotechnol.* **12,** 13–23.
2. Berman, H. M., Westbrook, J., Feng, Z., et al. (2000) The Protein Data Bank. *Nucleic Acids Res.* **28,** 235–242.
3. Grimes, J. M., Burroughs, J. N., Gouet, P., et al. (1998) The atomic structure of the bluetongue virus core. *Nature* **395,** 470–478.
4. Wikoff, W. R., Liljas, L., Duda, R. L., Tsuruta, H., Hendrix, R. W., and Johnson, J. E. (2000) Topologically linked protein rings in the bacteriophage HK97 capsid. *Science* **289,** 2129–2133.
5. Baumeister, W. and Steven, A. C. (2000) Macromolecular electron microscopy in the era of structural genomics. *Trends Biochem. Sci.* **25,** 624–631.
6. Zhou, Z. H., Dougherty, M., Jakana, J., He, J., Rixon, F. J., and Chiu, W. (2000) Seeing the herpesvirus capsid at 8.5 A. *Science* **288,** 877–880.
7. Zhou, Z. H., Baker, M. L., Jiang, W., et al. (2001) Electron cryomicroscopy and bioinformatics suggest protein fold models for rice dwarf virus. *Nat. Struct. Biol.* **8,** 868–873.
8. Bottcher, B., Wynne, S. A., and Crowther, R. A. (1997) Determination of the fold of the core protein of hepatitis B virus by electron cryomicroscopy. *Nature* **386,** 88–91.
9. Conway, J. F., Cheng, N., Zlotnick, A., Wingfield, P. T., Stahl, S. J., and Steven, A. C. (1997) Visualization of a 4-helix bundle in the hepatitis B virus capsid by cryo-electron microscopy. *Nature* **386,** 91–94.
10. Mancini, E. J., Clarke, M., Gowen, B. E., Rutten, T., and Fuller, S. D. (2000) Cryo-electron microscopy reveals the functional organization of an enveloped virus, Semliki Forest virus. *Mol. Cell* **5,** 255–266.
11. Trus, B. L., Roden, R. B., Greenstone, H. L., Vrhel, M., Schiller, J. T., and Booy, F. P. (1997) Novel structural features of bovine papillomavirus capsid revealed by a three-dimensional reconstruction to 9 A resolution. *Nat. Struct. Biol.* **4,** 413–420.
12. Grimes, J. M., Jakana, J., Ghosh, M., et al. (1997) An atomic model of the outer layer of the bluetongue virus core derived from X-ray crystallography and electron cryomicroscopy. *Structure* **5,** 885–893.

13. San Martín, C., Burnett, R. M., de Haas, F., et al. (2001) Combined EM/X-ray imaging yields a quasi-atomic model of the adenovirus- related bacteriophage PRD1 and shows key capsid and membrane interactions. *Structure* **9,** 917–930.

14. Conway, J. F., Wikoff, W. R., Cheng, N., et al. (2001) Virus maturation involving large subunit rotations and local refolding. *Science* **292,** 744–748.

15. Canady, M. A., Tihova, M., Hanzlik, T. N., Johnson, J. E., and Yeager, M. (2000) Large conformational changes in the maturation of a simple RNA virus, nudaurelia capensis omega virus (NωV). *J. Mol. Biol.* **299,** 573–584.

16. Modis, Y., Trus, B. L., and Harrison, S. C. (2002) Atomic model of the papillomavirus capsid. *EMBO J.* **21,** 4754–4762.

17. San Martín, C., Huiskonen, J. T., Bamford, J. K., et al. (2002) Minor proteins, mobile arms and membrane-capsid interactions in the bacteriophage PRD1 capsid. *Nat. Struct. Biol.* **9,** 756–763.

18. Stewart, P. L., Fuller, S. D., and Burnett, R. M. (1993) Difference imaging of adenovirus: bridging the resolution gap between X-ray crystallography and electron microscopy. *EMBO J.* **12,** 2589–2599.

19. Drenth, J. (1999) *Principles of Protein X-Ray Crystallography,* 2nd ed. Springer Advanced Texts in Chemistry. Springer, New York.

20. Frank, J. (1996) *Three-Dimensional Electron Microscopy of Macromolecular Assemblies.* Academic Press, San Diego.

21. Baker, T. S., Olson, N. H., and Fuller, S. D. (1999) Adding the third dimension to virus life cycles: three-dimensional reconstruction of icosahedral viruses from cryo-electron micrographs. *Microbiol. Mol. Biol. Rev.* **63,** 862–922.

22. Tagari, M., Newman, R., Chagoyen, M., Carazo, J. M., and Henrick, K. (2002) New electron microscopy database and deposition system. *Trends Biochem. Sci.* **27,** 589.

23. Peek, J. D., Todino, G., and Strang, J. (2001) *Learning the UNIX Operating System,* 5th ed. O'Reilly, Sebastopol, CA.

24. Baumeister, W., Grimm, R., and Walz, J. (1999) Electron tomography of molecules and cells. *Trends Cell Biol.* **9,** 81–85.

25. Jones, T. A., Zou, J.-Y., Cowan, S. W., and Kjeldgaard, M. (1991) Improved methods for building protein models in electron density maps and the location of errors in these models. *Acta Crystallogr. A* **47,** 110–119.

26. Brünger, A. T. (1992) *X-PLOR, Version 3.1: A System for X-Ray Crystallography and NMR.* Yale University Press, New Haven, CT.

27. Brünger, A. T., Adams, P. D., Clore, G. M., et al. (1998) Crystallography & NMR system: A new software suite for macromolecular structure determination. *Acta Crystallogr.* **D54,** 905–921.

28. Collaborative Computational Project, Number 4. (1994) The CCP4 suite: programs for protein crystallography. *Acta Crystallogr.* **D50,** 760–763.

29. Kleywegt, G. J., Zou, J. Y., Kjeldgaard, M., and Jones, T. A. (2001) Around O, in *International Tables for Crystallography* (Rossmann, M. G. and Arnold, E., eds.), Vol. F. *Crystallography of Biological Macromolecules.* Chapter 17.1, Kluwer Academic, Dordrecht, The Netherlands, pp. 353–356.

30. Frank, J., Radermacher, M., Penczek, P., et al. (1996) SPIDER and WEB: processing and visualization of images in 3D electron microscopy and related fields. *J. Struct. Biol.* **116,** 190–199.
31. Fuller, S. D., Butcher, S. J., Cheng, R. H., and Baker, T. S. (1996) Three-dimensional reconstruction of icosahedral particles—the uncommon line. *J. Struct. Biol.* **116,** 48–55.
32. Crowther, R. A., Henderson, R., and Smith, J. M. (1996) MRC image processing programs. *J. Struct. Biol.* **116,** 9–16.
33. Bamford, D. H., Caldentey, J., and Bamford, J. K. H. (1995) Bacteriophage PRD1: a broad host range dsDNA tectivirus with an internal membrane. *Adv. Virus Res.* **45,** 281–319.
34. San Martín, C. and Burnett, R. M. (2003) Structural studies on adenoviruses, in *Adenoviruses: Model and Vectors in Virus Host Interactions,* vol. 272, *Current Topics in Microbiology and Immunology* (Doerfler, W. and Böhm, P., eds.)., Springer-Verlag, Heidelberg, pp. 57–94.
35. Bamford, J. K., Cockburn, J. J., Diprose, J., et al. (2002) Diffraction quality crystals of PRD1, a 66-MDa dsDNA virus with an internal membrane. *J. Struct. Biol.* **139,** 103–112.
36. Cockburn, J. J., Bamford, J. K., Grimes, J. M., Bamford, D. H., and Stuart, D. I. (2003) Crystallization of the membrane-containing bacteriophage PRD1 in quartz capillaries by vapour diffusion. *Acta Crystallogr. D. Biol. Crystallogr.* **59,** 538–540.
37. Mindich, L., Bamford, D., McGraw, T., and Mackenzie, G. (1982) Assembly of bacteriophage PRD1: particle formation with wild-type and mutant viruses. *J. Virol.* **44,** 1021–1030.
38. Mindich, L., Bamford, D., Goldthwaite, C., Laverty, M., and Mackenzie, G. (1982) Isolation of nonsense mutants of lipid-containing bacteriophage PRD1. *J. Virol.* **44,** 1013–1020.
39. Benson, S. D., Bamford, J. K. H., Bamford, D. H., and Burnett, R. M. (1999) Viral evolution revealed by bacteriophage PRD1 and human adenovirus coat protein structures. *Cell* **98,** 825–833.
40. Benson, S. D., Bamford, J. K. H., Bamford, D. H., and Burnett, R. M. (2002) The X-ray crystal structure of P3, the major coat protein of the lipid-containing bacteriophage PRD1, at 1.65 Å resolution. *Acta Crystallogr. D* **58,** 39–59.
41. Butcher, S. J., Bamford, D. H., and Fuller, S. D. (1995) DNA packaging orders the membrane of bacteriophage PRD1. *EMBO J.* **14,** 6078–6086.
42. Caspar, D. L. D. and Klug, A. (1962) Physical principles in the construction of regular viruses. *Cold Spring Harbor Symp. Quant. Biol.* **27,** 1–24.
43. McRee, D. E. and David, P. R. (1999) *Practical Protein Crystallography,* 2nd ed. Academic Press, San Diego, CA.
44. Aldroubi, A., Trus, B. L., Unser, M., Booy, F. P., and Steven, A. C. (1992) Magnification mismatches between micrographs: corrective procedures and implications for structural analysis. *Ultramicroscopy* **46,** 175–188.
45. Reddy, V. S., Natarajan, P., Okerberg, B., et al. (2001) Virus Particle Explorer (VIPER), a website for virus capsid structures and their computational analyses. *J. Virol.* **75,** 11943–11947.

46. Wriggers, W., Milligan, R. A., and McCammon, J. A. (1999) Situs: a package for docking crystal structures into low-resolution maps from electron microscopy. *J. Struct. Biol.* **125,** 185–195.
47. Volkmann, N. and Hanein, D. (1999) Quantitative fitting of atomic models into observed densities derived by electron microscopy. *J. Struct. Biol.* **125,** 176–184.
48. Roseman, A. M. (2000) Docking structures of domains into maps from cryoelectron microscopy using local correlation. *Acta Crystallogr. D Biol. Crystallogr.* **56,** 1332–1340.
49. Chacón, P. and Wriggers, W. (2002) Multi-resolution contour-based fitting of macromolecular structures. *J. Mol. Biol.* **317,** 375–384.
50. Cheng, R. H., Kuhn, R. J., Olson, N. H., et al. (1995) Nucleocapsid and glycoprotein organization in an enveloped virus. *Cell* **80,** 621–630.
51. Rossmann, M. G., Bernal, R., and Pletnev, S. V. (2001) Combining electron microscopic with x-ray crystallographic structures. *J. Struct. Biol.* **136,** 190–200.
52. He, Y., Chipman, P. R., Howitt, J., et al. (2001) Interaction of coxsackievirus B3 with the full length coxsackievirus-adenovirus receptor. *Nat. Struct. Biol.* **8,** 874–878.
53. Kuhn, R. J., Zhang, W., Rossmann, M. G., et al. (2002) Structure of dengue virus: implications for flavivirus organization, maturation, and fusion. *Cell* **108,** 717–725.
54. Nandhagopal, N., Simpson, A. A., Gurnon, J. R., et al. (2002) The structure and evolution of the major capsid protein of a large, lipid-containing DNA virus. *Proc. Natl. Acad. Sci. USA* **99,** 14758–14763.
55. Kraulis, P. J. (1991) MOLSCRIPT: a program to produce both detailed and schematic plots of protein structures. *J. Appl. Crystallogr.* **24,** 946–950.
56. Esnouf, R. M. (1999) Further additions to MolScript version 1.4, including reading and contouring of electron-density maps. *Acta Crystallogr. D* **55,** 938–940.
57. Merritt, E. A. and Bacon, D. J. (1997) Raster3D: photorealistic molecular graphics. *Methods Enzymol.* **277,** 505–524.

9

Herpes Simplex Virus–Cell Interactions Studied by Low-Fading Contrasted Immunofluorescence

Helle Lone Jensen and Bodil Norrild

Summary

The low-fading immunofluorescence with propidium iodide contrast described here is recommended for light and confocal viral antigen identification and other cell biology studies because: (1) it is a simple, rapid, sensitive, and reproducible technique; (2) phase-contrast microscopy is unnecessary; (3) contrast is optimal without blurring the fluorescent labeling; (4) autofluorescence is minimal, even in fixed cells; (5) background staining is minimal; (6) fading is invisible for at least 5-min exposures, even in preparations with weak antigen presentation; (7) fluorescence is stable after storage in the dark at −20°C; (8) fluorochromes are small-sized markers without steric hindrance; and (9) there is no need for silver enhancement or substrate solutions, which increase the risk of diffusion and other artifacts.

Key Words: Immunofluorescence; low fading; counterstain; whole cells; semithin cryosections; HSV-1.

1. Introduction

Viruses do not have their own metabolism *(1)* and rely on host cell protein synthesis to replicate efficiently, which is why the propagation of animal viruses takes place only after its genome has been integrated in a eukaryotic and viable host cell. Advances in cell culture, immunocytochemistry, and microscopy facilities have intensified studies of virus–cell interactions, resulting in important discoveries of both viral morphogenesis and molecular cell biology.

All immunocytochemical methods depend on: (1) the cells or tissues; (2) the nature and accessibility of the antigen; (3) the extraction, redistribution, and conformation of the antigen after fixation and preparation; (4) the need for permeabilization; (5) the nature and reactivity of the antibody; (6) the need for

From: *Methods in Molecular Biology, vol. 292: DNA Viruses: Methods and Protocols*
Edited by: P. M. Lieberman © Humana Press Inc., Totowa, NJ

blocking agents; (7) the nature of the marker; (8) the nature of necessary enhancement techniques; (9) the background staining; (10) cross-labeling; (11) diffusion; (12) fading; and (13) various artifacts. Naturally, the more complicated the technique becomes, the greater the risk of unspecific, unreproducible results. In any case, many of the reflections on immunocytochemistry apply to any immunological method.

Reactions between antigens and antibodies are valuable biological tools, but the reaction is useless without a suitable marker. Optimum labeling requires a marker of small size with good penetration of the cells and sections. The steric hindrance decreases, and the labeling efficiency, penetration, and precision increase with decreasing probe size *(2–6)*. Fluorochromes are light microscopy dyes and markers of small size with good precision and minimal steric hindrance *(5)*. When fluorochromes are excited by light of the appropriate wavelength, they emit light of longer wavelength, which can be detected at very low concentrations. The main disadvantage of fluorochromes *(5,7)* is fading (the intensity of the fluorescence declines). The promising approach of insertion of an open reading frame of green fluorescent protein in, for instance, herpesvirus genes, allowing studies of fluorescent tagged proteins in living cells *(8,9)*, is beyond the scope of the present chapter.

Immunofluorescence analyses of herpes simplex virus type 1 (HSV-1) morphogenesis in intact cells have previously been carried out extensively, with drawbacks such as fading and the need for phase contrast. However, semithin cryosections are invisible in phase contrast, and the antigens are in small quantities, making the fluorescence highly vulnerable to fading *(5)*. Conventional immunofluorescence labeling of semithin cryosections fades within 10 s when exposed to epifluorescence, leaving an almost totally black field, which is useless for examining and photographing. Our goal has been to achieve an easy, rapid, sensitive, and reproducible immunohistochemical light microscopic technique with minimal artifacts and application for whole cells and semithin cryosections of HSV-1-infected cells. Low-fading immunofluorescence with propidium iodide contrast has proved valid for several light microscopy and confocal laser scanning microscopy studies of HSV-1 morphogenesis *(3–5,10–13)*, as well as in daily pathology service. *p*-Phenylenediamine retards fading, and propidium iodide provides counterstaining, which results in brilliant fluorescence and contrast, minimal autofluorescence, and invisible fading for at least 5-min exposures, even in preparations with weak antigen presentations *(5)*.

2. Materials

Chemicals are obtained from Merck (Darmstadt, Germany) unless otherwise stated.

2.1. Serum

1. Sterile fetal bovine serum (FBS; cat. no. 29-101, Flow Laboratories, McLean, VA).
2. Human IgG (The State Serum Institute, Copenhagen, Denmark).

2.2. Microscope Slides

1. Microscope slides (cat. no. 02 1201, Menzel, Braunschweig, Germany).
2. 1% HCl in 70% alcohol.
3. Distilled water.

2.3. Low-Fading, Contrasted Mountant

1. Phosphate-buffered saline (PBS): 10 mM Na$_2$HPO$_4$/KH$_2$PO$_4$, 150 mM NaCl, pH 7.4. Store at 4°C for up to 2 wk.
2. p-Phenylenediamine (cat. no. P-6001, Sigma, St. Louis, MO). **Caution:** toxic.
3. Glycerol (cat. no. 4094, Merck), fresh and stored at –20°C.
4. 0.5 M Sodium carbonate-bicarbonate buffer, pH 9.0.
5. Propidium iodide (cat. no. 81845; Fluka Chemie, Buchs, Switzerland). **Caution:** injurious to health and mutagenic.

2.4. Viral Stock

1. Prototype HSV-1 strain F (a gift from B. Roizman, University of Chicago, Chicago, IL). The viral stock of HSV-1 strain F is replaced every 6 mo.

2.5. Cells

2.5.1. Cultures

1. MRC-5 human embryonic lung cells delivered at passages 26–28 (cat. no. 02-021-83, Flow Laboratories), and Vero cells (African Green Monkey cells, cat. no. 03-230, Flow Laboratories), passage below 145, or another cell line permissive to the virus studied.
2. EBME (Flow Laboratories), pH 7.4: 50 mL Eagle's essential basal medium (modified) with Earle's salts (cat. no. 14-000) supplemented with 5 mL nonessential amino acids (cat. no. 16-810), 11 mL sodium bicarbonate (cat. no. 16-883), 5 mL glutamine (cat. no. 16-801), 5 mL penicillin + streptomycin (cat. no. 16-700), 50 mL (to 10% FBS) or 5 ml (to 1% FBS) heat-inactivated FBS (cat. no. 29-101) and 430 mL sterile water. Store the medium at 4°C, renew every 2 wk, and add glutamine just before use.
3. Trypsin (cat. no. 16-891, Flow Laboratories).
4. *Mycoplasma* Gen-Probe (Gen-Probe, San Diego, CA).
5. Hoechst stain (cat. no. 33258, Hoechst Pharmaceuticals, Sommerville, NJ).
6. 25-cm^2 Cell culture flasks (cat. no. 163371, Nunc, Roskilde, Denmark).
7. Glass cover slips (Smethwick, Warley, UK).

2.5.2. Preparation

1. Absolute methanol, stored at –20°C.
2. Paraformaldehyde (PFA; cat. no. P 026, TAAB Laboratories Equipment, Aldermaston, UK). **Caution:** injurious to health.

3. Glutaraldehyde (GA; cat. no. G 002, TAAB Laboratories Equipment). **Caution:** injurious to health.
4. NaN$_3$ (cat. no. 6688, Merck). **Caution:** Toxic.
5. 20% Gelatin (cat. no. 4078, Merck) in PBS. Store at 4°C, and renew every 2 wk.
6. Rubber policeman.
7. 2.3 *M* Sucrose (cat. no. 7651, Merck) in PBS, pH 7.4. Store at 4°C, and renew every 4 wk.
8. Sharp hobby knife, needle or awl, and razor blades.
9. Dental wax.
10. Tissue-Tek (cat. no. O.C.T. 4583, Ames Division, Miles Laboratories, IN).
11. Silver sticks.
12. Liquid nitrogen.
13. Glass for ultramicrotomy: Alkar UMBO 25, strips 8 × 25 × 400 mm (Glass Ultra Micro, Sweden).
14. Bromma, LKB 2178 knifemaker II (LKB, Sweden).
15. RMC MT 6000 XL cryo-ultramicrotome.
16. Wood sticks, each mounted with a hair from a Dalmatian dog or an eyelash.
17. Smallest possible 0.25-mm platinum wire loop.

2.6. Immunofluorescence Labeling

1. NP$_{40}$ (Nonidet P-40, cat. no. N-3516, Sigma). **Caution:** injurious to health.
2. Monoclonal antibodies (MAbs) are stored as aliquots of stock at –80°C, and the dilutions are made just before use. Avoid repeated thawing and freezing of the antibodies. Any antiviral or anticytoskeleton MAb reactive in the method may be used.
   a. MAbs *(14–19)* against HSV-1 glycoprotein D (gD-1): Fd 138-80 (a gift from S. Chatterjee, University of Alabama, Birmingham, AL), DL6 (a gift from R. J. Eisenberg and G. H. Cohen, University of Philadelphia, Philadelphia, PA), and HD1 (a gift from L. Pereira, University of California, San Francisco, CA). These are all type-common and of isotype IgG$_{2a}$.
   b. MAb *(16)* against HSV-1 glycoprotein C (gC-1): HCl (a gift from L. Pereira, University of California, San Francisco, CA). HC1 is type-specific and of subclass IgG$_{2a}$.
   c. MAb against the cytoskeleton: anti-β-tubulin, subclass IgG$_{1k}$ light chains (no. N 357, Amersham, UK).
3. Fluorescein isothiocyanate (FITC) conjugated rabbit antimouse (cat. no. F 232, DAKO, Glostrup, Denmark).
4. Bovine serum albumin (BSA; cat. no. A 4503, Sigma). Store at –20°C for up to 6 mo.

2.7. Control Antibody

1. Monoclonal antihuman T-cell UCHL1, subclass IgG$_{2ak}$ (cat. no. M 742, DAKO).

2.8. Microscopy

1. Microscope equipped for epifluorescence.
2. Photomicrography equipment.

3. Methods

The methods described below outline (1) decomplementation of serum, (2) preparation of microscope slides, (3) production of low-fading, contrasted mountant, (4) the viral stock, (5) the cell culture and preparation of whole cells and frozen cell culture, (6) two-layer indirect immunofluorescence labeling, (7) controls, (8) storage, and (9) microscopy and photomicrography.

Attention must be given to controlled conditions through all procedures to obtain reproducible results. All experiments must be repeated at least twice, and slides must be made in duplicate to ensure reproducibility.

3.1. Serum

FBS is decomplemented by heat inactivation for 30 min at 56°C in a water bath to avoid complement-mediated nonspecific binding *(6)*.

3.2. Microscope Slides

1. Scratch circles with a diamond on the back side of the slides to mark the location of the sections and to avoid glass dust on the top of the slide.
2. Rinse the slides overnight in running water.
3. Dip the slides several times in 1% HCl in 70% alcohol.
4. Rinse three times in distilled water, air-dry, and store the slides protected against dust and dirt.

3.3. Low-Fading, Contrast Mountant of Immunofluorescence Microscopy

1. Dissolve 100 mg *p*-phenylenediamine in 10 mL PBS.
2. Add 90 mL glycerol.
3. Protect the solution against light, and stir properly at 4°C.
4. Adjust pH to 8.0 with 0.5 *M* sodium carbonate-bicarbonate buffer, pH 9.0.
5. Add propidium iodide to 1 µg/mL mountant.
6. Store the solution in small aliquots protected against light at –20°C. The optimal solution is translucent, very light red, and prevents the fading of FITC immunofluorescence for at least 10 min when whole cells are continuously exposed to epifluorescence. Discard the mounting medium when thawed once or when it begins to discolor to brown.

3.4. Viral Stock

1. Make the viral stock phenotypically concordant by two passages in MRC-5 cells below passage 35 (or any other used cell line permissive to the virus studied) at a multiplicity of infection (MOI) of 0.01 PFU per cell in EBME supplemented with 1% FBS.
2. Carefully scrape the clearly HSV-1-infected cells off the culture flasks with a rubber policeman.

3. Centrifuge at 455g for 10 min.
4. Resuspend the pellet from 10^7 infected cells in 1.5 mL EBME medium with 1% FBS.
5. Sonicate the cell suspension at 14 μm four times for 5 s, and keep the solution on ice in between.
6. Centrifuge at 455g for 5 min to remove cell debris.
7. Store the supernatant in aliquots of 500 μL at $-80°C$.
8. Determine the viral titer of this viral stock by plaque assay after 1 h of adsorption of 1 mL diluted viral suspension in $\log_{10}$ steps on confluent Vero cell monolayers overlaid with 5 mL medium containing 0.2% human IgG. Fix the plaques with formaldehyde, and stain with crystal violet.

3.5. Cells

*3.5.1. Cultures (see **Note 1**)*

1. Culture MRC-5 cells (or any other low-passage cell line permissive to the virus studied) as a monolayer at 37°C in EBME medium with 10% FBS and 95% air + 5% CO_2.
2. Trypsinate the cells and passage twice a week (split-ratio 1:2 for MRC-5) just at confluence, corresponding to 1.5×10^6 MRC-5 cells in a 25-cm^2 cell culture flask.
3. Three days before the HSV-1 infection, plate the cells at 8×10^4 cells/cm^2 on sterile glass cover slips for investigation of whole cells or in 25-cm^2 cell culture flasks in preparation for cryosections.
4. Infect low-passage, subconfluent or just confluent cells (MRC-5 cells at passage below 35) with HSV-1 strain F at an MOI of 30 PFU/per cell in EBME supplemented with 1% FBS.
5. After 1 h of adsorption at 37°C, remove the inoculum, and add EBME containing 1% FBS.
6. At 12 h post infection at 37°C, remove the medium, wash the cells briefly in sterile 37°C PBS, and fix.

*3.5.2. Whole Cells (see **Note 2**)*

1. Fix in permeabilizing absolute methanol at $-20°C$ for 10 min for optimal preservation and presentation of the cytoskeleton. Otherwise, fix the whole cells in freshly prepared 3% PFA in PBS for 10 min.
2. Store the cells at 4°C in PBS containing 0.1% (v/w) NaN_3.
3. Perform the immunolabeling within 4 wk.

*3.5.3. Frozen Cell Cultures (see **Note 3**)*

1. Fix the cells for 10 min in PBS-buffered 3% freshly prepared PFA, to which is added 2% GA.
2. Overlay the monolayer of cells with 20% gelatin dissolved in PBS and preheated to 37°C.
3. Scrape the cells off the culture flask with a rubber policeman.

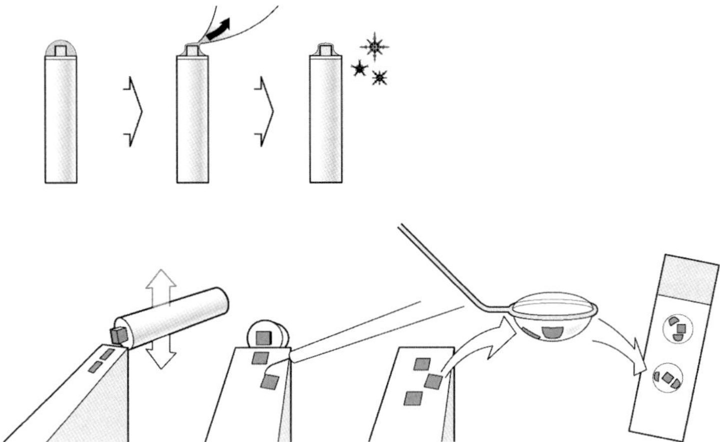

© Per Diemer 2003, Grafisk Værksted, Rigshospitalet afs. 7551 - tlf. 3545 3970, mail grafik@rh.dk

Fig. 1. Preparation of semithin cryosection. Mount the specimen in Tissue-Tek and sucrose on the tip of a silver stick. Remove excess sucrose with filter paper, but leave a thin film of sucrose. Afterward, plug the specimen holder into liquid nitrogen with shaking. Mount the silver stick in the cryochamber cooled with liquid nitrogen, and cut sections of about 500 nm with a glass knife at about –40°C. The sections can be rearranged on the knife using a wood stick mounted with a hair or eyelash. Remove the cryosections from the dry knife with a droplet of 2.3 *M* sucrose in PBS in a loop. The sections will fly to the approached cooled but not frozen droplet of sucrose. It is important not to let the knife touch the sucrose. The cryosections will spread during the thawing, resulting in thinner sections with fewer wrinkles. The sectioning and this collecting maneuver in particular require practice. By a slight touch, the sections and a part of the sucrose are placed on the microscope slide in the marked circles. The slide is now ready for immunoincubation. Throughout the procedure it is important that the sections never dry.

4. Transfer the sample without air bubbles to Eppendorf tubes, and after 30 min at 37°C, centrifuge at 300*g* for 5 min.
5. After 1 h at 4°C, use a sharp hobby knife to cut off the bottom of the tube containing the cell pellet.
6. Now coax the cells pelleted in 20% gelatin out of the tube bottom with a needle or awl, and place the pellet in a drop of 2.3 *M* sucrose on a microscope slide or dental wax cooled with ice.
7. Then cut the sample with razor blades into cubes of maximum 1 mm^3, which are cryoprotected overnight (~18 h) at 4°C in 2.3 *M* PBS-buffered sucrose (prevents formation of ice crystals) with 1% PFA.
8. The cubes with a little cover of sucrose are then mounted in Tissue-Tek on silver sticks and frozen by rapid immersion in liquid nitrogen (**Fig. 1**). For storage in liquid nitrogen, use cryoplastic tubes with screw caps and small holes made to ensure that the specimen is immersed in liquid nitrogen.

9. To make semithin cryosections (**Fig. 1**), prepare a glass knife according to the manufacturer's instructions. Alternatively, a diamond knife might be used. Mount the silver stick in the cryochamber cooled with liquid nitrogen in the cryo-ultramicrotome, cut sections of about 500 nm with the knife at about –40°C, and transfer the sections from the knife by means of a sucrose droplet in a wire loop to the microscope slides prepared as described in **Subheading 3.2.**
10. If convenient, the slides can be stored at 4°C in PBS-buffered 1% PFA overnight before immunolabeling.

3.6. Two-Layer Indirect Immunofluorescence Labeling

Immunoincubations should be carried out at room temperature in a humid chamber, and air-drying of the specimen must be prevented. PFA-fixed whole cells must be permeabilized to visualize intracellular antigens. Otherwise (*see* **Note 4**), the procedure is identical for whole cells (**Fig. 2**) and semithin cryosections (**Fig. 3**).

1. Wash the specimen with PBS four times for 5 min.
2. If necessary, permeabilize the PFA-fixed whole cells with 0.1% NP_{40} for 30 min, and wash in PBS.
3. Incubate with MAb anti-gD-1 Fd 138-80 diluted 1:500, HD1 1:1000, or DL6 1:4000; anti-gC-1 HC1 1:1600; or anti-β-tubulin 1:800 in PBS with 1% BSA for 45 min.
4. Rinse with PBS four times for 5 min.
5. FITC-conjugated rabbit antimouse antibody 1:25 in PBS with 1% BSA for 30 min.
6. Wash with PBS four times for 5 min.
7. Fix with 3% PFA in PBS for 10 min.
8. Rinse with PBS four times for 5 min.
9. Mount the specimens with low-fading, contrast mounting medium:
   a. Mount the immunofluorescence-labeled whole cells by placing the cover slip upside down in one drop of mounting medium on a clean microscope slide.
   b. Mount the immunofluorescence-labeled cryosections by placing a clean cover slip in one drop of mounting medium on the sections.
   In both cases, remove the excess mounting medium with filter paper by light pressure on the cover slip.
10. Store the specimens.

3.7. Controls

1. Immunostaining of non-virus-infected cells.
2. Replacement of one of the antibodies with a type-matched unrelated MAb such as, e.g., UCHL1, or with PBS buffer.
3. Omission of the FITC conjugate.

3.8. Storage

The cryopreparations can be stored in liquid nitrogen for at least 5 yr without loss of immunoreactivity. Slides stored in the dark at –20°C preserve the flu-

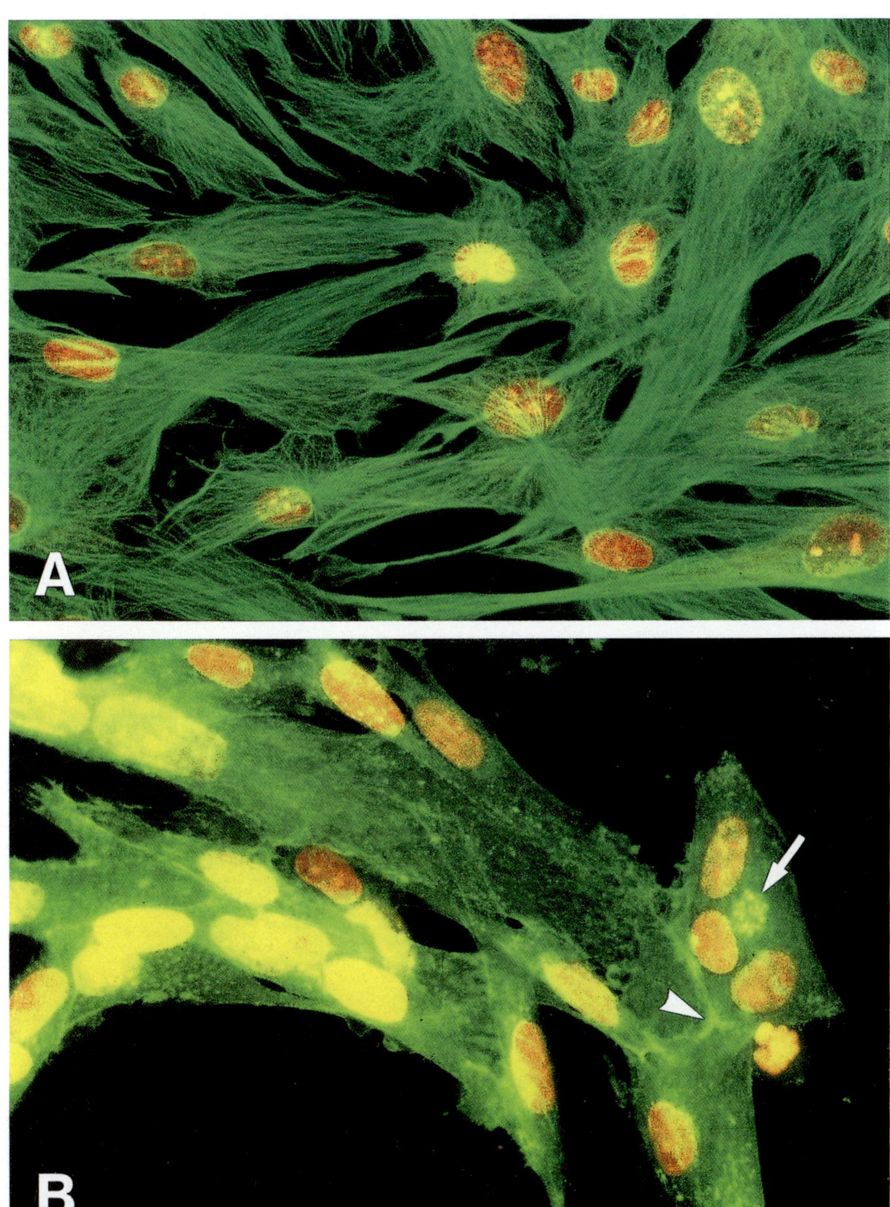

Fig. 2. (A) Nonvirus infected, methanol-fixed whole MRC-5 cells immunofluorescence-labeled for tubulin fibers. Original magnification ×25. **(B)** Permeabilized, HSV-1-infected whole MRC-5 cells immunofluorescence-labeled for gC situated in cytoplasmic vesicles, in a Golgi-like region (arrow) close to the nucleus and in cellular adhesion areas (arrowhead). Original magnification ×25. (Reprinted from **ref. 3** with permission from Kluwer Academic Publishers.)

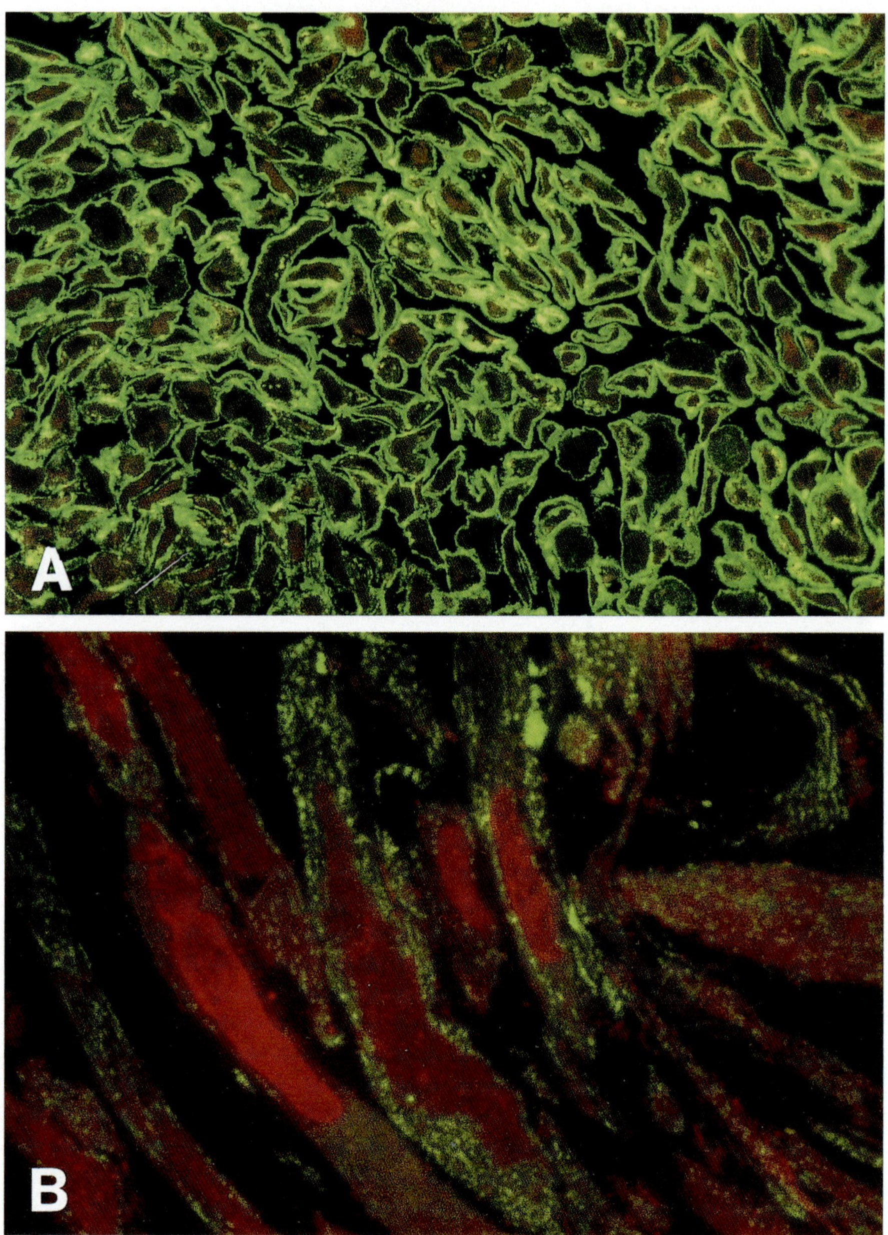

Fig. 3. (A) Semithin cryosections of HSV-1-infected L-fibroblasts. Low-fading immunofluorescence with suboptimal propidium iodide contrast shows gD-1 mainly at the cell surface. Original magnification ×25. **(B)** Well-contrasted low-fading immuno-fluorescence light microscopy of semithin cryosections with well-defined nucleus and cytoplasm. The tubulin stands out mostly as globules instead of fibers in these HSV-1-infected aging MRC-5 fibroblasts at passage 40. Original magnification ×100.

orescence with no significant diffusion for at least 1 yr. The mounting medium is liquid, which is why the slides must be stored horizontally in one layer.

3.9. Microscopy and Photomicrography

1. The Microscopy observations presented were made on a Leitz Dialux 20 microscope equipped for epifluorescence with a high-pressure mercury vapor lamp (HBO 50W; Osram, Tåstrup, Denmark). The lamp is used for a maximum of 50 h.
2. Contrast low-fading fluorescence studies were performed with water immersion 25/0.60 W or 100/1.20 W fluorescence objectives (*see* **Note 5**).
3. Photomicrography was performed with a Wild MP 551 camera and a Wild MPS 45 Photoautomat.
4. The color fluorescence photos were taken on Kodak Ektachrome EL ISO 400/27 (*see* **Note 6**).
5. Black-and-white fluorescence pictures were taken with Kodak Tmax professional film TMY ISO 400/27.
6. Suitable exposure times are 60–90 s with use of these films and low-fading mounting medium.

4. Notes

1. Attention must be given to careful handling and standardized culturing of the cells. It is recommended to examine cell growth by light microscopy every day, preferably by the same investigator. The cells must not be allowed to grow too dense. Every month the cell cultures should be proved *Mycoplasma*-free by means of cultivation (The State Serum Institute, Copenhagen, Denmark), using the Hoechst DNA staining method *(20)* and Gen-Probe.

   Any other cell type properly cultured and permissive to HSV or other viruses can be studied by the low-fading contrast immunofluorescence technique. It is important to use low-passage cells, because cellular aging causes cytoskeleton disorganization, reduced expression of viral proteins, and reduced capability to produce virus particles *(11)*.

   The American Type Culture Collection (ATCC) details the permissiveness to virus and biology of cell cultures (http://www.atcc.org).
2. To avoid contaminants such as formic acid and methanol *(21)*, the buffered formaldehyde solution is freshly prepared from PFA powder. Fixation with PFA and especially GA contributes to disturbing background fluorescence, but this, together with autofluorescence, is eliminated by use of propidium iodide.

   Permeabilization of intact cells and pre-embedding staining may not lead to equal accessibility of antibodies and markers to all compartments, which is why the staining can be difficult to interpret *(6)*.
3. The MOI, the required time of infection, the fixation, and the embedment in gelatin have to be optimized *(4)* and may differ depending on the aim of the study, the virus, the cells, the antigen, and the antibodies. Treatment of the cells with, e.g., brefeldin A blocks and accumulates glycoproteins and viral particles in the trans-

port systems anterior to the post-Golgi compartments, and the transport can be examined when brefeldin A is washed away *(10,13)*.

The about 500-nm-thick cryosections for light microscopy of fibroblasts represent approx 5% of the whole-cell antigen amount *(4)*. Semithin cryosections are invisible in phase contrast, and the level of HSV antigens is so low that the fluorescence disappears within 10 s in conventional immunofluorescence staining *(5)*. The low-fading immunofluorescence, with application of *p*-phenylenediamine to delay fading and propidium iodide to provide counterstaining, results in excellent, stable fluorescence and contrast both in whole cells (**Fig. 2**) and in semithin cryosections (**Fig. 3**), which allows identification of the nucleus and cytoplasmic compartment without phase-contrast microscopy. The low-fading contrast immunofluorescence technique has been compared with and proved superior to conventional immunofluorescence mounted in 90% PBS-buffered glycerol and silver-enhanced (Intense M, Amersham) colloidal 5-nm gold light microscopy *(5)*.

4. The optimal incubation time and dilution of antisera, defined as the highest dilution producing maximal staining and minimal background, must be determined for all batches of antisera and conjugates.

   Information on the many new Alexa Fluor dyes (Molecular Probes, Eugene OR) is available at the Web site http://www.probes.com. Nanoprobes (NY) has developed small uncharged gold markers and a dual immunoprobe containing both fluorescence dye and a gold marker (http://www.nanoprobes.com).

5. The advantages of using water-immersion objectives are obvious even when a high-power objective lens is used. No oil contaminates the preparation, the inherent fluorescence of oil is avoided, it is easy to clean both the objective and the glass slide, and the refractive index is optimal between immersion and the mounting medium.

6. High-speed films are convenient to prevent fading of fluorescent labeled specimens. However, films of more than 400 ASA result in gritty pictures and are not recommended. The recent explosive development in computerized data registration and image analysis is welcome for studies of virus–cell interactions, but it must be hoped that the new technologies will not be abused to manipulate the data.

References

1. Van Regenmortel, M. H. V. (2000) Introduction to the species concept in virus taxonomy, in *Virus Taxonomy. Classification and Nomenclature of Viruses. Seventh Report of the International Committee on Taxonomy of Viruses* (van Regenmortel, M. H. V, Fauquet, C. M., Bishop, D. H. L., et al., eds.), Academic, San Diego, CA, pp. 3–16.
2. Yi, H., Leunissen, J. L. M, Shi, G.-M., Gutekunst, C.-A., and Hersch, S. M. (2001) A novel procedure for pre-embedding double immunogold-silver labeling at the ultrastructural level. *J. Histochem. Cytochem.* **49,** 279–283.
3. Jensen, H. L. and Norrild, B. (1999) Easy and reliable double-immunogold labelling of herpes simplex virus type-1 infected cells using primary monoclonal antibodies and studied by cryosection electron microscopy. *Histochem. J.* **31,** 525–533.

4. Jensen, H. L. and Norrild, B. (1998) Herpes simplex virus type 1-infected human embryonic lung cells studied by optimized immunogold cryosection electron microscopy. *J. Histochem. Cytochem.* **46,** 487–496.

5. Jensen, H., Broholm, N., and Norrild, B. (1995) Low-fading immunofluorescence with propidium iodide contrast compared with immunogold light microscopy in whole cells and semi-thin cryosections. *J. Histochem. Cytochem.* **43,** 507–513.

6. Larsson, L.-I. (ed.) (1988) *Immunocytochemistry: Theory and Practice.* CRC, Boca Raton, FL.

7. Ono, M., Murakami, T., Kudo, A., Isshiki, M., Sawada, H., and Segawa, A. (2001) Quantitative comparison of anti-fading mounting media for confocal laser scanning microscopy. *J. Histochem. Cytochem.* **49,** 305–311.

8. Smith, G. A., Gross, S. P., and Enquist, L. W. (2001) Herpesviruses use bidirectional fast-axonal transport to spread in sensory neurons. *Proc. Natl. Acad. Sci. USA* **98,** 3466–3470.

9. Martin, A., O'Hare, P., McLauchlan, J., and Elliott, G. (2002) Herpes simplex virus tegument protein VP22 contains overlapping domains for cytoplasmic localization, microtubule interaction, and chromatin binding. *J. Virol.* **76,** 4961–4970.

10. Jensen, H. L., Rygaard, J., and Norrild, B. (1995) A time-related study of Brefeldin A effects in HSV-1 infected cultured human fibroblasts. *APMIS* **103,** 530–539.

11. Jensen, H. L. and Norrild, B. (2000) The effects of cell passages on the cell morphology and the outcome of herpes simplex virus type 1 infection. *J. Virol. Methods* **84,** 139–152.

12. Jensen, H. L. and Norrild, B. (2002) Morphologic, immunohistochemical, immunologic, ultrastructural, and time-related study of herpes simplex virus type 1-infected cultured human fibroblasts. *AIMM* **10,** 71–81.

13. Jensen, H. L. and Norrild, B. (2002) Temporal morphogenesis of herpes simplex virus type 1-infected and Brefeldin A-treated human fibroblasts. *Mol. Med.* **8,** 210–224.

14. Koga, J., Chatterjee, S., and Whitley, R. J. (1986) Studies on herpes simplex virus type 1 glycoproteins using monoclonal antibodies. *Virology* **151,** 385–389.

15. Eisenberg, R. J., Long, D., Pereira, L., Hampar, B., Zweig, M., and Cohen, G. H. (1982) Effect of monoclonal antibodies on limited proteolysis of native glycoprotein gD of herpes simplex virus type 1. *J. Virol.* **41,** 478–488.

16. Pereira, L., Klassen, T., and Baringer, J. R. (1980) Type-common and type-specific monoclonal antibody to herpes simplex virus type 1. *Infect. Immun.* **29,** 724–732.

17. Eisenberg, R. J., Long, D., Ponce de Leon, M., et al. (1985) Localization of epitopes of herpes simplex virus type 1 glycoprotein D. *J. Virol.* **53,** 634–644.

18. Isola, V. J., Eisenberg, R. J., Siebert, G. R., Heilman, C. J., Wilcox, W. C., and Cohen, G. H. (1989) Fine mapping of antigenic site II of herpes simplex virus glycoprotein D. *J. Virol.* **63,** 2325–2334.

19. Lausch, R. N., Staats, H., Oakes, J. E., Cohen, G. H., and Eisenberg, R. J. (1991) Prevention of herpes keratitis by monoclonal antibodies specific for discontinuous and continuous epitopes on glycoprotein D. *Invest. Ophthalmol. Vis. Sci.* **32,** 2735–2740.

20. McGarrity, G. J., Steiner, T., and Vanaman, V. (1983) Detection of mycoplasmal infection of cell cultures by DNA fluorochrome staining. *Methods Mycoplasmol.* **2,** 183–190.
21. Larsson, L.-I. (1993) Tissue preparation methods for light microscopic immunohistochemistry. *Appl. Immunohistochem.* **1,** 2–16.

10

Herpes Simplex Virus–Cell Interactions Studied by Immunogold Cryosection Electron Microscopy

Helle Lone Jensen and Bodil Norrild

Summary

A technique is presented for high-resolution postembedding immunolocalization of one or two (or several) antigens in the same ultrathin cryosection using primary monoclonal antibodies from the same species. The optimized three-layer indirect immunogold-labeled cryosection electron microscopy described is recommended for studies of virus–cell interactions, because: (1) it is a simple and reproducible method; (2) colloidal gold markers are electron-dense, stable, and easy to recognize; (3) the membraneous ultrastructure and immunolabeling are well preserved; (4) immunolabeling is less in the two-layer method; (5) silver-enhanced gold particles vary in size and shape; (6) it is possible to demonstrate herpes simplex virus type 1 glycoproteins gC-1 and gD-1 in the nuclear membranes and gC-1- and gD-1-labeled viral particles in the perinuclear space and to observe virions in the endoplasmic reticulum and Golgi area. The use of buffered 3% paraformaldehyde plus 2% glutaraldehyde for 2 h at room temperature effectively destroys free anti-IgG binding sites on the secondary antibodies in double-labeling immunogold cryosection electron microscopy and is recommended because: (1) inactivation is obtained through buffered primary fixative; (2) the method is simple and reproducible; (3) cross-labeling is effectively avoided; (4) silver-intensification, high temperature, and methyl cellulose cover of ultrathin cryosections are avoided between the staining sequences; and (5) ultrastructure and antigenicity are well preserved.

Key Words: Cryosection; electron microscopy; glycoproteins; HSV-1; immunogold; monolabeling; double labeling.

1. Introduction

By definition, viruses are unable to replicate on their own but must enter a host cell and use the host-cell macromolecular machinery and energy supplies in order to replicate *(1)*. Viruses are too small to be seen by ordinary light microscopy. Hence, the development of immunotechniques and electron

From: *Methods in Molecular Biology, vol. 292: DNA Viruses: Methods and Protocols*
Edited by: P. M. Lieberman © Humana Press Inc., Totowa, NJ

microscopy has been a breakthrough in the examination of viruses and viral diseases. The attractions of herpes simplex viruses (HSVs) are their ability to remain latent in their host for life, their reactivation, the variety of potential fatal infections, and the fact that there is still no final cure or vaccination available *(2)*. On the other hand, the intimate herpesvirus–cell interaction makes them attractive therapeutic couriers. However, the pathways for HSV entry, maturation, and egress are contentious issues that inspired our immunofluorescence light microscopy *(3)*, confocal laser scanning microscopy *(4)*, and immunogold cryosection electron microscopy *(5–9)* studies of the interaction between HSV type 1 (HSV-1) and the host cell.

The outcome of immunohistochemical methods depends not only on the antigen and the antibody but also on the character of the marker. Comparative studies and weak antigen presentation make special demands on the method. Simultaneous microscopic detection of two or more antigens is essential for studying spatial and possible functional relationships between different biomolecules. Electron microscopy is necessary to identify the fine structure of the cell, to observe the viruses, and to determine the precise localization of the labeled antigens. The advantages of colloidal gold markers at the electron microscopic level include their electron density and stability and the facts that they are easy to prepare in different and even small diameters *(10–13)* and are easy to recognize. The steric hindrance increases, and the precision decreases with increasing particle diameters.

There are, however, several essential problems in immunoelectron microscopy of HSV-1-infected cells *(5)*: (1) HSV-1 glycoproteins might not be immunodemonstrable after the process of Epon embedding; (2) one of the limitations of electron microscopy is sampling, and many immunolabeled cells are needed, because only minor areas of the tissue or the cell culture are investigated; (3) an ultrathin section represents only about 1% of the whole cell antigen amount; (4) the intensity of immunostaining is an issue; (5) the reproducibility of immunolabeling is an issue; (6) cross-labeling may occur, caused by free anti-IgG binding sites in multiple-staining techniques using several primary antibodies of the same species; (7) nonspecific staining may occur through each additional layer and staining cycle; (8) inhomogeneous gold particles with overlapping size are seen; (9) steric hindrance may exist; (10) preservation of labeling and ultrastructure is an issue; and (11) sections represent 2D, static formations of 3D, dynamic structures. The specific objectives *(5,6)* were to improve upon mono- and double-immunogold cryosection electron microscopy *(10,14)* with the purpose of obtaining reproducible results on the ultrastructure, the cellular distribution of HSV-1 glycoproteins, and virus–cell interactions *(5–9)*. The effects of cellular aging on HSV-1 infection *(7)* and the significance of the cytoskeleton in virus–cell interaction *(6–9)* have been demonstrated and studied by this technique. Furthermore, the data *(5–9)* support the now widely accepted view *(15,16)* that herpesvirus envelopment and maturation occur by a multiple-step pathway, at least in some cell types, and with a common route of HSV-1 virion and glycoprotein transport. Additionally, the immunogold cryosection electron microscopy method described may, with potential minor corrections, be useful for many cell biology studies.

2. Materials

All chemicals are from Merck (Darmstadt, Germany) unless otherwise indicated.

2.1. Serum

1. Sterile fetal bovine serum (FBS; cat. no. 29-101, Flow Laboratories, McLean, VA).
2. Human IgG (The State Serum Institute, Copenhagen, Denmark).
3. Goat serum (The State Serum Institute).

2.2. Viral Stock

1. Prototype HSV-1 strain F (a gift from B. Roizman, University of Chicago, Chicago, IL). The viral stock of HSV-1 strain F is replaced every 6 mo.

2.3. Cells

2.3.1. Cultures

1. MRC-5 human embryonic lung cells delivered at passages 26–28 (cat. no. 02-021-83, Flow Laboratories), and Vero cells (African Green Monkey cells, cat. no. 03-230, Flow Laboratories) passage below 145, or another cell line permissive to the virus studied.
2. EBME (Flow Laboratories), pH 7.4: 50 mL Eagle's essential basal medium (modified) with Earle's salts (cat. no. 14-000) supplemented with 5 mL nonessential amino acids (cat. no. 16-810), 11 mL sodium bicarbonate (cat. no. 16-883), 5 mL glutamine (cat. no. 16-801), 5 mL penicillin + streptomycin (cat. no. 16-700), 50 mL (to 10% FBS) or 5 mL (to 1% FBS) heat-inactivated FBS (cat. no. 29-101), and 430 mL sterile water. Store the medium at 4°C, renew every 2 wk, and add glutamine just before use.
3. Trypsin (cat. no. 16-891, Flow Laboratories).
4. *Mycoplasma* Gen-Probe (Gen-Probe, San Diego, CA).
5. Hoechst stain (cat. no. 33258, Hoechst Pharmaceuticals, Sommerville, NJ).
6. 25-cm^2 Cell culture flasks (cat. no. 163371, Nunc, Roskilde, Denmark).

2.3.2. Preparation

1. Paraformaldehyde (PFA; cat. no. P 026; TAAB Laboratories Equipment, Aldermaston, UK). **Caution:** injurious to health.
2. Glutaraldehyde (GA; cat. no. G 002, TAAB Laboratories Equipment). **Caution:** injurious to health.

3. Phosphate-buffered saline (PBS): 10 mM Na_2HPO_4/KH_2PO_4, 150 mM NaCl, pH 7.4. Store at 4°C, and renew every 2 wk.
4. 20% Gelatin (cat. no. 4078, Merck), in PBS. Store at 4°C and renew every 2 wk.
5. 2.3 M Sucrose (cat. no. 7651, Merck), in PBS, pH 7.4. Store at 4°C, and renew every 4 wk.
6. Rubber policeman.
7. Sharp hobby knife.
8. Needle or awl.
9. Razor blades.
10. Dental wax.
11. Tissue-Tek (cat. no. O.C.T. 4583, Ames Division, Miles Laboratories, IN).
12. Silver sticks.
13. Liquid nitrogen.
14. Glass for ultramicrotomy: Alkar UMBO 25, strips 8 × 25 × 400 mm (Glass Ultra Micro, Sweden).
15. Bromma, LKB 2178 knifemaker II (LKB, Sweden).
16. RMC MT 6000 XL cryo-ultramicrotome.
17. Wood sticks, each mounted with a hair from a Dalmatian dog or an eyelash.
18. Smallest possible 0.25-mm platinum wire loop.
19. 1% Solution of formvar (cat. no. F 005, TAAB Laboratories Equipment) in pure chloroform (cat. no. 2445, Merck).
20. 200-Mesh nickel grids (HR 24, Ni-3-0 mm, Graticules, Tunbridge, UK).

2.4. Methyl Cellulose With Uranyl Acetate

Store at 4°C and renew every 4 wk.

1. Methyl cellulose (64610 Mithocel MC 25 mPa.s USP, cat. no. 18.804-2, Aldrich, Steinheim, Germany).
2. Uranyl acetate (cat. no. 8473, Merck). **Caution:** highly toxic, radioactive material.

2.5. Immunogold Labeling

1. Monoclonal antibodies (MAbs) are stored as aliquots of stock at –80°C, and the dilutions are made just before use, avoiding repeated thawing and freezing.
   Any antiviral or anticytoskeleton MAb that works well in the method may be used.
   a. Monoclonals *(17–22)* against HSV-1 glycoprotein D (gD-1): Fd 138-80 (a gift from S. Chatterjee, University of Alabama, Birmingham, ALA), DL6 (a gift from R. J. Eisenberg and G. H. Cohen, University of Philadelphia, Philadelphia, PA), and HD1 (a gift from L. Pereira, University of California, San Francisco, CA). These are all type-common and of isotype IgG_{2a}.
   b. MAb *(19)* against HSV-1 glycoprotein C (gC-1): HC1 (a gift from L. Pereira, University of California, San Francisco, CA). This is type-specific and of subclass IgG_{2a}.
   c. MAb against the cytoskeleton: Anti-β-tubulin, subclass $IgG_{1\kappa}$ light chains (cat. no. N 357, Amersham, UK).

2. GAR-G5: goat antirabbit IgG (H+L) conjugated to 5-nm colloidal gold particles (cat. no. RPN 420, Amersham).
3. GAR-G15: goat antirabbit IgG (H+L) conjugated to 15-nm colloidal gold particles (cat. no. RPN 422, Amersham).
4. Affinity-purified RAM (rabbit antimouse) immunoglobulin (cat. no. Z 259, DAKO, Glostrup, Denmark).
5. 0.02 *M* Glycine (cat. no. 4201, Merck), in PBS, pH 7.4. Store at –20°C for 6 mo.
6. Bovine serum albumin (BSA; cat. no. A 4503, Sigma, St. Louis, MO). Store at –20°C for 6 mo.
7. IGSS quality gelatin (cat. no. RPN 416, Amersham).

2.6. Control Antibody

1. Monoclonal antihuman T cell UCHL1, subclass $IgG_{2a\kappa}$ (cat. no. M 742, DAKO).

2.7. Microscopy

1. Electron microscope with photomicrography equipment.

3. Methods

The methods described below include (1) decomplementation of serum, (2) the viral stock, (3) the cell culture and preparation of frozen cell culture, (4) the production of methyl cellulose and uranyl acetate, (5) postembedding indirect three-layer mono- and double- (or multiple) immunogold labeling with primary MAbs, (6) controls, and (7) electron microscopy and photomicrography.

Attention must be given to controlled conditions through all procedures, including preparation of solutions and controlled temperature, to obtain reproducible results. At no time during the entire procedure should sections be allowed to dry. In particular, the preparation of frozen cell cultures, the cutting of ultrathin frozen sections, the immunogold labeling, and the use of methyl cellulose with uranyl acetate require practice and experience. All experiments must be repeated at least twice, and grids must be made in duplicate to ensure reproducibility. In each experiment, 200–300 cells should be examined in the electron microscope.

3.1. Serum

Decomplement the goat serum and FBS by heat inactivation for 30 min at 56°C in a water bath to avoid complement-mediated nonspecific binding *(12)*.

3.2. Viral Stock

1. Make the viral stock phenotypically concordant by two passages in MRC-5 cells below passage 35 (or any other used cell line permissive to the virus studied) at a multiplicity of infection (MOI) of 0.01 PFU per cell in EBME supplemented with 1% FBS.

2. Carefully scrape the obvious HSV-1-infected cells off the culture flasks with a rubber policeman.
3. Centrifuge at 455g for 10 min.
4. Resuspend the pellet from 10^7 infected cells in 1.5 mL EBME medium with 1% FBS.
5. Sonicate this cell suspension at 14 μm four times for 5 s, and keep the solution on ice in between.
6. Centrifuge at 455g for 5 min to remove cell debris.
7. Store the supernatant in aliquots of 500 μL at –80°C.
8. Determine the viral titer of this viral stock by plaque assay after 1 h of adsorption of 1 mL diluted viral suspension in $\log_{10}$ steps on confluent Vero cell monolayers overlaid with 5 mL of medium containing 0.2% human IgG. Fix the plaques with formaldehyde, and stain with crystal violet.

3.3. Cells

*3.3.1. Cultures (see **Note 1**)*

1. Culture *Mycoplasma*-free MRC-5 cells (or any other low-passage cell line permissive to the virus studied) as a monolayer at 37°C in EBME medium with 10% FBS and 95% air + 5% CO_2.
2. Trypsinate the cells and passage twice a week (split ratio 1:2 for MRC-5 cells) just at confluence, corresponding to 1.5×10^6 MRC-5 cells in a 25-cm^2 cell culture flask.
3. Infect low-passage, just confluent cells (MRC-5 cells at passage below 35) with HSV-1 strain F at an MOI of 30 PFU per cell in EBME supplemented with 1% FBS.
4. After 1 h of adsorption at 37°C, remove the inoculum, and add EBME containing 1% FBS.
5. After 12 h at 37°C, remove the medium, wash the infected cells briefly in sterile 37°C PBS, and fix.

*3.3.2. Frozen Cell Cultures (see **Note 2**)*

1. Fix the cells for 10 min in PBS-buffered 3% freshly prepared PFA, to which is added 2% GA.
2. Then overlay the monolayer of cells with 20% gelatin dissolved in PBS and preheated to 37°C.
3. Scrape off the cells from the culture flask with a rubber policeman.
4. Transfer the sample without air bubbles to Eppendorf tubes, and after 30 min at 37°C, centrifuge at 300g for 5 min.
5. After 1 h at 4°C, cut off the bottom of the tube with the cell pellet using a sharp hobby knife.
6. Coax the cells pelleted in 20% gelatin out of the tube bottom with a needle or awl, and place the pellet in a drop of 2.3 *M* sucrose on a microscope slide or dental wax cooled with ice.
7. Cut the sample with a razor blade into cubes of maximum 1 mm^3, which are cryoprotected overnight (~18 h) at 4°C in 2.3 *M* PBS-buffered sucrose (prevents formation of ice crystals) with 1% PFA.

8. Then mount the cubes with a little sucrose cover in Tissue-Tek on silver sticks, and freeze by rapid immersion in liquid nitrogen (**Fig. 1**). For storage in liquid nitrogen, use cryoplastic tubes with screw caps and small holes made to ensure the cubes are immersed in liquid nitrogen.
9. To make ultrathin cryosections (**Fig. 1**), prepare a glass knife according to the manufacturer's instructions, immediately before use. Control the quality of the glass knife in a stereomicroscope. Mount the silver stick in the cryochamber cooled with liquid nitrogen in the cryo-ultramicrotome, and cut sections of about 100 nm (blue cellophane-like sections) with a glass knife at –75°C to –90°C. The optimal cutting temperature depends on the desired thickness of the sections and the material. Then transfer the sections from the knife to formvar-covered and carbon-coated 200-mesh nickel grids by means of a sucrose droplet in a wire loop.
10. If necessary, the grids with the sections downside in PBS-buffered 1% PFA can be stored in small Petri dishes at 4°C and immunolabeled within 24 h.

3.4. Methyl Cellulose With Uranyl Acetate (see Note 3)

1. For a final concentration of 1.35%, add methyl cellulose powder to distilled water preheated to 95°C.
2. Mix with a magnetic stirrer at 95°C for a few minutes, and put the solution on ice.
3. Stir for at least 4–8 h on ice and preferably 24–72 h at 0–4°C.
4. Leave for further 3–4 d at 4°C.
5. Centrifuge at 362,000g for 90 min at 4°C.
6. Store the tubes at 4°C for up to 4 wk without disturbing the pellet produced.
7. Carefully pipet aliquots of methyl cellulose from the surface layer for each experiment, and add filtered stock of aqueous uranyl acetate to a final concentration of 0.4%.

3.5. Immunogold Labeling

Transfer the ultrathin cryosections on the grids by means of the smallest possible 0.25-mm platinum wire loop through large drops on a sheet of Parafilm (**Fig. 1**) in a humid chamber at room temperature. Between each drop the loop may be rinsed in PBS and wiped on filtering paper. In this way contamination from one drop to another is minimized, and the wash optimized. It is important that only the one side with sections on the grid be wet, or the grid will drown.

3.5.1. Postembedding, Indirect Three-Layer Mono-Immunogold Labeling (*Fig. 1;* see *Notes 2–5*)

1. Wash the specimen two times for 5 min with PBS, pH 7.4.
2. Add 0.02 *M* glycine in PBS for 10 min to inactivate any remaining aldehyde.
3. Rinse with PBS two times for 5 min.
4. Rinse for 5 min in PBS containing 0.8% BSA and 0.1% IGSS quality gelatin.
5. Incubate for 45 min with the primary antibody (Fd 138–80 1:250, DL6 1:1500, HD1 1:1000, HC1 1:800, or anti-β-tubulin 1:600–1:1000 in different batches) diluted in PBS + BSA + gelatin solution with 1% goat serum.

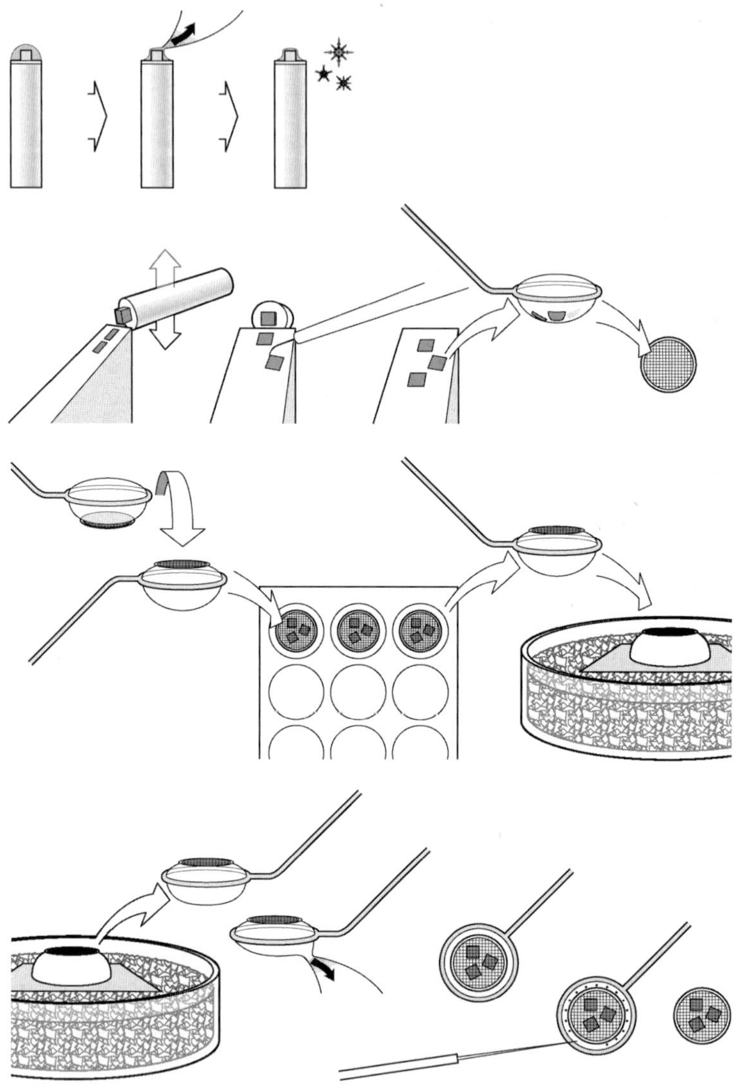

6. Wash off the primary antibody with PBS + BSA + gelatin solution (six times for 4 min).
7. Incubate for 30 min with 2.5 µg/mL affinity-purified RAM in PBS + BSA + gelatin solution containing 1% goat serum.
8. Wash with PBS + BSA + gelatin solution (six times for 4 min).
9. Incubate for 20 min with GAR-G5 at a suitable concentration of 1:25–1:75 diluted in PBS + BSA + gelatin buffer with 1% goat serum.
10. Then wash with PBS + BSA + gelatin buffer (six times for 4 min).

Fig. 1. Preparation of ultrathin cryosections. Mount the specimen in Tissue-Tek and sucrose on the tip of a silver stick. Remove excess sucrose with a filter paper, but leave a film of 0.5–1.0 mm on the specimen. Afterward, plug the specimen holder into liquid nitrogen with shaking. Mount the silver stick in the cryochamber cooled with liquid nitrogen, and cut sections of about 100 nm with a glass knife at about –90°C. The sections can be rearranged on the knife by using a wood stick mounted with a hair or eyelash. Remove the cryosections from the dry glass knife with a droplet of 2.3 *M* sucrose in PBS in a loop. The sections will fly to the approached cooled but not frozen droplet of sucrose. It is important not to let the knife touch the sucrose. The cryosections will spread during the thawing, resulting in thinner sections with fewer wrinkles. The sectioning and this collecting maneuver in particular require practice. By a slight touch, the sections and a part of the sucrose are placed on the grid. Turn the grid upside down, and immediately transfer the grid to a dish with the sections facing down in PBS-buffered 1% PFA. If necessary, the grids can now be stored at 4°C before incubation in drops on a piece of Parafilm in a humid chamber. In this and the following steps, make sure that the side of the grid without sections remains dry. Now pass each grid from one large drop of solution to another by means of the smallest possible platinum wire loop. Finally, float the grid on ice-cold methyl cellulose with uranyl acetate for 10 min. Catch the grid with a loop, and remove excess methyl cellulose with a piece of filter paper. Dry the grid; the sections are thereby covered with a film of methyl cellulose. Take care to remove only enough methyl cellulose to give an interference color from gold to blue after drying. Use a needle or awl to loosen the grid from the loop. The grid is now ready for examination in the electron microscope or storage in a grid box.

11. Rinse with PBS (six times for 4 min).
12. Postfix for 10 min with 2% GA in PBS.
13. Wash with PBS (four times for 4 min).
14. Wash with glass-distilled water (four times for 4 min).
15. Then stain and protect the sections against air-drying artifacts by a solution of ice-cold 1.35% methyl cellulose containing 0.4% aqueous uranyl acetate.
16. Remove any excess of methyl cellulose after 10 min with a piece of filter paper.
17. Dry the grids, and store them in grid boxes at room temperature, protected against dust and damage.

*3.5.2. Postembedding Indirect Three-Layer Double (or Multiple) Immunogold Labeling With MAbs (**Fig. 1;** see **Notes 2–6**)*

1. Perform the first staining sequence as for mono-immunogold labeling, **Subheading 3.5.1., steps 1–11,** with one of the antiviral monoclonals as the primary antibody.
2. Inactivate free anti-IgG binding sites on the antibodies in the first staining sequence with PBS-buffered, freshly prepared 3% PFA + 2% GA for 2 h at room temperature.

3. If convenient, store the grids with the sections down in PBS-buffered 1% PFA in small Petri dishes overnight at 4°C.
4. Perform the second staining sequence like the first staining sequence, but now with the primary antibody anti-β-tubulin for 45 min, and then with 2.5 µg/mL RAM for 30 min, and render the reaction visible with GAR-G15 (1:25–1:50) for 20 min. The anti-β-tubulin, RAM immunoglobulin, and GAR-G15 are all diluted in PBS + BSA + gelatin buffer with 1% goat serum.
5. Wash with PBS + BSA + gelatin buffer (six times, 4 min each).
6. Wash with PBS (six times for 4 min).
7. Postfix with 2% GA in PBS for 10 min.
8. Wash with PBS (four times, 4 min each).
9. Wash with glass-distilled water (four times for 4 min).
10. Stain and protect the sections with a solution of ice-cold 1.35% methyl cellulose containing 0.4% aqueous uranyl acetate.
11. Remove any excess of methyl cellulose after 10 min with a piece of filter paper.
12. Dry and store the grids at room temperature protected in grid boxes against dirt and damage.

3.6. Controls

Immunocytochemical controls are carried out as follows, in one or both staining sequences:

1. Immunostaining of nonvirus-infected cells.
2. Omission of the primary or secondary antibody and replacement with buffer or type-matched unrelated MAb such as, e.g., UCHL1 1:25.
3. Anti-β-tubulin serves as a positive control.

The specificity of the monoclonals and the reliability of the immunogold labeling are controlled by immunoblotting and low-fading, contrast immuno-fluorescence light microscopy.

3.7. Electron Microscopy and Photomicrography

The observations presented electron microscopic (**Figs. 2** and **3**) are made by examining the ultrathin sections in a Zeiss electron microscope 900, operating at 80 kV. Photomicrographs are taken on Agfa Scientia EM film (3.25 × 4 inches, HJQ7B).

4. Notes

1. The cell culture conditions are standardized. The cultures are proved *Mycoplasma*-free once a month by means of cultivation (The State Serum Institute), by the Hoechst DNA staining method *(23)*, and by Gen-Probe; light microscopic examination of the cells is performed nearly every day by the same investigator. The cells have to be handled carefully and must not be allowed to grow too dense.

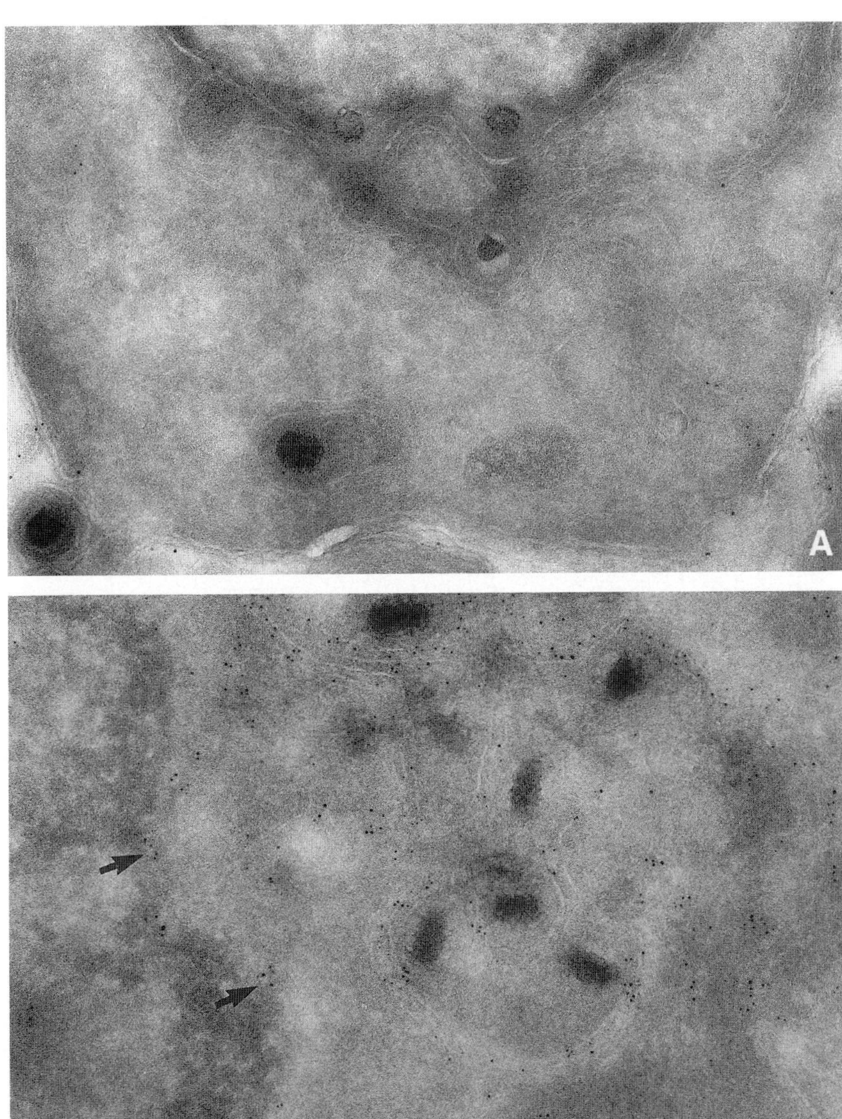

Fig. 2. (A) Mono-immunogold (GAR-G5) staining of viral glycoprotein in HSV-1-infected fibroblasts at 12 h post infection: capsids push aside the inner nuclear membrane, and viral particles can be seen in the endoplasmic reticulum. An extracellular virus particle is found in the lower left corner. Original magnification ×20,000. (Reprinted from **ref. 8** with permission from Lippincott Williams & Wilkins.) **(B)** Ultrathin cryosection of HSV-1-infected and brefeldin A-treated fibroblasts with brefeldin A effects removed for 3 h. The gD is identified with 5-nm gold particles both in the nuclear membranes (arrows) and in the endoplasmic reticulum containing enveloped and unenveloped viral particles. Original magnification ×30,500. (Reprinted from **ref. 9,** with permission from The Johns Hopkins University Press.)

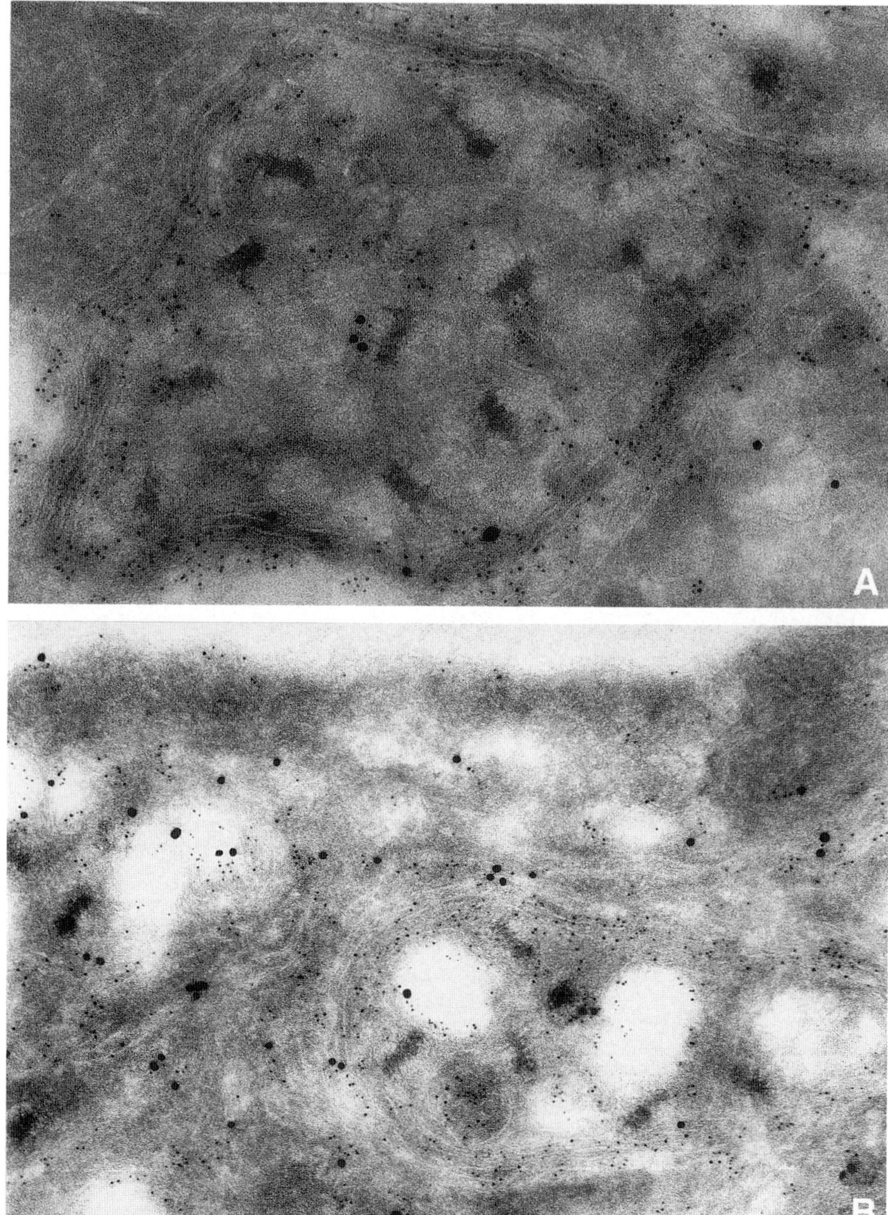

Fig. 3. (A) and **(B)** Double-immunogold-labeled ultrathin cryosection of HSV-1-infected MRC-5 cells. The gD is identified with 5-nm gold particles in cytoplasmic membranes around capsids. Tubulin is marked with 15-nm gold particles. (B) represents a brefeldin A-treated cell after 9 h washout allowing gD transport to the plasma membrane (top). Original magnification ×30,500. (B, reprinted from **ref. 9,** with permission from Johns Hopkins University Press.)

Any other cell type properly cultured and permissive to HSV or another virus studied can be examined by the immunogold labeled cryosection electron microscopy technique described. It is important to use low-passage cells, because cellular aging changes the functions of the cells, including cytoskeleton disorganization, reduced expression of viral proteins, and reduced capability to produce virus particles *(7)*. The American Type Culture Collection (ATCC) details the permissiveness to virus and biology of cell cultures (http://www.atcc.org).

2. The advantages of postembedding, indirect immunostaining is that the method is inexpensive and simple, and the immunolabeling can be repeated. However, there are at least three essential problems *(5)* in mono-immunoelectron microscopy of HSV-1-infected cells: (1) the number of labeled cells, because only minor areas are investigated; (2) the intensity of the immunostaining; and (3) the reproducibility of immunolabeling. Any variation can be eliminated by studying 100–200 cells *(24)*. An ultrathin section of about 100 nm represents approx 1% of the whole MRC-5 cell antigenicity *(5)*, which calls for special considerations. Moreover, the results may be influenced by the thickness of the sections and the asynchronous infection of different cells. Artifacts must be borne in mind. Furthermore, the speed of the fixation is a limiting factor in time-related studies like HSV-1 envelopment and transport of viral glycoproteins. In addition, the antigens gC-1 and gD-1 are destroyed by heat, which may in part explain the fact that with the antibodies used in our studies, these viral glycoproteins were not detectable in Epon-embedded material *(5,6)*. Fortunately, cryopreparations can be stored in liquid nitrogen for years without loss of immunoreactivity *(5)*.

The MOI, the required time of infection, the fixation, and the embedment in gelatin have to be optimized *(5)*, but they may differ depending on the aim of the study, the virus, the cells, the antigens, and the antibodies. Treatment of the cells with, e.g., brefeldin A, blocks and accumulates glycoproteins and viral particles in the transport systems anterior to post-Golgi compartments *(4,9)*, and the transport can be examined when brefeldin A is washed away (**Figs. 2B** and **3B**). To avoid contaminants such as formic acid and methanol *(25)*, the buffered formaldehyde solution should be freshly prepared from PFA powder. The optimal incubation time and dilution, defined as the highest dilution producing maximal, saturated staining and minimal background, must be determined for all batches of antisera and conjugates. Gold conjugates are titrated by anti-β-tubulin immunostaining of non-virus-infected cells. To obtain reliable results (**Figs. 2** and **3**), care must be taken with the antigen, the antibody, and the marker. The antigen can be destroyed in the processing, for instance by the fixative, heat, or embedding media *(5,6)*. The antibodies must be well characterized, and in our studies *(3–9)* only mouse monoclonals *(17–22)* are used. It should be noted that the gC-1 antibody used, HC-1 *(5–9)*, exhibits a prozone character. The prozone phenomenon *(12)* occurs only in indirect immunocytochemical procedures when too high concentrations of primary antibodies are used, so that the antibodies may become so crowded that their Fc regions hinder access for the immunomarker, and consequently no staining or only partial staining can be observed. Additionally, contrast staining of the sections must not reduce the immunolabels and the cell structures.

Problems such as chatter marks and static electricity in cryosectioning may trouble even experienced investigators, but these may be overcome by the correct angle of the knife in relation to the specimen block, by preventing any vibrations, by avoiding plastics, by wearing cotton materials, and so on. In preparation for easier frozen-sectioned specimens, it has been suggested to infuse the samples with a mixture of 10–30% poly(vinylpyrrolidone) and 2.07–1.61 *M* sucrose *(26)*. Embedment with resins may be advantageous if good preservation of the ultrastructure has priority. Nevertheless, the epoxy embedment in particular is carcinogenic and time-consuming and may destroy or reduce many antigenic epitopes *(5)*, whereas immunocytochemistry on ultrathin frozen sections is superior in all these aspects, with more than 50% higher labeling efficiency *(5,27)*. However, the ultrastructure of the cryosection differs from that of resin-embedded material.

3. Methyl cellulose and uranyl staining of cryosectioned gelatine-embedded material (**Figs. 2** and **3**) produce good delineation of membraneous structures *(5–9)*, but the cytoplasmic filaments and other nonmembraneous ultrastructures are not depicted as in Epon-embedded material and are barely visible without immunodetection. The final thickness of the uranyl–methyl cellulose is critical with respect to both contrast and fine structure. The optimal final thickness of the dried uranyl–methyl cellulose film has a gold to blue interference color. Thicker films improve the preservation of fine structure at the expense of contrast. Specimens with very thin films have air-drying artifacts such as holes and are vulnerable in the electron beam. Making use of both sucrose and methyl cellulose (and if convenient, further uranyl acetate) as cryoprotectants might further benefit the ultrastructure *(28)*, because most damage to the sections occurs by overstretching when the sections are thawed and transferred from the cryochamber of the microtome. Another initiative to preserve the ultrastructure, but with respect to the immunoreactivity, could be to cryoprotect the cultured cells with sucrose and methyl cellulose after fixation but before embedment in gelatin and sectioning.

4. Colloidal gold *(10)* is excellent for immunoelectron microscopy, giving an electron-dense, particulate, permanent signal allowing multiple staining using gold of different sizes *(5,6)*, but the labeling sensitivity is inversely proportional to the size of the marker. However, to reduce background immunogold staining, we recommend that the immunoglobulins be affinity-purified, that the serum be decomplemented by heat inactivation, that blocking agents be avoided, that only one buffer as PBS be used, and that careful washing steps be used. Many different buffers increase the risk of errors, precipitations, and other artifacts. Fortunately, gelatin is an effective inhibitor of nonspecific gold binding, at least to fibroblasts *(29)*. However, the labeling efficiency can be related to both the fixation and the gelatin concentration *(5,30)*.

It is possible to prepare homogeneous gold markers even of small size *(11,13)*, but the coefficient of variance often is as high as 10–20% *(6,11)*. Recent advances in gold technology have led to small and stable probes with higher labeling density, better sensitivity, and greater penetration into tissues *(31)*. Silver enhancement of gold particles is applicable at both the light and electron microscopic levels *(3,5,32,33)*. However, the background staining, the self-nucleation, and the size of enhanced gold particles vary significantly even though the silver enhancement is carefully performed, and this method cannot be recommended *(3,5)*. Without silver enhancement, the 1-nm gold probe needs great magnification, which gives rise to an inconveniently small visual field. By virtue of its particulate nature, gold-labeling facilitates a semiquantitative analysis of the antigen densities on ultrathin sections. However, both relative and absolute quantitation *(12,30,33,34)* presupposes reproducible and well-defined ultrastructure, prepared standards, and the biological relevance related to errors in the detection procedure. Quantitative ultrastructural analysis of gold particles has been omitted from our studies *(5,6)*, because (in addition to our reservations concerning the method whereby the sections represent only a minor part of the antigenicity) the thickness of the cryosection and the level of sectioned cells can vary, the infection of different cells varies, glycoproteins occur in various phases of processing during biosynthesis and with various immunoreactivity, and even the most optimized procedure allows antigens to be destroyed to various degrees.

The precision of the immunolabeling depends on the size of the marker, the conjugated macromolecule, enhancement, and diffusion *(5,6)*. The precision suffers, and there is a risk of introducing nonspecific labeling and artifacts by each staining sequence and additional layer *(5,6,12)*; the advantage of the indirect technique is greater sensitivity. Furthermore, the validity of the immunotechnique depends on the efficiency with which the first antibody can be saturated with the subsequent antibodies and markers *(3,5,6)*. For these reasons, labeling of more than two antigens in the same section may often be unreliable.

Nanoprobes has developed small uncharged gold markers and a dual immunoprobe containing both fluorescence dye and gold marker (http://www.nanoprobes.com).

5. Quality control of the immunogold-labeled ultrathin cryosections includes well-preserved ultrastructure, no precipitations or clumping of the gold probes, and minimal but just visible background staining with few gold particles over the nuclei to ensure saturation of the antigen–antibody reaction (**Fig. 2**). Additionally, it is required that cross-sectioned tubulin in non-virus-infected cells be labeled with about 10 G5 particles and that longitudinally sectioned tubulin show gold particles with a linear disposition. Cross-staining between the first- and second-staining sequences can be estimated by means of the extracellular HSV-1 virus particles, because intracellular closeness of GAR-G5 and GAR-G15 might be the biological coincidence of viral glycoprotein and tubulin (**Fig. 3**). It is advantageous to perform glycoprotein staining in the first staining cycle to make it possible to evaluate cross-labeling. Background staining is disclosed on the nuclei.

6. Two (or more) antigens are allowed to be labeled in the same section (**Fig. 3**), provided that the markers do not overlap in size and there is no crossreactivity or exchange of probes between the different targets. However, there are several essential problems in multiple-labeling immunogold electron microscopy of HSV-infect-

ed cells: (1) the level and availability of antigen; (2) the intensity of immunostaining in both first and second staining sequences; (3) the reproducibility of the labelings; (4) cross-labeling; (5) increasing nonspecific staining by each additional layer; (6) obtaining homogenous gold particles with nonoverlapping size; (7) steric hindrance; (8) preservation of labeling and the ultrastructure; and finally (9) sections representing 2D information of 3D cells, viral particles, and tubulin network. In addition, as in our studies *(6–9)*, primary antibodies of the same species require obstruction of free anti-IgG binding sites to prevent cross-labeling. Harsh inactivation procedures are avoided *(6)* by using the primary PBS-buffered fixative 3% PFA plus 2% GA for 2 h at room temperature between the first and second staining sequences. Also, to minimize the risk of steric hindrance, the smallest gold particles should be included exclusively in the first staining cycle. This approach will probably be successful for several staining sequences with any antigen and MAb.

References

1. Van Regenmortel, M. H. V. (2000) Introduction of the species concept in virus taxonomy, in *Virus Taxonomy. Classification and Nomenclature of Viruses. Seventh Report of the International Committee on Taxonomy of Viruses* (van Regenmortel, M. H. V., Fauquet, C. M., Bishop, D. H. L., et al., eds.), Academic, San Diego, CA.
2. Roizman, B. and Knipe, D. M. (2001) Herpes simplex viruses and their replication, in *Fundamental Virology,* 4th ed. (Knipe, D. M., et al.), Lippincott Williams & Wilkins, Philadelphia, PA, pp. 1123–1183.
3. Jensen, H., Broholm, N., and Norrild, B. (1995) Low-fading immunofluorescence with propidium iodide contrast compared with immunogold light microscopy in whole cells and semi-thin cryosections. *J. Histochem. Cytochem.* **43,** 507–513.
4. Jensen, H. L., Rygaard, J., and Norrild, B. (1995) A time-related study of Brefeldin A effects in HSV-1 infected cultured human fibroblasts. *APMIS* **103,** 530–539.
5. Jensen, H. L. and Norrild, B. (1998) Herpes simplex virus type 1-infected human embryonic lung cells studied by optimized immunogold cryosection electron microscopy. *J. Histochem. Cytochem.* **46,** 487–496.
6. Jensen, H. L. and Norrild, B. (1999) Easy and reliable double-immunogold labelling of herpes simplex virus type-1 infected cells using primary monoclonal antibodies and studied by cryosection electron microscopy. *Histochem. J.* **31,** 525–533.
7. Jensen, H. L. and Norrild, B. (2000) The effects of cell passages on the cell morphology and the outcome of herpes simplex virus type 1 infection. *J. Virol. Methods* **84,** 139–152.
8. Jensen, H. L. and Norrild, B. (2002) Morphologic, immunohistochemical, immunologic, ultrastructural, and time-related study of herpes simplex virus type 1-infected cultured human fibroblasts. *AIMM* **10,** 71–81.
9. Jensen, H. L. and Norrild, B. (2002) Temporal morphogenesis of herpes simplex virus type 1-infected and Brefeldin A-treated human fibroblasts. *Mol. Med.* **8,** 210–224.

10. Faulk, W. P. and Taylor, G. M. (1971) An immunocolloid method for the electron microscope. *Immunochemistry* **8,** 1081–1083.
11. Slot, J. W. and Geuze, H. J. (1985) A new method of preparing gold probes for multiple-labeling cytochemistry. *Eur. J. Cell Biol.* **38,** 87–93.
12. Larsson, L.-I. (ed.) (1988) *Immunocytochemistry: Theory and Practice.* CRC, Boca Raton, FL.
13. Baschong, W., Lucocq, J. M., and Roth, J. (1985) "Thiocyanate gold": small (2–3 nm) colloidal gold for affinity cytochemical labeling in electron microscopy. *Histochemistry* **83,** 409–411.
14. Tokuyasu, K. T. (1973) A technique for ultracryotomy of cell suspensions and tissues. *J. Cell Biol.* **57,** 551–565.
15. Granzow, H., Klupp, B. G., Fuchs, W., Veits, J., Osterrieder, N., and Mettenleiter, T. C. (2001) Egress of alphaherpesviruses: comparative ultrastructural study. *J. Virol.* **75,** 3675–3684.
16. Skepper, J. N., Whiteley, A., Browne, H., and Minson, A. (2001) Herpes simplex virus nucleocapsids mature to progeny virions by an envelopment → deenvelopment → reenvelopment pathway. *J. Virol.* **75,** 5697–5702.
17. Koga, J., Chatterjee, S., and Whitley, R. J. (1986) Studies on herpes simplex virus type 1 glycoproteins using monoclonal antibodies. *Virology* **151,** 385–389.
18. Eisenberg, R. J., Long, D., Pereira, L., Hampar, B., Zweig, M., and Cohen, G. H. (1982) Effect of monoclonal antibodies on limited proteolysis of native glycoprotein gD of herpes simplex virus type 1. *J. Virol.* **41,** 478–488.
19. Pereira, L., Klassen, T., and Baringer, J. R. (1980) Type-common and type-specific monoclonal antibody to herpes simplex virus type 1. *Infect. Immun.* **29,** 724–732.
20. Eisenberg, R. J., Long, D., Ponce de Leon, M., et al. (1985) Localization of epitopes of herpes simplex virus type 1 glycoprotein D. *J. Virol.* **53,** 634–644.
21. Isola, V. J., Eisenberg, R. J., Siebert, G. R., Heilman, C. J., Wilcox, W. C., and Cohen, G. H. (1989) Fine mapping of antigenic site II of herpes simplex virus glycoprotein D. *J. Virol.* **63,** 2325–2334.
22. Lausch, R. N., Staats, H., Oakes, J. E., Cohen, G. H., and Eisenberg, R. J. (1991) Prevention of herpes keratitis by monoclonal antibodies specific for discontinuous and continuous epitopes on glycoprotein D. *Invest. Ophthalmol. Vis. Sci.* **32,** 2735–2740.
23. McGarrity, G. J., Steiner, T., and Vanaman, V. (1983) Detection of mycoplasmal infection of cell cultures by DNA fluorochrome staining. *Methods Mycoplasmol.* **2,** 183–190.
24. Gundersen, H. J. G., Bendtsen, T. F., Korbo, L., et al. (1988) Some new, simple and efficient stereological methods and their use in pathological research and diagnosis. *APMIS* **96,** 379–394.
25. Larsson, L.-I. (1993) Tissue preparation methods for light microscopic immunohistochemistry. *Appl. Immunohistochem.* **1,** 2–16.
26. Tokuyasu, K. T. (1989) Use of poly(vinylpyrrolidone) and poly(vinyl alcohol) for cryoultramicrotomy. *Histochem. J.* **21,** 163–171.

27. Griffiths, G. (ed.) (1993) *Fine Structure Immunocytochemistry.* Springer-Verlag, Berlin.

28. Liou, W., Geuze, H. J., and Slot, J. W. (1996) Improving structural integrity of cryosections for immunogold labeling. *Histochem. Cell Biol.* **106,** 41–58.

29. Behnke, O., Ammitzbøll, T., Jessen, H., et al. (1986) Non-specific binding of protein-stabilized gold sols as a source of error in immunocytochemistry. *Eur. J. Cell Biol.* **41,** 326–338.

30. Posthuma, G., Slot, J. W., and Geuze, H. J. (1987) Usefulness of the immunogold technique in quantitation of a soluble protein in ultrathin sections. *J. Histochem. Cytochem.* **35,** 405–410.

31. Hainfeld, J. F. and Powell, R. D. (2000) New frontiers in gold labeling. *J. Histochem. Cytochem.* **48,** 471–480.

32. Lackie, P. M. (1996) Immunogold silver staining for light microscopy. *Histochem. Cell Biol.* **106,** 9–17.

33. Gilbert, R. and Ghosh, H. P. (1993) Immunoelectron microscopic localization of herpes simplex virus glycoprotein gB in the nuclear envelope of infected cells. *Virus Res.* **28,** 217–231.

34. Griffiths, G. and Hoppeler, H. (1986) Quantitation in immunocytochemistry: correlation of immunogold labeling to absolute number of membrane antigens. *J. Histochem. Cytochem.* **34,** 1389–1398.

11

FTIR Microscopy Detection of Cells Infected With Viruses

Vitaly Erukhimovitch, Marina Talyshinsky, Yelena Souprun, and Mahmoud Huleihel

Summary

Fourier-transform infrared (FTIR) microscopy is considered a comprehensive and sensitive method for detection of molecular changes in cells. The advantage of FTIR microspectroscopy over conventional FTIR spectroscopy is that it facilitates inspection of restricted regions of a cell culture or a tissue. We have shown that it is possible to apply FTIR microscopy as a sensitive and effective assay for the detection of cells infected with various members of the herpes family of viruses and retroviruses. Detectable and significant spectral differences between normal and infected cells were evident at early stages of the infection. Impressive changes in several spectroscopic parameters were seen in infected compared with uninfected cells. It seems that the change in spectral behavior is specific to the infecting virus, because cells infected with herpesviruses showed different spectral changes compared with cells infected with retoviruses.

Key Words: Viruses; FTIR microscopy; malignant cells; retroviruses; herpesviruses; viral infection.

1. Introduction

Early detection and diagnosis of viral infections play important roles in therapy and treatment strategy. Various in vitro assays (such as immunoassays, cell culture, and polymerase chain reaction) are currently in use for the detection of viruses or viral infections *(1,2)*. These methods are based largely on immunological and molecular characteristics of the virus or on the morphological effects of the virus on the infected cells. The major drawbacks of most of these assays are that they are time-consuming and expensive. One solution to these drawbacks lies in the use of Fourier-transform infrared (FTIR) spectroscopy, a noninvasive method that has already been applied for identifying various bio-

From: *Methods in Molecular Biology, vol. 292: DNA Viruses: Methods and Protocols*
Edited by: P. M. Lieberman © Humana Press Inc., Totowa, NJ

molecular components of the cell *(3,4)*. Indeed, one application of FTIR spectroscopy in biomedicine is the detection and monitoring of characteristic changes in the molecular compositions and structures that accompany cellular changes resulting from viral infections or transformations from a normal to a malignant state *(5–11)*.

Several features of IR techniques indicate that FTIR may be applied as an accurate and sensitive method for the diagnosis and study of different diseases: (1) IR, having a longer wavelength than UV or visible radiation, penetrates to a greater depth and is absorbed with less scattering by the tissue or cells; (2) many vibration bands in the IR region are well resolved, and therefore subtle changes in molecular structure may be monitored during the development of a disease *(12–15)*; (3) only small amounts of the sample to be tested are required for analysis; and (4) there is no need for expensive reagents. With the introduction of microscopy into modern FTIR instrumentation, which allows successful examination of heterogeneous samples, spectroscopic analysis of specific regions of cells and tissues has become a reality. In recent years, there has been increasing interest in applying FTIR as a tool in the diagnosis of cancer. Successful diagnoses of lung *(16)*, breast *(17)*, cervical *(18)*, prostate *(19)*, and colon *(20)* cancers have been reported in the literature, as have a range of in vitro studies on cells *(21–23)*.

In addition to the use of IR in cell and tissue diagnosis, Naumann is group has applied FTIR spectroscopy to the classification of different classes of bacteria *(24)*. Furthermore, its role in the examination of body fluids has gained importance in the past few years. FTIR spectroscopy in the mid-IR region has been used, for example, for identification of a disease pattern in human serum *(25)* and in quantification of serum components such as glucose, total protein, cholesterol, and urea *(26)*. Concentrations of urea, glucose, protein, and ketones in human urine have been determined by near-IR spectroscopy *(27)*.

The present authors have successfully used FTIR microscopy for the detection and characterization of malignant cells transformed by retroviruses and cells infected with various members of the herpes family of viruses. FTIR microscopy enables us to focus solely on specific relevant regions of a tissue or cell culture section and hence to avoid undesirable signals from unsuitable regions. The data obtained showed FTIR microscopy to be a sensitive and reliable method for the detection and characterization of malignant cells transformed by retroviruses *(23,24)* and for the early detection of cells infected with herpesviruses *(28,29)*. Our studies showed impressive differences in spectra between control cells and cells infected with either herpesviruses or retroviruses. Cluster analysis of FTIR spectra yielded 100% accuracy in the classification of both control and infected cells. Our ongoing research is directed toward examining the potential of FTIR spectroscopy for the detection of viral infections in vivo.

2. Materials

1. Zinc sellenide crystals (PIKE Technologies).
2. 0.9% NaCl solution (sterile).
3. Cell culture medium (RPMI-1640, Biological Industries, Israel); store at 4°C.
4. Fetal calf serum (FCS; Biological Industries); store at –20°C.
5. Trypsin-EDTA solution (Biological Industries); store at –20°C.
6. Penicillin-streptomycin-nystatin solution (Biological Industries); store at –20°C.

2.1. Cells and Viruses

1. African green monkey kidney (Vero) cells were purchased from the American Type Culture Collection (ATCC; Rockville, MD). Cells were grown in RPMI medium containing 10% FCS, 1% glutamine, and antibiotic solution and incubated at 37°C in humidified air containing 5% CO_2. Vero cells were used for infection with herpesviruses.
2. A mouse fibroblast cell line (NIH/3T3; obtained from ATCC) and primary cells obtained from different organs of different newborn animals were used for infection with murine sarcoma virus (MuSV). These cells were grown in the same medium and conditions as Vero cells.
3. Herpes simplex virus type 1 (HSV-1), type 2 (HSV-2) and varicella-zoster virus (VZV) were obtained from ATCC. These viruses were propagated to >10^7–10^8 PFU per mL in Vero cells, and concentrations were estimated by a standard plaque reduction assay, as described by Huleihel et al. *(30)*.
4. Clone 124 of TB cells (mouse fibroblast cells), chronically releasing Moloney murine sarcoma virus (MuSV-124), was used to prepare the appropriate virus stock. TB cells were grown in a minimum amount of medium containing 2% serum for 24 h. This medium was then spun at 400 *g* for 5 min to remove cells and cell debris (virus particles are released from the TB cells into the medium) and used for infection. This virus belongs to the retrovirus family and can transform cells in culture into malignant cells.

3. Methods

FTIR microscopy was applied for detection, in various cell cultures, of cells infected with herpesviruses or cells transformed by retroviruses.

3.1. Preparation of Slides

Since ordinary glass slides exhibit strong absorption in the wavelength range of interest to us, we used zinc sellenide crystals, which are highly transparent to IR radiation. Two different methods were used for preparing the slides with the cells to be tested, as follows.

3.1.1. Method 1

1. A zinc sellenide crystal, which had been sterilized by autoclaving, was placed in a 9.6-cm^2 cell culture plate, and 2.5 mL of medium (with 10% FCS and the antibiotic solution) containing 1×10^6 target cells was added.

2. After a confluent cell culture was obtained on the zinc sellenide crystal (usually 24 h after plating), the cell culture was infected with the appropriate virus (as detailed in **Subheading 3.2.** below).
3. At 24 h post infection (p.i.) or later, the sellenide crystal was picked out of the culture plate, washed gently (by soaking) twice with physiological saline (0.9% NaCl solution), dried in air for 2 h at room temperature (or for only 1 h by air-drying in laminar flow), and examined by FTIR microscopy.

3.1.2. Method 2

1. Confluent cells in cell culture were infected with the appropriate virus.
2. At 24 h p.i. or later, the cells were washed twice with saline and picked out of the cell culture plates after treatment with trypsin (0.25%) for 1 min at 37°C (*see* **Note 1**).
3. The cells were pelleted by centrifugation at 400 *g* for 5 min.
4. The pellet was washed twice with physiological saline and resuspended in 100 μL of physiological saline.
5. The number of cells was counted with hemacytometer, and the tested sample was pelleted again and resuspended in an appropriate volume of physiological saline to give a concentration of 1000 cells/μL.
6. A drop of 1 μL of the sample was placed on a certain area on the zinc sellenide crystal (*see* **Note 2**), air-dried for 2 h at room temperature (or 1 h by air-drying in a laminar flow), and examined by FTIR microscopy. The radius of such a 1-μL drop was about 1 mm.

3.2. Cell Infection

3.2.1. Herpesviruses

1. Monolayers of cells grown in 9-cm^2 tissue culture plates were incubated at 37°C for 2 h with HSV or VZV at various multiplicities of infection (MOI) in RPMI medium containing 2% NBCS.
2. The unabsorbed virus particles were removed, and fresh medium containing 2% NBCS was added.
3. The monolayers were incubated at 37°C.
4. At various times p.i., the infected cells were examined by FTIR microscopy.
5. The cells were examined under an inverted light microscope for the appearance of the cytopathic effect (CPE), defined as areas of complete destruction of cells or of morphologically modified cells in the fields inspected.

3.2.2. MuSV

1. A monolayer of target cells grown in 9-cm^2 tissue culture plates was treated with 8 μg/mL of polybrene (a cationic polymer required for neutralizing the negative charge of the cell membrane) for 24 h before infection with the virus.
2. Excess polybrene was then removed, and the cells were incubated at 37°C for 2 h with the infecting virus (MuSV-124) at various concentrations in RPMI medium containing 2% NBCS.

3. The unabsorbed virus particles were removed, fresh medium containing 2% NBCS was added, and the monolayers were incubated at 37°C.
4. Control cells were also treated with polybrene using the same procedure as for the MuSV-infected cells.

3.3. FTIR Spectra Measurement

1. FTIR measurements were performed in the transmission mode with a liquid-nitrogen-cooled MCT detector of the FTIR microscope (Bruker IRScope II) coupled to an FTIR spectrometer (BRUKER EQUINOX model 55/S, OPUS software).
2. The spectra were obtained in the wave number range of 600–4000 cm^{-1} in the mid-IR region.
3. Spectral resolution was set at 4 cm^{-1} with Backman Harris 4-Term apodization.
4. To increase the signal-to-noise ratio, a spectrum was taken as an average of 128 scans.
5. Since the samples to be analyzed were often heterogeneous, appropriate regions were chosen by FTIR microscopy so as to eliminate different impurities (salts, medium residuals, and so on) (*see* **Note 3**).
6. The aperture used in this study was 100 µm, since this aperture gave the best signal-to-noise ratio (*see* **Notes 4** and **5**). At lower apertures, the quality of the spectra was bad owing to the high noise level. In addition, at apertures lower than 20 µm, there was diffraction of the IR beam.
7. Baseline correction and normalization were obtained for all the spectra by OPUS software. Baseline correction was performed by the rubber band method as follows:
   a. Each spectrum was divided up into ranges of equal size.
   b. In each range, the minimum *y*-value was determined.
   c. The baseline was then created by connecting the minima with straight lines.
   d. Starting from "below," a rubber band stretched over this curve constituted the baseline.
   e. The baseline points that did not lie on the rubber band were discarded.
8. Normalization was performed by a vector method, as follows:
   a. The average *y*-value of the spectrum was first calculated.
   b. This average value was then subtracted from the spectrum so that the middle of the spectrum was pulled down to $y = 0$.
   c. The sum of the squares of all the *y*-values was then calculated, and the spectrum was divided by the square root of this sum.
   d. The vector norm of the resulting spectrum was 1.
9. Peak positions were determined by means of a second derivation method by OPUS software. For each cell type, the spectrum was taken as the average of five different measurements at various sites of the sample.
10. Each experiment with each cell type was repeated five times.
11. It is important to mention that there were no significant differences in the spectra from various sites (SD did not exceed 0.005).

3.4. FTIR Spectra Analysis

The spectra obtained were analyzed for specific regions that showed distinct differences between normal and infected cells. The main peaks considered were as follows.

3.4.1. Peak at 861 cm⁻¹

This peak is attributed to *N*-type sugars. Our results showed a gradual shift in the position of this peak from 861 cm^{-1} in normal samples to about 854 cm^{-1} in correspondence with the development of herpes viruses infection *(29)*. For retrovirus-transformed cells, there was no shift in this peak.

3.4.2. Peak at 1023 cm⁻¹

Our data showed a gradual disappearance of the peak at 1023 cm^{-1} over time with the development of herpesviruses infection *(29)*. This spectral peak can be attributed to carbohydrates, as was previously reported *(31)*. There was also a notable decrease in this peak in cells infected with retroviruses (**Fig. 1**).

3.4.3. Peak at 1081 cm⁻¹

This peak is attributed to PO_2 symmetric stretching vibration. A significant and detectable shift of the peak at 1080–1081 cm^{-1} for normal cells to 1086–1087 cm for retrovirus-transformed cells was observed *(23)*, whereas there was no shift in this peak in cells infected with herpesviruses.

3.4.4. Peaks at 1200–1400 cm⁻¹

The peaks in the region 1200–1400 cm^{-1} represent PO_2^- asymmetric stretching vibrations. Our results showed a statistically significant reduction in the intensity of the absorbance owing to these PO_2^- vibrations for retrovirus-transformed cells compared with normal cells *(23)* and a significant increase in the intensity of absorbance for cells infected with herpesviruses *(28)* (**Fig. 2**).

These differences may be considered key parameters for the detection of cells infected with these different viruses.

3.5. Cluster Analysis

Cluster analysis is an unsupervised statistical technique that examines the interpoint distances between all the samples and represents that information in the form of a 2D plot, known as a dendrogram *(32)*. Such dendrograms present the data from high-dimensional row spaces in a form that facilitates the use of human pattern recognition abilities. To generate the dendrogram, cluster analysis methods form clusters of samples based on their nearness in row space. A common approach is to treat every sample initially as a cluster and to join the closest clusters. This process is repeated until only one cluster remains.

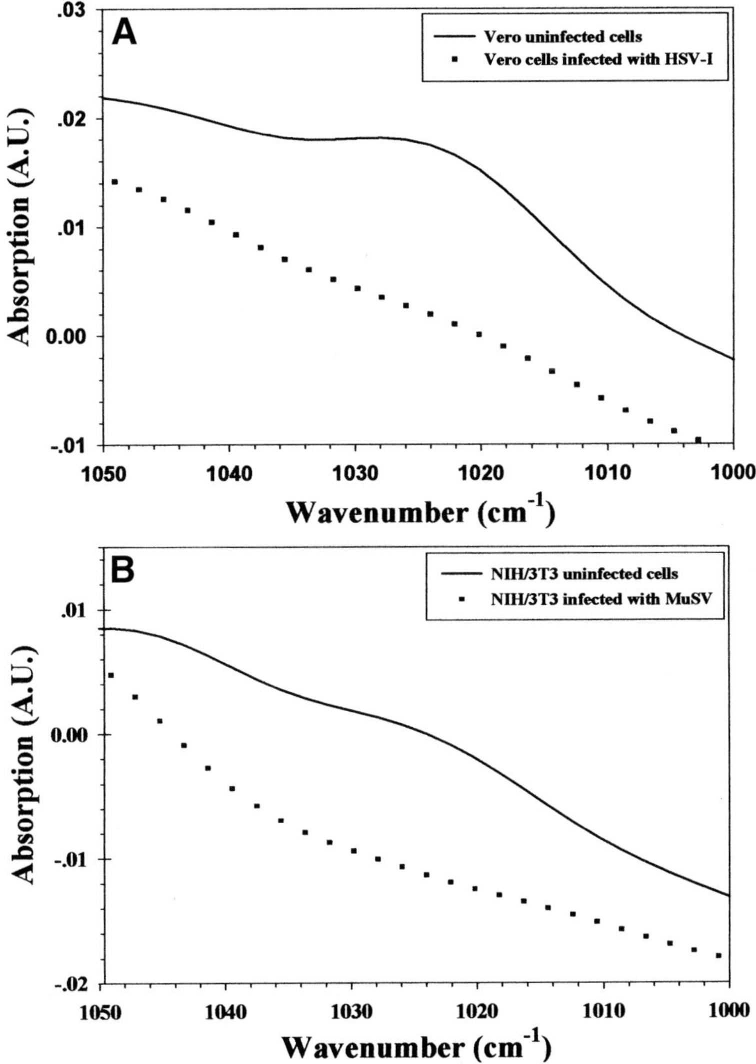

Fig. 1. FTIR spectra in the region of 1000–1050 cm^{-1} of **(A)** noninfected Vero cells and cells infected with 1 MOI of HSV-1 and **(B)** noninfected NIH/3T3 cells and cells infected with 1 FFU/cell of MuSV. Results are means of five different and separate experiments for each cell culture. The SD for these means was ≤0.001. FFU, focus-forming unit; HSV, herpes simplex virus; MuSV, murine sarcoma virus; AU, asymmetric unit.

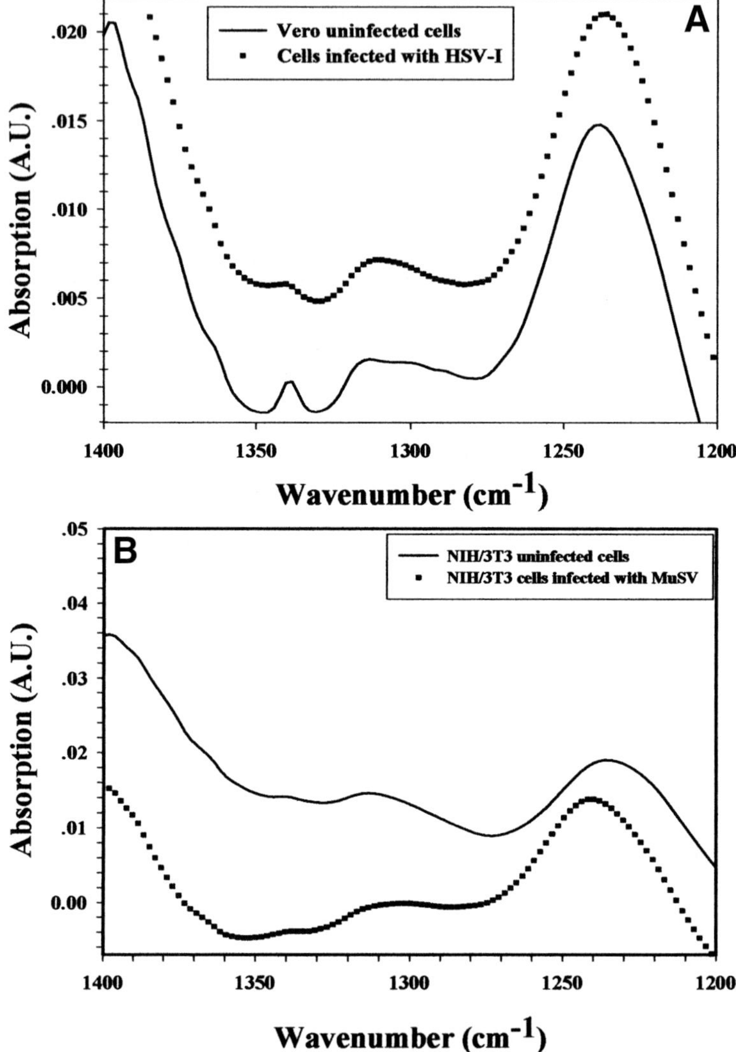

Fig. 2. FTIR spectra in the region of 1200–1400 cm⁻¹ of (**A**) noninfected Vero cells and cells infected with 1 MOI of HSV-1 and (**B**) noninfected NIH/3T3 cells and cells infected with 1 FFU/cell of MuSV. Results are means of five different and separate experiments for each cell culture. The SD for these means was ≤0.001. FFU, focus-forming unit; HSV, herpes simplex virus; MuSV, murine sarcoma virus; AU, asymmetric unit.

4. Notes

1. Treatment of the cells with trypsin should be performed carefully and for a short time (about 1 min) so as to avoid destroying the cells.
2. The 1-μL drop of sample (cells) should be placed on the zinc sellenide crystal as a concentrated drop.
3. Be careful not to choose for scanning possible contaminants, such as salt, rather than cells.
4. When choosing by microscope the region of the cells to be scanned, it is important to choose a region with confluent cells to obtain a better signal-to-noise ratio.
5. Set the condenser position of the FTIR microscope according to the thickness of the ZnSe crystals for obtaining maximum signal.

References

1. Dworkin, L. L., Gibler, T. M., and Van Galder, R. N. (2002) Real-time quantitive polymerase chain reaction diagnosis of infectious posterious uveitis. *Arch. Ophthalmol.* **120,** 1534–1539.
2. Coiras, M. T., Perez-Brena, P., Garcia, M. L., and Casas I. (2003) Simultaneous detection of influenza A, B, and C viruses, respiratory syncytial and adenoviruses in clinical samples by multiplex reverse transcription nested-PCR assay. *J. Med. Virol.* **69,** 132–141.
3. Mantsch, H. H. and Chapman, D. (1996) *Infrared Spectroscopy of Biomolecules.* John Wiley, New York.
4. Jackson, M., Tetteh, J., Manfield, B., et al. (1998) Cancer diagnosis by infrared spectroscopy: methodology aspects. *SPIE* **3257,** 24–34.
5. Diem, M., S., Boydston, W., and Chiriboga, L. (1999) Infrared spectroscopy of cells and tissues: shining light onto a novel subject. *Appl. Spectrosc.* **53,** 148–161.
6. Franck, P., Nabet, P., and Dousset, B. (1998) Applications of infrared spectroscopy to medical biology. *Cell. Mol. Biol.* **44,** 273–275.
7. Huleihel, M., Talyshinsky, M., and Erukhimovitch, V. (2001) FTIR microscopy as a method for detection of retrovirally transformed cells. *Spectroscopy* **15,** 57–64.
8. Wong, P., Goldstein, S., Grekin, R., Godwin, A., Pivik, C., and Rigas, B. (1993) Distinct infrared spectroscopic pattern of human basal cell carcinoma of the skin. *Cancer Res.* **53,** 762–765.
9. Afanasyeva, N. I., Kolyakov, F. S., Artjushenko, S. G., Sokolov, V. V., and Frank, G. A. (1998) Minimally invasive and *ex vivo* diagnostics of breast cancer tissues by fiber optic evanescent wave Fourier transform IR (FEW-FTIR) spectroscopy. *SPIE* **3250,** 140–146.
10. Lasch, P. and Naumann, D. (1998) FT-IR microspectroscopic imaging of human carcinoma thin sections based on pattern recognition techniques. *Cell. Mol. Biol.* **44,** 189–202.
11. Mordechai, S., Argov, S., Salman, A., et al. (2000) FTIR microscopic comparative study on normal, premalignant and malignant tissues of human intestine. *SPIE* **4129,** 231–242.

12. Yazdi, H. M., Bertrand, M. A., and Wong, P. T. (1996) Detecting structural changes at the molecular level with Fourier transform infrared spectroscopy. *Acta Cytol.* **40,** 664–668.

13. Benedetti, E., Bramanti, E., and Rossi, I. (1997) Determination of the relative amount of nucleic acids and proteins in leukemic and normal lymphocytes by means of FT-IR microspectroscopy. *Appl. Spectrosc.* **51,** 792–797.

14. Chiriboga, L., Xie, P., Zarou, D., Zakim, D., and Diem, M. (1998) Infrared spectroscopy of human tissue. IV. Detection of dysplastic and neoplastic changes of human cervical tissue via infrared microscopy. *Cell. Mol. Biol.* **44,** 219–229.

15. Yang, D., Castro, D., El-Sayed, I., El-Sayed, M., Saxton, R., and Nancy, Y. (1995) A Fourier-transform infrared spectroscopic comparison of cultured human fibroblast and fibrosarcoma cells. *SPIE* **2389,** 543–550.

16. Wang, H. P., Wang, H. C., and Huang, Y. J. (1997) Microscopic FTIR studies of lung cancer cells in pleural fluid. *Sci. Total. Environ.* **204,** 283–287.

17. Gao, T., Feng, J., and Ci, Y. (1999) Human breast carcinomal tissues display distinctive FTIR spectra: implication for the histological characterization of carcinomas. *Anal. Cell. Pathol.* **18,** 87–93.

18. Lowry, S. R. (1998) The analysis of exfoliated cervical cells by infrared microscopy. *Cell. Mol. Biol.* **44,** 169–177.

19. Malins, D. C., Polissar, N. L., and Gunselman, S. J. (1997) Models of DNA structure achieve almost perfect discrimination between normal prostate, benign prostatic hyperplasia (BPH) and adenocarcinoma, and have a high potential for predicting BPH and prostate cancer. *Proc. Natl. Acad. Sci. USA* **94,** 259–264.

20. Salman, A., Argov, S., Jagannathan, R., et al. (2001) FTIR microscopic characterization of normal and malignant human colonic tissues. *Cell. Mol. Biol.* **47,** 159–166.

21. Jagannathan, R., Salman, A., Hammody, Z., et al. (2001) FTIR microscopic studies on normal and H-Ras oncogene transfected cultured mouse fibroblasts. *Eur. Biophys. J.* **30,** 250–255.

22. Erukhimovitch, V., Talyshinsky, M., Souprun, Y., and Huleihel, M. (2002) Spectroscopic characterization of human and mouse primary cells, cell lines and malignant cells. *Photochem. Photobiol.* **76,** 446–451.

23. Huleihel, M., Erukhimovitch, V., Talyshinsky, M., and Karpasas, M. (2002) Spectroscopic characterization of normal primary and malignant cells transformed by retroviruses. *Appl. Spectrosc.* **56,** 640–645.

24. Naumann, D., Helm, D., and Labischinski, H. (1991) Microbiological characterizations by FT-IR spectroscopy. *Nature* **351,** 81–82.

25. Petrich, W. and Werner, G. (1999) Recognition of disease pattern in IR of blood serum. *Physikalische Blaetter* **55,** 49–51.

26. Shaw, R. A., Kotowich, S., Leroux, M., and Mantsch, H. H. (1998) Multianalyte serum analysis using mid-infrared spectroscopy. *Ann. Clin. Biochem.* **35,** 624–627.

27. Shaw, R. A., Kotowich, S., Mantsch, H. H., and Leroux, M. (1996) Quantitation of protein, creatinine and urea in urine by near-infrared spectroscopy. *Clin. Biochem.* **29,** 11–19.

28. Salman, A., Erukhimovitch, V., Talyshinsky, M., Huleihel, M., and Huleihel, M. (2002) FTIR-spectroscopic method for detection of cells infected with herpes viruses. *Biopolymers (Biospectroscopy)* **67,** 406–412.
29. Huleihel, M., Talyshinsky, M., Souprun, Y., and Erukhimovitch, V. (2003) Spectroscopic evaluation of cells infected with herpes viruses and treated with red microalgal polysaccharides. *Appl. Spectrosc.* **57,** 390–395.
30. Huleihel, M., Ishanu, V., Tal, J., and Arad, S. (2001) Antiviral effect of red microalgal polysaccharides on herpes simplex and varicella zoster viruses. *J. Appl. Phycol.* **13,** 127–134.
31. Dukor, R. K. (2001) Vibrational spectroscopy in the detection of cancer, in *Handbook of Vibrational Spectroscopy* (Chalmers, J. M. and Griffiths, P. R., eds.), John Wiley & Sons, New York, pp. 3335–3361.
32. Lasch, P. and Naumann, D. (1998) FTIR microspectroscopic imajing of human carcinoma thin sections based on pattern recognition techniques. *Cell. Mol. Biol.* **44,** 189–202.

III

VIRUS ENTRY

12

The JC Virus-Like Particle Overlay Assay

Hirofumi Sawa and Rika Komagome

Summary

JC virus (JCV) belongs to the family of double-stranded DNA polyomaviruses and in humans causes a demyelinating disease of the central nervous system, progressive multifocal leukoencephalopathy (PML). It has been reported that sialic acids play a pivotal role in hemagglutination of red blood cells and entry into host cells of JCV and that JCV can enter a wide variety of cell types and localize to the nuclei. The outer shell of the JCV virion comprises the major capsid protein VP1, and a virus-like particle (VLP) consisting of recombinant VP1 made from *Escherichia coli* exhibit a virion-like structure and physiological functions (cellular attachment and intracytoplasmic trafficking) similar to those of JCV virions. To examine the mechanism of cell attachment of JCV, an overlay assay using a VLP has been developed, revealing that sialoglycoproteins, including $\alpha 1$ acid-glycoprotein, fetuin, and transferrin receptor bind with VLP. In addition, VLPs bind to glycolipids, such as lactosylceramide and gangliosides including GM3, GD2, GD3, GD1b, GT1b, and GQ1b, and VLP weakly bind to GD1a. In this section, detailed procedures for the synthesis of VLP from *E. coli* and VLP overlay assay are described.

Key Words: JC virus; virus-like particle (VLP); thin-layer chromatography; SDS-polyacrylamide gel electrophoresis; VLP overlay assay; electron microscopy; hemagglutination assay.

1. Introduction

It has been reported that the receptor of JC virus (JCV) is a glycoprotein containing terminal $\alpha 2$-6-linked sialic acid, based on the finding that sialidase inhibits infection of glial cells by JCV. JCV belongs to a slow virus family, and it is difficult to obtain enough amounts of the virus to identify its receptor; we therefore use a virus-like particle (VLP) made from the recombinant JCV major capsid protein, VP1, which is synthesized in *E. coli* and have established an overlay assay using a VLP *(1)*. The methods for synthesis of VLP and the overlay assay are described.

From: *Methods in Molecular Biology, vol. 292: DNA Viruses: Methods and Protocols*
Edited by: P. M. Lieberman © Humana Press Inc., Totowa, NJ

2. Materials

1. pJC1-4→pJCV plasmid (HSRRB VG015, Osaka, Japan).
2. pBRMad1 plasmid *(2)*.
3. pET15-b prokaryotic expression vector (Novagen, Madison, WI).
4. *E. coli* BL21 (DE3) pLysS (Stratagene, La Jolla, CA).
5. SOC: 20 g bacto-tryptone, 5 g bacto-yeast extract, 0.5 g NaCl, and 2.5 mL of 1 *M* KCl are added to 900 mL of distilled water and the pH adjusted to 7.0 with NaOH. The volume is adjusted to 960 mL, and the solution is sterilized by autoclaving. Before use, the solution is mixed with 10 mL of 1 *M* $MgCl_2$, 10 mL of 1 *M* $MgSO_4$, and 20 mL of 1 *M* glucose.
6. Bacto-agar plate containing 50 µg/mL of ampicillin: 10 g bacto-tryptone, 5 g yeast extract, 10 g NaCl, and 15 g bacto-agar are added to the distilled water. The volume is adjusted to 1000 mL with distilled water, and the solution is sterilized by autoclaving for 20 min. Ampicillin (50 mg) is added to the mixture when the temperature of the solution is 50–60°C.
7. Terrific broth (TB) containing ampicillin (50 µg/mL): 18 g bacto-tryptone, 36 g yeast extract, and 6 mL glycerol are added to the 1300 mL of distilled water. The volume is adjusted to 1350 mL with distilled water and sterilized by autoclaving for 20 min (solution A). Then, 3.465 g KH_2PO_4 and 18.81 g K_2HPO_4 are mixed, dissolved with 150 mL of distilled water, and autoclaved for 20 min (solution B). Then 75 mg ampicillin is added to a mixture of solutions A and B, at a temperature of 50–60°C.
8. Isopropyl-β-D-thiogalactopyranoside (IPTG) (Takara, Tokyo, Japan).
9. Lysozyme (Sigma, St. Louis, MO)
10. DNase I (Roche Diagnostics, Indianapolis, IN).
11. Phenylmethylsulfonyl fluoride (PMSF) (Sigma).
12. Tris-buffered saline (TBS): 6.05 g Tris-HCl and 8.76 g NaCl, dissolved in 800 mL distilled water and adjusted to pH to 7.5 with 1 *N* HCl; adjust the volume to 1000 mL with distilled water, and sterilize by autoclaving.
13. Reassociation buffer: 1 m*M* $CaCl_2$ in TBS.
14. CsCl solution: CsCl (Sigma) is dissolved with distilled water at the appropriate concentration, for instance, 2.7 g of CsCl is dissolved with distilled water, and the volume is adjusted to 2 mL for 1.35 g/mL of CsCl solution.
15. 2X sodium dodecyl sulfate (SDS) sample buffer: 10 mL of 1.5 *M* Tris-HCl, pH 6.8, 6 mL of 20% SDS, 30 mL of glycerol, 15 mL of β-mercaptoethanol (Sigma), and 1.8 mg of bromophenol blue (Sigma); mix and keep at 4°C.
16. SDS running buffer: 1 g SDS, 3.03 g Tris-HCl, and 14.41 g glycine, dissolved in distilled water. Adjust the total volume to 1000 mL.
17. Blotting buffer: 1.5 g Tris-HCl, 7.2 g glycine, and 100 mL of methanol; add to distilled water. Adjust the total volume to 1000 mL with distilled water.
18. TBS-T buffer: TBS buffer containing 0.05% Tween-20.
19. ECL solution: freshly mixed with solution I and II, ECL Plus (Amersham Biosciences, Piscataway, NJ).
20. Phosphate-buffered saline (PBS), pH 7.15: 1.2 g Na_2HPO_4, 0.7 g KH_2PO_4, and 6.8 g NaCl dissolved with distilled water; adjust the pH to 7.15 with diluted HCl. Adjust the volume to 1000 mL with distilled water and sterilize by autoclaving.

21. Bovine serum albumin (BSA; Sigma).
22. 96-Well microplate for hemagglutination (HA) assay (Costar, Cambridge, MA).
23. Alsever solution: 2.05 g glucose, 0.8 g sodium citrate, 0.42 g sodium chloride, and 0.055 g citric acid, dissolved with distilled water; adjust the volume to 100 mL.
24. 0.5% Chloroform in polyvinyl formal solution (OkenShoji, Tokyo, Japan).
25. Parafilm (American National Can, Chicago, IL).
26. 2.5% Phosphotungstic acid: dissolve 2.5 g phosphotungstic acid (Sigma) with distilled water and keep at 4°C.
27. Immobilon-P (polyvinylidene fluoride [PVDF] membrane; Millipore, Billerica, MA).
28. Blocking buffer 1: PBS containing 0.1% Tween-20 and 1% BSA.
29. PBS-T: PBS containing 0.1% Tween-20.
30. Horseradish peroxidase (HRP)-conjugated F(ab′)$_2$ goat antirabbit immunoglobulins (BioSource International, Camarillo, CA).
31. Silica gel plastic plates (Polygram Sil G; Macherey-Nagel, Düren, Germany).
32. Orcinol/H_2SO_4 reagent: 200 mg of orcinol dissolved in 11.4 mL of H_2SO_4. This solution is carefully added to 80 mL distilled water. Adjust the total volume to 100 mL with distilled water and keep at 4°C.
33. Blocking solution 3: PBS containing 1% ovalbumin and 1% polyvinylpyrrolidone.
34. Immunostaining reagent for VLP overlay assay for lipid: add 1 μL of H_2O_2, 200 μL of 0.06 *M* *N,N*-diethyl-*p*-phenylenediamine dihydrochloride dissolved with acetonitrile, 200 μL of 4-chloro-1-naphthol in sequential order to 10 mL of 100 m*M* citrate buffer, pH 6.0.
35. Sialidase (*Arthrobacter ureafaciens;* Nacalai tesque, Kyoto, Japan).
36. α2-3 Sialidase (*Salmonella typhimurium;* Takara, Tokyo, Japan).

3. Methods

The methods described below outline (1) construction of the plasmids, (2) purification of VLPs, and (3) the VLP overlay assay.

3.1. Construction of Plasmid and Transformation of E. coli

The construction of plasmids for making JCV VLP is described in **Subheading 3.1.1.,** and the transformation of *E. coli* with the expression plasmid for JCV VP1 is described in **Subheading 3.1.2.**

3.1.1. Construction of Plasmids for Making JCV VLP

Sequence analysis of the pJC1-4→pJCV plasmid (HSRRB, Osaka, Japan) containing JCV genomic DNA revealed a 4-bp insertion between the 98-bp tandem repeat in the regulatory region compared with the sequence of Mad1 *(3)*. Two DNA segments flanking the insertion were amplified by polymerase chain reaction (PCR), subcloned into a cloning vector, pBR322 (*see* **Note 1**), and designated pBRMad1 *(2)*. From the pBRMad1 plasmid, the major capsid protein, VP1 encoding the *JCV* gene, is amplified by PCR and subcloned into the pET15b (*see* **Note 2**).

3.1.2. Transformation of VP1 Encoding Plasmid

The pET15b plasmid including the JCV VP1 DNA is transformed into *E. coli,* BL21 (DE3) pLysS (*see* **Note 3**) by the following procedure:

a. Plasmid (5 ng) is mixed with 50 µL of BL21 (DE3) pLysS, incubated on ice for 10 min, heated at 42°C for 45 s, and incubated on ice for 5 min.
b. After incubation with 500 µL of SOC for 30 min with vigorous shaking, the mixture is plated onto the bacto-agar plate containing 50 µg/mL of ampicillin.
c. After 16 h of incubation at 37°C, the *E. coli* colony is picked up by a yellow tip and precultured in TB at 37°C with vigorous shaking for 16 h.
d. The culture containing *E. coli* is transferred to 300 mL of TB and cultured at 37°C with vigorous shaking overnight.
e. The 300 mL of solution is added to 1200 mL of TB, cultured at 37°C with vigorous shaking for 3 h, and thereafter cultured at 30°C with vigorous shaking for 30 min.
f. IPTG is added to the 1500 mL of TB containing *E. coli* (final concentration: 1 mmol/L), and the mixture is cultured at 30°C with vigorous shaking for 4 h (*4*).

3.2. Purification of VLPs

The next steps in this process involve the purification of VP1 from *E. coli* described in **Subheading 3.2.1.** and the isolation of VLP described in **Subheading 3.2.2.** The confirmation assay for isolated VLP is given in **Subheadings 3.2.3.–3.2.5.**

3.2.1. Purification of VP1 From E. coli

1. The grown *E. coli* is collected by centrifugation at 4000g for 10 min at 4°C.
2. The precipitated *E. coli* is resuspended with 20 mL of reassociation buffer.
3. Lysozyme is added to the mixture at a concentration of 1 mg/mL and incubated on ice for 30 min.
4. Sodium deoxycholate is added to the mixture at a final concentration of 0.2% and the mixture is incubated on ice for 10 min and then sonicated for 15 s five times.
5. DNase I (final concentration: 50 µg/mL) and PMSF (final concentration: 1 mmol/L) are added to the sonicated sample and incubated at 30°C for 30 min.
6. The treated sample is centrifuged at 12,000g, 4°C, for 10 min.

3.2.2. Isolation of VLPs

1. The 20 mL supernatant of the centrifuged sample is gently overlaid at the top of 2 mL of 20% sucrose/reassociation buffer in the ultracentrifugation tube (30 mL) and centrifuged at 100,000g, 4°C, for 2 h.
2. Precipitates are dissolved with 4 mL of the reassociation buffer and sonicated for 15 s five times.
3. Each 2 mL of the CsCl solutions (concentrations: 1.35, 1.32, 1.29, 1.25, and 1.20 g/mL) are sequentially overlaid to the ultracentrifugation tube (13 mL) from the

bottom to the top, and 2 mL of sonicated sample is overlaid at the top of the 1.20 g/mL of CsCl solution.

4. The sample is ultracentrifuged at 200,000*g* at 16°C for 16 h.
5. After centrifugation, each 500 μL of sample is fractionally collected.
6. The fractionated samples are analyzed by immunoblotting with anti-VP1 antibody, HA assay, and electron microscopy.

3.2.3. Immunoblotting With Anti-VP1 Polyclonal Antibody

1. Five microliters of fractionated sample is mixed with 5 μL of 2X SDS sample buffer, boiled for 5 min, and kept on ice for 5 min.
2. The sample is applied to 10% of the polyacrylamide gel and electrophoresed in SDS running buffer.
3. The gel is blotted to the PVDF membrane in the presence of blotting buffer at 100 V for 1 h.
4. After blocking with 5% skim milk in TBS-T buffer for 30 min, the transferred membrane is incubated with diluted anti-VP1 antibody (1:2000) in TBS-T buffer at room temperature for 3 h.
5. After three washes with TBS-T buffer for 10 min each, the membrane is incubated with the secondary antibody (peroxidase-conjugated antirabbit immunoglobulin antibody) at room temperature for 3 h.
6. After three washes with TBS-T buffer, the membrane is soaked in the ECL solution (freshly mixed with solutions I and II, ECL Plus, Amersham Biosciences) for 1 min and visualized with a Luminoimage analyzer (LAS-1000, Fujifilm, Tokyo, Japan).

3.2.4. Hemagglutination Assay

The HA assay is performed to estimate the amounts of virion for fractionated samples *(5)*.

1. Fifty microliters of PBS, pH 7.15, containing 0.2% BSA (*see* **Note 4**) are added to each well of the 96-well V-shaped microplate.
2. Fifty microliters of diluted VLP sample are added to the first well, and mixed. Then 50 μL of mixture are added to the second well, and 50 μL of mixture are added to the third well. In this way, serially diluted samples is prepared.
3. Next, 1 mL of type-O blood is prepared and mixed with 9 mL of the Alsever solution. The blood can be kept at 4°C for 1 mo as a diluted solution from the Alsever solution.
4. The mixture is centrifuged at 500*g* at 4°C for 5 min and resuspended with 10 mL of the Alsever solution following centrifugation at 500*g* at 4°C for 5 min.
5. Thereafter, the mixture is resuspended with 10 mL of the Alsever solution and centrifuged at 500*g* at 4°C for 10 min.
   Then 50 μL of precipitates are diluted with 10 mL of PBS, pH 7.15, and 50 μL of the mixture (0.5% type-O red blood cells) are added to each well, into which serially diluted VLP is poured. The 96-well plate is kept at 4°C for 3 h.
6. The HA titer is defined as the reciprocal of the greatest dilution of the VLP suspension with which complete HA is observed **(Fig. 1)**.

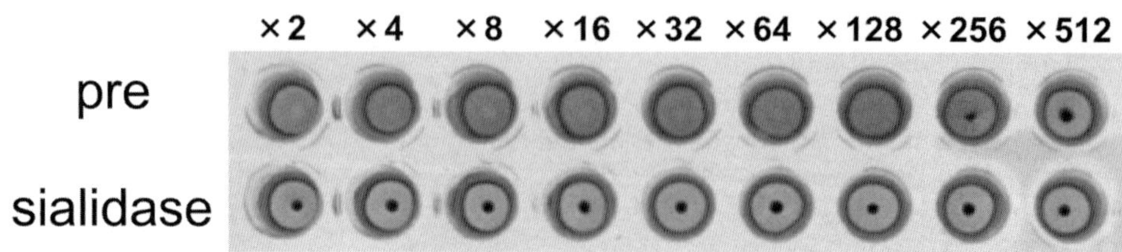

Fig. 1. Results of the HA assay. Serially diluted samples are prepared on the 96-well plate. This figure demonstrates the 128 HA titers of sample (upper lanes, pre). The HA activity of the sample is completely inhibited using red blood cells incubated with sialidase (0.1 U/mL) at 37°C for 30 min (lower lanes, sialidase), suggesting that the HA activity is dependent on the sialic acid on the surface of red blood cells.

3.2.5. Electron Microscopy for Observation of VLP (Negative Staining)

1. To eliminate the remaining lipid on the surface, the microscope slide is prewashed by ethanol and completely dried.
2. The slide is soaked in 0.5% polyvinyl formal in chloroform and immediately removed.
3. After drying the slide, the polyvinyl formal membrane on the surface of the slide is scored in a rectangle along the edge of the slide with a razor blade. The edge of the slide with the scored polyvinyl formal membrane is gently dipped in distilled water in the 10-cm- culture dish at an angle of approx 30° to the surface of the water. With gentle progression of the slide into the water, the film will float off.
4. The grids for electron microscopy are gently placed onto the floating film (**Fig. 2**).
5. To remove the grids and the film from the water, the film with grids is gently covered with adequately sized Parafilm.
6. After the film is stuck to the Parafilm, the Parafilm with the polyvinyl formal film and grids is gently removed from the water by a forceps.
7. The Parafilm is placed onto the desk upside down, i.e., the grids are placed onto the Parafilm and covered with the polyvinyl formal film.
8. The polyvinyl formal film is gently scored around each grid with a small forceps for electron microscopy.
9. Then 5 μL of the VLP solution are dripped onto the polyvinyl formal-coated grids and left for 5 min.
10. After removal of residual solution by filter paper, 10 drops of 2.5% phosphotungstic acid are placed onto each grid.
11. After air drying, the VLP is observed with an electron microscope (H-800, Hitachi, Tokyo, Japan) (**Fig. 3**).

3.2.6. Dialysis With Reassociation Buffer

1. After analysis by immunoblotting, HA assay, and electron microscopic examination, the optimal fractions are collected and dialyzed with 1000 mL of reassociation buffer at 4°C for 12 h twice.

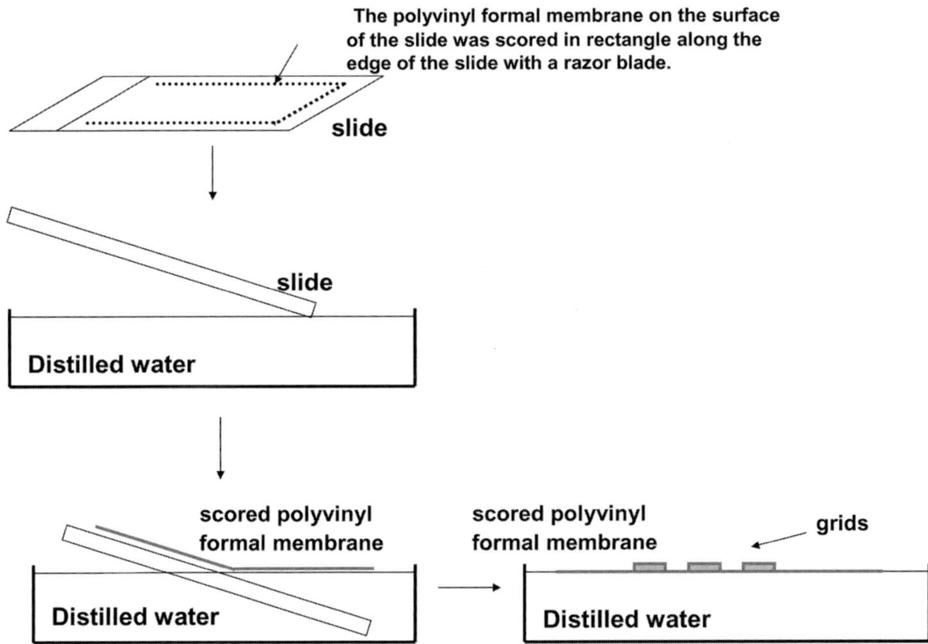

Fig. 2. Preparation strategy of grids coated with polyvinyl formal membrane for electron microscopy. The polyvinyl formal membrane on the surface of the slide is scored in a rectangle along the edge of the slide with a razor blade. The edge of the slide is gently dipped in distilled water at an angle of approx 30° to the surface of the water. With gentle progression of the slide into the water, the film will float off. The grids for electron microscopy are gently placed onto the floating film.

2. After measurement of HA activity of the dialyzed samples, they are kept at 4°C as VLPs.

3.3. VLP Overlay Assay

Described below are the steps involved in the VLP overlay assay for proteins (**Subheading 3.3.1.**) and lipids (**Subheading 3.3.2.**). Sialidase treatment is given in **Subheading 3.3.3.**

3.3.1. VLP Overlay Assay for Proteins

1. The protein is mixed with the equal volume of 2X SDS sample buffer, boiled for 5 min, kept on ice for 5 min, and separated by 10% SDS-polyacrylamide gel in the SDS running buffer.
2. After blotting of the gel onto the PVDF membrane (Immobilon, Millipore), the membrane is treated with the blocking buffer for 30 min.

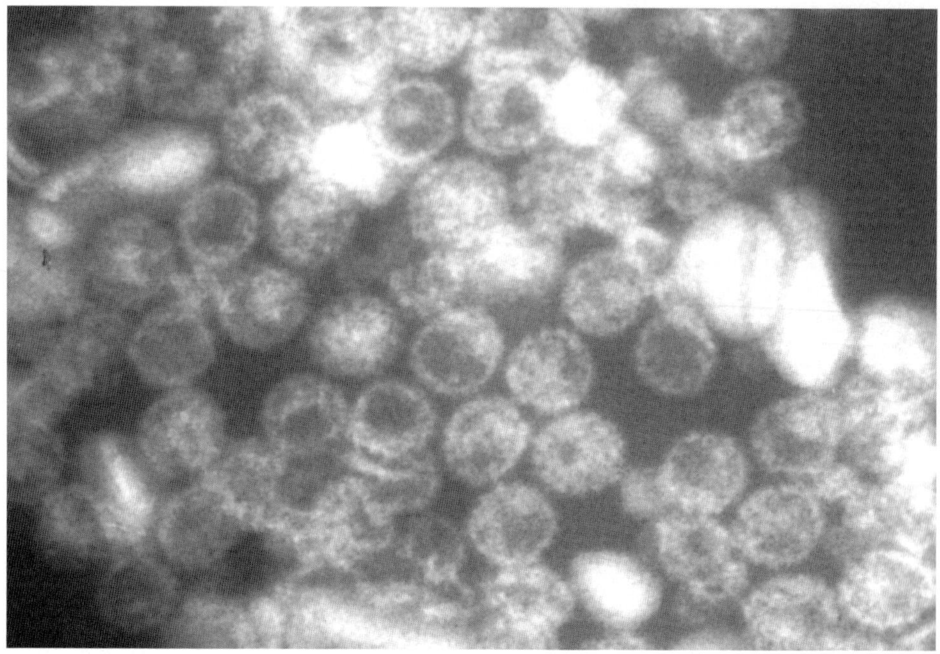

Fig. 3. Electron micrograph of negatively stained virus-like particles (VLPs). The diameters of the VLPs are approx 40–50 nm.

3. The membrane is incubated with diluted VLP solution (at a concentration of 5 µg/mL in blocking buffer) at 4°C for 2 h.
4. After washing with ice-cold PBS-T four times, the membrane is incubated at 4°C for 1 h with anti-VP1 antibody (diluted 1:1000 in PBS-T).
5. After two ice-cold PBST washes, the membrane is incubated at 4°C for 1 h with HRP-conjugated F(ab′)$_2$ goat antirabbit immunoglobulins (diluted 1:5000 in PBS-T).
6. The membrane is then incubated with equal amounts of ECL plus substrates and incubated for 5 min at room temperature, and the bound VP1 is visualized by a fluoroimage analyzer (LAS System, Fuji Film).

3.3.2. VLP Overlay Assay for Lipids

1. Gangliosides or other glycolipids (100 pmol) are subjected to thin-layer chromatography (TLC) on silica gel plastic plates in a solvent system of chloroform/methanol/12 m*M* MgCl$_2$ at a volume ratio of 5:4:1.
2. The overlay and immunochemical detection of VLP on the TLC plate are performed using a modification of a method described previously *(6)*.
3. The glycolipids are visualized by spraying the plates with orcinol reagent (*see* **Note 5**).
4. The chromatogram is blocked with blocking solution 3 at room temperature for 1 h.

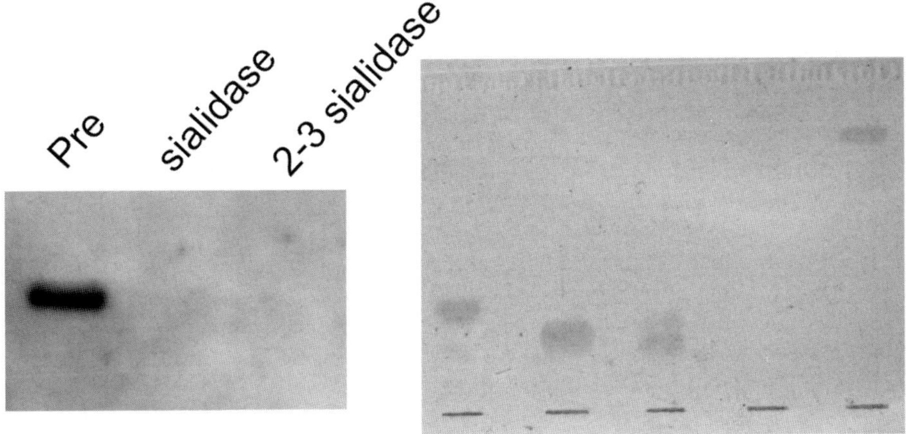

Fig. 4. The VLP overlay assay of proteins and glycolipids. Left, VLP overlay assay using each 1 μg of fetuin, which is a glycoprotein containing sialic acids. Fetuin in the second (sialidase) and third lanes (2–3 sialidase) is treated with sialidase and 2–3 sialidase, respectively. Right, VLP assay of glycolipids. Gangliosides (100 pmol), including GD1b, GT1b, GQ1b, GalCer, and LacCer, are analyzed by the TLC-VLP overlay assay. Intense positive signals are detected in GD1b, GT1b, GQ1b, and LacCer.

5. After four washes with PBS, the plate is incubated at 4°C for 2 h with diluted VLP (at a concentration of 5 μg/mL in blocking buffer 3).
6. After removal of VLP suspension with suction and four washes with ice-cold PBS to remove unbound VLP, the plate is incubated at 4°C for 1 h with the anti-VP1 antibody (diluted 1:1000 in PBS containing 1% polyvinylpyrrolidone).
7. Following four ice-cold PBS washes, the plate is incubated at 4°C for 1 h with HRP-conjugated F(ab′)$_2$ goat antirabbit immunoglobulins (diluted 1:5000 in PBS containing 1% polyvinylpyrrolidone) and then washed four times with ice-cold PBS.
8. The VLP bound to the plates is visualized by incubation with the immunostaining reagent containing the *N,N*-diethyl-*p*-phenylenediamine with dihydrochloride and 4-chloro-1-naphthol *(7)*.

3.3.3. Sialidase Treatment of Glycoproteins and Glycolipids

Optionally, glycoprotein may be treated by sialidase as follows;

1. Glycoproteins are incubated with 100 m*M* acetate buffer, pH 5.5, containing 0.1 U/mL sialidase or 0.5 U/mL α2-3 sialidase at 37°C for 1 h.
2. Then binding of VLP to the enzyme-treated proteins is examined by the VLP overlay assay.

3. For glycolipids, the samples are initially subjected to TLC and then developed and blocked as described above.
4. The TLC plate is incubated at 37°C for 16 h with 10 m*M* acetate buffer, pH 5.0, in the presence of 0.1 U/mL sialidase.
5. The plate is washed four times with PBS, blocked with blocking buffer 3 for 30 min, and incubated with VLP following the VLP overlay assay.

4. Notes

1. The 4361-bp pBR322 has often been utilized as an *E. coli* cloning vector, including the *rep* gene responsible for the replication of plasmid, the *rop* gene promoting a stable RNA complex, the *bla* gene encoding for β-lactamase for resistance to ampicillin, and the *tet* gene for tetracycline resistance protein. The GenBank/EMBL sequence accession number is J01749 *(8)*.
2. pET15b (Novagen) has the N-terminal His tag followed by a thrombin cleavage site that can separate the object protein and the His-tag protein. The coding strand is transcribed by T7 RNA polymerase.
3. BL21 (DE3) pLysS (Stratagene) from *E. coli* competent cells is useful for expression of recombinant toxic proteins derived from the T7 promoter-driven vector.
4. BSA is very difficult to dissolve. For making 1% BSA, 1 g of BSA powder is gently placed onto 95 mL of distilled water with the stirrer bar in the beaker. The stirrer bar is gently rotated after the BSA powder is placed onto distilled water. After it is completely dissolved, the volume of the solution is adjusted to 100 mL.
5. The orcinol reagent is sprayed onto the TLC plate after chromatography. The plate is covered with a clean glass plate, and then the silica surface of the plate is placed upward. The plate is kept at 110°C for 5 min.

Acknowledgments

The authors thank Drs. Takashi Suzuki, Yasuo Suzuki, Chizuka Henmi, Shingo Semba, Qiumin Qu, Tadaki Suzuki, Yuki Okada, Yasuko Orba, Shinya Tanaka, and Kazuo Nagashima for numerous suggestions. We also thank Ms. Mami Satoh and Mayumi Sasada for their excellent technical support. This study is supported in part by grants from the Ministry of Education, Science, Sports, and Culture of Japan.

References

1. Komagome, R., Sawa, H., Suzuki, T., et al. (2002) Oligosaccharides as receptors for JC virus. *J. Virol.* **76,** 12992–13000.
2. Okada, Y., Sawa, H., Tanaka, S., et al. (2000) Transcriptional activation of JC virus by human T-lymphotropic virus type I Tax protein in human neuronal cell lines. *J. Biol. Chem.* **275,** 17016–17023.
3. Frisque, R. J., Bream, G. L., and Cannella, M. T. (1984) Human polyomavirus JC virus genome. *J. Virol.* **51,** 458–469.
4. Suzuki, S., Sawa, H., Komagome, R., et al. (2001) Broad distribution of the JC virus receptor contrasts with a marked cellular restriction of virus replication. *Virology* **286,** 100–112.

5. Padgett, B. L. and Walker, D. L. (1973) Prevalence of antibodies in human sera against JC virus, an isolate from a case of progressive multifocal leukoencephalopathy. *J. Infect. Dis.* **127,** 467–470.
6. Suzuki, Y., Nakao, T., Ito, T., et al. (1992) Structural determination of gangliosides that bind to influenza A, B, and C viruses by an improved binding assay: strain-specific receptor epitopes in sialo-sugar chains. *Virology* **189,** 121–131.
7. Conyers, S. M. and Kidwell, D. A. (1991) Chromogenic substrates for horseradish peroxidase. *Anal. Biochem.* **192,** 207–211.
8. Watson, N. (1988) A new revision of the sequence of plasmid pBR322. *Gene* **70,** 399–403.

13

Analysis of Fusion Using a Virus-Free Cell Fusion Assay

Marisa P. McShane and Richard Longnecker

Summary

For enveloped viruses, such as viruses within the herpesvirus family, of which Epstein-Barr virus (EBV) is a member, infection of target cells includes two distinct steps. The first is characterized by the binding of viral envelope glycoproteins to host cellular receptors. After binding, the viral membrane and the cellular membrane fuse. Without both binding and fusion, the virus is not able to enter the host target cell efficiently. Combined with the specific tropism of EBV for primarily two cell types, B lymphocytes and epithelial cells, and the difficulty in inducing lytic replication of EBV in vitro, there is a lack of a good experimental model to study EBV-induced viral fusion. To study fusion more efficiently and effectively, we have employed a virus-free cell–cell fusion assay. In the effector cell, the viral glycoproteins and a plasmid containing the T7 promoter, driving the luciferase gene, are expressed. In the target cell type, T7 RNA polymerase is transfected. Fusion is quantitated by the amount of luciferase expression after mixing of the two cell types. Alongside the fusion assay, a CELISA is performed to determine glycoprotein expression on the effector cells. This methodology has been useful in studying membrane fusion induced by other herpesvirus family members.

Key Words: Glycoproteins; membrane fusion; viral entry; cellular receptors; epithelial cells; B lymphocytes; CELISA.

1. Introduction

Viral infection begins with the entry of the virus into the host target cell. In the absence of viral entry, productive infection cannot proceed, and there is an absence of viral replication. Epstein-Barr virus (EBV) infects over 90% of the human population and therefore, the virus has evolved an efficient manner to gain access to target cells within the human population to establish latency.

Entry of EBV is in part mediated by viral surface glycoproteins and results in the transfer of the viral genome into cells at the beginning of infection. Several of the glycoproteins, including gH, gL, gB, gM, and gN, are conserved

From: *Methods in Molecular Biology, vol. 292: DNA Viruses: Methods and Protocols*
Edited by: P. M. Lieberman © Humana Press Inc., Totowa, NJ

in the herpesvirus family. gH, gL, and gB have a role in entry/fusion for all human herpesviruses studied to date *(1–6)*, but a function has not been ascertained for several of the glycoproteins. Studying entry has primarily involved antibodies directed against the glycoproteins or viruses null for individual glycoproteins. These studies have provided a great deal of information about EBV entry of B cells and whether specific glycoproteins are essential for the process, but overall information is still lacking in the general mechanism of EBV-induced membrane fusion.

By using a virus-free cell fusion assay *(7)*, viral fusion can be analyzed more directly. The fusion assay utilizes a luciferase reporter gene activation system to quantify fusion. In effector cells, which mimic the virion, EBV glycoproteins of interest (in addition to a plasmid containing the luciferase reporter gene under the control of the T7 RNA polymerase) are transiently transfected. CHO-K1 cells are utilized since they are not susceptible to EBV infection. The effector cells are then mixed with target cells transfected with a plasmid containing T7 RNA polymerase. In order for the T7 RNA polymerase to turn on expression of the luciferase gene, the contents of the cell types must combine, providing an accurate measure of cell fusion.

Using this assay, several aspects of fusion can be analyzed including the protein requirements for fusion, putative receptors, and specific fusion domains within the glycoproteins. In regard to the tropism of EBV, the glycoproteins necessary and sufficient to cause fusion in a wide variety of cells, including primary cells, can be determined. For example, EBV-induced cell fusion mediated by B cells requires gp42, gB, gH, and gL *(1)*. Since there may be differences in the requirements between cell types and even cell lines, other cell types can be used as the target cell in the fusion assay.

Once the requirements are known for a particular cell line, mutations can be made in the required glycoproteins to map important functions further. With the presence of over 10 glycoproteins in the viral genome, a modulatory function may be ascribed to any of the other glycoproteins, such as gM/gN, by adding them to the minimal requirements for the specific cell type and looking for a decrease or increase in fusion. This modulatory effect in fusion may not be seen using antibodies or with a null virus, as is the case with gM/gN of EBV *(8)*.

By expressing a limited subset of glycoproteins, the individual role that each of the viral proteins contributes to fusion can be assessed. For example, peptides or small-molecule inhibitors can be added to the fusion assay and tested for their inhibition. Information gained from these studies could further lead to the identification of drugs and/or inhibitory molecules to inhibit infection.

When studying cell fusion in vitro, the surface expression of glycoproteins is also important. Specific levels of expression may be required for appropriate

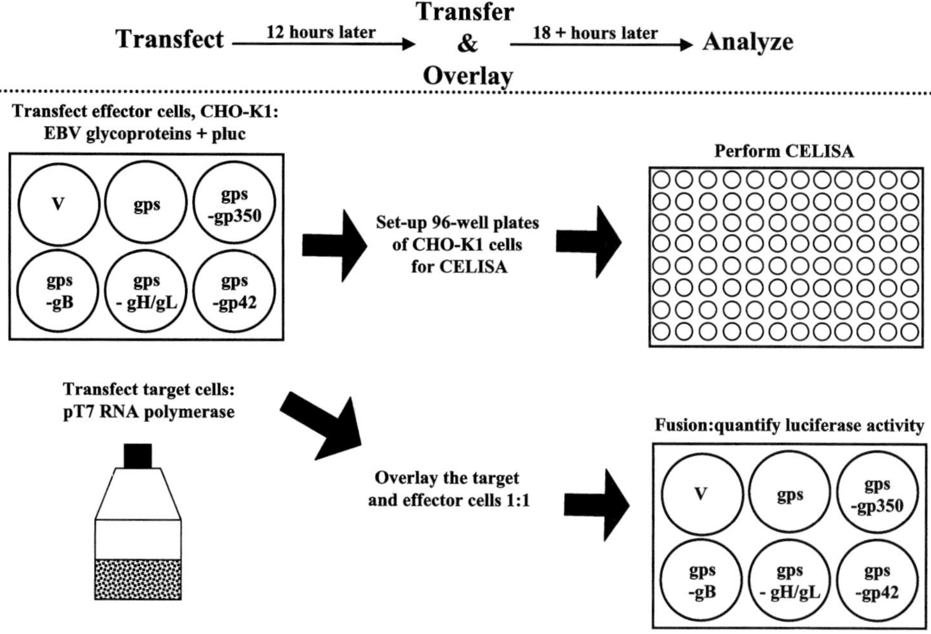

Fig. 1. Schematic diagram of fusion assay performed simultaneously with cell enzyme-linked immunosorbent assay (CELISA). Effector cells (in the example shown CHO-K1 cells are used) are transfected with various Epstein-Barr virus (EBV)-encoded glycoproteins. Similar mutant forms of specific EBV-encoded glycoproteins can be substituted. A plasmid encoding a T7 RNA polymerase-driven luciferase gene is included in the transfection. Twelve hours post transfection, the effector cells are split and either used to perform a CELISA to confirm glycoprotein expression or used in a fusion assay by overlaying the effector cells with target cells. In the example shown, Daudi, a human Burkitt's lymphoma cell line that has been transfected by a plasmid encoding a constitutively expressed T7 RNA polymerase, is used. Previous studies have shown that fusion occurs only in the well-expressing gp350, gB, gH/gL, and gp42 (gps). Interestingly, fusion is not dramatically effected by the absence of gp350 (gps, –gp350), whereas the omission of any of the other glycoproteins results in a complete absence of fusion.

fusion to occur. To ascertain whether differences within the fusion assay are owing to differences in expression levels of the glycoproteins, a cell enzyme-linked immunosorbent assay (CELISA) is performed alongside the fusion assay **(Fig. 1)**. After transfecting the effector cells with the glycoproteins, the cells are split, with some of the cells being plated for the CELISA and the remaining used in the fusion assay. This assay can provide both the amount of protein expressed on the surface and total protein expression by either permeabilizing or not permeabilizing the cells prior to addition of antibody.

Overall, the fusion assay and CELISA described in this chapter provide an efficient and quantitative manner to study various aspects of viral fusion.

2. Materials

2.1. Transient Transfection

1. Cell lines: effector cell line, CHO-K1; target cell lines, daudi, PEAK, AGS, and other cell lines as well as primary cells (*see* **Note 1**).
2. Media appropriate for cell lines: CHO-K1 cells require Ham's F-12 containing 10% fetal calf serum (FCS) and penicillin/streptomycin.
3. Opti-MEM (minimum essential medium).
4. Lipofectamine 2000 (Invitrogen).
5. 1X Phosphate-buffered saline (PBS), sterile.
6. Trypsin-versene (Gibco).
7. DNA: pCAGGS.MCS vector, viral glycoproteins in pCAGGS.MCS vector, T7 RNA polymerase plasmid, and plasmid containing the T7 RNA promoter upstream of the luciferase gene (pluc).
8. Plastic tissue-culture-treated 12-well plates and 10-cm^2 dishes (*see* **Note 2**).
9. 37°C, 5% CO_2 Humidified incubator.

2.2. CELISA

1. Plastic tissue-culture-treated 96-well flat-bottomed plates.
2. Fixative: 2% formaldehyde, 0.2% glutaraldehyde in PBS-A (make fresh each time)
3. PBS-ABC.
   a. A: to make 10 L, add 80 g NaCl, 2 g KCl, 11.5 g Na_2HPO_4, 2 g KH_2PO_4.
   b. B: to make 1 L, add 1 g $MgCl_2.6H_2O$.
   c. C: to make 1 L, add 1 g $CaCl_2$. The pH should be 7.2–7.4.

Make each part separately and autoclave. Cool overnight to room temperature. Under sterile conditions, add 50 mL solution B and 50 mL solution C to 400 mL solution A.

4. Permeabilization solution: 0.01% DOC (deoxycholic acid), 0.02% NP-40 in H_2O.
5. 3% Bovine serum albumin (BSA) in PBS-ABC; store at 4°C.
6. PBS + 0.1% Tween; store at 4°C.
7. Primary antibody: dilute in PBS-BSA, 50 µL/well (*see* **Note 3**).
8. Secondary antibody: dilute 1:500 in PBS-BSA; biotinylated antimouse or antirabbit IgG (cat. no. B7389 or B7264, Sigma), 100 µL/well (*see* **Note 3**)
9. Tertiary antibody: 1:20,000 in PBS-BSA, AMDEX streptavidin-conjugated horseradish peroxidase (cat. no. RPN 4401, Amersham), 100 µL/well (*see* **Notes 3** and **4**).
10. Substrate: TMB peroxidase (cat. no. TMBW-0100-01, BioFX TMB Microwell).
11. Plate reader.

2.3. Cell Fusion Assay

1. Plastic tissue-culture-treated 24-well plates (*see* **Note 5**).
2. PBS.
3. Versene.
4. Promega Luciferase 5X Lysis Buffer.
5. Promega Luciferase Substrate.
6. Plate reader (*see* **Note 6**).

3. Methods

3.1. Transient Transfection (see Note 7)

1. Seed CHO-K1 cells and target cells so the following day they are at 60–80% confluency (*see* **Notes 8** and **9**).
2. Transfect cells following the Lipofectamine 2000 protocol.
   a. Tube 1: Opti-MEM + DNA (*see* **Note 10**).
      i. Effector cells (12-well plate): glycoproteins (gH, gL, gB=0.25 µg, gp42=1 µg) and T7 promoter-luciferase (luc, 0.4 µg).
      ii. Target cells (10-cm^2 dish): T7 polymerase, 20 µg.
   b. Tube 2: Opti-MEM + Lipofectamine 2000 (5 µL/250 µL Opti-MEM)
3. Incubate the tubes for 5 min.
4. Combine tube 1 and tube 2 by aliquoting an equal amount of tube 2 into the glycoprotein DNA + Opti-MEM tubes. Incubate for 25 min at room temperature.
5. Wash cells twice with PBS, and add Opti-MEM to each well.
6. Add Opti-MEM + DNA + Lipofectamine 2000 mixture to cells.
7. Incubate at 37°C for 12 h.
8. After 12 h, the CHO-K1 (target) cells will be split for the CELISA and fusion assay as described below and as depicted in **Fig. 1**.

3.2. CELISA

3.2.1. Total Intracellular and Surface Expression

1. Pour off media from the 96-well plate, and gently blot the plate by turning upside down onto a paper towel.
2. Wash cells gently once with PBS-ABC (100 µL/well).
3. Fix with 0.2% glutaraldehyde, 2% formaldehyde in PBS-A 100 µL/well at RT for 10 min (*see* **Note 11**).
4. Wash cells three times with PBS-BSA. (The only time you wash with BSA.)
5. Permeabilize cells with permeabilization solution (100 µL/well) at RT for 10 min.
6. Wash cells three times with PBS-ABC.
7. Aspirate and add 50 µL/well of primary antibody. Incubate at room temperature for 30 min (*see* **Note 3**).
8. Wash cells gently five times with PBS-ABC.
9. Incubate the cells with secondary antibody (100 µL/well) at room temperature for 30 min on the rocker.

10. Wash five times with PBS-ABC.
11. Incubate the cells with tertiary antibody (100 µL/well) at room temperature for 30 min on a rocker.
12. Wash cells gently five times with PBS-ABC + 0.1% Tween (*see* **Note 4**).
13. Add TMB substrate (50 µL/well) (*see* **Note 12**).
14. Read cells at 370 nm on a plate reader for 0.1 s/well.

3.2.2. Surface Expression Alone

1. Pour off media from the 96-well plate, and gently blot the plate by turning upside down onto a paper towel.
2. Wash cells gently once with PBS-ABC (100 µL/well).
3. Aspirate and add 50 µL/well of primary antibody. Incubate at room temperature for 30 min (*see* **Note 3**).
4. Wash cells gently five times with PBS-ABC.
5. Fix with 100 µL/well at room temperature for 10 min (*see* **Note 11**).
6. Wash cells three times with PBS-BSA. (The only time you wash with BSA.)
7. Incubate the cells with secondary antibody (100 µL/well) at room temperature for 30 min on the rocker.
8. Wash five times with PBS-ABC.
9. Incubate the cells with tertiary antibody (100 µL/well) at room temperature for 30 min on a rocker.
10. Wash cells gently five times with PBS-ABC + 0.1% Tween (*see* **Note 4**).
11. Add TMB substrate (50 µL/well) (*see* **Note 12**).
12. Read cells at 370 nm on a plate reader for 0.1 s/well.

3.3. Cell Fusion Assay

1. Warm up PBS and Ham's media.
2. Wash target and effector (CHO-K1) cells once with PBS.
3. Detach cells using versene (*see* **Note 13**).
4. Count cells (*see* **Note 14**).
5. Spin down the amount of cells necessary to add the target and effector cells to each other in a 1:1 manner.
6. Aspirate media off the pelleted cells, and add complete Ham's F-12 media to each sample in order to add 0.5 mL/well for each cell type.
7. Combine 0.5 mL of the CHO-K1 cells and 0.5 mL of the target cells to each well of a 24-well plate.
8. Incubate at 37°C for 18–24 h (*see* **Note 15**).
9. Aspirate media off the overlaid cells. At this point, the work may be done on the benchtop.
10. Wash once with PBS.
11. Add 100 µL 1X Promega Passive Lysis Buffer (diluted with PBS) to each well.
12. At this time (or it can be done earlier), take out Promega Luciferase Substrate from the freezer (*see* **Note 16**).
13. Rotate the plate for 20 min at room temperature.

14. Transfer 20 µL of lysate to a 96-well plate in duplicate (*see* **Note 17**).
15. Immediately before analyzing, add 100 µL of substrate to each well, and place in plate reader (*see* **Note 6**). If the plate reader has a shaking option, shake the samples for 2 s before reading each well for 10 s.

4. Notes

1. The effector cell line used should be easily transfected and also not susceptible to infection, as interference in *cis* between the glycoproteins and receptor could potentially inhibit fusion. The target cells only need to express the T7 RNA polymerase efficiently. If you have trouble transfecting the target cells, alternative methods are retroviral or adenoviral approaches. Furthermore, stable cell lines expressing the T7 RNA polymerase will help decrease inconsistencies between experiments and, overall, ease the experimental procedure.
2. The size of the plates will vary depending on the transfection method and the size of the experiment. Just keep in mind you will need an equal amount of target and effector cells for the fusion assay as well as an additional amount of effector cells for the CELISA. If you are testing several cell types in one experiment, we recommend using the same effector cells for all cell types. Therefore, you should increase the number of effector cells transfected by increasing the size of plate/dish and splitting them between the cell types during the overlay step.
3. For each glycoprotein expressed, an antibody will be needed. Serial dilutions of the primary antibodies should be done first to determine the most accurate reading. The concentrations given for the secondary and tertiary antibodies are suggested starting points. Different dilutions should be tested to get the best readings.
4. If you experience high background, dilute the tertiary antibody in PBS-ABC + 0.1% Tween.
5. For the overlay step, the plate to use depends on the number of cells you are using; if you use 0.1×10^6 cells for each sample, use a 24-well plate, and if you use 2×10^6 cells for each sample, use a 6-well plate.
6. A tube luminometer reader may also be used to measure luminescence. The amount of lysate and substrate remain the same. If you are using a tube reader, remember to vortex the sample briefly.
7. This method uses Lipofectamine 2000 for efficient transfection. Other methods may be used as preferred.
8. Most lipid-based transfection procedures suggest a density of 60–90%. Transfection efficiency will vary between cell types and lines, so optimization is suggested using a GFP plasmid to quantitate prior to performing the entirety of the fusion assay. CELISA is also adequate for checking the transfection efficiency of the glycoproteins.
9. With increasing passage of the CHO-K1 cells, the background increases so it is recommended to use only lower passages.
10. The amount of DNA should be altered for the size of the well; if doing so, then change the amount of Opti-MEM and DNA accordingly. The transfection length is dependent on how sensitive the cells are to Lipofectamine. If the cells are dead after

12 h with the Lipofectamine-DNA, then the length of time should be decreased. An alternative is to decrease the amount of Lipofectamine as this is what is cytotoxic to the cells. It is important to remember to keep the amount of DNA equal for all transfection samples.

11. If you need to stop, the cells have been fixed and the experiment can be continued later. Add 100 µL/well of PBS-ABC, and store at 4°C for up to 1 wk. This may lower the sensitivity of the experiment.

12. Avoid bubbles when adding the TMB, as this will increase the reading of the well.

13. By detaching the cells with versene instead of trypsin, any disruption of the glycoproteins or the receptors necessary for fusion is alleviated.

14. There are two ways in which to perform the overlay step of this assay; both should work. The preferred way is outlined above in **Subheading 2.** The alternative method is to overlay the counted target cells on top of the undisrupted CHO-K1 (or effector population). This can be done if you have an accurate count of the number of effector cells in a well and the wells are consistent.

15. The length of the overlay for fusion should be consistent between experiments in order to compare. Keep in mind that at least several hours are necessary for luciferase gene expression to occur after fusion has taken place. Time may be optimized as deemed necessary.

16. It is crucial that the substrate be warmed to room temperature. Luciferase is temperature- and light-sensitive. The substrate should not be warmed up to 37°C, as this will degrade the enzyme.

17. Please avoid the lysate debris, as it can alter the reading. If you find it difficult to avoid, transfer the lysate to tubes, spin down, and then transfer 20 µL of the supernatant for analysis.

Acknowledgments

We wish to thank Peter Pertel and Keith Haan for their earlier experiments in the establishment of fusion assays for human herpesvirus-8 and EBV as well as members of the Spear and Longnecker laboratories for their technical support and guidance.

References

1. Haan, K. M., Lee, S. K., and Longnecker, R. (2001) Different functional domains in the cytoplasmic tail of glycoprotein B are involved in Epstein-Barr virus-induced membrane fusion. *Virology* **290**, 106–114.

2. Pertel, P. E. (2002) Human herpesvirus 8 glycoprotein B (gB), gH, and gL can mediate cell fusion. *J. Virol.* **76**, 4390–4400.

3. Pertel, P. E., Fridberg, A., Parish, M. L., and Spear, P. G. (2001) Cell fusion induced by herpes simplex virus glycoproteins gB, gD, and gH-gL requires a gD receptor but not necessarily heparan sulfate. *Virology* **279**, 313–324.

4. Klupp, B. G., Nixdorf, R., and Mettenleiter, T. C. (2000) Pseudorabies virus glycoprot in M inhibits membrane fusion. *J. Virol.* **74**, 6760–6768.

5. Muggeridge, M. I. (2000) Characterization of cell-cell fusion mediated by herpes simplex virus 2 glycoproteins gB, gD, gH and gL in transfected cells. *J. Gen. Virol.* **81,** 2017–2027.
6. Turner, A., Bruun, B., Minson, T., and Browne, H. (1998) Glycoproteins gB, gD, and gHgL of herpes simplex virus type 1 are necessary and sufficient to mediate membrane fusion in a Cos cell transfection system. *J. Virol.* **72,** 873–875.
7. Kobasa, D., Rogers, M. E., Wells, K., and Kawaoka, Y. (1997) Neuraminidase hemadsorption activity, conserved in avian influenza A viruses, does not influence viral replication in ducks. *J. Virol.* **71,** 6706–6713.
8. Lake, C. M. and Hutt-Fletcher, L. M. (2000) Epstein-Barr virus that lacks glycoprotein gN is impaired in assembly and infection. *J. Virol.* **74,** 11162–11172.

14

Pseudovirions as Specific Tools for Investigation of Virus Interactions With Cells

Martin Sapp and Hans-Christoph Selinka

Summary

This chapter outlines the generation and application of human papillomavirus type 33 (HPV33) pseudovirions. The method describes (1) the construction of vaccinia viruses recombinant for the major and minor HPV capsid proteins, L1 and L2, respectively; (2) the transfection of Cos7 cells with a marker plasmid replicating to high copy numbers; (3) the expression of L1 and L2 using the vaccinia virus expression system; (4) the extraction, purification, and analysis of HPV33 pseudovirions; and (5) their use in pseudoinfection assays. These pseudovirions are structurally indistinguishable from native virions and are therefore valuable tools for the study of papillomavirus–cell interactions. The methods described can be adopted for other nonenveloped DNA viruses and may be useful for gene transfer.

Key Words: Papillomavirus; vaccinia virus expression system; pseudovirus; pseudoinfection; HPV capsid protein; DNA encapsidation; virus-like particle; nuclear extract; density gradient purification; gene transfer.

1. Introduction

Since some viruses cannot be easily propagated in vitro owing to special requirements (e.g., regarding cell differentiation), surrogate systems have been developed to study the infection process of these viruses. In addition to genome-free virus-like particles, which allow the study of certain aspects of virus–cell interactions, pseudoviruses carrying foreign genes have been generated for this purpose *(1,2)*. These pseudoviruses have turned out to be valuable tools for analysis of virus internalization pathways and morphogenesis, for determination of neutralizing antibodies, and for testing drugs interfering with the virus infection. We describe here a method for generating pseudovirions of human papillomaviruses (HPV), to circumvent the strict dependence on termi-

From: *Methods in Molecular Biology, vol. 292: DNA Viruses: Methods and Protocols*
Edited by: P. M. Lieberman © Humana Press Inc., Totowa, NJ

nally differentiating keratinocytes for their replication *(3)*. HPV are nonenveloped viruses, which are composed of 360 copies of the major capsid protein L1, probably 12 copies of the minor capsid protein L2, and an 8-kb circularized double-stranded DNA genome in the form of chromatin *(4)*. We have chosen the approach of expressing the capsid proteins in cells harboring high copy numbers of a marker plasmid to generate HPV pseudovirions. In contrast to alternative procedures, which package naked DNA in or outside virus-like particles, these pseudovirions are structurally indistinguishable from native virions and allow the complete packaging of any histone-associated DNA up to 8 kb *(5,6)*. They allow the study of single-cell infections and a fast and easy quantitative analysis of infection events.

2. Materials

2.1. Basic Technical Equipment

1. Ultracentrifuge and rotors (e.g., Beckman Vti65 and SW40).
2. Electroporator and electroporation chambers (Gibco Cellporator®).
3. Fluorescence microscope.
4. Sonifier.
5. Density refractometer.
6. Polymerase chain reaction (PCR) equipment.
7. Agarose gel equipment.
8. Cell culture equipment.
9. Sodium dodecyl sulfate polyacrylamide gel electrophoresis (SDS-PAGE) equipment.

2.2. Viruses, Cells, and Plasmids

1. Vaccinia viruses (VV): wild-type VV strain WR; temperature-sensitive VV strain ts7; T7 RNA polymerase recombinant VV strain vTF7-3 *(7)*.
2. Plasmids: pTM1 *(7)*; pEGFP-C1 (Clontech); pHPV33 *(8)*.
3. Cell lines: HuTK-143 B (ATCC: CRL-8303); Cos7 (ATCC: CRL-1651).
4. Bacterial cells: *E. coli* DH5α; ELECTROMAX DH5α-E cells.

2.3. Buffers, Chemicals, and Other Materials

1. Hypotonic Dounce buffer: 10 m*M* HEPES, 10 m*M* KCl, 1.5 m*M* MgCl$_2$, 0.5% NP40, pH 7.6.
2. HBS buffer: 21 m*M* HEPES, 137 m*M* NaCl, 5 m*M* KCl, 0,7 m*M* Na$_2$HPO$_4$, 6 m*M* glucose, pH 7.05.
3. TE buffer: 10 m*M* Tris-HCl, 1 m*M* EDTA, pH 8.0.
4. Phosphate-buffered saline buffer: Dulbecco's PBS without calcium and magnesium (Invitrogen).
5. Proteinase K.
6. Restriction endonucleases.

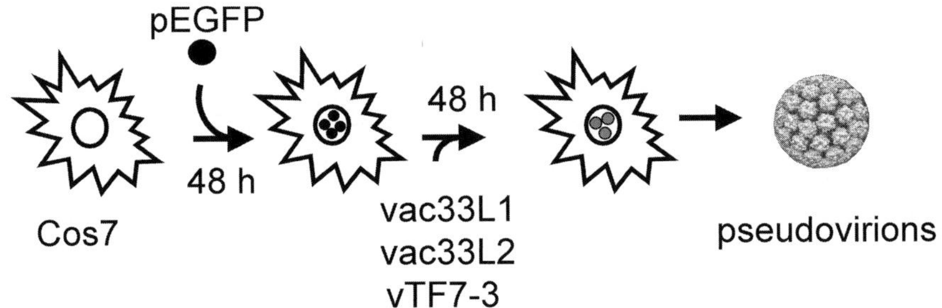

Fig. 1. Schematic drawing outlining the important steps in pseudovirus production.

7. DNA modifying enzymes.
8. Oligonucleotide primers.
9. 5-Bromo-2′-deoxyuridine (BrdU).
10. HPV capsid protein-specific antibodies.

3. Methods

We generate HPV pseudovirions by expression of HPV33 L1 and L2 using recombinant VVs in Cos7 cells harboring high copy numbers of a marker plasmid (**Fig. 1**). The methods described below outline (1) the construction of the transfer plasmids for (2) the generation and purification of recombinant VV-expressing HPV33 L1 and L2, (3) the generation of HPV33 pseudovirions bearing a green fluorescent protein coding sequence, (4) their purification and analysis, and (5) infection assays performed with these pseudovirions.

3.1. Transfer Plasmids for Generation of L1 and L2 Recombinant Vaccinia Viruses

Because of its size, recombinant VV can only be generated by homologous recombination. This requires the use of transfer vectors carrying the target genes surrounded by VV sequences. Construction of the transfer plasmids for L1 and L2 of HPV33 is described here. This includes (1) a description of the transfer vector pTM1; (2) amplification of the L1 and L2 genes by PCR and their cloning into pTM1; and (3) in vitro transcription/translation to control for the functionality of the expression cassette.

3.1.1. pTM1 Transfer Plasmid

The pTM1 transfer vector is a pUC-based plasmid, which was constructed and kindly provided by Bernhard Moss *(7,9)*. The vector contains the VV thymidine kinase (TK) coding sequence, whose open reading frame was destroyed by insertion of the bacteriophage T7 promoter and terminator, which enclose an EMC sequence, to allow cap-independent translation, and a multiple cloning site **(Fig. 1)**.

3.1.2. Amplification of the HPV33 L1 and L2 Open Reading Frames

Plasmid pHPV33, harboring the complete HPV33 genome cloned via its singular *Bgl*I site (position 2796) into the *Bgl*I site of pBR322link, was kindly provided by Gerard Orth *(8)*.

1. The open reading frame of HPV33 L1 is amplified by PCR with oligonucleotides 5'-GGCC**TCATGA**CCGTGTGGCGGCCTAGTG-3' and 5'-GGCC**GGATCC**ACA-CAATT ACACAAAGTG-3' using proofreading thermostable DNA polymerase.
2. The resulting fragment is cloned via the *Bsp*H1 and *Bam*H1 sites (boldfaced letters) into the *Nco*I and *Bam*H1 sites of pTM1 using standard molecular biology techniques.
3. The L2 gene is amplified from pHPV33 by PCR with oligonucleotides 5'-GGC-**GAATTC**ATGAGACACAAAG ATCT-3' and 5'-GAC**GGATCC**AGGTACACT-GTGGCC-3' as primers.
4. The resulting fragment is cut with *Eco*RI and *Bam*HI and cloned into the *Eco*RI and *Bam*HI sites of pTM1.
5. The L1 and L2 genes of the resulting plasmids pTM33L1 and pTM33L2 have to be sequenced to confirm the absence of mutations.

3.1.3. In Vitro Transcription/Translation

Since the generation of recombinant VVs is a time-consuming step, it is recommended to control the correctness of the expression cassette by combined in vitro transcription/translation using ^{35}S-labeled methionine (*see* **Note 1**). This is done according to the T7 Coupled Reticulocyte Lysate System supplied by Promega. The analysis involves an SDS-10% PAGE and exposure of a film overnight (*see* **Note 2**).

3.2. Generation of Recombinant Vaccinia Viruses

This section describes (1) purification of wild-type (wt) VV DNA, (2) infection of cells defective for the *TK* gene with temperature sensitive helper virus, (3) cotransfection of cells with transfer vector and wt VV DNA, (4) selection for recombinant VVs using BrdU, and (5) plaque purification of recombinant viruses.

3.2.1. Preparation of Wild-Type Vaccinia Virus DNA

1. Grow ten 150-mm dishes of confluent HuTK$^-$143 B cells in Dulbecco's modified eagle's medium (DMEM) supplemented with 10% fetal calf serum (FCS). After washing each plate with 20 mL serum-free DMEM, infect the plates with wt VV in 5 mL serum-free DMEM at a multiplicity of infection (MOI) of 0.1 at room temperature (RT) for 1 h under repeated agitation. Replace the medium with supplemented DMEM and grow for 24 h at 37°C in a CO_2 incubator.

2. Wash the cells with PBS and scrape them off the plate in 1.5 mL PBS. For further treatment, two 1.5-mL aliquots are required. Additional tubes can be stored for up to 2 yr at –70°C.

3. To each 1.5 mL cell suspension, add 30 µL 10% Triton X-100, 1.5 µL β-mercaptoethanol, and 48 µL EDTA (250 m*M*, pH 8.0). Invert the tube three to six times slowly.

4. Centrifuge the cell lysate for 2.5 min at 850 *g* in a tabletop centrifuge. Transfer the supernatant into a new 1.5-mL reaction tube and spin for 10 min at 16,000 *g*. Discard the supernatant, and combine the pellets in 200 µL TE buffer (*see* **Note 3**).

5. Add 13.4 µL 3 *M* NaCl, 20 µL 10% SDS, 0.6 µL β-mercaptoethanol, and 3 µL proteinase K (10 mg/mL). Incubate overnight at 56°C (*see* **Note 4**).

6. Extract the solution twice with 400 µL buffered phenol, followed by two extractions with chloroform/isoamylalcohol (24:1), and precipitate the VV DNA with ice-cold ethanol.

7. Remove the DNA from the solution by winding around a glass dipstick (e.g., Pasteur pipet), wash it by dipping twice into 70% ethanol, and air-dry. Resuspend the DNA in up to 300 µL TE, and store at 4°C. Determine the DNA concentration.

3.2.2. In Vivo Recombination

1. Seed HuTK⁻143 B cells into 6-well plates (4×10^5/well), and grow them overnight. Wash cells with DMEM and infect with temperature-sensitive VV ts7-VV at an MOI of 0.1 for 1 h in a total volume of 1 mL. Replace virus with 2 mL supplemented DMEM, and incubate for 2 h at 33°C in a CO_2 incubator. Remove cells from the incubator, and increase temperature to 39°C.

2. Transfect cells with 1 µg each of transfer vector and wt VV DNA, using standard calcium phosphate precipitation.

3. Replace the calcium precipitate after 1 h at RT with 2 mL DMEM, and incubate for 2 h at 39°C. Wash cells twice with DMEM, add 5 mL supplemented DMEM, and grow cells for 48 h at 39°C to kill temperature-sensitive helper viruses.

3.2.3. Selection With BrdU

In recombinant VV the target genes replace the wt *TK*. This can be used for elimination of *TK*-positive wt VV owing to their incorporation of BrdU.

1. Discard the cell culture supernatant carefully, and scrape cells in 900 µL of 10 m*M* HEPES, pH 7.2, off the plates. Lyse cells by two cycles of freezing and thawing. Add 100 µL of 10-fold concentrated PBS.

2. Use the lysate to infect confluent HuTK⁻143 B cells in 6-well plates. Incubate for 48 h at 37°C in the presence of 0.2 mg/mL BrdU.

3. Repeat **steps 1** and **2** twice. Cell lysis induced by infection with recombinant VV should become obvious.

4. To control for the successful recombination, test the unpurified lysates for the presence of L1- or L2-containing recombinant VV by coinfection with the helper virus vTF7-3. Coinfect HuTK⁻143 B cells with 300 µL of the final lysate and vTF7-3

(MOI 1) for 1 h at RT, and grow cells for 24 h. Scrape cells off the plates in PBS, wash once with PBS, and lyse cells in 100 µL of Laemmli sample buffer. Analyze 10 µL for the synthesis of L1 or L2 by Western blot.

3.2.4. Plaque Purification of Recombinant Vaccinia Virus

1. Prepare 10-fold serial dilutions of the lysates (10^{-2}–10^{-7}) in PBS. Use these to infect HuTK⁻143 B cells grown in 6-well plates.
2. Remove the VV lysates, and overlay cells with 0.625% agarose in DMEM supplemented with 10% FCS following standard protocols.
3. Check for the formation of plaques daily.
4. Pick plaques using a 1-mL filtered pipet tip, and dispense the agarose into 1 mL DMEM. Incubate overnight at 4°C.
5. Amplify VV using HuTK⁻143 B cells grown in 6-well plates. Repeat amplification until complete lysis of the monolayer is seen after 24 h of incubation.
6. Test individual recombinant VV for expression of L1 and L2 by coinfection with vTF7-3 as described above.
7. Positive VV can now be amplified on a large scale and stored at –70°C.

3.2.5. Determination of Plaque-Forming Units

For generation of pseudovirions, it is important to determine the number of plaque-forming units (PFUs) of recombinant VV, which should be in the range of 5×10^7 to 1×10^9.

1. Prepare 10-fold serial dilutions of the lysates (10^{-3}–10^{-8}) in PBS. Use these to infect HuTK⁻143 B cells grown in 6-well plates, and grow cells for 24 h.
2. Carefully wash cells with PBS and stain with Coomassie brilliant blue.
3. Count the number of plaques and multiply with the dilution factor to obtain the PFU/mL.

3.3. Production of Human Papillomavirus Type 33 (HPV33) L1/L2 Pseudovirions

3.3.1. Construction of a GFP Marker Plasmid

Natural infections with HPV require delivery of the viral DNA genome into the nucleus. A green fluorescent protein (GFP)-encoding marker plasmid encapsidated by HPV pseudovirions allows identification of successfully infected cells by their bright green cytoplasmic fluorescence. To facilitate analysis, we use a dimeric GFP fused with a nuclear localization signal (pGFPGFPNLS). This plasmid is based on pEGFP-C1 (Clontech). The GFP fragment of plasmid pEGFP-C1 was amplified by PCR using a 5′NTR primer (5′-GCC**GAATTC**TATGGTGAGCAAGGGCGAG GAG-3′) containing an *Eco*RI restriction site and a reverse primer (5′-CTC**GGATCC**TTATTTTT-TAACCTTTTTGCGTTTCTTGTACAGCTCGTCCAT-3′) containing a *Bam*HI restriction site adjacent to the nuclear location signal sequence of HPV33 L1 (amino acid position 493–499). The PCR fragment was cut with *Eco*RI and *Bam*HI, purified, sequenced, and cloned adjacent to the original GFP sequence into the *Eco*RI/*Bam*HI-digested pEGFP-C1 vector by standard molecular biology techniques.

3.3.2. Electroporation of Cos7 Cells

This step requires large quantities of purified marker plasmid DNA (*see* **Note 5**). Transfection of the GFP reporter plasmid preceding the incorporation into pseudovirions can be achieved by various methods. The highest levels of transfection have been reported using electroporation, which we performed according to the following protocol (*see* **Note 6**).

1. For detachment, incubate Cos7 monolayer cells with 3 mL PBS/2.5 mM EDTA for 1 min at RT, followed by a 5-min incubation at 37°C after removal of PBS/EDTA.
2. Resuspend Cos7 cells at 5×10^6 cells/mL in HBS buffer.
3. Transfer cells into precooled (4°C) electroporation chambers.
4. Add 12 µg of the GFP marker plasmid resuspended in HBS, PBS, or TE buffer.
5. Close electroporation chamber, and mix carefully by inverting the chamber.
6. Electroporation conditions: 200–220 V, 333 µF, low Ohm, fast charge rate.
7. For optimal pulse recovery, cells should be kept on ice for at least 5 min.
8. Harvest cells with a Pasteur pipet, and resuspend in complete culture medium (e.g., DMEM/10% FCS)
9. Monitor transient GFP protein expression after 24–48 h of incubation at 37°C.

3.3.3. Infection of Transfected Cells With Recombinant Vaccinia Viruses

The GFP plasmid-transfected Cos7 cells are subsequently (*see* **Note 7**) infected with recombinant VV encoding the HPV33 L1 and HPV33 L2 capsid proteins. Expression of these capsid proteins by recombinant vaccinia viruses requires simultaneous expression of the vTF7-3 vaccinia helper virus encoding the phage T7 RNA polymerase (*see* **Note 8**).

1. Remove culture supernatants of transfected cells.
2. Add 5 mL of serum-free medium (DMEM) containing vac33L1, vac33L2, and vTF7-3 (each with an MOI of 1) to the monolayer cells (*see* **Note 9**).
3. Incubate for 1 h at RT, moving of the plates occasionally.
4. Replace supernatant by DMEM/10% FCS, and incubate for 40–48 h at 37°C.

3.4. Purification and Analysis of Pseudovirions

At 40–48 h post infection with the recombinant VVs, cells become rounded up and can easily be harvested by simple pipeting. Although some cells have already released pseudovirions into the culture supernatant at that time, the bulk of the DNA-containing pseudovirions is still captured in cell nuclei. DNA-containing pseudovirions and empty virus-like particles may be prepared from nuclear lysates.

3.4.1. Preparation of Nuclear Lysates

1. Harvest VV-infected cells by centrifugation (10 min, 300*g*).
2. Wash cells with PBS and resuspend in 10 mL hypotonic dounce buffer (*see* **Note 10**).
3. Disrupt cells in a tight-fitting dounce homogenizer (40–50 strokes, 4°C).
4. Sediment cell nuclei by centrifugation (e.g., 10 min, 1000 *g* in a Sorvall SS34 Rotor).
5. Resuspend pellet in dounce buffer (10 mL).
6. Disrupt cell nuclei by sonification (three times, 45 s; 40% output).
7. Remove nuclear debris by centrifugation at 4°C (10 min, 8000 *g*, Sorvall SS34 Rotor).
8. Save supernatant for further purification steps.

3.4.2. CsCl Density Gradient Purification

Pseudovirions are further purified from nuclear lysates by density gradient centrifugation. During this procedure, pseudovirions are separated from nuclear debris not only but also from the high content of DNA-free VLPs still present in these preparations **(Fig. 2)**.

1. Transfer supernatants (*see* **Subheading 3.2.5.**) to 15-mL conical tubes, and add 0.4 g CsCl per mL solution to obtain a density of 1.29 g/cm^3.
2. Incubate for 60–90 min at RT. (This step is required for complete destruction of contaminating VVs.)
3. Transfer solution to 5-mL polyallomer centrifugation tubes (e.g., OptiSeal®, Beckman).
4. Ultracentrifugation: Vti65, 350,000 *g*, 12°C overnight.
5. Collect 0.25-mL fractions from the bottom of the centrifuge tubes.

3.4.3. Analysis of Pseudovirus-Containing Fractions

CsCl density gradient fractions are analyzed for the presence of pseudovirions by refractive index determination, Western blot analysis, and pseudovirus infection assays **(Fig. 2)**. Determination of the fraction density by measuring the refraction index in a density refractometer allows separation of DNA-containing pseudovirus fractions (1.33 g/cm^3) and fractions containing VLPs (1.29 g/cm^3). This separation may be confirmed by Western blot analysis (5 μL of each fraction) using an HPV capsid protein L1-specific antibody. A detailed description of pseudovirus infection assays is presented in **Subheading 3.5.** The encapsidated marker plasmid can be quantified by the following procedure.

1. Dialyse 100 μL each of fractions with densities from 1.35 to 1.30 g/cm^3 for at least 6 h.

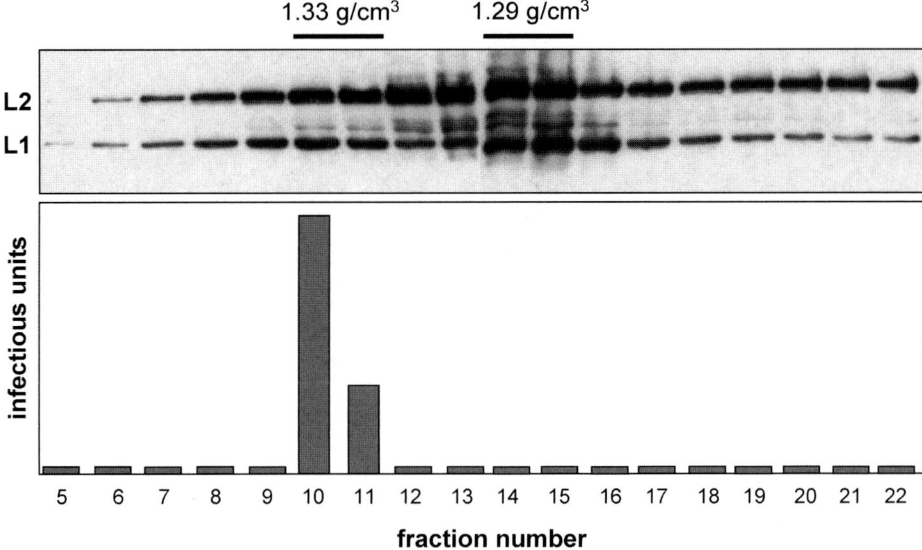

Fig. 2. Analysis of a buoyant cesium chloride density gradient centrifugation. Nuclear extracts containing HPV33 pseudovirions were subjected to cesium chloride gradient centrifugation. Fractions 5–22 were analyzed for the presence of L1 and L2 protein by Western blot *(top)*, using monoclonal antibodies 33L1-7 *(14)* and 33L2-1 *(15)* and infectivity assays *(bottom)*. The positions of pseudovirions and DNA-free virus-like particles at densities of 1.33 and 1.29 g/cm³, respectively, are indicated.

2. Adjust to 10 m*M* $MgCl_2$ and 10 U DNase I, and incubate for 1 h at 37°C.
3. Phenol/chloroform-extract and ethanol-precipitate DNA overnight at –20°C.
4. Resuspend the pellet in 15 μL TE, and transform *E. coli* by electroporation using ELECTROMAX DH5α-E cells.
5. Use a purified marker plasmid of known concentration as the standard for determination of transformation efficiency.
6. Plate onto LB plates containing kanamycin, grow overnight, and count the colonies.

3.4.4. Further Purification of Pseudovirons by Sucrose Step Gradients

Pseudovirions purified by CsCl gradient centrifugation may be used directly in pseudovirus infection assays, provided that the high salt concentration is reduced by dialysis (*see* **Note 11**). In addition to pseudovirions, significant amounts of unincorporated L1 and L2 capsid proteins are still present in these CsCl fractions. These may interfere in some assays, e.g., analyses of pseudovirus interactions with the cell surface. For further purification, pseudovirus-containing CsCl fractions may be subjected to sucrose step gradient centrifugation **(Fig. 3)**.

For this purpose, pseudovirus-containing CsCl fractions are diluted 1:5 with PBS to reduce the salt concentration of the sample. A sucrose step gradient is prepared in a siliconized 12-mL polyallomer ultracentrifugation tube (Beckman) by carefully pipeting 3 mL of 30% sucrose (diluted in PBS containing 50 µg/mL BSA) on top of a 2-mL layer of 70% sucrose. The gradient is loaded with up to 6.5 mL of diluted pseudovirus fractions, placed into an SW40 rotor (Beckman), and run at 12°C for 4 h at 285,000 g. Subsequently, the gradient is collected from the bottom in 0.5-mL aliquots, and each 5 µL of fractions 1–6 is tested in infectivity assays (*see* **Note 12**).

3.5. Pseudovirus Infection Assays

3.5.1. Infection Assay

1. Grow Cos7 cells (5×10^4 cells/well) in 24-well plates, and infect them with pseudovirions in a total volume of 250 µL DMEM. The volume of added pseudovirions is normally in the range of 0.5–5 µL and should preferably be titrated to yield 50–500 infectious units.
2. After 1 h at 4°C under constant agitation, replace the pseudovirions with 1 mL of supplemented culture medium, and continue incubation at 37°C for 72 h.
3. Determine infectious events by counting cells with nuclear bright green fluorescence **(Fig. 4)**.

3.5.2. Neutralization Assay

Pseudovirions are suitable tools in HPV diagnosis and in research for large-scale characterization of virus-specific neutralizing antibodies, which arise during natural HPV infection or in immunized individuals.

1. In neutralization assays, preincubate pseudovirions for 1 h at 4°C with serial dilutions of antiserum or purified antibodies in a total volume of 30 µL.
2. After addition of 220 µL of DMEM, add the sample to Cos7 cells, as outlined in **Subheading 3.5.1.**
3. In addition, a pseudovirus depletion assay, using virus-specific antibodies coupled to magnetic beads (e.g., Dynabeads®) may be used to determine the virus-binding capacities of nonneutralizing antibodies. This can be achieved by preincubation of pseudovirus with antibody-loaded magnetic beads (1 h at 4°C) prior to the addition to Cos7 cells.
4. Controls with unrelated antibodies and uncoupled magnetic beads need to be run in parallel.
5. Pseudovirions have also been used to measure virus–receptor interactions as well as virus internalization kinetics, using antibodies in postattachment pseudovirus neutralization assays *(10,11)*.

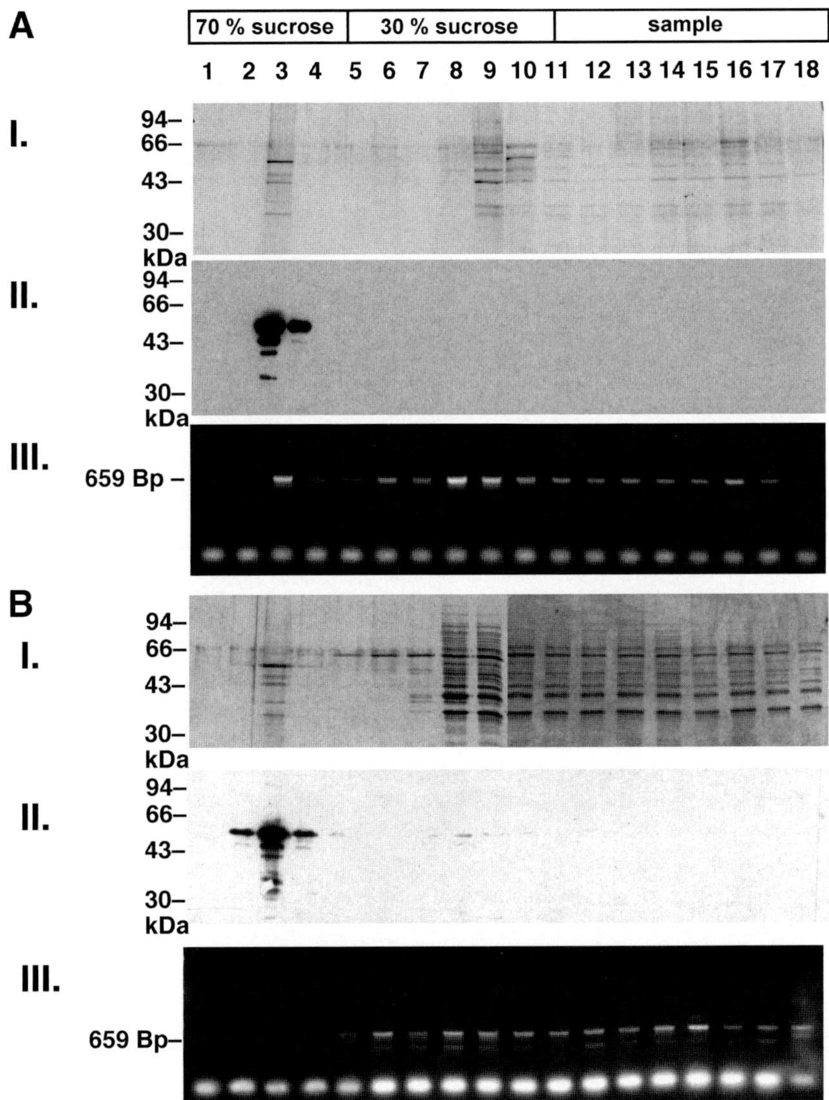

Fig. 3. Purification of pseudovirions by sucrose step gradient. (**A**) Pseudovirus containing fractions of cesium chloride gradients were combined, dialyzed, and subjected to sucrose step gradient centrifugation. Fractions were analyzed by silver staining (I), L1-specific Western blot (II), and marker gene-specific PCR without prior DNase I digestion (III). In this case, a pSVβgal marker plasmid was packaged. (**B**) As a control, fractions containing DNA-free virus-like particles were analyzed in parallel. Although marker plasmid is detectable in most fractions of the gradients, it is present in pseudovirus-containing but not virus-like particle-containing fractions.

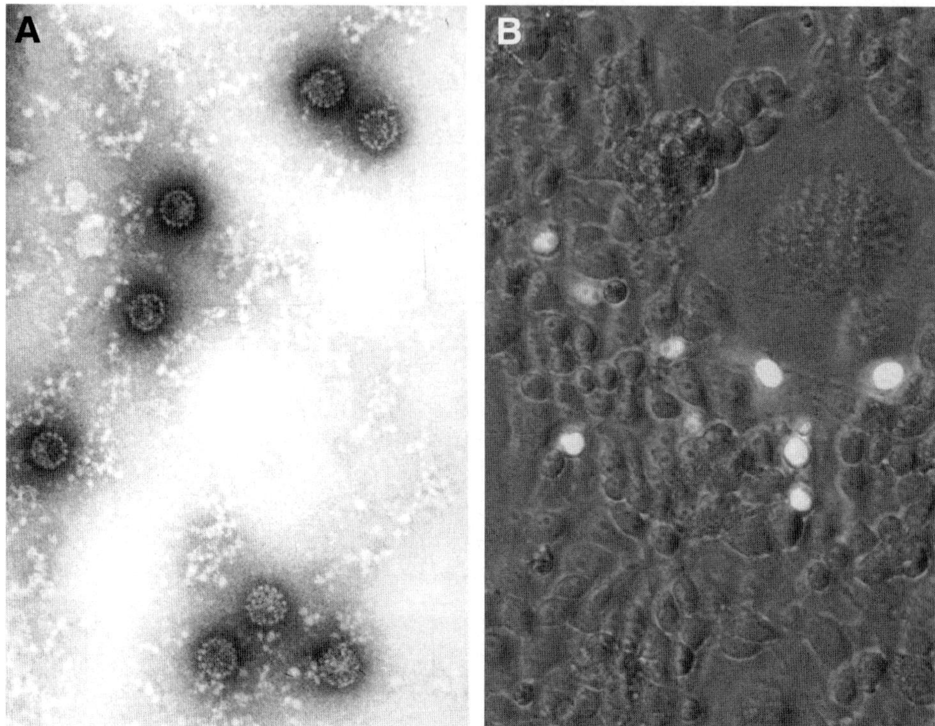

Fig. 4. (A) Electron micrograph of HPV33 pseudovirions negatively stained with uranyl acetate. **(B)** Readout of a pseudoinfection assay. Infected cells are detectable by their nuclear bright green fluorescence. A merged photograph of phase contrast and fluorescence microscopy is shown.

4. Notes

1. Alternatively, constructs can be tested by transient expression in mammalian cells. To this end, subconfluent monolayer cells are infected with helper virus vTF7-3 at an MOI of 5 (*see* **Subheading 3.2.1.**). Cells are subsequently transfected by lipofection with 4 µg of the recombinant plasmids. Five to 10 h later, cells are harvested and processed for Western blot with capsid protein-specific antibodies.
2. The L1 protein has an apparent molecular mass of 55 kDa, as determined by SDS-PAGE. The molecular mass of L2 protein is 72 kDa, which is considerably larger than the predicted 50.5 kDa. This seems to be an intrinsic property of L2 proteins from several papillomaviruses and is not caused by posttranslational modifications.
3. If problems arise from loose pellets by removing the supernatant, centrifuge the sample again at 2400 *g* for 2 min.
4. From now on the large VV DNA is sensitive to shearing. Strictly avoid vortexing, and always use wide-mouthed pipet tips.

5. The purity of DNA is considered a critical parameter for efficient transfection and is even more critical when transfection is followed by subsequent packaging into virus-like particles. In our hands, purification by commercially available DNA purification kits resulted in efficient but not reproducible transfection in regard to the formation of pseudovirions. The best results were obtained by purifying the marker plasmid using two subsequent cesium chloride (CsCl) density purification steps.

6. The quality of Cos7 cells is very important, for both generation of pseudovirions and pseudoinfection. Virus yields and pseudoinfection can vary dramatically. The best results were obtained using cells with low passage numbers. We always use freshly thawed cell aliquots for expansion and subsequent electroporation. Never use cells grown longer than 4 wk for pseudoinfection. We made the observation that high-level expression of syndecans reduces infectivity. For standard pseudovirus preparation, it is suggested to use about 5×10^8 cells. Although approx 10,000 marker plasmid-encapsidating particles are required for one infectious event, up to 10^6 infectious units can be generated. In addition to marker plasmid DNA, cellular DNA is being incorporated *(1)*.

7. Expression of GFP is often obvious 24 h after transfection. We suggest starting infection not before 48 h post transfection, since amplification of the marker plasmid by the simian virus 40 T antigen is important to achieve high yields of pseudovirions. If one waits longer, many transfected cells begin to die owing to the T-antigen-driven runoff replication.

8. Of course, expression systems other than the VV system, which allow high-level expression of L1 and L2, can be used. However, owing to the codon usage of papillomaviruses, codon-optimized (humanized) L1 and L2 genes have to be used for nonviral transient expression of capsid proteins *(12,13)*.

9. In our hands, we did not see significant differences in yields using MOIs between 0.1 and 2. This is probably owing to the spread of the virus to neighboring cells.

10. VVs partially copurify with pseudovirions. It is therefore crucial to eliminate the VVs, which interfere in pseudovirus infection assays. This is achieved by addition of the detergent NP40, which does not affect pseudovirions. NP40 is present throughout the purification, since extended incubation is required for complete destruction of VVs. At this step, disrupted cells may be stored at –20°C for 1 mo without any loss in pseudovirus recovery.

11. Dialysis for 1 h is sufficient to remove 90% of the cesium chloride, which is good enough for the use in pseudoinfection. At this stage, pseudovirions can be frozen in aliquots at –20°C. Avoid repeated freezing and thawing. Once thawed, the aliquots can be kept up to 1 mo at 4°C without significant loss in infectivity.

12. Although sucrose gradient centrifugation significantly increases the purity of the pseudovirus preparation, this may result in considerable reduction in yield. The use of siliconized tubes is highly recommended. The presence of BSA in the gradient also reduces the loss but may be a disadvantage for some assays, e.g., direct coupling to enzyme-linked immunosorbent assay plates.

Acknowledgments

The authors thank B. Moss for providing reagents for the T7 polymerase-based vaccinia virus expression system and G. Orth for pHPV33. We are grateful to former and current members of the lab, especially F. Unckell, T. Giroglou, F. Schäfer, and L. Florin for their contributions to the development and improvement of the system. This work was supported by grants to M. Sapp from Deutsche Forschungsgemeinschaft and Stiftung Rheinland-Pfalz für Innovation.

References

1. Unckell, F., Streeck, R. E., and Sapp, M. (1997) Generation and neutralization of pseudovirions of human papillomavirus type 33. *J. Virol.* **71,** 2934–2939.
2. Stauffer, Y., Raj, K., Masternak, K., and Beard, P. (1998) Infectious human papillomavirus type 18 pseudovirions. *J. Mol. Biol.* **283,** 529–536.
3. McMurray, H. R., Nguyen, D., Westbrook, T. F., and McAnce, D. J. (2001) Biology of human papillomaviruses. *Int. J. Exp. Pathol.* **82,** 15–33.
4. Howley, P. M. (1996) Papillomavirinae: the viruses and their replication, *in Fields Virology,* vol. 2 (Fields, B. N., Knipe, D. M., Howley, P. M., et al., eds.), Lippincott-Raven, Philadelphia, pp. 2045–2076.
5. Fligge, C., Schäfer, F., Selinka, H. C., Sapp, C., and Sapp, M. (2001) DNA-induced structural changes in the papillomavirus capsid. *J. Virol.* **75,** 7727–7731.
6. Schäfer, F., Florin, L., and Sapp, M. (2002) DNA binding of l1 is required for human papillomavirus morphogenesis in vivo. *Virology* **295,** 172–181.
7. Moss, B., Elroy-Stein, O., Mizukami, T., Alexander, W. A., and Fuerst, T. R. (1990) New mammalian expression vectors. *Nature* **348,** 91–92.
8. Beaudenon, S., Kremsdorf, D., Croissant, O., Jablonska, S., Wain-Hobson, S., and Orth, G. (1986) A novel type of human papillomavirus associated with genital neoplasias. *Nature* **321,** 246–249.
9. Fuerst, T. R., Niles, E. G., Studier, F. W., and Moss, B. (1986) Eukaryotic transient-expression system based on recombinant vaccinia virus that synthesizes bacteriophage T7 RNA polymerase. *Proc. Natl. Acad. Sci. USA* **83,** 8122–8126.
10. Giroglou, T., Florin, L., Schäfer, F., Streeck, R. E., and Sapp, M. (2001) Human papillomavirus infection requires cell surface heparan sulfate. *J. Virol.* **75,** 1565–1570.
11. Selinka, H. C., Giroglou, T., and Sapp, M. (2002) Analysis of the infectious entry pathway of human papillomavirus type 33 pseudovirions. *Virology* **299,** 279–287.
12. Zhou, J., Liu, W. J., Peng, S. W., Sun, X. Y., and Frazer, I. (1999) Papillomavirus capsid protein expression level depends on the match between codon usage and tRNA availability. *J. Virol.* **73,** 4972–4982.
13. Leder, C., Kleinschmidt, J. A., Wiethe, C., and Müller, M. (2001) Enhancement of capsid gene expression: preparing the human papillomavirus type 16 major structural gene L1 for DNA vaccination purposes. *J. Virol.* **75,** 9201–9209.

14. Sapp, M., Kraus, U., Volpers, C., et al. (1994) Analysis of type-restricted and cross-reactive epitopes on virus-like particles of human papillomavirus type 33 and in infected tissues using monoclonal antibodies to the major capsid protein. *J. Gen. Virol.* **75,** 3375–3383.
15. Volpers, C., Sapp, M., Snijders, P. J., Walboomers, J. M., and Streeck, R. E. (1995) Conformational and linear epitopes on virus-like particles of human papillomavirus type 33 identified by monoclonal antibodies to the minor capsid protein L2. *J. Gen. Virol.* **76,** 2661–2667.

IV

GENE EXPRESSION

15

Simultaneous *In Situ* Detection of RNA, DNA, and Protein Using Tyramide-Coupled Immunofluorescence

Brian A. Van Tine, Thomas R. Broker, and Louise T. Chow

Summary

The use of tyramide-coupled immunofluorescence at the single cell level provides expedient, clean, and sensitive signals for detection of DNA, RNA, or proteins. The principle is based on the ability of horseradish peroxidase (HRP) to cleave tyramides into a free radical species with a very short diffusion radius. The free radicals are then covalently bound to electron-rich moieties such as tyrosine in proteins proximal to the targets. Here we present protocols for tyramide fluorescent *in situ* hybridization (T-FISH), which detects unique DNA species using DNA probes as short as approx 300–500 bp, or unique RNA species with probes as small as an oligonucleotide. We also present a protocol for tyramide immunofluorescence (T-IF) to detect protein antigens. By combining these protocols with several tyramide-coupled fluorophores, multiple targets can be detected simultaneously *in situ,* which is ideal for in-depth analyses at the molecular and cellular levels. Finally, we describe the detection of nascent viral RNA transcripts simultaneously with integrated viral genomes or chromosomal domains in single cells or tissue sections.

Key Words: Fluorescence microscopy; tyramide; fluorophores; T-FISH; T-IF; HPV; DNA; RNA; protein; chromosome paints; antibody; hybridization; *in situ* analyses; tissue sections; single cell.

1. Introduction

In creating strategies for enzyme-linked immunosorbent assay-based techniques, Bobrow et al. *(1)* were the first to report the use of tyramides for signal amplification. This technology has been further developed for use with tyramide fluorescent *in situ* hybridization (T-FISH) for nucleic acids *(2–7)*, tyramide immunofluorescence (T-IF) for protein antigens *(8–14)*, and combination applications *(15–19)*. In this technique, the targeted DNA, RNA, or protein is reacted with the primary probe, which can be a biotinylated nucleic acid (or other tagged probes, to be described) or a primary antibody (*see* **Note 1**). The

From: *Methods in Molecular Biology, vol. 292: DNA Viruses: Methods and Protocols*
Edited by: P. M. Lieberman © Humana Press Inc., Totowa, NJ

primary probe is then detected by streptavidin–horseradish peroxidase (HRP; or the alternative and equivalent peroxidase [POD]) or a secondary antibody coupled to streptavidin–HRP, in conjunction with a tyramide fluorophore. The enzymatic activity of HRP generates tyramide free radicals from tyramide. The free radicals then react with tyrosine and other electron-rich amino acid residues in proteins in the vicinity of the target.

By conjugating tyramide to various fluorophores, the sensitivity and speed of detection are significantly increased above and beyond the traditional immuno-histochemical detection of chromagens or autoradiographic detection of radiolabeled probes with a light microscope. The enhanced sensitivity confers the ability to detect targets of low abundance as well as protein antigens with diluted primary antibodies. Indeed, tyramides have been used to localize chromosomal genes in metaphase with a probe in the range of 1 kb *(2,7)*. Because tyramides are covalently bonded, one can inactivate the HRP with hydrogen peroxide and remove the original nucleic acid probe and/or target by denaturation without dramatically affecting the strength or localization of the signals already deposited. This allows additional rounds of probing for other targets by using exactly the same chemistry but different tyramide fluorophores. Consequently, this probe strategy allows simultaneous detection at the single cell level of multiple targets, be they DNA, RNA, or protein.

One slight limitation relative to conventional detection methods is the diminished ability to view the histology or morphology of the tissues or cells in which the signals are revealed. However, this drawback is more than compensated for by the simultaneous localization and detection of multiple targets (*see* **Note 2**). One can always examine serial sections or parallel specimens for histology and morphology, to which the conventional detection methods also resort. A good pathologist or histologist should be able to retrain themselves to read nuclear stain and autofluorescence to discern tissue morphology. In certain applications, we note a drawback in that the signals could be larger than the targets when the latter is of high copy numbers. One can try adjusting tyramide concentration or shortening reaction times to seconds instead of minutes to reduce the signal size, but the simultaneous detection of the same target of low copy number would be hampered. Such is the case in the investigation of viral genomes that integrated into host chromosomes at multiple loci at variable copy numbers *(15)*. However, for most applications, one is not likely to encounter this situation. Of course, a prerequisite is access to a good fluorescence microscope with appropriate filters, which is much more expensive than an upright light microscope.

As will be discussed, there are multiple methods for labeling the probes for *in situ* hybridization. Dinitrophenol (DNP), digoxigenin, fluorescein isothiocyanate (FITC), and biotin are all good building blocks for amplification. We personally like the biotin tag and have found it to give the least background. We have been able to detect successfully multiple targets, each based on the association of biotin–streptavidin *(7)*. In fact, one biotinylated probe can be successfully utilized to detect sequentially RNA and DNA of the same gene, as long as one removes the first-round probe/HRP for RNA before applying the probe/HRP for the next round probing for DNA *(7)*.

Two commercial sources presently exist for fluorescent tyramides. The original source was Perkin Elmer (Boston, MA), which bases their labeling schema on the CyDye fluorophores FITC and tetramethylrhodamine. A newer source for tyramides is Molecular Probes (Eugene, OR), which bases the conjugation on the Alexa dyes. The choice of fluorophores depends on the available filter systems for the fluorescence microscope. There are two types of commercially available tyramides, general reagents and the tyramide signal amplification (TSA)+ reagents. The general reagents are relatively time- and temperature-independent with respect to signal deposition and hence are easier to use. Although the TSA+ reagents are much more sensitive, they are more difficult for the novice to use, as it takes more effort to optimize conditions to achieve a low background (*see* **Note 3**). In addition, primary antibodies should be even more diluted when using the TSA+ reagents than with the regular reagents, an additional advantage for these TSA+ products over the general reagents.

The manual that follows is based on the first-generation tyramides. Protocols for probing DNA, RNA, and proteins are presented. Investigators are encouraged to fine-tune their protocols for their specific applications. When protocols are properly conducted, there should be little or no background. A recent advancement in detection techniques utilizing tyramides incorporated Quantum Dots to enhance sensitivity further *(20)*. The ability to conduct multiplex assays with tyramide-coupled fluorophores described here will have considerable applicability to investigating diseases, transgenic stem cells and animals, and potentially gene replacement therapy.

2. Materials

2.1. Slides

1. 2-Well chamber slides (cat. no. 12-525-16, Fisher) (*see* **Note 4**).
2. Superfrost Plus slides (Fisher).
3. Cover slips, sizes 22 × 22, 22 × 40, and 24 × 50 mm, all no. 1.5 in thickness.

2.2. Reagents and Solutions

1. 20X SSC: 3 M NaCl, 0.3 M sodium citrate, 1 L H_2O, pH. 7.0.
2. 10X phosphate-buffered saline (PBS) 80 g NaCl, 2 g KCl, 11.5 g $Na_2HPO_4 \cdot 7H_2O$.
3. CSK solution: 10 mM PIPES, pH 7.8, 100 mM NaCl, 0.3 M sucrose, 3 mM $MgCl_2$. Filter through a 0.45-µm cellular acetate filter, and store at 4°C. Before use, add Triton X-100 to 0.5% and vanadyl ribonucleoside complex (VRC; NE Biolabs) to 5%.

4. Acetylation buffer: 0.1 *M* triethanolamine/0.9 *M* NaCl, pH 7.0 (made fresh before use).
5. Xylene.
6. Ethanol (EtOH).
7. RNAse (100 µg/mL in 2X SSC).
8. 0.05 mg/mL 4,6-Diamino-2-phenylindole (DAPI) in either 4X SSC or 1X PBS.
9. Antifade: various commercial sources (*see* **Note 5**).
10. Triton X-100.
11. Tween-20 (polyethylene glycol sorbitan monolaurate).
12. H_2O_2.
13. Proteinase K (cat. no. S3020, DAKO).
14. Triethanolamine.
15. Formamide.
16. Dextran sulfate.
17. Acetic anhydride.
18. Cot-1 DNA (Invitrogen).
19. Salmon sperm DNA.
20. Yeast t-RNA (Sigma).
21. Antigen retrieval solution (cat. no. S1700, DAKO).
22. Goat serum for blocking for antigen detection (50% goat serum, 1X PBS).
23. Casein blocking solution for biotinylated nucleic acid probes (1% casein, included as Blocking Agent from any tyramide kit), in 4X SSC or 1X PBS). Heat at 65°C for 1 h, and then filter with a 0.45-µm cellulose acetate filter.
24. Tyramide kits
    Perkin Elmer: http://las.perkinelmer.com/catalog/Category.aspx?CategoryName =Tyramide+Signal+Amplifi cation+(TSA)+Systems
    Molecular Probes: www.probes.com/handbook/tables/0397.html.

2.2.1. Paraformaldehyde Preparation (see **Note 6**)

1. Place 4 g of paraformaldehyde (PFA) in 86 mL of water at room temperature.
2. Add 1.5 mL of 10 *N* NaOH, and mix until solution clears (about 5 min).
3. Add 10 mL of 10X PBS.
4. Add 1 mL of 1 *M* $MgCl_2$. The solution turns cloudy again.
5. Add concentrated HCl until pH is 7.0–7.5.
6. Add water to a final volume of 100 mL.
7. Filter through a 0.45-µm cellulose acetate filter if used for RNA *in situ* hybridization (*see* **Note 7**).

2.3. Probe Labeling (see **Note 8**)

1. DNP labeling kit: direct labeling kit (cat. no. NEL655001KT, Perkin Elmer).
2. Digoxigenin: nick translation kit (cat. no. 1745816, Roche); PCR labeling (cat. no. 1585550, Roche).
3. Biotin: nick translation kit (cat. no. 1745824, Roche).
4. FITC labeling: PCR labeling (cat. no. 1636154, Roche).
5. Poly(A) tailing: (cat. no. NEL624001KT, Perkin Elmer).

2.4. Primary and Secondary Antibodies

1. EPOS++ (enhanced polymer one-step staining): anti-mouse K4004; anti-rabbit K4008 (DAKO) (*see* **Note 1**).
2. Anti-digoxin (Dig)–POD (cat. nos. 1207733 or 1633716, Roche).
3. Anti-FITC–POD (cat. no. 1426346, Roche).
4. Antirabbit-POD (cat. no. AP307P, Chemicon).
5. Antimouse-POD (cat. no. AP308P, Chemicon).
6. Streptavidin-HRP (comes with tyramide kits).
7. Anti-DNP–HRP (comes with the labeling kit).

2.5. Chromosome Paints (see Note 9)

1. Vysis: http://www.vysis.com/.
2. Cambio: http://www.cambio.co.uk/.

2.6. Hybridization Solutions

1. Denaturation buffer: 70% formamide, 2X SSC, pH 7.0–7.5.
2. Washing buffers: 50% formamide, 2X SSC, pH 7.0–7.5 for probes other than oligonucleotides. For oliognucleotide probe, use 20% formamide, 2X SSC, pH 7.0–7.5.
3. Hybridization buffer for nucleic acid probes other than oligonucleotides: Hybrisol VII (cat. no. S1390-10, Qbiogene).
4. Oligonucleotide probe hybridization solution: 20% formamide, 2X SSC, 10% dextran sulfate, pH 7.0–7.5.
5. Rubber cement: Wal-Mart.

2.7. Large and Small Equipment

1. Fluorescence microscope with appropriate filters and camera.
2. Coplin jars.
3. Forceps.
4. Circulating water baths that hold ±0.1°C at 37°C and at 74°C.
5. A 37°C oven.
6. A 65°C oven.
7. A microcentrifuge.
8. Slide warmer.
9. Cytospin.
10. Cryostat or microtome for tissue sections.
11. Micropipetors (p10, p20, p100, p200, p1000).

3. Methods
3.1. Slide Preparation for Adherent Cells

1. Grow adherent cells on chamber slides.
2. Aspirate off the media, and wash once with 1X PBS before permeabilization or fixation with PFA (*see* **Note 10**).

3. If cells detach easily, remove the chamber, and place the slide directly into PFA or a permeabilization solution, depending on the application.

3.2. Slide Preparation for Nonadherent Cells

3.2.1. Cytospin

1. Cytospin 250,000 cells onto SuperFrost® Plus slides at 20 g for 2 min.
2. Store the slides at –80°C, and fix or permeabilize/fix depending on the application.

3.2.2. Dry Cell Spread (see **Note 11**)

1. Spin down 1–10×10^6 suspension cells or adherent cells that have been trypsinized at 1000g for 5 min.
2. Wash once with PBS to remove serum/medium residues.
3. Resuspend cells in PBS so that the final volume is less than 50 µL.
4. Pipet cells to the middle of a slide within a small area. Using the side of a p100 micropipet tip, spread suspension to the desired area of use, depending on the number of cells and the size of the cover slips (22×22 or 22×40 mm).
5. Dry the slides by placing them on the air flow grid at the outer edge of a laminar flow hood for approx 15 min. Drying time may vary, depending on the humidity of the room.
6. Store the slides at –80°C, and fix or permeabilize/fix the cells depending on the application. Except for nascent transcription detection (go to **Subheading 3.8.**), fix slides for 10 min at room temperature with PFA and then permeabilize with 1X PBS, 0.5% Triton X-100 for 10 min.

3.3. Slide Preparation for Buffered Formalin-Fixed and Paraffin-Embedded Tissue Sections

1. Adhere 4–10-µm sections to Superfrost Plus microscope slides.
2. Stabilize and age the sections for 24 h before deparaffinizing in an oven at 65°C for 1 h.
3. From the oven, place the slides directly into 100% xylene for three washes, 5 min each (*see* **Note 12**).
4. Wash and rehydrate for 5 min each (*see* **Note 13**) in 100% EtOH twice; 85, 70, and 50% EtOH once; and then 1X PBS three times for 2–5 min each. Optional procedure (**steps 5** and **6**) for *in situ* protocols (*see* **Note 14**):
5. Treat with proteinase K (*see* **Note 15**) for 6 min at room temperature.
6. Wash once in 1X PBS.
    Additional optional steps (**steps 7** and **8**): acetylation (*see* **Note 16**):
7. Make a fresh solution of 0.1 M triethanolamine/0.9 M NaCl, pH 7.0.
8. Within 5 min of placing the slides in the above solution, add 625 µL of acetic anhydride. Acetylate the slides for 10 min.
9. For *in situ* hybridization, dehydrate for 2 min each in 70% EtOH, 85% EtOH, and 100% EtOH. For protein detection, wash three times for 2 min each in PBS, 0.1% Tween-20.

3.4. Slide Preparation for Frozen Sections

1. Cut frozen tissue in a cryostat into 4–10-μm section onto Superfrost Plus slides.
2. Fix immediately in PFA for 10 min at room temperature.
3. Wash once in 1X PBS for 2 min.
4. Treat with proteinase K for exactly 6 min (*see* **Note 15**).
5. Wash in once 1X PBS for 1 min to stop enzymatic activity.
6. Either dehydrate or equilibrate in PBS depending on the application.

3.5. Probe Synthesis for T-FISH

3.5.1. DNA Probes Longer Than 500 bp for RNA or DNA Detection (see **Note 17**)

1. Nick-translate 1.3 μg of plasmid DNA in the volume recommended by the kit (which contains one tagged deoxyribonucleoside triphosphate) for 90 min at 15°C (kit, Roche). The product should be 50–500 bp in length, as confirmed by agarose gel electrophoresis of 5–10% of the reaction (*see* **Note 18**).
2. Precipitate with 0.3 *M* sodium acetate/70% EtOH in the presence of 40 μg of cot-1 DNA, 60 μg of salmon sperm DNA, and 20 μg of yeast t-RNA. Place on dry ice for 15 min. Spin in a microfuge for 15 min.
3. Wash the precipitate with 250 μL of 70% EtOH. Spin for 5 min in a microfuge.
4. Decant, and dry the precipitate.
5. Resuspend the probe mixture in 100 μL Hybridsol VII. Store at –20°C.
   This amount of probe mixture is sufficient for 10 hybridizations at 10 μL per slide.

3.5.2. DNA Probes Shorter Than 500 bp for DNA or RNA Detection

1. Perform PCR on the fragment of interest using standard conditions.
2. Nick-translate 1 μg of gel-purified PCR product. The kit contains one tagged deoxyribonucleoside triphosphate (for instance, biotinylated dUTP).
3. Go to **Subheading 3.5.1., steps 2–5** to complete the probe preparation.

3.5.3. Oligonucleotide Probes for RNA Detection

1. Oligonucleotides can be either internally labeled during their synthesis (Oligos Etc.) or poly(A)-tailed to incorporate labeled nucleotides such as biotin (kit, Perkin Elmer).
2. Precipitate the reaction product with sodium acetate/70% EtOH in the presence of 60 μg of salmon sperm DNA and 20 μg of yeast t-RNA. Place on dry ice for 15 min. Spin in a microfuge for 15 min.
3. Wash with 250 μL of 70% EtOH. Spin for 5 min in a microfuge. Decant and dry.
4. Resuspend the probe mixture in 100 μL of 20% formamide, 10% dextran sulfate, 2X SSC. Store at –20°C.
   This produces a 10X probe solution, sufficient for approx 100 hybridizations at 10 μL per slide (*see* **Note 19**).

3.6. T-FISH

3.6.1. Probe Hybridization to Detect RNA

1. For freshly prepared cells on slides, start from **step 2.** For fixed frozen tissues, start at **step 3,** and skip **step 4.** For deparaffinized tissue sections, start from **step 5.**
2. Wash slides once with 1X PBS.
3. Fix slides in PFA at room temperature for 10 min.
4. Permeabilize cells with 1X PBS, 0.5% Triton X-100 for 10 min at 4°C. **Steps 5–9** are optional.
5. Treat tissues with proteinase K for exactly 6 min (*see* **Note 15**).
6. Wash once in 1X PBS for 1 min to stop the proteinase.
7. Equilibrate in acetylation buffer for 2 min.
8. In a new solution, acetylate tissue for 10 min.
9. Wash slides three times for 2 min each in PBS.
10. Dehydrate for 2 min each in 70% EtOH, 85% EtOH, and 100% EtOH.
11. Ten minutes before the last wash, denature the probe at 74°C for 10 min, and then chill on ice.
12. Dry the slides on a slide warmer (45–50°C), and add 10 µL of probe. Seal with cover slips and rubber cement, and hybridize overnight (12–16 h) in a humid box (with wet paper towel) at 37°C.
13. Remove rubber cement and soak off cover slips in 2X SSC at room temperature.
14. Wash three times for 10 min each in 50% formamide, 2X SSC, pH 7.0, at 37°C.
15. Wash in 1X SSC for 30 min at 37°C.
16. Wash in 1X SSC/3% H_2O_2 for 15 min at room temperature.
17. Wash in 1X SSC for 15 min at room temperature. If background is high after tyramide deposition (*see* **Subheading 3.6.3.**), increase **step 16** and decrease **step 17.** Total time for **steps 16** and **17** is 30 min.
18. Go to **Subheadings 3.6.3.** or **3.6.4.**

3.6.1.1. Hybridization of Oligonucleotide Probes to Detect RNA

1. For freshly prepared cells on slides, start from **step 2.** For fixed frozen tissues, start at **step 3** and skip **step 4.** For deparaffinized tissue sections, start from **step 5.**
2. Wash slides once with 1X PBS.
3. Fix slides in PFA at room temperature for 10 min.
4. Permeabilize cells with 1X PBS, 0.5% Triton X-100 for 10 min at 4°C.
5. **Steps 6–10** are optional.
6. Treat tissues with proteinase K for exactly 6 min (*see* **Note 15**).
7. Wash once in 1X PBS for 1 min to stop the proteinase.
8. Equilibrate in acetylation buffer for 2 min.
9. In a new solution, acetylate tissue for 10 min.
10. Wash slides three times for 2 min each in PBS.
11. Dehydrate for 2 min each in 70% EtOH, 85% EtOH, and 100% EtOH.
12. Dry the slides on a slide warmer (45–50°C), and add 10 µL of probe. Seal with cover slips and rubber cement and hybridize overnight (12–16 h) in a humid box (with wet paper towel) at 37°C.

13. Remove rubber cement and soak off cover slips in 2X SSC at room temperature.
14. Wash once for 15 min in 2X SSC, 3% H_2O_2 at room temperature.
15. Wash three times for 15 min in 2X SSC at room temperature.
16. Wash once for 15 min in 0.5X SSC at room temperature.
17. Go to **Subheading 3.6.3.** or **3.6.4.**

3.6.2. Probe Hybridization to Detect DNA

1. For freshly prepared cells on slides, start from **step 2.** For deparaffinized tissue sections, start at **step 5.** For fixed frozen tissues, start from **step 3** and skip **step 4.**
2. Wash slides once with 1X PBS.
3. Fix slides with PFA at room temperature for 10 min.
4. Permeabilize cells with 1X PBS, 0.5% Triton X-100 for 10 min at 4°C.
5. Treat the tissues/cells with 100 µg/mL of RNase (Sigma) in 2X SSC for 1 h at 37°C.
   Steps 6–9 are optional.
6. Treat the tissues/cells with proteinase K for exactly 6 min (*see* **Note 15**).
7. Wash slides in 1X PBS to stop the proteinase.
8. Equilibrate slides in acetylation buffer for 2 min.
9. In a new solution, acetylate slides for 10 min.
10. Wash slides with 2X SSC three times for 2 min each.
11. Dehydrate for 2 min each in 70% EtOH, 85% EtOH, and 100% EtOH.
12. Dry slides on a slide warmer at 45–50°C.
13. Denature slides at 74°C in 70% formamide, 2X SSC, pH 7.0–7.5, for exactly 2 min (*see* **Note 20**).
14. Dehydrate for 2 min each in 70% EtOH (–20°C), 85% EtOH, and 100% EtOH at room temperature.
15. Ten minutes before the last wash, denature the probe at 74°C for 10 min and then place on ice.
16. Dry the slides on a slide warmer (45–50°C), and add 10 µL of probe mixture. Seal with cover slip and rubber cement and hybridize overnight (12–16 h) in a humid box at 37°C.
17. Remove rubber cement and soak off cover slip in 2X SSC at room temperature.
18. Wash three times for 10 min each in 50% formamide, 2X SSC, pH 7.0, at 37°C.
19. Wash in 1X SSC for 30 min at 37°C.
20. Wash in 1X SSC/3% H_2O_2 for 15 min at room temperature.
21. Wash in 1X SSC for 15 min at room temperature.
    If background is high after coupling to tyramide fluorophore (*see* **Subheading 3.7.3.**), increase **step 20** and decrease **step 21.** Total time for **steps 20** and **21** is 30 min.
22. Go to **Subheading 3.6.3.** or **3.6.4.**

3.6.3. Detection of a Biotinylated Nucleic Acid Probe With Tyramide Fluorophore

1. Equilibrate the slides in 4X SSC, 0.1% Triton X-100 three times for 5 min each.
2. Wash once for 5 min in 4X SSC.

3. Block in 4X SSC, 1% casein for 30 min at 37°C
4. Using the blocking buffer, make a 1:50–1:100 dilution (use 1:100 first) for strepta-vidin-HRP included with the kit. Apply 80 µL per slide (for a 24 × 50mm cover slip).
5. Invert the slide, and use capillary action to pick up the cover slip. This keeps air bubbles from forming.
6. Incubate the slides for 1 h at 37°C.
7. Remove cover slip by soaking in 4X SSC, 0.1% Triton X-100.
8. Wash the slides in 4X SSC, 0.1% Triton X-100 three times for 5 min each.
9. Wash once for 5 min in 4X SSC.
10. Dilute tyramide fluorophore 1:100 in the diluent provided by the manufacturer. Apply 80 µL per slide, cover slip, and incubate for 10 min at room temperature (*see* **Note 20**).
11. Soak off the cover slip in 4X SSC, 0.1% Triton X-100.
12. Wash the slides in 4X SSC, 0.1% Triton X-100 three times for 5 min each.
13. For additional rounds of probing, continue onto multiplexing, **Subheading 3.10.**
14. For mounting, go to **Subheading 3.11.**

3.6.4. Detection of a Nonbiotinylated Nucleic Acid Probe With Tyramide Fluorophore

1. Equilibrate the slides in 1X PBS, 0.1% Triton X-100 three times for 2 min each.
2. Wash once for 5 min in 1X PBS.
3. Dilute antibodies to the probe tag in 50% goat serum, 1X PBS. For instance, 1:500 anti-Dig–POD, or 1:100 anti-FITC–POD, and so on.
4. Cover slip, and incubate slides for 1–2 h at 37°C.
5. Remove cover slip by soaking in 1X PBS, 0.1% Triton X-100.
6. Wash three times in 1X PBS, 0.1% Triton X-100, 5 min each.
7. Wash once for 5 min in 1X PBS.
8. Dilute tyramide 1:100 in the provided diluent. Apply 80 µL per slide and incubate for 10 min.
9. Wash the slides in 1X PBS, 0.1% Triton X-100 three times for 5 min each.
10. For mounting, go to **Subheading 3.11.**
11. For a second round of probing, go to multiplexing, **Subheading 3.10.**

3.7. Detection of a Protein Antigen With Tyramide Fluorophore

1. Fix and permeabilize cells on slides (*see* **Subheadings 3.1.–3.4.** above).
2. Soak slides in 1X PBS, 3% H_2O_2 for 15 min at room temperature.
3. Wash slides three times for 2 min each in 1X PBS/0.1% Tween 20.
4. Perform optional antigen retrieval using the antigen retrieval reagent from DAKO (S1700) for 20 min at 95°C followed by a 20-min cool down at room temperature.
5. Equilibrate in 1X PBS, and then block in 50% goat serum in 1X PBS at 37°C for 30 min.
6. Apply primary antibody at optimal concentration (to be determined for each anti-body). Dilute the antibody in the same blocking buffer as **step 5.**

7. Cover slip, and incubate either overnight at 4°C or for 1–2 h at 37°C. Soak off the cover slip in 1X PBS.
8. Wash slides three times for 5 min each in 1X PBS/0.1% Tween-20.
9. If the signal is suboptimal after signal deposition using the antirabbit/mouse POD from Chemicon at 1:100, one can enhance the signals by using the EPOS+ antirabbit or mouse-HRP conjugate from DAKO.
10. Incubate for 1 h at 37°C.
11. Wash three times for 5 min each in 1X PBS/0.1% Tween-20.
12. Wash once with 1X PBS for 5 min at room temperature.
13. Detect the protein with a tyramide (80 µL of a 1:100 dilution of tyramide in the buffer that comes with the Perkin Elmer kit) for 10 min at room temperature.
14. Wash three times for 5 min each in 1X PBS/0.1% Tween-20.
15. If only a single protein is to be detected, go to mounting, **Subheading 3.11.**
    If more than one antigen is to be detected, continue to **step 16** or go to **Subheading 3.10.** for multiplexing.
16. Soak slides in 1X PBS, 3% H_2O_2 for 15 min at room temperature.
17. Wash once in 1X PBS for 2 min.
18. Repeat **steps 6–16.**

3.8. Detection of Nascent RNAs

1. Permeabilize the cells, and remove finished transcripts in the cytoplasm or nucleus by placing slides in the CSK buffer *(21)* at 4°C for 3–5 min (depending on the cell type).
2. Fix cells with PFA for 10 min at room temperature.
3. Follow the protocols for *in situ* hybridization to detect RNA: go to **Subheading 3.6.1., step 5.**

3.9. Detection of Interphase Chromosomal Territory With Chromosome Paints

This protocol can be used in conjunction with the detection of a protein, nascent RNA transcript of a gene, or DNA locus.

1. After detection of the first target just stated above, treat slides with RNAse, if they have not been treated already. Apply 100 µg/mL of RNase (Sigma) in 2X SSC for 1 h at 37°C.
2. Dehydrate for 2 min each in 70% EtOH, 85% EtOH, and 100% EtOH.
3. Dry slides on a slide warmer at 45–50°C.
4. Denature slides at 74°C in 70% formamide, 2X SSC, pH 7.0–7.5, for exactly 2 min (*see* **Note 20**).
5. Dehydrate for 2 min each in 70% EtOH (–20°C), 85% EtOH, and 100% EtOH at room temperature.
6. In general, 10 min before the last wash, denature the paint (follow probe-specific directions for denaturation and prehybridization, as some need reannealing time) at 74°C for 10 min, and then place on ice.

7. After an overnight incubation, follow paint-specific washing conditions as provided by the suppliers (*see* **Note 21**).
8. Go to mounting, **Subheading 3.11.** or multiplexing, **Subheading 3.10.**

3.10. General Multiplexing Strategies

Multiplexing with tyramide fluorophores involves using any of the above protocols with an intervening incubation with 3% H_2O_2 for 15 min. This step inactivates HRP and allows for one or more rounds of signal enhancement with HRP-tyramides in the subsequent protocols.

1. In general, RNA is detected first, as it is more labile than DNA or proteins. When detecting RNA and DNA, the RNAse treatment step in the DNA *in situ* hybridization protocol also necessitates detecting RNA target first.
2. Proteins are usually detected before DNA, but this order is not as critical and is application-specific.

3.11. Nuclear DNA Staining and Mounting of Slides

1. Stain cells with DAPI (0.05 mg/mL; Sigma) in either 1X PBS or 4X SSC for 5 min, depending on the solutions that have been used right before the staining (*see* **Note 22**). Other nuclear DNA stains are available from Molecular Probes at a variety of wavelengths. Mount them in an antifade (home-made or purchased; *see* **Note 23**).
2. Overlay with a no. 1 cover slip of appropriate size, and seal with clear nonchip fingernail polish.

3.12. Fluorescence Filters

The choice of filters for the fluorescence microscope can affect the ability to detect low-level signals present in fixed cells or tissue sections. Use green (rather than Texas Red [TR], Cy3, Cy5, and so on) fluorescence on targets that are small or of low abundance in a tissue sections or fixed cells. FITC-tyramide is best viewed through an FITC/TRITC (tertamethylrhodamine isothiocyanate) or FITC/TR filter (*see* **Note 2**). For instance, nascent RNA signals are small, but highly recognizable by eye both on tissue sections and in single cells. We also suggest the use of narrow-bandpass filters such as the Speicher filter set from Chroma Technology. Filter sets should be purchased with the detector(s) in mind: human eye, black and white or color film (rarely used now), color digital camera, high-sensitivity monochrome digital camera, and/or confocal microscope. Single-band filter sets are best for specificity; wide-bandpass filter sets are excellent for looking at several colors by eye or color digital camera; double and triple-pass tend to be dimmer but have good color separation.

3.13. Added Control for the Tyramides

In addition to the usual controls for *in situ* and immunohistochemistry to verify probe specificity, one more controls must be run to demonstrate signal specificity. Control for the H_2O_2 treatment of the cells and tissue sections to ensure that all endogenous peroxidase activity has been quenched. To do so, treat the slides with a tyramide after they have been washed with 4X SSC following 3% H_2O_2 treatment. No signal deposition should occur.

4. Notes

1. When coupled to a highly specific primary antibody, our experience with T-IF has been that protein antigen can be detected virtually free of background. Precautions must be taken to inactivate endogenous peroxidases. We also recommend using Chemicon POD or Dako EPOS+ antibodies.

2. The inherent drawback to this technique is that the signal deposition does not allow 3D localization, as one would want with confocal microscopy. On the other hand, confocal laser technology is very useful for separating multiple fluorophores.

3. If one needs added sensitivity, dilute the TSA+ product an extra 10-fold over the manufacturer's directions and apply it at 4°C overnight.

4. The two-well chamber slides are the best to work with and the easiest not to destroy. We find that the NUNC Nalgene Lab-Tek II CC chamber slide system (the slides with the blue ink around the wells) are problematic. They adhere firmly to the rubber cement used to seal the cover slips, making it difficult to unseal the slides during *in situ* hybridization procedures. They also require additional volumes for hybridization because of their construction.

5. To make your own antifade, we suggest: http://www.cshl.org/labs/spector/fluores-cence_medium.html. In our experience, the antifade from Vector quenches cyanine-3.

6. A fresh PFA preparation is essential each time for optimal results. The protocol only takes 5 min. We suggest making this solution right before you use it and then discarding it so that no one would reuse it. Aged PFA solution can result in higher backgrounds. PFA preparation and cell fixation are both conducted at room temperature.

7. Filters are expensive and, empirically, it only seems to be important for RNA detection.

8. You can assemble your own kits for these and other labels.

9. There are many washing protocols for chromosome paints, but we only get clean/clear nuclear domains utilizing formamide washing procedures. The salt wash protocols for paints recommended by the supplier for hybridization to metaphase chromosomes appear to be too stringent for interphase work, as they remove most of the signals.

10. Working with solutions directly in the chamber helps minimize volumes of solutions needed. Cells such as 293 are not adherent enough for this application. In such cases, slides are handled in staining dishes after removal of the chamber but not the gasket. The gasket is then removed after fixation.

11. The spread-and-dry technique allows for a greater number of cells to be analyzed in a larger area.

12. Sheeting action is very important for the removal of wax from paraffin-embedded tissue sections. Take slides in and out of the xylene solution five times at each washing step.

13. Two minutes is adequate for all the washes for most applications and for smaller numbers of slides.

14. Prior proteinase K treatment dramatically enhances our ability to detect RNA and DNA but could negatively impact on the ability to detect certain proteins, if antigenic epitopes are eliminated. However, the detection of certain other proteins may require protease treatment to expose the antigen. Regardless, once the tyramide fluorophores are deposited, no protease can be used, as it removes the tyramide fluorophore conjugates.

15. DAKO has optimized this kit for 6 min exactly. In our time-course trials, a deviation of as short as 15 s alters the results.

16. Prior acetylation of the tissue, which reduces nonspecific retention of probes owing to electrostatic charges, is rarely useful. Include this application when one has high background in the negative controls.

17. The use of the same DNA probe for RNA and DNA detection makes long-term planning easier. We generally have not had a need to use riboprobes in our applications. Also, riboprobes are prone to RNAase contamination and require significant precautions in preparation and application. In contrast, DNA probes are more stable.

18. For ease, we usually nick-translate the entire plasmid. We have never had a background issue owing to plasmid backbone, and, as a matter of fact, we generally use the vector as one of our negative controls. For viral work, we usually probe with a different or unrelated virus of approximately the same size as an additional control. With the exception of oligonucleotide probes, we add cot-1 DNA, yeast tRNA, and salmon sperm DNA to all nucleic acid probes to reduce background. The cocktail lasts for years at –20°C.

19. One should optimize the oligonucleotide probe dilution by performing hybridizations to find an optimal signal. Yeast tRNA must be added to the hybridization buffer used to dilute the oligonucleotide probe to a final concentration of 20 µg/100 µL.

20. These protocols hinge on exact timing at the critical steps. Strict adherence to the protocols generates highly reproducible results.

21. Sources for chromosome paints include Vysis and Cambio (UK). Alternatively, these protocols will work with BACS and YACS.

22. DAPI (0.05 mg/mL) can be made up in a Coplin jar, which is shielded from light with aluminum foil. The solution can be used repeatedly over time. As the solution gets weaker, either spike with more DAPI or increase the staining time.

23. http://www.cshl.org/labs/spector/fluorescence_medium.html). Do not use the antifade from Vector on cyanine-3, as it reacts and quenches it.

Acknowledgments

The research conducted in the authors' laboratory was supported by USPHS grants CA36200 and CA83679. B.A.V.T. was partially supported by the MSTP Program of UAB. We thank George McNamara and Phil Moen, Jr. for their assistance and supportive guidance with these techniques. We thank George McNamara for review of this manuscript. We also acknowledge the 1998 Cold Spring Harbor class on *in situ* hybridization of which B.A.V.T. was a participant.

References

1. Bobrow, M. N., Harris, T. D., Shaughnessy, K. J., and Litt, G. J. (1989) Catalyzed reporter deposition, a novel method of signal amplification. Application to immunoassays. *J. Immunol. Methods* **125,** 279–285.
2. Schriml, L. M., Padilla-Nash, H. M., Coleman A., et al. (1999) Tyramide signal amplification (TSA)-FISH applied to mapping PCR-labeled probes less than 1 kb in size. *Biotechniques* **806,** 494–497.
3. Bobrow, M. N. and Moen, P. T. (2000) Tyramide Signal Amplification (TSA) systems for the enhancement of ISH signals in cytogenetics, in *Current Protocols in Cytometry* (J. P. Robinson, et al., eds.), John Wiley & Sons, New York, pp. 8.9.1–8.9.16.
4. Speel, E. J., Ramaekers, F. C., and Hopman, A. H. (1997) Sensitive multicolor fluorescence in situ hybridization using catalyzed reporter deposition (CARD) amplification. *J. Histochem. Cytochem.* **45,** 1439–1446.
5. Raap, A. K., van de Corput, M. P., Vervenne, R. A., van Gijlswijk, R. P., Tanke, H. J., and Wiegant, J. (1995) Ultra-sensitive FISH using peroxidase-mediated deposition of biotin- or fluorochrome tyramides. *Hum. Mol. Genet.* **4,** 529–534.
6. Kerstens, H. M., Poddighe, P. J., and Hanselaar, A. G. (1994) Double-target in situ hybridization in brightfield microscopy. *J. Histochem. Cytochem.* **42,** 1071–1077.
7. Van Tine, B. A., Knops, J. F., Butler, A., Deloukas, P., Shaw, G. M., and King, P. H. (1998) Localization of HuC (ELAVL3) to chromosome 19p13.2 by fluorescence in situ hybridization utilizing a novel tyramide labeling technique. *Genomics* **53,** 296–299.
8. Buki, A., Walker, S. A., Stone, J. R., and Povlishock, J. T. (2000) Novel application of tyramide signal amplification (TSA): ultrastructural visualization of double-labeled immunofluorescent axonal profiles. *J. Histochem. Cytochem.* **48,** 153–162.
9. Teramoto, N., Szekely, L., Pokrovskaja, K., et al. (1998) Simultaneous detection of two independent antigens by double staining with two mouse monoclonal antibodies. *J. Virol. Methods* **73,** 89–97.
10. Wang, G., Achim, C. L., Hamilton, R. L., Wiley, C. A., and Soontornniyomkij, V. (1999) Tyramide signal amplification method in multiple-label immunofluorescence confocal microscopy. *Methods* **18,** 459–464.
11. van Gijlswijk, R. P., Zijlmans, H. J., Wiegant, J., et al. (1997) Fluorochrome-labeled tyramides: use in immunocytochemistry and fluorescence in situ hybridization. *J. Histochem. Cytochem.* **45,** 375–382.

12. Shindler, K. and Roth, K. (1996) Double immunofluorescent staining using two unconjugated primary antisera raised in the same species. *J. Histochem. Cytochem.* **44,** 1331–1335.

13. Macechko, P. T., Krueger, L., Hirsch, B., and Erlandsen, S. L. (1997) Comparison of immunologic amplification vs enzymatic deposition of fluorochrome-conjugated tyramide as detection systems for FISH. *J. Histochem. Cytochem.* **45,** 359–364.

14. Strappe, P. M., Huei Wang, T., McKenzie, C. A., Lowrie, S., Simmonds, P., and Bell, J. E. (1997) Enhancement of immunohistochemical detection of HIV-1 p24 antigen in brain by tyramide signal amplification. *J. Virol. Methods* **67,** 103–112.

15. Van Tine, B. A., Knops, J., Broker, T. R., Chow, L. T., and Moen, P. T., Jr. (2001) In situ analysis of the transcriptional activity of integrated viral DNA using tyramide-FISH, in *Developmental Biology: Evolving Scientific and Regulatory Perspectives on Cell Substrates for Vaccine Development,* vol. 106 (F. Brown, et al., eds.), Karger, Basel, Switzerland, pp. 381–385.

16. Jian, Y., Van Tine, B. A., Chien, W. M., Shaw, G. M., Broker, T. R., and Chow, L. T. (1999) Concordant induction of cyclin E and p21cip1 in differentiated keratinocytes by the human papillomavirus E7 protein inhibits cellular and viral DNA synthesis. *Cell Growth Differ.* **10,** 101–111.

17. Swindle, C. S., Zou, N., Van Tine, B. A., Shaw, G. M., Engler, J. A., and Chow, L. T. (1999) HPV-11 DNA replication compartments in a transient DNA replication system. *J. Virol.* **73,** 1001–1009.

18. Ma, T., Van Tine, B. A., Wei, Y., et al. (2000) Cell cycle-regulated phosphorylation of p220(NPAT) by cyclin E/Cdk2 in Cajal bodies promotes histone gene transcription. *Genes Dev.* **14,** 2298–2313.

19. Zaidi, A. U., Enomoto, H., Milbrandt, J., and Roth, K. A. (2000) Dual fluorescent in situ hybridization and immunohistochemical detection with tyramide signal amplification. *J. Histochem. Cytochem.* **48,** 1369–1375.

20. Ness, J. M., Akhtar, R. S., Latham, C. B., and Roth, K. A. (2003) Combined tyramide signal amplification and quantum dots for sensitive and photostable immunofluorescence detection. *J. Histochem. Cytochem* **51,** 981–987.

21. Fey, E. G., Krochmalnic, G., and Penman, S. (1986) The nonchromatin substructures of the nucleus: the ribonucleoprotein (RNP)-containing and RNP-depleted matrices analyzed by sequential fractionation and resinless section electron microscopy. *J. Cell Biol.* **102,** 1654–1665.

16

Identification and Characterization of Herpesviral Immediate-Early Genes

Yan Yuan

Summary

Immediate-early (IE) genes are the first class of viral genes expressed after primary infection or reactivation. As transcription of IE genes does not require prior viral protein synthesis, this class of genes is experimentally defined by their transcription following primary infection or reactivation in the presence of inhibitors of protein synthesis. This chapter describes an approach to identify IE genes in a novel herpesvirus genome. Transcription of IE genes is selectively induced with sodium butyrate in the presence of the protein synthesis inhibitor cycloheximide. The transcripts of the induced genes are identified by using a cDNA subtraction-based method of gene expression screening.

Key Words: Immediate-early transcripts; cDNA subtraction; KSHV; γ-herpesvirus; reactivation.

1. Introduction

Herpesviruses have two modes of replication, latent and lytic. In latently infected cells that contain a limited number of the herpesvirus genomes, there is no infectious virus. Only a limited number of viral genes are expressed during latency and are referred to as latent genes. When latency is disrupted, the virus can switch to a lytic cycle (also referred to as a productive cycle). In the lytic life cycle, herpesviruses express their lytic genes in a temporal and sequential order. A few viral genes are expressed independently of *de novo* protein synthesis and are classified as immediate-early (IE) genes. Delayed-early genes are expressed slightly later, and their expression is not affected by inhibition of viral DNA replication. Late genes are expressed after viral DNA synthesis, and their expression is, in general, blocked in the presence of inhibition of viral DNA *(1)*.

From: *Methods in Molecular Biology, vol. 292: DNA Viruses: Methods and Protocols*
Edited by: P. M. Lieberman © Humana Press Inc., Totowa, NJ

Reactivation of herpesviruses is initiated and controlled by a small number of regulatory proteins encoded by viral IE genes. In general, viral IE gene products function in one of the following three ways: (1) to activate a cascade of viral gene expression; (2) to alter host cell physiological status and make host cells ready for supporting lytic viral replication; and (3) to subvert host antiviral immune defense. In many cases, IE gene products are multifunctional proteins that may involve different biological processes. For example, the EBV major IE protein Zta is a transcriptional activator that activates the viral lytic gene expression cascade and initiates the switch of EBV from a latent to a lytic life cycle *(2)*. Zta is also found to influence cell cycle progression *(3)* and inhibit interferon-γ (IFN-γ) signaling *(4)*.

IE genes are the first class of viral genes expressed after primary infection or reactivation. As transcription of IE genes does not require prior viral protein synthesis, this class of genes is experimentally defined by their transcription following primary infection or reactivation in the presence of inhibitors of protein synthesis. Therefore, inhibitors of protein synthesis, such as cycloheximide and puromycin, have been employed as efficient tools in studies of IE genes of a variety of herpesviruses, such as herpes simplex type 1 *(5)*, human cytomegalovirus *(6)*, Kaposi's sarcoma-associated herpesvirus (KSHV) *(7)*, and murine γ-herpesvirus 68 *(8)*. This chapter describes a procedure that was successfully used for identification of IE genes of KSHV.

In general, identification of herpesvirus IE genes includes three steps. First, transcription of IE genes is selectively induced in either *de novo* viral infection or reactivation. This is usually achieved by infection of permissive cells by the viruses or induction of reactivation of virus in latently infected cells in the presence of protein synthesis inhibitors. Second, these specifically induced viral transcripts are isolated, as candidates for viral IE mRNA. In this chapter, a cDNA subtraction-based gene expression screen method is described. Recently a DNA microarray technique has been successfully employed for identifying IE transcriptions in cycloheximide-treated cells *(8)*. Finally, the kinetics of a putative IE gene transcription are analyzed by using Northern blotting or other techniques to confirm its IE transcription nature.

2. Materials

2.1. Cell Culture

1. BC-1 and BCBL-1 cells. BC-1 *(9)* cells were purchased from the American Type Culture Collection (ATCC) and grown in RPMI-1640 medium (Gibco-BRL, Gaithersburg, MD) supplemented with 15% fetal bovine serum (FBS; Gibco-BRL). BCBL-1 *(10)* cells were obtained from the NIH AIDS Research and Reference Reagent Program and grown in RPMI-640 medium supplemented with 10% FBS.

2. 0.6 *M* Sodium butyrate (Sigma, St. Louis, MO).
3. Cycloheximide.
4. Trypan blue solution (0.4%).
5. ^{35}S-methionine.
6. Methionine-free RPMI-1640 medium.

2.2. DNA, RNA, Southern, and Northern Analyses

1. Cosmid DNA: six KSHV cosmid clones, namely, GB11, GA29, GB22, GA1, GA-2, and GB1, prepared from BC-1 cells, were kindly provided by Drs. Ren Sun and George Miller (Yale University, New Haven, CT).
2. Nytran membranes (Schleicher & Schuell, Keene, NH).
3. [α-^{32}P]dCTP.
4. TRIzol reagent for total RNA isolation (Promega, Madison, WI).
5. PolyAtract mRNA isolation system (Promega).
6. Seakem GTG agarose (FMC).

2.3. Subtractive Hybridization

1. Photoactivatable biotin (Vector).
2. Mercury Vapor bulb (Vector).

2.4. Enzymes and Reagents

1. Restriction endonucleases and 10X buffers supplied (New England Biolabs).
2. T4 DNA ligase and 10X ligation buffer: 500 m*M* Tris-HCl, pH 7.8, 100 m*M* MgCl$_2$, 100 m*M* dithiothreitol (DTT), 10 m*M* ATP, 250 μg/mL bovine serum albumin (BSA; New England Biolabs).
3. Polynucleotide kinase and 10X kinase buffer: 700 m*M* Tris-HCl, pH 7.6, 100 m*M* MgCl$_2$, 50 m*M* DTT (New England Biolabs).
4. Polymerase chain reaction (PCR) system including 10X PCR buffer: 200 m*M* Tris-HCl, pH 8.4, 2.5 m*M* MgCl$_2$, 500 m*M* KCl, 1 mg/mL BSA (Perkin Elmer Cetus).

2.5. Solutions and Buffers

1. Diethyl pyrocarbonate (DEPC)-treated H$_2$O: double-distilled water mixed with 0.1% DEPC and stirred overnight. The DEPC is inactivated by autoclaving for 20 min.
2. 20X SSC: 3 *M* NaCl, 0.3 *M* trisodium citrate.
3. 1 *M* Tris-HCl, pH 7.5.
4. 1 *M* HEPES buffer, pH 7.5.
5. 1 *M* MOPS buffer, pH 7.0.
6. 1 *M* Sodium phosphate buffer, pH 6.8.
7. Annealing buffer: 10 m*M* Tris-HCl, pH 7.5, 10 m*M* Mg$_2$Cl, 30 m*M* NaCl.
8. 4X Hybridization buffer: 200 m*M* HEPES, pH 7.5, 0.8% sodium dodecyl sulfate (SDS), 8 m*M* EDTA, 2 *M* NaCl.
9. HE buffer: 10 m*M* HEPES pH 7.5, 1 m*M* EDTA.
10. Streptavidin solution: 2 μg/μL in 0.15 *M* NaCl, 10 m*M* HEPES, 1 m*M* EDTA, pH 7.4.

11. 10 mg/mL Ethidium bromide. **Caution:** Toxic and carcinogenic; handle with gloves.
12. Phenol saturated with 10 m*M* Tris-HCl, pH 7.5, 1 m*M* EDTA.
13. Chloroform/isoamyl alcohol (24:1).
14. Southern hybridization buffer: 6X SSC, 2X Denhardt's solution (1X Denhardt's solution is 0.1% BSA, 0.1% polyvinylpyrrolidone, 0.1% Ficoll), 0.5% SDS, and 50 µg/mL denatured sheared salmon sperm DNA.
15. Northern hybridization buffer (100 mL): 1 g BSA, 43 mL H_2O, 7 g SDS, 50 mL 1 *M* sodium phosphate, pH 6.8, 0.2 mL 0.5 *M* EDTA.
16. Northern mix A: 60 µL 10X MOPS, 135 µL H_2O.
17. RNA gel loading buffer: 0.5 mL glycerol, 2 mL 0.5 *M* EDTA, 5 mg xylene cyanol,5 mg bromphenol blue.
18. Formaldehyde/formamide mix: 105 µL 37% formaldehyde, 300 µL formamide. This mix should be freshly prepared just before use.
19. Northern washing solution (1 L): 5 g BSA, 50 g SDS, 40 mL 1 *M* sodium phosphate, pH 6.8, 2 mL 0.5 *M* EDTA.
20. Northern washing solution II (1 L): 10 g SDS, 40 mL 1 *M* sodium phosphate, pH 6.8, 2 mL 0.5 *M* EDTA.

2.6. Equipment

1. Automated thermal cycler.
2. Water bath.
3. Molecular Dynamics PhosphorImager.
4. Oligonucleotide synthesis facilities.
5. Tissue culture facility.
6. DNA sequencing facility

3. Methods

3.1. Induction of KSHV IE Gene Expression

Theoretically, herpesvirus IE genes can be selectively induced in the presence of inhibition of protein synthesis. Since protein synthesis inhibitors, such as cycloheximide, are in general toxic to host cells and cause apoptosis in these cells, a range of cycloheximide concentrations has to be determined in which protein synthesis is completely inhibited but cell viability rate remains high at least for 8–12 h.

1. BC-1 cells are grown in RPMI-1640 medium supplemented with 15% FBS.
2. Cycloheximide is added to BC-1 cells in T75 flasks to the following concentrations: 0, 10, 50, and 100 µg/mL. After 8 or 12 h, the cell viability of each treatment is assessed by trypan blue (0.4%) staining followed by microscopic examination (*see* **Note 1**).
3. Protein synthesis inhibition by cycloheximide is determined by [^{35}S]methionine incorporation.

   a. BC-1 cells are collected, washed with 1X phosphate-buffered saline (PBS), suspended in methionine-free growth medium (10^7/mL), and seeded in 24-well plates (5×10^6/well).

   b. Cells are incubated with cycloheximide at various concentration (0, 10, 50, and 100 µg/mL) for 30 min at 37°C prior to the addition of [^{35}S]methionine (50 µCi/well).

   c. Protein synthesis inhibition is determined after various periods by comparing the percentage of trichloroacetic acid-insoluble [35S]methionine from cycloheximide-treated cells with that from untreated cells (*see* **Note 2**).

4. Once the window of cycloheximide dosage that ensures a complete blockade of protein synthesis and maximal cell survival is determined, BC-1 cells are induced for KSHV reactivation in the presence of cycloheximide within the window.

   a. Exponentially growing BC-1 cells (10^6/mL) are induced by treatment with 3 m*M* sodium butyrate in the presence of 50 µg/mL cycloheximide for 4 h.

   b. To ensure a maximal blockade of protein synthesis upon virus reactivation, cells are induced with sodium butyrate 1–4 h after cycloheximide addition.

3.2. Mapping of the Regions That Are Transcriptionally Active in the IE Stage

1. Total RNAs are purified using TRIzol reagent (Gibco-BRL) from 10^8 BC-1 cells that have been treated with 3 m*M* sodium butyrate for 4 h in the presence of 50 µg/mL cycloheximide, as well as from the same number of cells that are not treated with sodium butyrate but incubated with cycloheximide for 8 h. Poly(A$^+$) RNAs are purified using the PolyAtract mRNA isolation system (Promega).

2. The double-stranded cDNAs are synthesized using the Universal riboclone cDNA synthesis system (Promega).

3. These cDNAs are labeled using a random priming method with [α-^{32}P]dCTP (Amersham, Arlington Heights, IL).

4. Six overlapping KSHV cosmid DNAs, which represent the whole KSHV genome, are digested with *Eco*RI, *Bam*HI, or both and separated in 0.8% agarose gels (**Fig. 1**). DNAs are transferred onto a Nytran membrane and probed with ^{32}P-labeled cDNA prepared from sodium butyrate-induced and -uninduced BC-1 cells in the presence of cycloheximide. As illustrated in **Fig. 1,** a number of bands that are only present in the blot hybridized with the cDNA pool from the induced cells (**Fig. 1D**) is observed. Examples include the 2.2-kb *Eco*RI/*Bam*HI fragment in cosmid 1 (22,409–24,637); the 2.6-kb *Eco*RI/*Bam*HI fragments in cosmids 2 and 3 (66,444–69,094); the 3.1-kb *Eco*RI/*Bam*HI fragment in cosmid 2 (47,518–50,637), and the 3.8-kb *Eco*RI/*Bam*HI fragment in cosmid 3 (69,095–72,888) (**Fig. 1D**). The transcripts originating from these DNA fragments may correspond to KSHV genes induced by sodium butyrate in the presence of cycloheximide and thus are candidates for the IE mRNAs of KSHV (*see* **Note 3**).

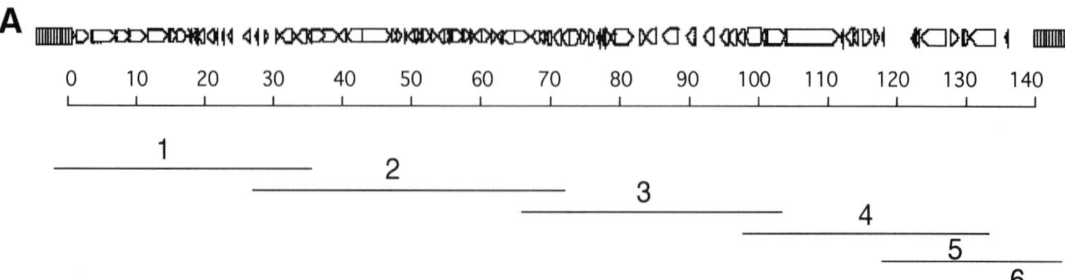
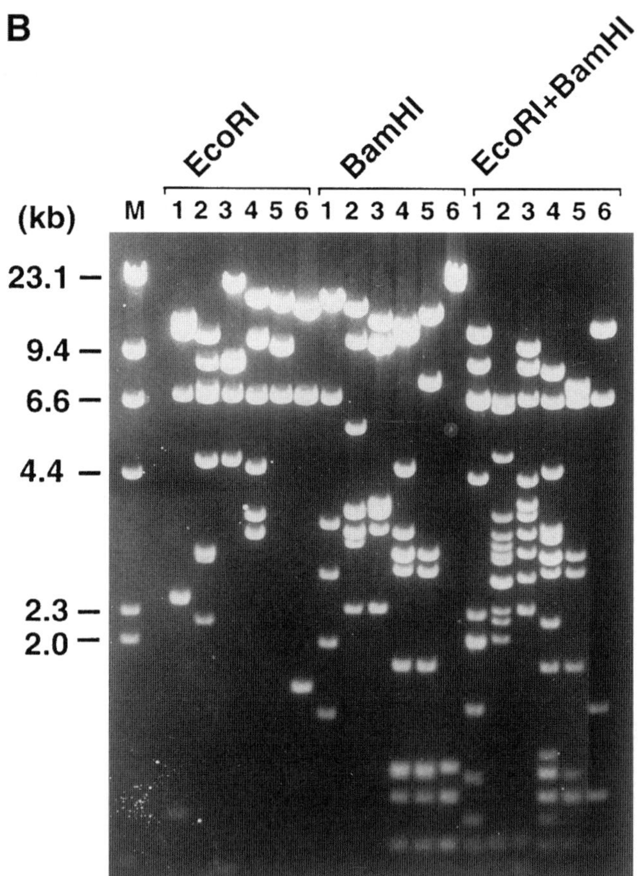

Fig. 1. Identification of regions that are actively transcribed in the KSHV genome before and after induction with sodium butyrate in the presence of cycloheximide. **(A)** Summary of overlapping cosmid clones of KSHV DNA. Cosmids GB11, GA29, GB22, GA1, GA-2, and GB1 are referred to as cosmids 1, 2, 3, 4, 5, and 6, respectively. **(B)** Ethidium bromide-stained agarose gel (0.8%) of KSHV cosmid DNAs digested with *Eco*RI, *Bam*HI, or both as indicated. **(C)** Southern hybridization of restricted KSHV cosmid DNAs with cDNA probe prepared from uninduced BC-1 cells. **(D)** Southern hybridization of the restricted DNAs with cDNA probe prepared from sodium butyrate induced BC-1 cells. (Reproduced with permission from **ref.** *7*.)

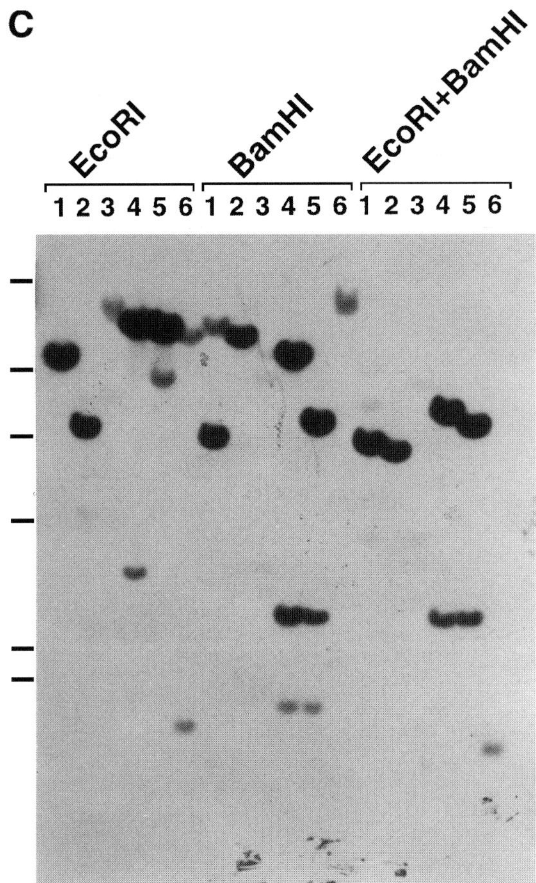

Fig. 1. (*Continued*)

3.3. Identification of Induced Transcripts by cDNA Subtractive Selection

To isolate the cDNAs of KSHV IE mRNAs from a complex cDNA pool that contains cellular mRNA sequences as well as numerous KSHV sequences that can be from latent viral gene expression or low-level lytic gene expression, we employed a gene expression screening method developed from cDNA subtractive hybridization (*7,11*). This method is designed to isolate mRNAs that differ in abundance between two RNA populations, so that the KSHV transcripts whose amounts were dramatically increased after induction in the presence of cycloheximide can be obtained. **Figure 2** depicts the strategy used in this study, and a detailed protocol is as follows.

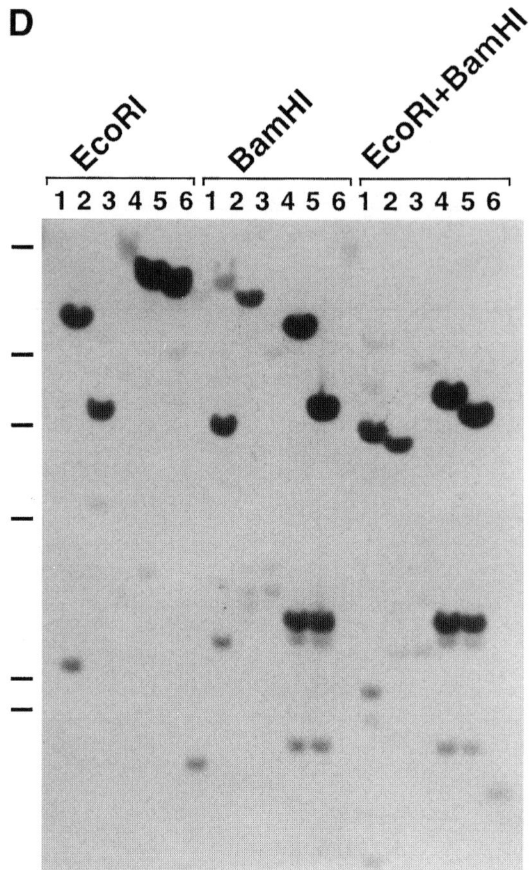

Fig. 1. (*Continued*)

3.3.1. Restriction Enzyme Digestion and Preparation of cDNA Fragments Suitable for PCR Amplification

1. 0.1–1.0 µg Double-stranded cDNA is digested with the restriction endonuclease *Alu*I overnight. The digested cDNA is extracted with an equal volume of phenol/chloroform/isoamyl alcohol and precipitated with ethanol.
2. For linker ligation, the oligonucleotides are:

   Ad-1A, 5′-GATCCCAGTCACGACGAATTCC-3′
   Ad-1B, 5′-GGAATTCGTCGTGACTGG-3′

3. 20 µg Oligo Ad-1A and 18 µg phosphorylated oligo Ad-1B in 100 µL of 1X annealing buffer are heated to 95°C for 5 min and then cooled down slowly to room temperature.
4. *Alu*I-digested cDNA is ligated with 0.5 µg (14 µL) of double-stranded phosphorylated oligodeoxyribonucleotide linker in 50 µL 1X T4 DNA ligase buffer, 3 µL (1 U/µL) T₄ ligase at 16°C overnight.

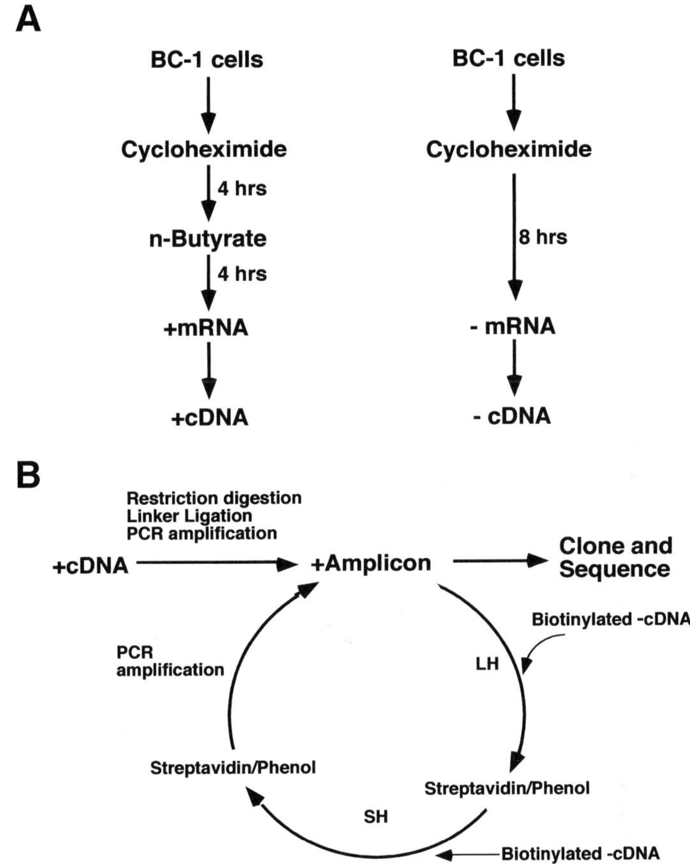

Fig. 2. Scheme for chemical induction of KSHV reactivation in latently infected BC-1 cells **(A)** and subtractive cDNA cloning of differentially expressed KSHV mRNAs in the immediate-early stage of reactivation **(B)**. A plus sign (+) refers to the mRNA or cDNA prepared from the BC-1 cells induced with sodium butyrate. A minus sign (−) refers to mRNA or cDNA from uninduced BC-1 cells. LH, long hybridization (20 h); SH, short hybridization (2 h). (Reproduced with permission from **ref. 7**.)

5. Ligating is extracted with phenol/cholorform/isoamyl alcohol. Then 10 µL of 6X DNA dye is added, and the sample is loaded on a 1.4% low-melting agarose gel. The gel is run for a short distance. The linker-ligated cDNA fragments in the size range of 0.2–2 kb are collected.

3.3.2. PCR Amplification

1. Twenty 100 µL PCRs are performed for each of the (−) and (+) cDNA samples using oligo Ad-1A as primer (94°C, 1 min; 50°C, 1 min; 72°C, 2 min; 30 cycles). Each PCR reaction includes 2 µL of melted agarose that contains linker-ligated cDNA fragments.

2. PCR reactions are extracted with an equal volume of phenol, followed by an additional extraction with phenol/choloroform.
3. DNA (100 µL, 5 µg) is subjected to a spin column (Sephedex G50) and ethanol-precipitated. The DNA will serve as tester (tracer) DNA.
4. The rest of the DNA (1900 µL, 100 µg) is digested with *Eco*RI after ethanol precipitation. DNAs are extracted with phenol/chloroform/isoamyl alcohol and precipitated with ethanol.

3.3.3. Biotinylation of Driver DNAs

1. Photoactivatable biotin, purchased as a dry powder, is dissolved in 500 µL of distilled water in the dark.
2. Driver DNA (100 µg) in 100 µL 0.1X TE is mixed with an equal volume of Photoprobe biotin solution in the dark.
3. The mixture is irradiated in an ice bath 10 cm below a sunlamp (mercury vapor bulb) for 20 min.
4. After photoirradiation, 200 µL of 0.1 M Tris-HCl, pH 9.5 is added, to deprotonate the unreactive Photoprobe biotin.
5. An equal volume (400 µL) of 2-butanol (water satured) is added, vortexed, and centrifuged for 1 min. The upper organic phase is discarded.
6. The 2-butanol extraction is repeated twice more.
7. The biotinylated DNA is precipitated by adding NaCl to 100 mM and a 2.5 vol of ethanol.

3.3.4. Subtraction Hybridization

3.3.4.1. Long Hybridization

1. Biotinylated driver DNA is dissolved in H_2O at concentration of 20 µg/10 µL.
2. Then 1–1.5 µg of tracer DNA (+) with 20 µg of biotinylated driver DNA (–) are mixed, ethanol-precipitated, and dissolved in 15 µL H_2O.
3. The DNA mixture is heated at 100°C for 3 min and then briefly centrifuged to collect condensed water.
4. Then 5 µL of 4X hybridization buffer is added. The DNA solution is overlaid with mineral oil and boiled again for 3 min.
5. The denatured DNA sample is incubated in a 68°C water bath for 20 h.
6. Eighty microliters of HE buffer, prewarmed at 55°C, is added to the hybridization mixture to bring the final NaCl concentration to 100 mM. The tube is then incubated at 55°C for 5 min.
7. The aqueous phase is transferred into a fresh tube. Then 20 µL of streptavidin solution is mixed with the hybridized DNA solution and incubated at room temperature for 20 min.
8. The solution is extracted with an equal volume of phenol/chloroform.
9. Repeat **steps 7** and **8** three more times, followed by an addition two phenol/chloroform extractions.

3.3.4.2. SHORT HYBRIDIZATION (*SEE* **NOTE 4**)

1. The subtracted tracer (+) cDNA is mixed with 15 μg of biotinylated driver (–) DNA. DNAs are precipitated with ethanol.
2. DNAs are denatured and hybridized as described above (**steps 3–5**), except that the hybridization is carried out for just 2 h.
3. Biotinylated DNAs are removed by using streptavidin–phenol extraction as before (**steps 6–9**), and enriched tracer DNA is ethanol-precipitated.
4. The DNA pellet is rinsed with 75% ethanol, dried, and resuspended in 20 μL of 0.1X TE buffer. Then 2 μL of subtracted cDNA is amplified in two 100-μL PCR reactions with the Ad-1A primer (94°C, 1 min; 50°C, 1 min; 72°C, 2 min; 25 cycles). The amplified cDNA is extracted with phenol/choloroform, followed by ethanol precipitation.
5. The resultant DNAs will be used for the next cycle of the subtraction experiment.

3.3.5. Cloning and Sequencing

1. After four or five cycles of subtraction, enriched DNA is digested with *Eco*RI.
2. *Eco*RI-digested DNA is ligated to *Eco*RI-digested and dephosphorylated pBluescript vector for transformation into competent *E. coli* DH5-α cells.
3. Colonies are screened by hybridization with the ^{32}P-labeled endriched cDNA pool.
4. The colonies that can be hybridized detectably are picked up for miniassay and sequencing analysis.

3.4. Northern Analyses of Putative IE mRNA

To characterize the mRNAs corresponding to the isolated cDNAs and determine whether these mRNAs are indeed IE transcripts of KSHV, Northern analyses are usually performed.

3.4.1. Preparation of mRNA Sample

1. Total RNAs are isolated from cells using TRIzol reagent (Gibco-BRL), and poly(A$^+$) mRNAs are purified using the PolyAtract mRNA isolation system (Promega).
2. In general, each RNA sample contains 20 μg total RNA or 1–2 μg poly(A$^+$) mRNA. RNA is dissolved in 6.5 μL mix A and 13.5 μL formaldehyde/formamide mix. RNA samples are denatured at 70°C for 10 min. Then 2 mL gel loading buffer are added.

3.4.2. Agarose Gel Electrophoresis and Northern Transfer

1. Cast 1% agarose/6% formaldehyde gel (for 300 mL), as follows: 3 g agarose (Seakem GTG from FMC), 221.4 mL DEPC-treated H$_2$O. Melt the agarose in a microwave oven. Add 30 mL 1 *M* MOPS pH 7.0 and 13 mL 37% formaldehyde.
2. Load mRNA in the gel. Each lane receives mRNA from 2×10^7 cells.
3. Run the gel at 80 V overnight in 1X running buffer (1.5 L): 150 mL 1 *M* MOPS, 243 mL 37% formaldehyde, and 1107 mL ddH$_2$O.

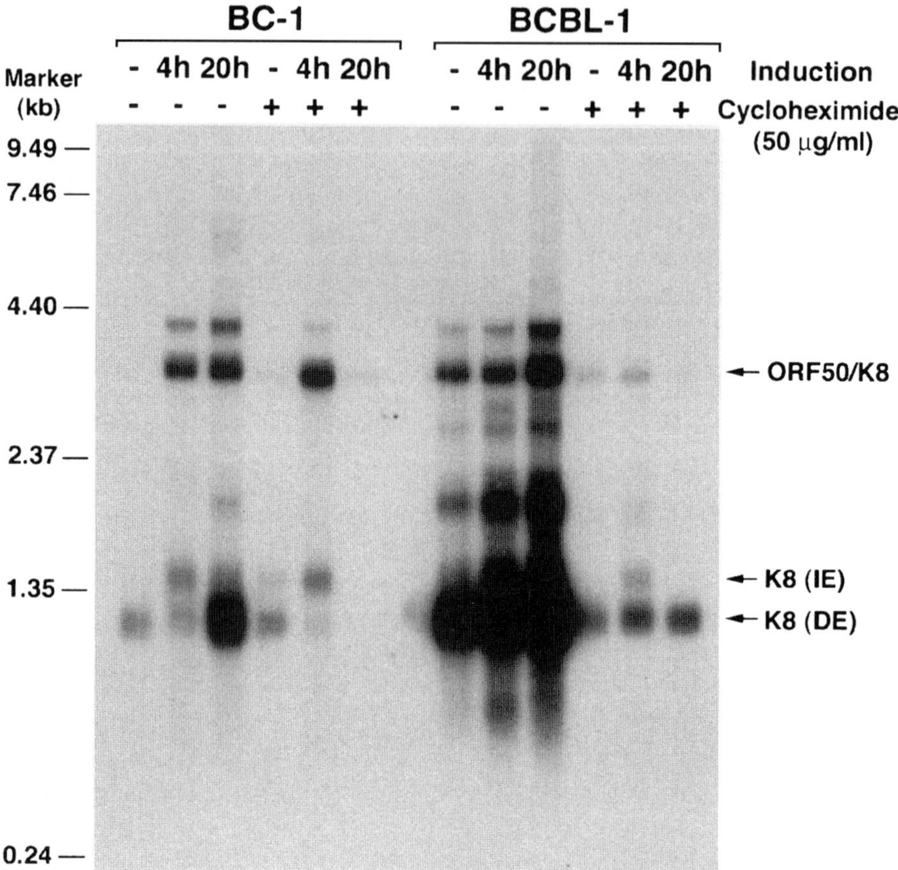

Fig. 3. Northern analysis of transcripts of KSHV ORF50K8. Poly(A$^+$) RNAs were isolated from BC-1 cells that had been treated with sodium butyrate for 4 or 20 h and from BCBL-1 cells that had been induced with TPA in the absence or presence of cycloheximide, as indicated above each lane. These RNAs were separated on a 1.0% agarose–formaldehyde gel and transferred onto nytran membranes. The membranes were probed with ^{32}P-labeled single-stranded DNA probes complementary to the K8 coding sequence. Molecular marker: 0.24–9.5 kb RNA ladder.

4. The gel is treated with the following solutions and times:
   a. ddH$_2$O for 10 min.
   b. 50 m*M* NaOH, 10 m*M* NaCl for 20 min.
   c. 0.1 *M* Tris-HCl, pH 7.5, for 20 min.
   d. ddH$_2$O for 10 min.
5. Transfer RNA onto Nytran membranes with 20X SSC via upward capillary transfer.

3.4.3. Northern Hybridization

1. Wet-blot in 0.5 *M* sodium phosphate buffer, pH 6.8.
2. Prehybidize the blot in Northern prehybridization mix at 68°C for 4 h.
3. Single-stranded DNA probes are prepared using asymmetric PCR with linearized plasmid templates and specific oligonucleotide primers, which are either a plasmid vector primer (i.e., KS or SK primers in pBluecript) or a primer specific to an insert sequence. The labeling reactions are performed in 15 µL of reaction solution (1X *Taq* polymerase buffer, 16.67 µ*M* each dATP, dGTP, dTTP, 1.67 µ*M* dCTP, 5 µL [α-^{32}P]dCTP (800 Ci/mmol, 10 µCi/µL [Amersham]), 100 ng DNA, 20 pmol primer and 2.5 U *Taq* polymerase). The PCR is initiated with a denaturing step of 2 min at 94°C, followed by 15 cycles of sequential steps of 1 min at 94°C, 1 min at 50°C, and 3 min at 74°C. Finally, the reaction is extended for 10 min at 74°C.
4. DNA probes are denatured at 100°C for 10 min and then added to hybridization tubes. Hybridization usually takes 16 h at 68°C.
5. Wash the filters three times (15 min each time) with Northern washing solution I at 68°C and two times with Northern washing solution II at 68°C.
6. A 0.24–9.5-kb RNA ladder (Gibco-BRL) is included in each agarose–formaldehyde gel and detected in Northern blots by hybridization with labeled λ DNA.
7. The filters are exposed to X-ray films or the phsophorimager screen. **Figure 3** is an image of a Northern blot hybridized with a KSHV ORF50-K8 probe.

4. Notes

1. We found that after an 8-h incubation in 50 and 100 µg/mL of cycloheximide, BC-1 cell viability rates were 84 and 74%, respectively, in comparison with the cell viability rate of 82% in the absence of cycloheximide. In contrast, most of the BCBL-1 cells died in this condition, indicating that BCBL-1 is much more sensitive to the toxicity of cycloheximide.
2. It was shown in our laboratory that cycloheximide at 50 and 100 µg/mL can block protein synthesis by over 95% as measured by [^{35}S]methionine incorporation.
3. Some induced transcripts may not be detected by Southern hybridization if they originated from the same DNA fragments (especially large restriction fragments) to which abundant latent transcripts also hybridized.
4. Long hybridization (20 h) is designed to suppress the highly complex, low abundant common DNA fragments. However, long hybridization does not efficiently reduce the abundant common mRNA sequences and can actually suppress some differentially expressed cDNAs that have a baseline level in the driver cDNA. Thus, short hybridization (2 h) is used to remove the abundant common mRNA sequences *(11)*.

References

1. Roizman, B. and Pellett, P. (2002) The family Herpesviridae: a brief introduction, in *Fields Virology,* 4th ed. (Knipe, D. M., Howley, P. M., Griffin, D. E., et al., eds.), Lippincott Williams & Wilkins, Philadelphia, pp. 2381–2397.

2. Miller, G. (1990) The switch between latency and replication of Epstein-Barr virus. *J. Infect. Dis.* **161,** 833–844.

3. Wu, F. Y., Chen, H., Wang, S. E., et al. (2003) CCAAT/enhancer binding protein alpha interacts with ZTA and mediates ZTA-induced p21CIP-1 accumulation and G1 cell cycle arrest during the Epstein-Barr virus lytic cycle. *J. Virol.* **77,** 1481–1500.

4. Morrison, T. E., Mauser, A., Wong, A., Ting, J. P., and Kenney, S. C. (2001) Inhibition of IFN-gamma signaling by an Epstein-Barr virus immediate-early protein. *Immunity* **15,** 787–799.

5. Honess, R. W. and Roizman, B. (1974) Regulation of herpesvirus macromolecular synthesis. I. Cascade regulation of the synthesis of three groups of viral proteins. *J. Virol.* **14,** 8–19.

6. Wathen, M. W., Thomsen, D. R., and Stinski, M. F. (1981) Temporal regulation of human cytomegalovirus transcription at immediate early and early times after infection. *J. Virol.* **38,** 446–459.

7. Zhu, F. X., Cusano, T., and Yuan, Y. (1999) Identification of the immediate-early transcripts of Kaposi's sarcoma-associated herpesvirus. *J. Virol.* **73,** 5556–5567.

8. Ebrahimi, B., Dutia, B. M., Roberts, K. L., et al. (2003) Transcriptome profile of murine gammaherpesvirus-68 lytic infection. *J. Gen. Virol.* **84,** 99–109.

9. Cesarman, E., Moore, P. S., Rao, P. H., Inghirami, G., Knowles, D. M., and Chang, Y. (1995) In vitro establishment and characterization of two acquired immunodeficiency syndrome-related lymphoma cell lines (BC-1 and BC-2) containing Kaposi's sarcoma-associated herpesvirus-like (KSHV) DNA sequences. *Blood* **86,** 2708–2714.

10. Renne, R., Zhong, W., Herndier, B., et al. (1996) Lytic growth of Kaposi's sarcoma-associated herpesvirus (human herpesvirus 8) in culture. *Nat. Med.* **2,** 342–346.

11. Wang, Z. and Brown, D. D. (1991) A gene expression screen. *Proc. Natl. Acad. Sci. USA* **88,** 11505–11509.

V

REPLICATION AND GENOME MAINTENANCE

17

Methods for Measuring the Replication and Segregation of Epstein-Barr Virus-Based Plasmids

Priya Kapoor and Lori Frappier

Summary

Plasmids containing the Epstein-Barr virus (EBV) latent origin of replication, *OriP*, are stably maintained in human cells expressing the viral EBNA-1 protein. This stable maintenance is owing to the ability of EBNA-1 to activate DNA replication from *OriP* and to facilitate the segregation of the plasmids during cell division. Methods for quantifying the replication and stable maintenance of EBV-based plasmids in human cells are presented here, as is a reconstituted segregation system in yeast that enables the segregation activity of EBNA1 to be measured independently from its replication activity.

Key Words: Epstein-Barr virus; DNA replication; segregation; EBNA-1; plasmid loss.

1. Introduction

The genomes of DNA viruses such as Epstein-Barr virus (EBV), papillomavirus, and Kaposi's sarcoma-associated herpesvirus (KSHV) are stably maintained in proliferating cells as double-stranded circular DNA plasmids or episomes. This stable maintenance occurs because of systems that ensure the replication of the viral genomes and their efficient segregation during cell division *(1–3)*. Techniques used to measure the stability of EBV-based plasmids in mammalian and yeast cells are the focus of this chapter. Specifically, the methods used to measure replication (transient assay) and maintenance (long-term assay, which serves as a measurement for both replication and segregation) of plasmids in mammalian cells and the method used to measure plasmid segregation in yeast are discussed.

The stable persistence of EBV episomes requires two viral components; the Epstein-Barr nuclear antigen-1 (EBNA-1) protein and the *cis*-acting *OriP* sequence, which consists of the dyad symmetry (DS) and family of repeats (FR)

From: *Methods in Molecular Biology, vol. 292: DNA Viruses: Methods and Protocols*
Edited by: P. M. Lieberman © Humana Press Inc., Totowa, NJ

elements separated by a 1-kb intervening sequence *(4–6)*. The interaction of EBNA-1 with the EBNA-1 binding sites in the DS element is required for the replication of the EBV episomes, which occurs once per cell cycle; the interaction of EBNA-1 with the EBNA-1 binding sites in the FR element is required for the segregation of the episomes *(5–9)*. EBNA-1 mediates the segregation of the EBV episomes by tethering the episomes (via FR) to host mitotic chromosomes *(10–15)*. Evidence suggests that EBNA-1 attaches to host human chromosomes in mitosis by binding to the human EBP-2 (hEBP-2) protein, which is a component of the mitotic chromosomes *(13,16,17)*. Recent studies have demonstrated that EBNA-1 can also segregate FR-containing plasmids in budding yeast, providing that hEBP-2 is present and that the function of hEBP-2 in this system is in tethering EBNA-1 to the yeast chromosomes in mitosis *(16,18)*. This reconstituted yeast system enables the segregation function of EBNA-1 to be studied without complications from its replication function, since replication occurs via a yeast origin of replication and yeast replication proteins.

2. Materials

2.1. Transient Replication and Long-Term Maintenance Assays in Mammalian Cells

1. pcDNA3 (Invitrogen).
2. EBV *OriP* element and EBNA-1 gene.
3. Plasmid cloning reagents.
   a. Restriction enzymes.
   b. Polymerases.
   c. Ligases.
   d. Kinases.
   e. Oligonucleotide primers
   f. Agarose gel equipment and reagents
   g. DH5α.
   h. LB media.
   i. Ampicillin.
   j. LB/AMP plates.
   k. Plasmid isolation reagents and kits.
4. Human C33A cells.
5. Tissue culture equipment and reagents. All reagents are stored at 4°C.
   a. Dulbecco's minimal essential media (DMEM) containing 100 U penicillin and 0.1 mg streptomycin per mL.
   b. Fetal bovine serum (FBS).
   c. Phosphate-buffered saline (PBS).
   d. 10X Trypsin-EDTA solution.
   e. G418 (50 mg/mL stock).

6. Hemocytometer.
7. Calcium phosphate transfection reagents: 2X HBS (0.28 *M* NaCl, 0.05 *M* HEPES, 1.5 m*M* Na$_2$HPO$_4$, pH 6.95), 2.5 *M* CaCl$_2$. Store at 4°C.
8. Hirt's solution: 2X Hirt's solution (1.2% sodium dodecyl sulfate [SDS], 20 m*M* EDTA, 20 m*M* Tris-HCl, pH 7.5), 5 *M* NaCl, phenol/chloroform, chloroform, ethanol, 0.3 *M* NaOAc, pH 7.5, 10 m*M* Tris-HCl, pH 8.5.
9. Southern blotting apparatus and reagents.
   a. Nylon membrane.
   b. 0.25 *N* HCl.
   c. 0.4 *N* NaOH/0.6 *M* NaCl.
   d. Blocking buffer: 50% formamide, 5X Denhardt's solution 0.5% SDS, 6X SSC, 100 µg/mL herring sperm DNA, Make fresh.
   e. 20X SSC: 3 *M* NaCl, 0.15 *M* sodium citrate, pH 7.0.
   f. 2X SSC/1% SDS.
   g. 0.2X SSC/0.1% SDS.
   h. Reagents for radiolabeling by random priming.
   i. Film.
   j. Phosphoimager screen and apparatus.
10. SDS-PAGE apparatus and reagents.

2.2. Yeast Plasmid Loss Assays

1. YRp7, p416MET25, and p425PGK plasmids.
2. FR element, EBNA-1, and hEBP2 cDNAs.
3. Plasmid cloning reagents (*see* **Subheading 2.1., item 3**).
4. KY320 yeast strain.
5. Yeast media and plates: YPD and synthetic complete media/plates containing lysine, adenine, and histidine but lacking uracil, tryptophan, and leucine (SC-Ura, Trp, Leu) or containing lysine, adenine, histidine, and tryptophan but lacking uracil, and leucine (SC-Ura, Leu).
6. Yeast transformation reagents.
   a. 0.1 *M* LiAc.
   b. 1 *M* LiAc.
   c. 50% PEG 3500.
   d. 2 mg/mL single-strand (ss) DNA.
7. Cell counting apparatus (hemocytometer or spectrometer).
8. SDS-polyacrylamide gel electrophoresis (PAGE) apparatus and reagents.

3. Methods

3.1. Transient Replication and Plasmid Maintenance Assays in Mammalian Cells

The replication and maintenance of plasmids containing *OriP* (negative control) or containing *OriP* and expressing the EBNA-1 gene (positive control) is measured in human cells as described below in the transient replication and plasmid maintenance assays. The plasmid maintenance assay measures a combination of DNA replication and segregation efficiency and reflects segregation ability when DNA replication occurs normally. The main differences in the transient replication and plasmid maintenance assays are in the amount of DNA transfected, the length of time that the plasmids are left in the cells, and the fact that the plasmid maintenance assay involves selecting for the plasmid. The steps describe (1) construction of the plasmids used in the assays, (2) transfection of a human cell line with the plasmids, (3) growth of the transfected cells, (4) extraction of the plasmids from the cells, and (5) Southern blot analysis of the extracted plasmids to determine whether they replicate and/or can be maintained in the cell line used.

3.1.1. Plasmids

Subheadings 3.1.1.1.–3.1.1.3. describe the backbone plasmid used to construct the plasmids used in the transient replication and maintenance assays, the DNA fragments containing *OriP* or the EBNA-1 gene, and the strategy used to clone these DNA fragments into the backbone plasmids.

3.1.1.1. pCDNA3

pcDNA3 (**Fig. 1**) is a 5.4-kb vector that contains a bacterial origin of replication (colE1) and the ampicillin resistance marker, allowing its propagation in bacteria; it also contains the SV40 origin of replication, allowing its propagation in mammalian cells expressing the SV40 T antigen. The plasmid expresses the neomycin resistance gene, which allows selection of cells containing the plasmid by growth in the antibiotic G418 (*see* **Note 1**). It also contains the mammalian cytomegalovirus (CMV) promoter, with a downstream multicloning site, allowing cloning/expression of a gene of interest in mammalian cells.

3.1.1.2. ORIP AND EBNA 1

The *OriP* element was obtained from the plasmid pGemoriP *(19)* as a 2.0-kb *Rsa*I–*Bam*HI DNA fragment. The EBNA-1 gene used in the assays lacks most of the Gly-Ala repeat region (amino acids 101–324), which is not required for the replication and segregation activities of EBNA-1. The fragment containing the EBNA-1 gene was obtained by PCR amplification of the EBNA-1 gene in plasmid p205 *(4)* using an N-terminal primer containing an *Nde*I site and a C-terminal primer containing a *Bam*HI site. The amplified fragment was digested with *Nde*I, the *Nde*I site was filled in with Klenow, and the fragment was then digested with *Bam*HI.

3.1.1.3. CLONING

To generate plasmids used to measure replication and maintenance activities, standard cloning methods were used. Plasmid containing the *OriP* element

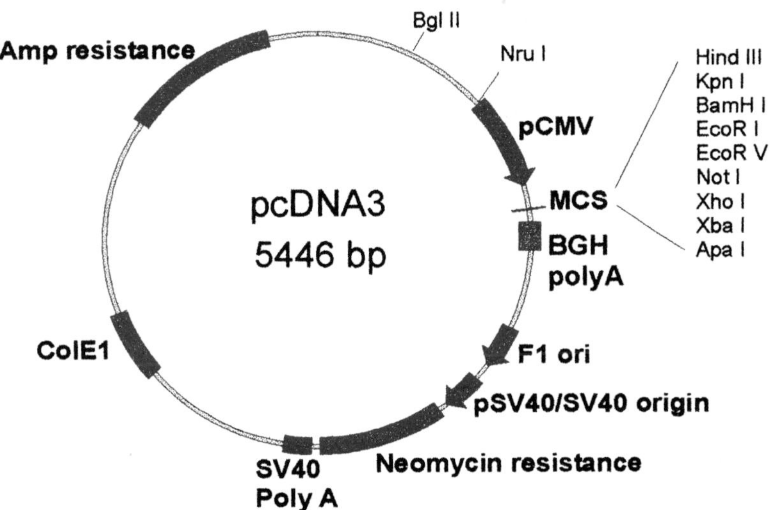

Fig. 1. pcDNA3, the backbone vector for the plasmids used in the transient replication and long-term maintenance assays. The EBV *OriP* sequence was cloned into the *Bgl*II and *Nru*I sites of pcDNA3 to generate pc3oriP. The EBNA-1 gene was cloned into the *Hind*III and *Bam*HI sites located in the multiple cloning site (MCS) of pc3oriP to give pc3oriPE.

(pc3oriP) *(20)* was constructed by ligation of the *Bam*HI–*Rsa*I *OriP* fragment into the *Bgl*II and *Nru*I sites of pcDNA3. To construct pc3oriPE *(20)*, which contains both *OriP* and the EBNA-1 gene, the *Nde*I (filled-in)–*Bam*HI fragment consisting of the EBNA gene was inserted between the *Hind*III (blunted by mung bean nuclease digestion) and *Bam*HI sites of pc3oriP (*see* **Note 2**). Ligation products for each construct were transformed into *E. coli* DH5α, and the resulting cells were plated on LB/ampicillin plates. Ampicillin-resistant colonies were grown in LB/ampicillin medium, and plasmids were isolated by standard miniplasmid purification methods and screened for positive clones through restriction enzyme digestion and DNA sequencing. Plasmids representing the right clones were retransformed into DH5α and purified by a large-scale plasmid purification method that would yield clean transfectable DNA (such as CsCl gradient or Qiagen maxiplasmid purification).

3.1.2. Introduction of Plasmids Into Human Cells

Described below is the introduction of the EBV-based plasmids used in the transient replication and maintenance assays into C33A cells by calcium phosphate transfection (*see* **Note 3**).

3.1.2.1. CELL LINE

C33A is a human cervical carcinoma cell line that grows as adherent cells.

1. Grow C33A cells in culture plates in DMEM supplemented with 10% FBS, until cells are 80% confluent.
2. To split cells, remove medium, wash cells with PBS, add trypsin to the cells, and incubate at 37°C until cells lift off the plates.
3. Replate 1/6 of the cells in tissue culture plates containing fresh medium.
4. Repeat when cells are 80% confluent (usually every 2 d).

3.1.2.2. TRANSFECTION OF C33A

1. At 24 h prior to transfection, plate 1×10^6 cells in a 10-cm tissue culture dish with 10 mL of DMEM/FBS. It is a good idea to plate enough dishes such that the assays can be conducted in duplicate for each plasmid sample to ensure reproducibility of the results.
2. For the transient replication assay, mix 10 μg of pc3oriP or pc3oriPE with 50 μL 2.5 M CaCl$_2$ and 450 μL dH$_2$O. For the plasmid maintenance assay, do the same using 1 μg of the plasmids.
3. Add the 500-μL mixture from **step 2** to 500 μL of 2X HBS, and mix by pipeting up and down seven times.
4. Incubate the resulting mixture at room temperature for 25 min.
5. Add the 1-mL mixture dropwise to the cells from **step 1,** and incubate for 16 h in a 37°C/5% CO$_2$ incubator. The precipitate in the cell medium should become visible within 10 min of incubation.
6. After the 16-h incubation, remove medium containing the precipitate, wash the cells twice with PBS, trypsinize, and transfer to 15-cm plates containing 20 mL fresh medium. Add 400 mg/mL G418 to the plates that will be used for the plasmid maintenance assay. Place all plates back into the incubator.

3.1.3. Growth of the Transfected Cells

1. Replication and maintenance assays.
   a. For the transient replication assay, grow C33A cells transfected with pc3oriP or pc3oriPE for 72 h post transfection.
   b. For the plasmid maintenance assay, grow C33A cells transfected with pc3oriP or pc3oriPE for 2 wk post transfection. Change media/G418 selection every 3 d. Cells lacking the plasmids will start to die off within 4–5 d of growth in G418. If G418-resistant cells within colonies start to crowd, then harvest cells by trypsinization, and replate all the cells to spread them out.
2. At the end of the growth period, harvest and count cells (use a hemocytometer).
   a. For the transient replication assay, remove as many cells as possible, ensuring that equal cell numbers are obtained from pc3oriP- and pc3oriPE-transfected plates.
   b. For the plasmid maintenance assay, remove 5×10^6 cells from pc3oriP- and pc3oriPE-transfected plates (*see* **Note 4**).

Pellet cells by centrifugation at 350 *g*, and wash with PBS.

3.1.4. Extraction and Digestion of Plasmids

Only some of the cells harvested for the transient replication assay will contain the plasmid of interest, since these cells were not placed under G418 selection. The number of cells containing the plasmid will be dependent on transfection efficiency and on whether the plasmid can replicate in the cells. Cells harvested for the plasmid maintenance assay will contain the plasmid of interest, since these cells were placed under G418 selection. However, the plasmid can either be maintained as an episome or can be integrated into the C33A genome. The plasmid will integrate in cells where it cannot either replicate or segregate efficiently. A plasmid that can replicate and segregate in cells can be recovered from the cells in both the replication and maintenance assays. To determine whether pc3oriP and pc3oriPE can transiently replicate and be maintained in C33A, these plasmids are extracted from the harvested cells by Hirt's extraction method *(21)* as described below.

1. Resuspend cells harvested in **Subheading 3.1.3.** in 350 µL of PBS.
2. Add 350 µL of 2X Hirt's solution, and incubate at room temperature for 10 min.
3. Add 140 µL 5 *M* NaCl to the 700-µL lysate, mix by gently inverting the tube 10 times, and incubate overnight at 4°C.
4. Centrifuge the lysate for 30 min at top speed in a microcentrifuge at 4°C.
5. Keep the supernatant and extract with an equal volume of phenol/chloroform and then with chloroform; in each case, keep the upper layer.
6. Add absolute ethanol (twice the volume of the upper layer), mix, and incubate overnight at –20°C to precipitate DNA.
7. Pellet DNA for 30 min at top speed in a microcentrifuge at 4°C.
8. Resuspend the pellet in 300 µL 0.3 *M* NaOAc, pH 7.5, and then add 600 µL of absolute ethanol and incubate at –20°C overnight to reprecipitate the DNA.
9. Pellet DNA as in **step 7,** and dry the pellet at 37°C.
10. To linearize the plasmid DNA with a unique enzyme (for Southern analysis), resuspend in the pellet in: 17 µL 10 m*M* Tris-HCl, pH 8.5, 2 µL *Xho*I restriction digest buffer (NEB buffer 2), 1 µL *Xho*I (20 U; NEB). Incubate at 37°C for 4 h.
11. To the digestion mixture, add 1 µL *Dpn*I (20 U; NEB) to remove nonreplicated pc3oriP or pc3oriPE. Incubate further at 37°C for 3 h.
12. Add DNA loading buffer, and load onto 0.8% TAE agarose gel. Also include 100 pg of linearized pc3oriP and/or pc3oriPE on the gel as marker. Run the gel overnight at 30 V.

3.1.5. Determining Replication and Maintenance of Plasmids by Southern Blotting

1. Place the gel in 0.25 *N* HCl depurination buffer, and gently shake on a rocker for 10 min; rinse with dH₂O and then incubate in 0.4 *N* NaOH/0.6 *M* NaCl denaturation buffer for 30 min with shaking.

2. Transfer the DNA samples in the gel to a nylon membrane (GeneScreen Plus, NEN Life Sciences) using standard Southern blotting techniques (*see* **Note 5**) *(22)*.

3. Rinse the membrane in 2X SSC, and incubate with Denhardt's blocking buffer for 2 h at 42°C in a hybridization oven.

4. During the 2-h incubation, radiolabel 300 ng of linearized pc3oriP with [α-^{32}P]dCTP using standard techniques for random oligonucleotide primed synthesis *(22)*. Determine the counts per minute (cpm)/mL of the probe.

5. After the 2-h incubation, discard the blocking buffer, and incubate the nylon blot in 10 mL of fresh blocking buffer containing 5×10^6 cpm of the probe that has been denatured (by incubation at 100°C for 5 min followed by incubation on ice for 5 min). Incubate overnight at 42°C in the hybridization oven.

6. Wash the probed nylon: (1) twice with 100mL of 2X SSC/1% SDS for 10 min at room temperature; (2) twice with 100 mL of 2X SSC/1% SDS for 30 min at 65°C; (3) once with 100 mL of 0.2X SSC/0.1% SDS for 10 min at room temperature.

7. Wrap the blot in Saran wrap, and expose on film or a phosphoimager screen for qualitative or quantitative (quantify by ImageQuant software from Molecular Dynamics) measurements, respectively, of the replication and maintenance capabilities of the plasmids. In the experiment described here, the negative control pc3oriP should not be recovered in the transient replication and plasmid maintenance assays, whereas the positive control pc3oriPE should be recovered in both the assays, giving a signal comparable to the 100-pg marker if the transfection was efficient (**Fig. 2**).

3.2. Plasmid Loss Assays in Yeast

A protocol to determine the stability of a plasmid in yeast is described below, using the EBV-based plasmid segregation system as an example. The protocol steps outlined involve (1) the description of the plasmids used; (2) the introduction of the plasmids into yeast and the growth of the resulting yeast to allow loss, if any, of the plasmid whose stability is in question; and (3) the plating of the yeast to determine plasmid stability both qualitatively and quantitatively.

3.2.1. Plasmids

The segregation test plasmid, which is the plasmid whose stability is measured in the yeast EBV-based plasmid segregation system, and the plasmids used to express EBNA-1 and hEBP2, which are proteins required for efficient plasmid segregation in the system, are described below in **Subheadings 3.2.1.1.–3.2.1.3.** The description includes (1) an explanation of the backbone vectors; (2) a description of the EBV segregation element FR and the EBNA-1 and hEBP2 cDNAs; and (3) the strategy used to clone the FR element, hEBP2, and EBNA1 into the backbone vectors.

3.2.1.1. YRp7, p416MET25, and p425PGK Backbone Vectors

To determine whether *cis*- and *trans*-acting factors of interest can confer stability upon a plasmid, it is important that the backbone plasmid selected for sta-

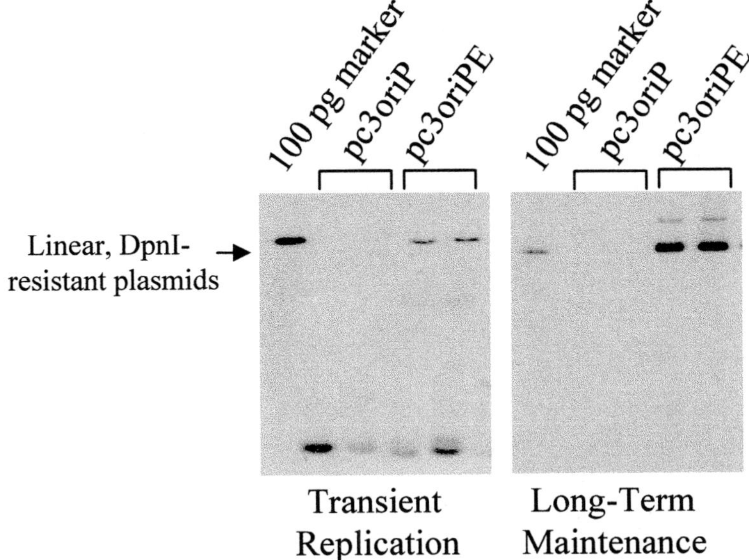

Fig. 2. Plasmid replication and maintenance by the *OriP*/EBNA1 system. C33A cells were transfected with a plasmid containing *OriP* (pc3oriP) or a plasmid containing *OriP* and expressing EBNA-1 (pc3oriPE) and grown for 3 d without selection (transient replication assay) or for 2 wk with selection (long-term maintenance assay) for the plasmid. At the end of the growth period, cells were harvested and plasmids were extracted, *Xho*I/*Dpn*I-digested, and run on agarose gel, along with a 100-pg marker. Southern blot analysis was conducted on the gel to determine the ability of the plasmids to replicate and/or be maintained in the cells.

bility measurements be unstable to begin with. The YRp series of yeast vectors *(23)* are well suited for this purpose. These vectors have a bacterial pBR322 plasmid backbone, which allows their propagation in *E. coli,* and contain sequences for the yeast origin of replication (termed the *ARS* element) and a yeast-selectable marker. Since these vectors have the yeast replication origin, they can undergo autonomous replication and transform yeast at high frequency. However, they lack elements that are required for segregation in yeast, such as the *CEN* element. As a result, YRp plasmids are very unstable and are completely lost from cells within a few generations in nonselective medium. In contrast, plasmids containing both the *CEN* and *ARS* sequences, such as pRS314 (**Fig. 3A**) *(24)*, are very stable and can serve as a positive control in the plasmid loss assays.

The YRp7 vector (**Fig. 3B**) *(23)*, which can serve as a negative control and is the backbone for the segregation test plasmid in the EBV segregation system described here, contains an *Eco*RI fragment consisting of the *ARS1* replication

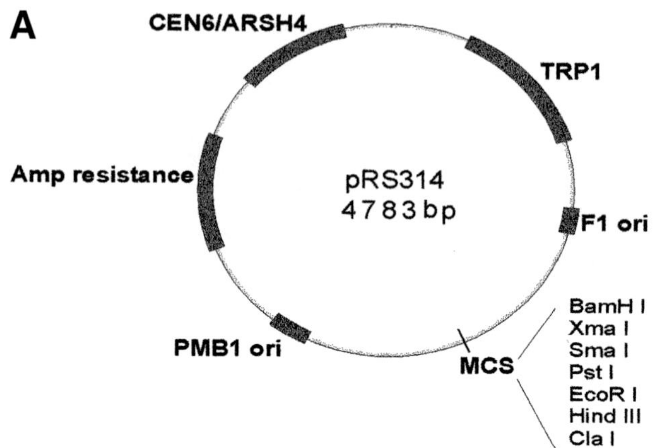
A
CEN6/ARSH4
TRP1
Amp resistance
pRS314
4783 bp
F1 ori
PMB1 ori
MCS
BamH I
Xma I
Sma I
Pst I
EcoR I
Hind III
Cla I

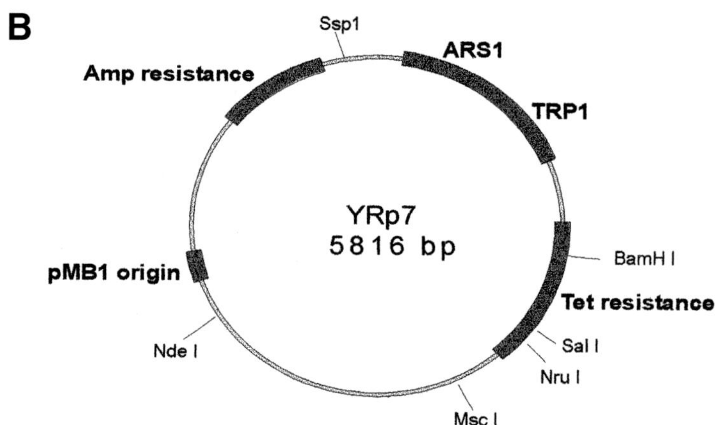
B
Ssp1
ARS1
Amp resistance
TRP1
YRp7
5816 bp
pMB1 origin
BamH I
Tet resistance
Nde I
Sal I
Nru I
Msc I

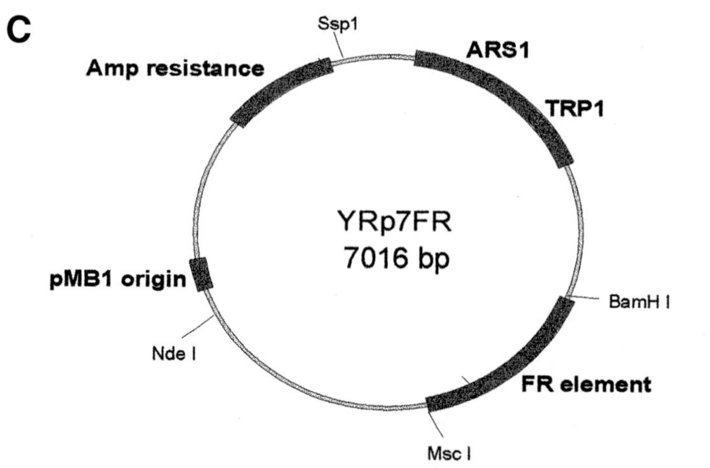
C
Ssp1
ARS1
Amp resistance
TRP1
YRp7FR
7016 bp
pMB1 origin
BamH I
Nde I
FR element
Msc I

Fig. 3. Segregation plasmids. Stability of three plasmids, each containing an *ARS* element and a *TRP1* selectable marker, are tested in the yeast plasmid loss assay. (**A**) The pRS314 plasmid contains the yeast *CEN* element and serves as a positive control. (**B**) The YRp7 plasmid lacks a segregation element and serves as a negative control. (**C**) The YRp7FR plasmid contains the EBV segregation element (FR) and is the experimental plasmid.

element and the *TRP1* gene inserted into the *Eco*RI site of pBR322. The *TRP1* gene serves as a selectable marker, allowing cells containing YRp7 to grow in medium lacking tryptophan. Although YRp7 does not contain a defined multiple cloning region for insertion of sequences, fragments can be cloned into any of the unique sites present within the plasmid, as long as the yeast sequences and the bacterial origin of replication (pMB1 ori) are not disrupted and at least one of the bacteria-selectable markers (ampicillin or tetracycline) is intact.

Proteins to be expressed in the plasmid loss assay should be expressed from plasmids that are mitotically stable, such as yeast centromeric or episomal plasmids (YCp or YEp, respectively) *(24,25)*. YCp are low-copy-number plasmids (one to two copies per cell) that contain the *ARS* and *CEN* elements and therefore can replicate and segregate in a manner similar to the yeast chromosomes. YEp are high-copy-number plasmids (20–100 copies per cell) containing the 2-μm sequence that maintains plasmids through a copy number amplification system and a partition mechanism. Both YCp and YEp are efficiently maintained in cells for over 20 generations, even under nonselective conditions.

The p4XXprom series *(26)* of yeast expression plasmids can be conveniently used to express proteins of interest (*see* **Note 6**). It is a family of vectors that is based on the bacterial Bluescript plasmid (contains ampicillin-selectable marker for propagation in bacteria). These vectors contain the *CEN6/ARSH4* or 2-μm system for plasmid maintenance, a yeast-selectable marker, and a promoter that allow expression of proteins (*see* **Note 7**). In the yeast segregation system described in this chapter, the expression plasmids p416MET25 and p425PGK were used (**Fig. 4A** and **B**) *(26,27)*. p416MET25 contains the *CEN6/ARSH4* sequence, the *URA3* marker for selection of plasmid-bearing cells in medium lacking uracil, a MET25 promoter that can be turned off with 2 *mM* methionine and a CYC1 terminator. An MCS is present downstream to the promoter for insertion of the protein coding sequence. p425PGK is a 2-μm plasmid that contains the *LEU2* gene for selection of cells containing the plasmid in medium lacking leucine and contains the PGK promoter (strong and constitutive) and terminator. The PGK promoter–terminator fragment was inserted into the *Hin*dIII site of the p425 (YEp plasmid with no promoter) MCS, and the only site available for insertion of the protein coding sequence between the promoter and terminator is the *Bgl*II site.

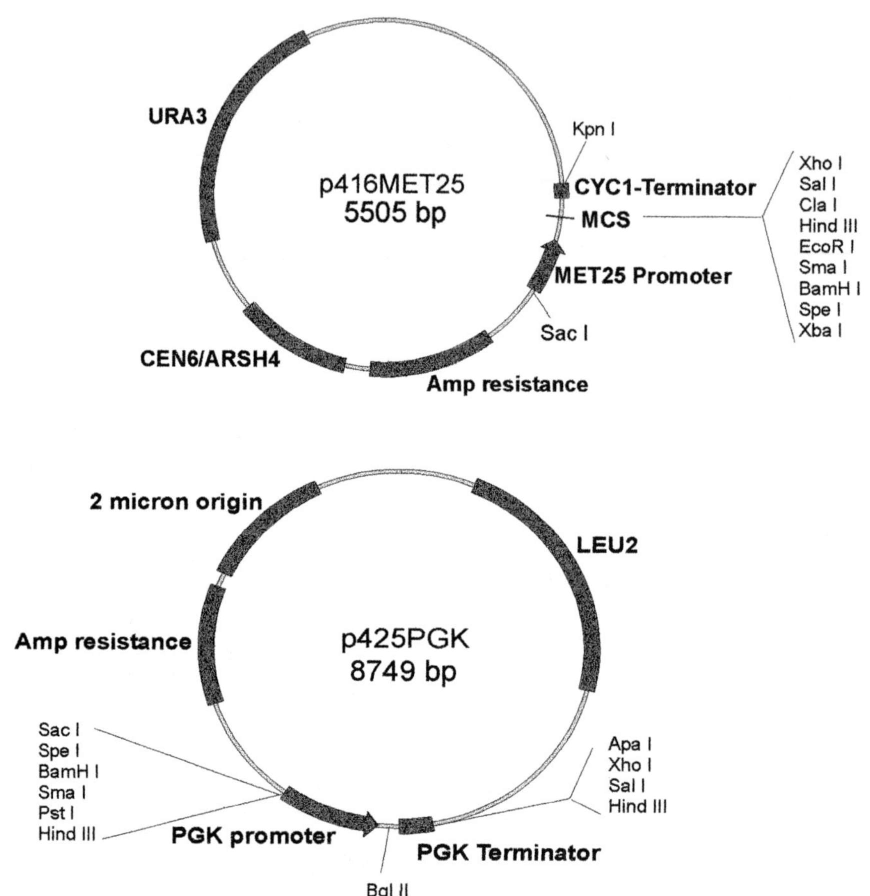

Fig. 4. Vectors used to express EBNA-1 and hEBP2 in yeast. EBNA-1 was expressed from the MET25 promoter by insertion of its gene into the *Xba*I and *Bam*HI sites of p416MET25, a low-copy-number expression plasmid containing the *URA3* marker. hEBP2 was expressed from the PGK promoter by cloning its gene into the *Bgl*II site of p425PGK, a high-copy-number expression plasmid containing the *LEU2* marker.

3.2.1.2. THE FR ELEMENT AND CDNAS FOR EBNA1 AND HEBP2

The FR element was isolated as a 1200-bp *Bam*HI–*Msc*I fragment from pGEMoriP *(19)*. The EBNA-1 gene, as described in **Subheading 3.1.1.2.,** was obtained as a *Xba*I–*Bam*HI fragment from pEBNA-1 *(17)*. The hEBP2 gene (306 amino acids) was obtained from pVLEBP2 by PCR amplification (*see* **Note 8**) *(17)*. Standard molecular biology techniques were followed.

3.2.1.3. Cloning

The segregation test plasmid containing the FR element (YRp7FR, **Fig. 3C**) *(16)* was constructed by ligation of the FR element (*Bam*HI–*Msc*I fragment from pGEMoriP) to YRp7 that was digested with *Bam*HI and *Msc*I. The EBNA-1-expressing vector (p416MET25EBNA1) *(16)* was made by ligating the EBNA-1 fragment isolated from pEBNA-1 into the *Xba*I and *Bam*HI sites of p416MET25. To construct the hEBP2-expressing vector (p425PGKhEBP2) *(16)*, p425PGK was *Bgl*II-digested, filled in with Klenow, treated with calf intestinal phosphatase (CIP) to prevent plasmid religation, and ligated with the hEBP2 PCR product that had been phosphorylated with T4 nucleotide kinase. Ligation products for each construct were screened for positive clones as described in **Subheading 3.1.1.3.**

3.2.2. Preparation of Yeast Culture for Determination of Plasmid Stability

This section of the chapter describes the yeast strain used to conduct the plasmid loss assay, the transformation of the strain with the plasmids of interest, followed by the growth of the transformants in selective and nonselective liquid media to access plasmid stability.

3.2.2.1. Yeast Strain

KY320 (*MATa leu2-PET56 ura3-52 trp1-Δ1 lys2-801*am *ade2-101*oc *his3-Δ200 GAL⁺*) *(28)* is a yeast strain that is auxotrophic for all the amino acids that serve as standard selectable markers in yeast vectors (*see* **Note 9**). As a result, it allows selection of plasmids that are used in the yeast plasmid loss assay. KY320 is a slow-growing strain compared with most other yeast strains (doubling time of 6–8 h, depending on growth conditions), but it displays strong and consistent expression of full-length hEBP2 and EBNA-1.

3.2.2.2. Yeast Transformation

The transformation protocol outlined here is a modified version of that described in Gietz and Schiestl (1995) *(29)*. All centrifugation steps are at room temperature.

1. Inoculate 5 mL YPD with a fresh KY320 colony, and grow overnight in a 30°C shaker, shaking at 300 rpm.
2. The next morning, inoculate 25 mL of fresh YPD with all the overnight culture and grow in a 30°C shaker for 3 h (*see* **Note 10**).
3. Place culture in a 50-mL sterile conical tube, and harvest the cells by centrifugation (980 *g* in Beckman GS 6KR for 10 min).
4. Resuspend pellet in 25 mL dH₂O, and centrifuge as in **step 3** to wash cells.
5. Resuspend pellet in 1 mL 0.1 *M* LiAc, transfer to a microfuge tube, and centrifuge at maximum speed in a microcentrifuge for 1 min.

6. Resuspend pellet in 600 µL of 0.1 *M* LiAc, and aliquot 50 µL of the suspension into microfuge tubes (the number of microfuge tubes used depends on the number of transformations to be done; in the present example, aliquot 50 µL into three microfuge tubes); centrifuge as in **step 5**).

7. To the pellet in each tube, add in order, as quickly as possible: 240 µL 50% PEG, 36 µL 1.0 *M* LiAc, 25 µL 2.0 mg/mL ssDNA, 50 µL of dH$_2$O, and the necessary plasmids (0.2–0.5 µg). For the EBV-based yeast segregation system, add 44 µL of dH$_2$O and 2 µL each of the *TRP1, URA3,* and *LEU2* plasmids (0.1 mg/mL) as follows:
   Tube 1: YRp7FR, p416MET25EBNA1, and p425PGKhEBP2 (experimental).
   Tube 2: pRS314, p416MET25, and p425PGK (positive control).
   Tube 3: YRp7, p416MET25, and p425PGK (negative control).

8. Vortex each tube until pellet is completely resuspended.

9. Incubate the tubes at room temperature for 30 min and then at 42°C for 22 min.

10. Centrifuge the tubes for 15 s at 4500 *g* in microcentrifuge.

11. Resuspend pellet in 200 µL of dH$_2$O, and plate all on SC-Trp, Ura, Leu plates.

12. Incubate plates in a 30°C incubator until well-sized colonies appear. (This can take up to a week for KY320.)

3.2.2.3. Growth of Transformants in Liquid Media

1. For each transformation carried out in **Subheading 3.2.2.2.,** use one colony to inoculate (separately) 3 mL of selective SC-Trp, Ura, Leu medium. It is recommended that duplicates of the 3 mL of inoculated medium be set up for each transformation to access the reproducibility of the results obtained from the plasmid loss assay.

2. Incubate in a 30°C shaker for 24 h. This medium allows selection of cells that contain the segregation and expression plasmids.

3. Determine the cell density of each culture (*see* **Note 11**).

4. Use each culture to separately inoculate 5 mL of nonselective SC-Ura, Leu medium at a concentration of 1×10^4 cells/mL. This medium allows growth of cells that contain the expression plasmids but may or may not contain the segregation plasmid (i.e., both cells are transformed with an unstable segregation plasmid that is lost, and a stable segregation plasmid that is maintained can grow).

5. Grow the resulting cultures in the 30°C shaker until the concentration of each culture has doubled approx 10 times (i.e., for 10 generations; in the EBV segregation system, cultures have been grown anywhere from 10 to 15 generations). For KY320, this will take about 72 h, and the nonselective cultures can be diluted at 48 h in 5 mL of fresh SC-Ura, Leu medium (at a concentration of 1×10^5 cells/mL) so that high cell concentration/stationary phase does not slow down the growth of the cultures.

3.2.3. Determination of the Stability of the Segregation Test Plasmid

The liquid cultures grown for 10 generations are subsequently used in a spot assay to determine qualitatively the proportion of cells that have maintained their respective segregation plasmid. The cultures are also spread onto both

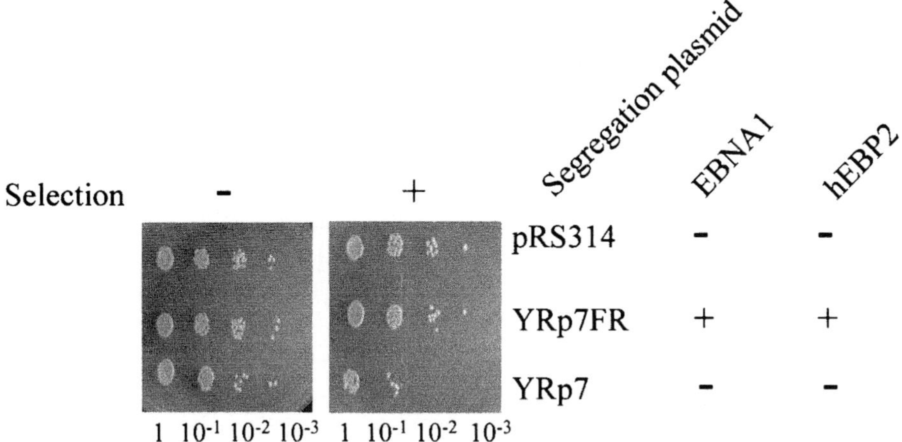

Fig. 5. Measuring EBV-based plasmid segregation in yeast. KY320 was transformed with a positive control segregation plasmid (pRS314), a negative control segregation plasmid (YRp7), or an experimental test plasmid (YRp7FR). KY320 was also transformed with a *URA3* plasmid expressing (+) or not expressing (−) EBNA-1 and a *LEU2* plasmid expressing (+) or not expressing (−) hEBP2. After growth for 11 generations without selection for the segregation plasmid, 10-fold dilutions of the cultures were plated on SC-Ura, Leu (no selection for the segregation plasmid) and SC-Ura, Leu, Trp (selection for the segregation plasmid) plates.

selective and nonselective plates to determine quantitatively the percentage of cells that still contain the segregation plasmid. An outline of these procedures is given in **Subheadings 3.2.3.1.** and **3.2.3.2.**

3.2.3.1. Spot Assays

1. Determine the cell concentration of the cultures (from **step 5** of **Subheading 3.2.2.3.**).
2. Remove 1×10^6 cells from each culture to check expression of EBNA-1 and hEBP2 by Western analysis. Cultures transformed with expression plasmids not expressing EBNA-1 and hEBP2 (those prepared from tubes 2 and 3, **Subheading 3.2.2.2.**) will serve as negative controls for the Western analysis (*see* **Note 12**).
3. Dilute a portion of each of the remaining cultures to a concentration of 1×10^6 cells/mL in sterile dH₂O.
4. Using a portion of the diluted cultures from **step 3** above, prepare three 10-fold dilutions (in dH₂O) to give cultures containing 1×10^5, 1×10^4, and 1×10^3 cells/mL.
5. Spot 10 μL of all dilutions on SC-Trp, Ura, Leu and SC-Ura, Leu plates as shown in **Fig. 5.** It is important that these plates be dried (for instance, place plates with the cover open in a tissue culture hood for 15 min) before the samples are spotted

on them, as plates that are still wet will result in spots that are too spread out and will run into each other.

6. Incubate the plates in the 30°C incubator until colonies appear. Determine stability of the positive control, negative control, and test segregation plasmids by comparing the growth of their respective cultures on SC-Trp, Ura, Leu and SC-Ura, Leu plates. It is expected that the culture corresponding to the positive control pRS314 segregation plasmid will exhibit almost equal growth on both SC plates, whereas the culture corresponding to the negative control YRp7 segregation plasmid will exhibit much reduced growth on the SC-Trp, Ura, Leu plate compared with the SC-Ura, Leu plate. The growth profile of the culture transformed with the segregation test plasmid YRp7FR and expressing EBNA-1 and hEBP2 should be similar to that of the positive control **(Fig. 5)**. This growth profile of YRp7FR in the presence of EBNA-1 and hEBP2 can be used as a positive control in experiments that measure the segregation capabilities of mutant forms of FR, EBNA-1, and hEBP2.

3.2.3.2. QUANTIFICATION OF PLASMID STABILITY

1. Dilute a portion of the SC-Ura, Leu cultures (from **step 5, Subheading 3.2.2.3.**) in dH$_2$O at a concentration of approx 150 cells/100 µL. (Serial dilutions will be required to obtain this concentration.)
2. Plate at least 600 µL of each diluted culture on SC-Ura, Leu and SC-Trp, Ura, Leu plates. (This will result in ~900 colonies for each culture on the SC-Ura, Leu plates, which is a large enough population to give accurate plasmid stability measurements.) Do this by plating 200 µL of each culture on three SC-Ura, Leu and SC-Trp, Ura, Leu plates. This will result in approx 300 colonies on each SC-Ura, Leu plate and will ensure that the colonies are well separated and easy to count.
3. Incubate the plates in the 30°C incubator until colonies appear.
4. Count the number of colonies on the plates. For a given culture, the ratio of colonies on SC-Trp, Ura, Leu vs SC-Ura, Leu plates will give the percentage of cells that have maintained their respective segregation plasmid (i.e., positive, negative, or test segregation plasmid).

4. Notes

1. If a plasmid is to be used in the transient replication but not the maintenance assay, then a G418-selectable marker is not required. It is only required that the plasmid can be propagated in bacteria, contain a mammalian promoter for protein expression, and does not contain the SV40 replication system (can contain the SV40 origin of replication or the SV40 T-antigen gene but cannot have both).
2. Mutants of EBNA-1 have also been previously ligated into pc3oriP by standard molecular biology techniques, and their replication and segregation activities have been compared relative to pc3oriP and pc3oriPE to map EBNA-1 sequences involved in these processes *(20,30)*. Elements and proteins other than *OriP* and EBNA-1 can be cloned into pcDNA3 to test their role(s) in plasmid replication and/or segregation.

3. The *OriP*/EBNA-1 system has been shown to replicate and segregate plasmids in several human cell lines, such as C33A, 293, HeLa, and B cells. The human cell line that is chosen for the assays should be EBV-negative (Raji cells are EBV-positive and cannot be used) and should not express the SV40 T-antigen (293T cells express T antigen and cannot be used). Plasmids containing the *OriP*/EBNA1 system can be introduced into the cells through a variety of transfection methods, such as calcium phosphate, electroporation, Lipofectamine 2000 (Invitrogen), and ESCORT (Sigma). The method of transfection that proves to be the most efficient will be dependent on the cell line used, and factors affecting transfection efficiency of a given method will have to be optimized for each cell line.

4. For both the transient replication and plasmid maintenance assays, EBNA-1 expression from pc3oriPE should be determined. This will ensure that an unsuccessful experiment did not result from the inability of EBNA-1 to express and will also indicate whether the transfection worked. To determine expression of EBNA-1, harvest 1×10^6 cells from each transfection (pc3oriP-transfected cells will serve as a negative control) 72 h post transfection, resuspend the cells in SDS loading buffer containing 0.5 *M* NaCl, boil the samples for 10 min, and load onto 12% SDS-PAGE gel for Western analysis.

5. The conventional method for transferring DNA samples from the gel to the nylon membrane is a overnight procedure that depends on capillary action and involves the use of a sandwich in which the gel and the membrane are held between stacks of filter paper. A faster and more efficient method of transfer can be achieved using a vacuum blot apparatus (Pharmacia Biotech VacuGene Pump apparatus). Here, the membrane is placed on the vacuum blotter, and the gel is then placed on the membrane. The edges of the membrane and the gel are sealed together to create a vacuum. Then 0.4 *N* NaOH/0.6 *M* NaCl is continuously added to the top of the gel under 50 mbar of vacuum for 30 min, followed by the continuous addition of 20X SSC for 1 h. The membrane is then removed from the apparatus, washed in 2X SSC, and air-dried.

6. Vectors in the p4XXprom series are named as follows: The second digit specifies the plasmid maintenance system present (1 = *ARS/CEN*, 2 = 2 µm), the third digit specifies the selectable marker (1 = *MET15*, 2 = *ADE2*, 3 = *HIS3*, 4 = *TRP1*, 5 = *LEU2*, 6 = *URA3*, 7 = *LYS2*), and the prom specifies the type of promoter present. Thus, p425GAL1 represents a 2-µm plasmid containing the *LEU2* marker and the inducible GAL1 promoter.

7. The promoter chosen will depend on the nature of the protein to be expressed. If the protein of interest negatively affects cell viability when overexpressed, then one may want to use a weak, inducible promoter. Alternatively, if the protein can be overexpressed without compromising cell viability, then a strong constitutive promoter can be used. p4XXprom series come with a wide variety of promoters, from weak promoters such as CYC1 to strong promoters such as ADH and from constitutive promoters such as GPD to inducible promoters such as GAL1.

8. Apart from full-length EBNA-1 and hEBP2, mutant forms of these proteins were also tested in the yeast plasmid loss assay to map the precise sequences required

for the EBV-based plasmid segregation system to function *(16,18)*. These mutants were obtained through standard molecular biology techniques (restriction digestion, PCR, site-directed mutagenesis) and will not be discussed here. The FR element, EBNA-1, and hEBP2 can be replaced in the plasmids by other factors to test the role of these factors in plasmid segregation.

9. In theory, any yeast strain that is auxotrophic for the selectable markers present on the plasmids used in the assay (yeast strain that is auxotrophic for leucine, tryptophan, and uracil for the segregation system described here) can be used. However, it is important to ensure that proteins to be tested or that are required in the assay are expressed efficiently and mostly as full length (minimal degradation of the proteins) in the strain chosen.

10. Diluting the overnight culture in 25 mL YPD and growing for 3 h as indicated has worked quite well for KY320 and typically yields a culture containing $1–4 \times 10^7$ cells/mL, which is enough for 12 transformations. Depending on the doubling time of the yeast strain used, the incubation time for the 25-mL culture can be adjusted.

11. The most convenient way to determine the concentration of a cell culture is to determine the absorbance of the culture at OD = 600 nm, where an absorbance of 0.1 is equal to 1×10^6 cells/mL. This relationship between absorbance and concentration is approximate and varies among yeast strains. However, this approximation is good enough for use in the yeast plasmid loss assays, in which it is more important that the relationship be constant among the samples assayed. (This should be true, given that the same strain is used for all transformations.) Alternatively, a hemocytometer can be used to give a more accurate cell concentration. Place 20 µL of cell culture between a cover slip and the hemacytometer slide and wait for a few seconds to allow the culture to settle onto the slide. Using a microscope, count the number of cells in the large square located at the top left-hand corner of the hemacytometer. (A hemacytometer grid consists of a large square at each of its four corners, with each square containing 16 smaller squares.) Multiply the number of cells counted by 10^4 to give the cell concentration per mL.

12. If the FR-containing segregation test plasmid is not maintained in the plasmid loss assay, then it may be because of problems in the expression of EBNA-1 and hEBP2. Thus, it is important to ensure that both these proteins are being expressed from their respective plasmids by Western blotting. Several techniques exist to lyse yeast cells for the preparation of protein lysates for Western analysis, such as spheroplasting of yeast followed by lysis in Triton X-100 and lysis by glass beads. A simple and quick method that has worked relatively well for the determination of protein expression involves harvesting the yeast culture, washing the cell pellet in cold dH_2O containing 1 m*M* phenylmethylsulfonyl fluoride (and any other protease inhibitors, if available), resuspending the pellet in protein loading buffer, centrifuging the sample, and loading the supernatant on a 12% SDS-PAGE protein gel.

Acknowledgments

The authors thank Dr. Hong Wu for providing the figure for the transient replication and long-term maintenance assays **(Fig. 2)**. This work was supported by grants to L.F. from the Canadian Institutes of Health Research and the National Cancer Institute of Canada (NCIC). P.K. is the recipient of an NCIC studentship.

References

1. Calos, M. P. (1998) Stability without a centromere. *Proc. Natl. Acad. Sci. USA* **95,** 4084–4085.
2. Collins, C. M. and Medveczky, P. G. (2002) Genetic requirements for the episomal maintenance of oncogenic herpesvirus genomes. *Adv. Cancer Res.* **84,** 155–174.
3. Frappier, L. (2004) Viral plasmids in mammalian cells, in *The Biology of Plasmids* (Funnell, B. E. and Phillips, G., eds.), ASM Press, Washington, D.C., 325–339.
4. Yates, J. L., Warren, N., and Sugden, B. (1985) Stable replication of plasmids derived from Epstein-Barr virus in various mammalian cells. *Nature* **313,** 812–815.
5. Lupton, S. and Levine, A. J. (1985) Mapping of genetic elements of Epstein-Barr virus that facilitate extrachromosomal persistence of Epstein-Barr virus-derived plasmids in human cells. *Mol. Cell. Biol.* **5,** 2533–2542.
6. Reisman, D., Yates, J., and Sugden, B. (1985) A putative origin of replication of plasmids derived from Epstein-Barr virus is composed of two *cis*-acting components. *Mol. Cell. Biol.* **5,** 1822–1832.
7. Krysan, P. J., Haase, S. B., and Calos, M. P. (1989) Isolation of human sequences that replicate autonomously in human cells. *Mol. Cell. Biol.* **9,** 1026–1033.
8. Gahn, T. A. and Schildkraut, C. L. (1989) The Epstein-Barr virus origin of plasmid replication, *OriP,* contains both the initiation and termination sites of DNA replication. *Cell* **58,** 527–535.
9. Yates, J. L. and Guan, N. (1991) Epstein-Barr virus-derived plasmids replicate only once per cell cycle and are not amplified after entry into cells. *J. Virol.* **65,** 483–488.
10. Grogan, E. A., Summers, W. P., Dowling, S., Shedd, D., Gradoville, L., and Miller, G. (1983) Two Epstein-Barr viral nuclear neoantigens distinguished by gene transfer, serology and chromosome binding. *Proc. Natl. Acad. Sci. USA* **80,** 7650–7653.
11. Harris, A., Young, B. D., and Griffin, B. E. (1985) Random association of Epstein-Barr virus genomes with host cell metaphase chromosomes in Burkitt's lymphoma-derived cell lines. *J. Virol.* **56,** 328–332.
12. Delecluse, H. J., Bartnizke, S., Hammerschmidt, W., Bullerdiek, J., and Bornkamm, G. W. (1993) Episomal and integrated copies of Epstein-Barr virus coexist in Burkitt's lymphoma cell lines. *J. Virol.* **67,** 1292–1299.
13. Wu, H., Ceccarelli, D. F. J., and Frappier, L. (2000) The DNA segregation mechanism of the Epstein-Barr virus EBNA-1 protein. *EMBO Rep.* **1,** 140–144.
14. Hung, S. C., Kang, M. S., and Kieff, E. (2001) Maintenance of Epstein-Barr virus (EBV) *OriP*-based episomes requires EBV-encoded nuclear antigen-1 chromosome-binding domains, which can be replaced by high-mobility group-I or histone H1. *Proc. Natl. Acad. Sci. USA* **98,** 1865–1870.
15. Kanda, T., Otter, M., and Wahl, G. M. (2001) Coupling of mitotic chromosome tethering and replication competence in Epstein-Barr virus-based plasmids. *Mol. Cell. Biol.* **21,** 3576–3588.

16. Kapoor, P., Shire, K., and Frappier, L. (2001) Reconstitution of Epstein-Barr virus-based plasmid partitioning in budding yeast. *EMBO J.* **20,** 222–230.

17. Shire, K., Ceccarelli, D. F. J., Avolio-Hunter, T. M., and Frappier, L. (1999) EBP2, a human protein that interacts with sequences of the Epstein-Barr nuclear antigen 1 important for plasmid maintenance. *J. Virol.* **73,** 2587–2595.

18. Kapoor, P. and Frappier, L. (2003) EBNA-1 partitions Epstein-Barr virus plasmids in yeast cells by attaching to human EBNA-1-binding protein 2 on mitotic chromosomes. *J. Virol.* **77,** 6946–6956.

19. Frappier, L. and O'Donnell, M. (1991) Overproduction, purification and characterization of EBNA1, the origin binding protein of Epstein-Barr virus. *J. Biol. Chem.* **266,** 7819–7826.

20. Ceccarelli, D. F. and Frappier, L. (2000) Functional analyses of the EBNA1 origin DNA binding protein of Epstein-Barr virus. *J. Virol.* **74,** 4939–4948.

21. Hirt, B. (1967) Selective extraction of polyoma DNA from infected mouse cell culture. *J. Mol. Biol.* **26,** 365–369.

22. Sambrook, J., Fritsch, E. F., and Maniatis, T. (eds.) (1989) *Molecular Cloning: A Laboratory Manual.* Cold Spring Harbor Laboratory Press, Cold Spring Harbor, NY.

23. Struhl, K., Stinchcomb, D. T., Scherer, S., and Davis, R. W. (1979) High-frequency transformation of yeast: autonomous replication of hybrid DNA molecules. *Proc. Natl. Acad. Sci. USA* **76,** 1035–1039.

24. Sikorski, R. S. and Hieter, P. (1989) A system of shuttle vectors and yeast host strains designed for efficient manipulation of DNA in *Saccharomyces cerevisiae. Genetics* **122,** 19–27.

25. Christianson, T. W., Sikorski, R. S., Dante, M., Shero, J. H., and Hieter, P. (1992) Multifunctional yeast high-copy shuttle vectors. *Gene* **110,** 119–122.

26. Mumberg, D., Muller, R., and Funk, M. (1994) Regulatable promoters of *Saccharomyces cerevisiae:* comparison of transcriptional activity and their use for heterologous expression. *Nucleic Acids. Res.* **22,** 5767–5768.

27. Marcus, S. L., Miyata, K. S., Rachubinski, R. A., and Capone, J. P. (1995) Transactivation by PPAR/RXR heterodimers in yeast is potentiated by exogenous fatty acid via a pathway requiring intact peroxisomes. *Gene Expr.* **4,** 227–239.

28. Chen, W. and Struhl, K. (1988) Saturation mutagenesis of a yeast his3 'TATA element': genetic evidence for a specific TATA-binding protein. *Proc. Natl. Acad. Sci. USA* **85,** 2691–2695.

29. Gietz, R. D. and Schiestl, R. H. (1995) Transforming yeast with DNA. *Methods Mol. Cell. Biol.* **5,** 255–269.

30. Wu, H., Kapoor, P., and Frappier, L. (2002) Separation of the DNA replication, segregation, and transcriptional activation functions of Epstein-Barr nuclear antigen 1. *J. Virol.* **76,** 2480–2490.

18

DNA Affinity Purification of Epstein-Barr Virus OriP-Binding Proteins

Constandache Atanasiu, Larissa Lezina, and Paul M. Lieberman

Summary

DNA affinity purification has been used to identify cellular and viral proteins associated with the Epstein-Barr virus origin of plasmid DNA replication. This approach allows for a one- or two-step purification scheme of high-affinity DNA binding proteins from crude nuclear extracts. Additionally, this approach may be useful for isolation of proteins that are found in the insoluble fractions of the nuclear matrix or scaffold.

Key Words: DNA affinity; magnetic beads; OriP; EBNA-1; telomere repeat binding factors.

1. Introduction

The isolation and identification of sequence-specific DNA binding proteins is essential for molecular and biochemical characterization of gene regulation. We describe here a simple DNA affinity technique that has been useful for identifying proteins in nuclear extracts that bind with high affinity and specificity to a DNA regulatory sequence. This technique has been applied specifically to the identification of cellular and viral proteins that bind to the Epstein-Barr virus (EBV) origin of plasmid replication (OriP) but should be of general utility for identification of any high-affinity protein–DNA complex.

Numerous studies have demonstrated the uses of DNA affinity methods for purification of sequence-specific DNA binding proteins (*1,2*; reviewed in **refs. 3** and **4***). The original methodologies involved the use of tandem repeats of specific DNA recognitions sites covalently crosslinked to sepharose beads. This method was best suited for identification of high-affinity DNA binding proteins that interact with a well-characterized genetic element. Many of these DNA binding activities were first identified by electrophoretic mobility shift assay

From: *Methods in Molecular Biology, vol. 292: DNA Viruses: Methods and Protocols*
Edited by: P. M. Lieberman © Humana Press Inc., Totowa, NJ

(EMSA) or DNase I footprinting assays, and these DNA binding activities could be easily monitored over several chromatographic steps, including DNA affinity chromatography. This approach was especially important when protein microsequencing required nanomolar quantities and was relatively expensive.

Several variations and modifications have been introduced to this method. The most significant technological change has been the increase sensitivity and affordability of matrix-assisted laser desorption ionization time of flight (MALDI-TOF) and tandem mass spectrometry (MS/MS) approaches to identify small amounts (picomolar) of unknown proteins. The liquid chromatography (LC)–MS/MS) protein identification provides a proteomic approach to DNA binding proteins and allows a much wider array of protein detection and post-detection verification.

In addition to the power of mass spectrometry, we have found that several other technical changes have made this approach simpler and more attractive. In particular, the use of magnetic beads coupled to biotinylated DNA allows for better separation of chromatographic support from crude starting materials of nuclear extracts *(5)*. Second, the use of multiple purification cycles can also help to enrich specific binding factors *(6)*. Third, the solubilization of nuclear pellets allows for purification of proteins that are typically enriched in the nuclear matrix or scaffold and are likely to be functionally relevant for DNA replication or chromosome maintenance functions. Finally, it needs to be emphasized that empirical determination of competitor DNA, salt, and detergent concentrations is critical for purification from contaminants.

In this application of the DNA affinity purification technique, we focus on the proteins associated with EBV oriP. Cells latently infected with EBV maintain their genome as a multicopy episome that replicates during the cellular S-phase (reviewed in **refs.** *7* and *8*). A virally encoded origin binding protein, Epstein-Barr nuclear antigen-1 (EBNA-1), is essential for replication and maintenance of OriP-containing plasmids *(9,10)*. EBNA-1 binds to multiple recognition sites in OriP, referred to as the family of repeats (FR) and the dyad symmetry (DS) elements *(10)*. The FR consists of twenty 30-bp repeats and is required for nuclear retention of DNA and plasmid maintenance. The DS consists of four EBNA-1 binding sites and three nonamer repeats and is essential for initiation of DNA replication. Here we describe in more detail methods used to isolate cellular proteins that bind to the DS element of OriP.

2. Materials

1. Raji, HeLa/EBNA-1+, and HeLa cells.
2. Dynabeads M-280 streptavidin (Dynal Biotech).
3. Magnetic Particle Concentrator (MPC_S; Dynal Biotech).
4. 2X B&W buffer: 10 mM Tris-HCl, pH 7.5, 2 M NaCl, 1 mM EDTA.

5. Buffer A: 10 m*M* HEPES, pH 7.9, 10 m*M* KCl, 1.5 m*M* MgCl$_2$, supplemented with 1 m*M* dithiothreitol (DTT), 1 m*M* phenylmethylsulfonyl fluoride (PMSF) before use.
6. Buffer C: 20 m*M* HEPES, pH 7.9, 420 m*M* NaCl, 1.5 m*M* MgCl$_2$, 0.2 m*M* EDTA, 25% glycerol, supplemented with 1 m*M* DTT, 1 m*M* PMSF before use.
7. Buffer D: 10 m*M* HEPES, pH 7.9, 20% glycerol, 0.4 m*M* EDTA, freshly supplemented with 0.05% NP40, 1 mm PMSF, 10 m*M* 2-mercaptoethanol, protease inhibitor cocktail for mammalian cells (Sigma P8340) before use. D150, D300, or D1000 represents buffer D plus 150, 300, or 1000 m*M* KCl.
8. Nuclear pellet solubilization (NPS) buffer: 10 m*M* Tris-HCl, pH 8.0, 140 m*M* NaCl, 1 m*M* DTT, protease inhibitor cocktail (Sigma P8340), 1% Triton X-100, 0.1% NaDeoxycholate.
9. 10 mg/mL Sonicated salmon sperm DNA (Invitrogen). Salmon sperm DNA (1 mL of 10 mg/mL) was sonicated three times for 20 s at 30% output in a Bransom Sonifier 250.
10. Biotinylated oligonucleotide for polymerase chain reaction (PCR) amplification. Biotin should be included in the synthesis at the 5′-end of one primer used for PCR amplification of template DNA.
11. PCR amplification reagents: dNTPs, primers, templates, *Taq* polymerase, and others.
12. 50% Trichloracetic acid (TCA).
13. Acetone (–20°).
14. Wheaton dounce tissue grinder (type 357544 or equivalent).
15. Qiaquick PCR purification kit (Qiagen or equivalent).
16. Colloidal Blue Staining Kit (Invitrogen).

3. Methods

DNA affinity purification is described for isolation of cellular and viral proteins that bind to EBV OriP. Raji cells are EBV-positive Burkitt's lymphoma cells that are defective for lytic cycle replication. They have been a useful source for isolation of cellular and viral proteins involved in the episomal maintenance and latent cycle replication of EBV. We describe methods for isolation of Raji nuclear proteins that bind with high affinity to the DS element of OriP. This method should be generalizable for characterization of proteins associated with other DNA regulatory elements.

3.1. Small-Scale Partial Purification for Western Blotting (see Note 1)

3.1.1. Preparation of the Nuclear Extract

Raji nuclear extract was prepared using a modified Dignam protocol *(11)*.

1. Harvest Raji cells (0.5–20 L) by centrifugation in a Beckman JA-10 rotor (or equivalent) for 10 min at 1200 rpm (250 *g*).
2. Wash cells twice with ice-cold PBS buffer. All following procedures should be performed at 4°C.

3. Resuspend cells in buffer A (hypotonic buffer).
4. Incubate cells on ice for 10 min.
5. Place cells in a Wheaton 357544 Dounce Tissue Grinder and dounce 10 times using pestle B.
6. Centrifuge for 10 min at 8000 rpm (7700 *g*; JA-20 rotor or equivalent).
7. Remove the supernatant (containing primarily cytoplasmic fraction).
8. Resuspend nuclei in buffer C (3 mL for each 10^9 cells).
9. Dounce 10 times using pestle A.
10. Transfer to mixing container (centrifuge or beaker) and incubate for 30 min at 4°C with mixing.
11. Centrifuge for 30 min at 18000 rpm (39,000 *g*; JA-20 rotor).
12. Collect soluble fraction (nuclear extract) and the pellet (nuclear pellet).
13. Dialyze nuclear extract in D150 buffer overnight at 4°C.
14. Aliquot the nuclear extract, freeze in dry ice, and keep it at –80°C.
15. Freeze nuclear pellet or proceed with purification steps below.

3.1.2. Solubilization of Nuclear Pellet

1. Resuspend nuclear pellet in NPS buffer (3 mL for each 10^9 cells).
2. Sonicate three times for 30 s at 30% output for each mg of Raji cells pellet.
3. Centrifuge for 30 min at 18,000 rpm (39,000 *g*; JA-20 rotor).
4. Collect supernatant containing partially soluble nuclear pellet fraction (*see* **Note 2**).

3.1.3. Primer Design

Primers are 20–30 bp long and terminate in one or more G or C bases. The forward primer has a biotin moiety attached at the 5′-end. The resulting PCR product has 300–500 bp (*see* **Note 3**).

3.1.4. PCR Conditions: Target and Control Templates

PCR is performed by standard procedure using the following cycles: 90 s at 94°C, 60 s at 52°C, 90 s at 72°C, 35 cycles and 3 μg/mL DNA template. In the experiments we describe, we amplify the DS region of EBV (120 bp containing four EBNA-1 sites and three nonamer repeats) in the pBSKII+ polylinker. A control plasmid containing a mutated insert or irrelevant DNA of similar size is an essential control for analysis of DNA binding proteins. It is best when both control and target template DNA are generated with the same primer sets. We used the M13 reverse and forward primers for inserts cloned into the pBSKII+ multiple cloning site. The PCR product is purified using the Qiaquick PCR Purification Kit (Qiagen) and checked for purity and quantity on 1.2% agarose gel.

3.1.5. Binding PCR Product to Magnetic Beads

For DNA affinity purification assay, we use superparamagnetic polystyrene beads with streptavidin covalently attached to the their surface (Dynabeads). To saturate the beads, about 50 pmol of biotinylated PCR product (~5 µg of 200-bp PCR product) per 35 µL of starting slurry of Dynabeads (~2.3 × 10^7 beads) are added. The following steps are followed for each 5 µg of PCR template:

1. Resuspend magnetic beads in solution by gently shaking the bottle.
2. Add 35 µL beads solution in an Eppendorf tube, and spin for 10 s at 14,000 rpm (16,000 *g*) in a table-top centrifuge (*see* **Note 4**).
3. Place the Eppendorf tube in an MPC-S, and leave the beads to settle for 2 min. Remove solution without touching the magnetic beads.
4. Resuspend magnetic beads in 1 mL 2X B&W buffer.
5. Wash the beads for 5 min with rotating on a Labquake shaker at room temperature.
6. Pulse-spin the beads for 10 s at 14,000 rpm (16,000 *g*), and place in the MPC-S for 2 min. Remove buffer.
7. Repeat **steps 4, 5,** and **6.**
8. Resuspend the magnetic beads in 50 µL 2X B&W buffer, and mix with 50 µL biotinylated PCR product (5 µg).
9. Incubate for 30 min to 2 h at room temperature with mixing (*see* **Note 5**).
10. Pulse-spin down, and place in the MPC-S for 2 min. Remove solution.
11. Wash twice with 1 mL 2X B&W buffer, by mixing, 5 min each wash.
12. Wash once with 1 mL D150 buffer (containing 0.05% NP40).
13. Pulse-spin, and place in the MPC-S for 2 min. Remove D150 buffer.

3.1.6. Preincubation of Nuclear Extract With Sonicated Salmon Sperm DNA

Mix Raji nuclear extract with solubilized nuclear pellet at a 1:1 ratio (8 mg/mL total protein concentration) (*see* **Note 6**). Add sonicated salmon sperm DNA (10 mg/mL stock) to the nuclear protein mixture to get 0.4 mg/mL final concentration, and incubate for 45 min at 4°C (*see* **Notes 7** and **8**).

3.1.7. Incubation of DNA Beads With Nuclear Extract

Split the nuclear protein mixture into equal portions for incubation with control and target templates that have been prebound to magnetic beads, as described in **Subheading 3.1.5.** Nuclear extracts should be mixed with DNA beads according to the following ratio: 1 mL nuclear extract mixture (0.5 mL nuclear extract + 0.5 mL solubilized nuclear pellet, as described in **Subheading 3.1.6.**) is combined with 50 µL bead slurry (3.7 × 10^7 beads (bound to 0.5 nmol DNA). (*see* **Note 9**)

Incubate extracts with beads for 45 min at 4°C with rotating or mixing.

3.1.8. Washing and Elution

1. Gently centrifuge (3000 rpm [730 *g*] in table-top Eppendorf microcentrifuge for 1 min) the sample to recover material from the lid (*see* **Note 10**).
2. Place sample on the MPC-S for 2 min. Remove solution.

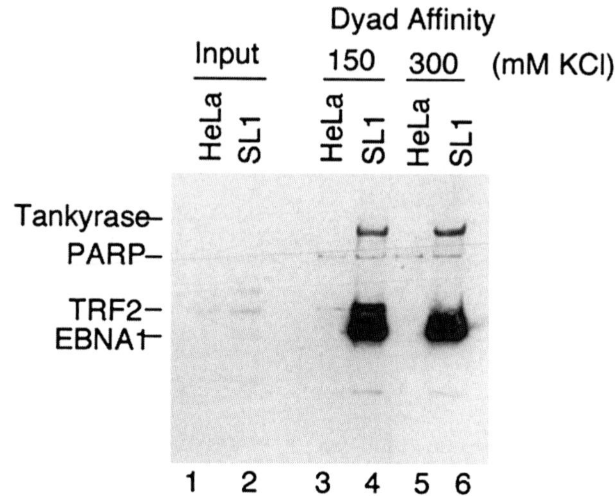

Fig. 1. Comparison of different cell types using DNA affinity purification. Nuclear extracts derived from HeLa or EBNA-1+ HeLa cells (HeLa/EBNA-1+) were subjected to DNA affinity purification with EBV DS DNA. Proteins were affinity-purified as described above and analyzed by Western blotting with antibodies to EBNA-1, Tankyrase, TRF2, and PARP1. Starting material (input, lanes 1 and 2) represents 10% of the binding reaction before affinity purification.

3. Resuspend well, and wash with 1 mL D150 buffer by rotating for 5 min at 4°C (*see* **Note 11**).

4. Repeat **steps 2 and 3** two more times (at least three washes with D150).

5. After the last wash, spin for another 10 s. Remove the last traces of buffer (*see* **Note 12**).

6. For Western blotting analysis, resuspend beads in 25 μL 2X Laemli buffer, and boil for 5 min.

7. Load samples on an 8–16% acrylamide gel, and analyze by Western blotting (**Figs. 1 and 2**).

3.2. Generation of Samples for LC–MS/MS Analysis

1. Scale up 5–10 times as described above. Typically, we use 10 mL Raji nuclear extract with 350 μL magnetic bead slurry combined with 500 μL PCR (50 μg of 200-bp fragment).

2. Perform the first round of purification as described in **Subheadings 3.1.1.–3.1.5.** Bound proteins are eluted two times with 100 μL D1000 buffer for 15 min at 4°C with mixing.

3. The second round of purification starts with diluting the combined eluted material (200 μL) with D0 buffer, on ice, until the final concentration of KCl becomes

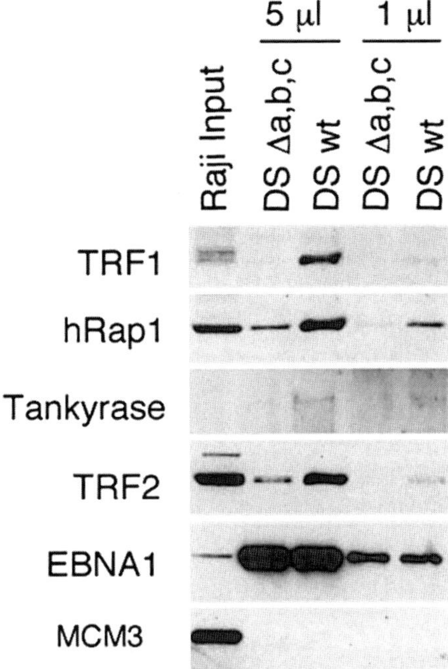

Fig. 2. Comparison of different DNA templates using DNA affinity purification. Nuclear extracts from Raji cells were subjected to DNA affinity purification with DNA templates derived from EBV DS or a defective DS lacking the nonamer repeats (DSΔa,b,c). Affinity-purified proteins were analyzed by chemiluminescence detection of Western blots.

150 m*M*. Dilution should be slow and stepwise (15 steps of dilution with 10 min between each step) (*see* **Note 13**).

4. After dilution is complete, add 10 µg/mL sonicated salmon sperm competitor DNA, and incubate for at least 15 min.
5. Add fresh magnetic beads coupled with biotin DNA. Use only 35 µL beads and 50 µL PCR. Incubate for 45 min at 4°C with rotating.
6. Recover the beads, and wash three times with 1 mL D150 as described in **Subheading 3.1.8.**
7. Resuspend the beads in 100 µL D300. Elute for 15 min with mixing.
8. Wash three times with 1 mL D300.
9. Resuspend the beads in 100 µL D1000. Elute for 15 min (*see* **Note 14**).
10. Concentrate protein by adding TCA to a 15% final concentration (*see* **Note 15**).
11. Add another 500 µL of 15% TCA. Incubate on ice for 20 min.
12. Spin down for 30 min at 14,000 rpm (16,000 *g*).
13. Remove solution. Wash the pellet briefly with 500 µL cold acetone.
14. Spin down. Remove acetone.

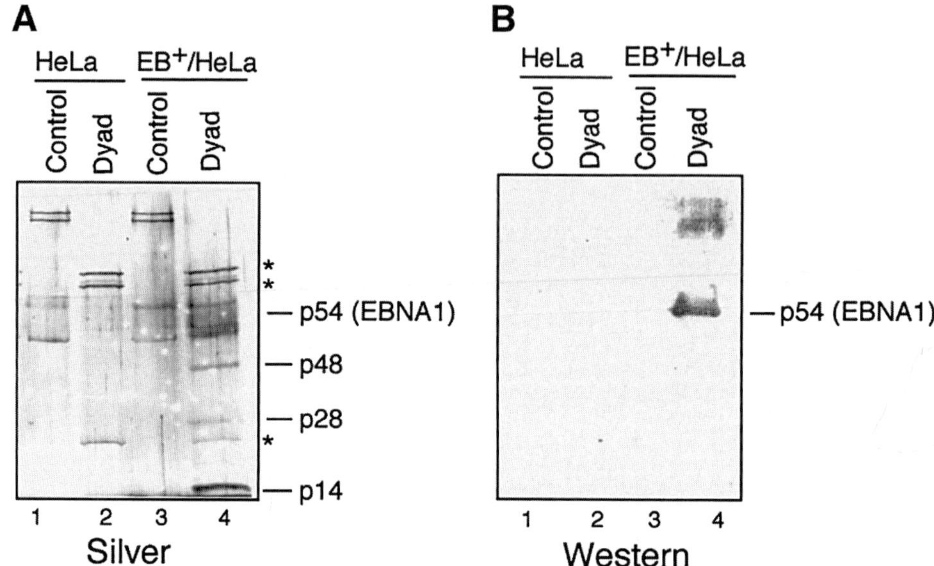

Fig. 3. (**A**) Silver staining analysis after two rounds of affinity purification. HeLa or EBNA-1-positive HeLa nuclear extracts were subject to two rounds of affinity purification (as described in **Subheading 3.2.**) and analyzed by silver staining of SDS-PAGE gels. Extracts were purified with DS DNA (Dyad) or with control BSKII DNA (control) as indicated above each lane. (**B**) The same material was subjected to Western blotting analysis with antibodies specific for EBNA-1.

15. Vacuum speed for 5 min.
16. Resuspend the pellet in 30 µL 2X Laemli buffer.
17. Incubate for 20 min at 65°C.
18. Boil for 10 min.
19. Load sample on an 8–16% Tris-glycine gel.
20. Stain with colloidal blue (Novagen), and identify specific protein bands of interest for submission to LC–MS/MS (**Fig. 3**) (*see* **Note 16**).

4. Notes

1. This approach is relatively fast and simple and can be quite powerful when combined with Western blotting antibodies. The limitations include the requirement for antibodies to candidate binding proteins and the relatively high level of contaminating proteins. This one-step method by itself is not recommended for protein identification by LC–MS/MS owing to the large background of contaminants.
2. The volume of the residual nuclear pellet after sonication should be greatly reduced relative to the starting pellet volume derived from the D buffer-extracted nuclei. The "soluble" extract derived from the supernatant after centrifugation remains tur-

bid but can still be effectively used with this protocol. It is this partially insoluble material that is highly enriched in nuclear matrix-associated components.

3. For PCR products larger than 2 kb, the Dynabeads kilobaseBINDER Kit (Dynal Biotech) is recommended by the manufacturer.

4. In experiments with multiple DNA templates, it is appropriate to prewash the Dynabeads together. We will wash up to 300 μL of Dynabead slurry in a 1 mL vol of 2X B&W buffer.

5. Binding of the biotinylated PCR product with streptavidin magnetic beads is nearly complete after 30 min but may vary for different length PCR products. There is some increase in binding with incubations up to 2 h.

6. We found that mixing the nuclear extract with solubilized nuclear pellet worked best for isolation of OriP binding proteins. However, under some circumstances it may be best to leave out the solubilized nuclear pellet, which contains high levels of histones and detergents that can interfere with some protein complex assembly on DNA.

7. Preincubation of nuclear extract with competitor DNA is critical for this procedure, and the precise amount of competitor may need to be titrated depending on the cell type and type of nuclear extract. The major reasons for failure of this technique include an overwhelming excess of nonspecific DNA binding proteins and DNA nucleases, which destroy template DNA. High concentrations of competitor DNA significantly reduces these inhibitory activities. Incubations on ice and elimination of the Mg^+ ion from the binding reaction may also protect DNA templates from degradation.

8. It is convenient to preincubate nuclear extracts with competitor DNA while simultaneously coupling DNA to magnetic beads.

9. The ratio of beads to extract can be altered depending on the goal of the experiment. To enrich for rare proteins, it is important to oversaturate the beads with excess extract relative to template. This appears to increase specificity. To minimize the use of nuclear extracts for Western blotting, the extract-to-template ratio can be decreased to ensure quantitative recovery of specific proteins. This may also increase background, but that may not interfere with Western blotting detection.

10. When solubilized nuclear pellet is included, it is best to avoid excessive centrifugation, which may enrich for insoluble components that can otherwise be separated by magnetic selection of beads, as opposed to density-dependent gravity and centrifugal forces.

11. Increasing the number, duration, and composition of the wash buffer can have profound effects on the purity of the protein components eluted. This can be determined empirically for each DNA target and depends on the stability of protein complexes assembled. EBNA-1 is resistant to 300 mM NaCl and 0.1% NP40, but this is not true for some other DNA binding proteins.

12. Since only 25 μL of 2X Laemli buffer is used for elution, it is very important to remove any trace of buffer that could dilute the sample. Usually, a brief spin at 14,000 rpm (16,000 g) after the last wash gives 20–40 μL buffer.

13. Dilution volumes should increase proportionately over the course of the diluting process, and additions should never be more than 10% of the total volume. This is important to avoid precipitation and denaturation of proteins.

14. If the eluted material is used for silver staining, boiling of beads is not recommended since large amounts of streptavidin protein are also eluted.

15. There are two reasons for precipitation with TCA. First, only 50 µL of protein can be loaded in one line, so the sample must be concentrated. Second, samples containing salts greater than 300 mM KCl interfere with loading and electrophoretic migration. TCA precipitation removes most of the salts.
16. Proteomics facilities may vary in their preferred preparation of proteins to be submitted for LC/MS/MS detection. At this stage, it is best to coordinate with a protein microsequencing facility.

Acknowledgments

C. Atanasiu was a fellow of the Leukemia Research Foundation. This work was funded by grants from the NIH (CA93606) to P.M.L.

References

1. Kadonaga, J. T. and Tjian, R. (1986) Affinity purification of sequence-specific DNA binding proteins. *Proc. Natl. Acad. Sci. USA* **83,** 5889–5893.
2. Rosenfeld, P. J. and Kelly, T. J. (1986) Purification of nuclear factor I by DNA recognition site affinity chromatography. *J. Biol. Chem.* **261,** 1398–1408.
3. Gadgil, H., Oak, S. A., and Jarrett, H. W. (2001) Affinity purification of DNA-binding proteins. *J. Biochem. Biophys. Methods* **49,** 607–624.
4. Kadonaga, J. T. (1991) Purification of sequence-specific binding proteins by DNA affinity chromatography. *Methods Enzymol.* **208,** 10–23.
5. Ranish, J. A., Yudkovsky, N., and Hahn, S. (1999) Intermediates in formation and activity of the RNA polymerase II preinitiation complex: holoenzyme recruitment and a postrecruitment role for the TATA box and TFIIB. *Genes Dev.* **13,** 49–63.
6. Kumar, R., Reynolds, D. M., Shevchenko, A., Shevchenko, A., Gladstone, S. D., Dalton, S. (2000) Forkhead transcription factors, Fkh1p and Fkh2p, collaborate with Mcm1p to control transcription required for M-phase. *Curr. Biol.* **10,** 896–906.
7. Sugden, B. and Leight, E. R. (2001) Molecular mechanisms of maintenance and disruption of virus latency, in *Epstein-Barr Virus and Human Cancer* (Takada, K., ed.), Springer, Heidelberg, pp. 3–11.
8. Leight, E. R. and Sugden, B. (2000) EBNA-1: a protein pivotal to latent infection by Epstein-Barr virus. *Rev. Med. Virol.* **10,** 83–100.
9. Yates, J. L. Warren, N., Reisman, D., Sugden, B., et al. (1984) A *cis*-acting element from Epstein-Barr viral genome that permits stable replication of recombinant plasmids in latently infected cells. *Proc. Natl. Acad. Sci. USA* **81,** 3806–3810.
10. Rawlins, D. R. Millman, G., Hayward, S. D., Howard, G. S. (1985) Sequence-specific DNA binding of the Epstein-Barr virus nuclear antigen (EBNA-1) to clustered sites in the plasmid maintenance region. *Cell* **42,** 859–868.
11. Dignam, J. D., Lebovitz, R. M., and Roeder, R. G. (1983) *Accurate transcription initiation by RNA polymerase II in a soluble extract from isolated mammalian nuclei. Nucleic Acids Res.* **11,** 1475–1489.

VI

PATHOGENESIS

19

Pre-B-Cell Colony Formation Assay

Masato Ikeda and Richard Longnecker

Summary

Latent membrane protein 2A (LMP-2A) of Epstein-Barr virus (EBV) mimics a constitutively active B-cell receptor (BCR) and plays a key role in viral latency and EBV pathogenesis. By functioning as a BCR mimic, LMP-2A drives B-cell development, resulting in the bypass of normal B-cell developmental checkpoints. To assess the function of LMP-2A, we have utilized a colony formation assay for the progenitor B cells that uses the B-cell proliferation factor IL-7.

Key Words: B-cell development; pre-B-cell; BCR; IL-7; methylcellulose; colony formation; LMP-2A, EBV.

1. Introduction

The generation of mature B cells from hematopoietic stem cells is a complex process regulated by a series of developmental checkpoints *(1)*. Among the most critical of these selection events are the production of the B-cell antigen receptor (BCR) and its precursor (pre-BCR). The pre-BCR is formed once a pro-B cell has completed a rearrangement of the immunoglobulin heavy chain (HC) locus yielding an HC protein that can associate with surrogate light chain (LC) proteins, λ5 and VpreB, and the signaling chains Igα and Igβ. Surface expression of the pre-BCR complex is required for the transition to the pre-B-cell stage. Subsequent rearrangement of the immunoglobulin LC locus in pre-B cells leads to the formation of the mature BCR, which mediates the progression to the immature and the mature B-cell stage. Other signals are also known to participate during B-cell development. The proliferation and differentiation of B-cell progenitors in the bone marrow depends on cytokines such as interleukin-7 (IL-7) produced by stromal cells within the bone marrow microenvironment *(2)*. IL-7 was originally isolated as a growth factor for B-cell precursors and plays a crucial role in B-cell development.

From: *Methods in Molecular Biology, vol. 292: DNA Viruses: Methods and Protocols*
Edited by: P. M. Lieberman © Humana Press Inc., Totowa, NJ

Hematopoietic progenitor cells are present in bone marrow at a low frequency. To determine the frequency of pre-B cells in the bone marrow, a pre-B-cell colony formation assay in methylcellulose medium containing IL-7 is commonly utilized. In IL-7-containing methylcellulose culture, large pre-B cells can selectively expand and form colonies *(3,4)*. This protocol is a cytokine bioassay based on the growth of a variety of hematopoietic progenitor cells as colonies in a semisolid culture medium. Progenitor cells are suspended in a combination of culture medium, growth factors, and a semisolid matrix that mimics extracellular matrix. In the course of culture, the suspended progenitors proliferate and differentiate, forming colonies that can be readily counted using an inverted or dissecting microscope. Originally, layers of stromal feeder cells were used for a source of growth factors and production of an extracellular matrix; however, they have been replaced by conditioned media and recombinant cytokines discovered in part through the use of a colony formation assay.

The ability of Epstein-Barr virus (EBV) latent membrane protein-2A (LMP-2A) to impart developmental and survival signals to developing mature B cells *(5)* has been highlighted by the construction of LMP-2A transgenic mice. LMP-2A causes a developmental alteration characterized by a bypass for the requirement that BCR be expressed, resulting in BCR-negative B cells in peripheral lymphoid organs *(6)*. Normally, B cells lacking a cognate BCR normally undergo apoptosis. The lack of BCR expression in LMP-2A transgenic mice results from the absence of immunoglobulin heavy chain rearrangement *(5)*. These results indicate that LMP-2A functions as a pre-BCR homolog as well as a BCR homolog during B-cell development, as demonstrated by the observation that LMP-2A-positive pre-B cells can produce colonies of Ig HC-negative progenitor B cells in IL-7-containing methylcellulose medium. The analysis of LMP-2A transgenic B cells indicates that LMP-2A mimics a constitutively active BCR and plays a key role in ensuring viral latency and EBV pathogenesis. Many DNA viruses encode homologs of cytokines and their receptors that play a crucial role in control of the host immune response. Construction of transgenic mice is a useful tool to examine the functions of these viral proteins and provides study models of pathogenesis in the human host. In particular, this technique provides a powerful approach when the transgenic proteins are related in overall biological functions such as cell development or cell differentiation.

Methylcellulose-based culture has become the standard tool for enumeration of hematopoietic progenitors characterized as colony-forming units (CFUs). Granulocyte/macrophage progenitors (CFU-GM), multipotential or granulocyte/erythroid/macrophage/megakaryocyte progenitors (CFU-Mix or CFU-GEMM), or erythroid blast-forming unit progenitors (BFU-E) can be detected in murine cells from bone marrow, spleen, blood, and fetal liver using methylcellulose medium containing stem cell factor (SCF), IL-3, IL-6, and erythropoietin. Therefore, this system can be also used to detect other progenitors (CFUs) by substituting cells or growth factors (*see* **Note 1**). In addition, this method can be used to determine whether pharmacological inhibitors can block LMP-2A function using pre-B-cell colony formation with bone marrow from LMP-2A transgenic mice *(7)*. This approach not only provides insights into the function of LMP-2A in viral latency but also may allow for the design of therapeutics to eradicate latently infected B cells.

2. Materials

2.1. Preparation of Murine Bone Marrow Cells

1. Mice.
2. Forceps and scissors. Keep in sterile beaker containing 70% ethanol.
3. 5-mL Syringes.
4. 26-Gage needles.
5. 15-mL Conical centrifuge tubes.
6. Frosted microslides (cat. no. 48312–013, VWR Scientific).
7. 60 × 15-mm Tissue culture dish.
8. Red blood cell (RBC) lysis buffer: 0.15 M NH$_4$Cl, 10 mM KHCO$_3$, 0.1 mM Na$_2$EDTA; adjust pH to 7.2–7.4 with 1 N HCl, and filter sterilize through a 0.2-μm filter.
9. Phosphate-buffered saline (PBS), sterile.
10. Fetal bovine serum (FBS).
11. Iscove's modified Dulbecco's medium (MDM; GIBCO).

2.2. Colony Assay Using Methylcellulose

1. MethoCult™: 0.9% methylcellulose in Iscove's MDM, 30% FBS, 100 μM 2-mercaptoethanol, 2 mM L-glutamine, 10 ng/mL IL-7 (cat. no. 03630, StemCell Technologies, Vancouver, Canada). Methylcellulose media should be aliquoted in 3-mL aliquots in 15-mL conical tubes and stored at –20°C. The media is stable for at least 1 yr at –20°C.
2. 35 × 10-mm Tissue culture dish or 6-well tissue culture plate.
3. 5% CO$_2$ incubator.
4. Inverted microscope.

3. Methods

3.1. Preparation of Murine Bone Marrow Cells

1. Euthanize mice using procedures approved by the institutional animal care review board. Minimize contamination by thoroughly wetting the pelt with 70% ethanol, and peel the skin to expose the hindlimbs.
2. Before cutting off the hind legs, remove most of the muscle tissue by scissors. Cut off the leg bones at the hip joint, and remove excess tissues using sterile scissors and Kimwipes. Sever the femur and tibia at the knee joint and ankle, respectively.

3. Attach a 5-mL syringe to a 26-gage needle, and fill with 5 mL of Iscove's MDM containing 1% FBS. Insert needle into bone marrow cavity of femur or tibia. Flush the marrow with approx 2 mL medium per leg or until the bone cavity appears white. Allow wash medium to collect in 15-mL conical centrifuge tubes on ice (*see* **Note 2**).

4. Centrifuge for 5 min at 600 g, and discard supernatant.

5. Resuspend cells in 10 mL RBC lysis buffer, and incubate for 5 min on ice (*see* **Note 3**). During the incubation time, count cells using a hemacytometer.

6. Wash cells for 5 min at 600 g, and discard supernatant. Resuspend pellet to 2×10^7 cells/mL in Incove's MDM with 1% FBS (*see* **Note 4**).

3.2. Colony Assay Using Methylcellulose

1. Take appropriate number of cells (a total from 1×10^5 to 1×10^6) and dilute into 0.3 mL of Incove's MDM with 1% FBS (*see* **Note 5**).

2. Add 0.3 mL of cell suspension to 3.0 mL of MethoCult methylcellulose media in 15-mL conical tube. Vortex the tube, and allow 5–10 min for any bubbles to dissipate.

3. Dispense 1.5 mL into each of two 35-mm tissue culture dishes.

4. Place the dishes in a 100-mm Petri dish with a third 35-mm dish containing 3–4 mL of sterile water. The lid must be removed from this water dish to supply enough moisture (*see* **Note 6**).

5. Incubate for 7 d at 37°C and 5% CO_2 in a humidified incubator.

6. Score colonies using an inverted microscope (*see* **Note 7**).

4. Notes

1. A variety of cell culture media based on methylcellulose medium to detect human and murine hematopoietic progenitors are available from StemCell Technologies.

2. Expect cell recovery as follows: $1–2 \times 10^7$ per femur, 0.6×10^7 per tibia, and total $3–5 \times 10^7$ per mouse.

3. Red blood cells are more efficiently lysed at room temperature than on ice.

4. Purified bone marrow cells can be used for other analysis. Flow cytometric analysis (FACS) is one of the most preferable and common experiments. The cell density of 2×10^7 cells/mL is convenient to carry out FACS at the same time. For performing FACS, mix 50 μL cell suspension (10^6 cells) and 50 μL appropriately diluted labeled primary antibody to 12×75-mm round-bottomed test tubes, and incubate at 4°C for 30 min in the dark.

5. Overplating of cells in the medium may result in increased myeloid colony growth and inhibition of CFU–pre-B.

6. Six-well plates can be substituted for 35-mm dishes.

7. Bone marrow cells at 5×10^5 should yield 300–500 pre-B colonies per dish. Pre-B-cell colonies are usually uniform in shape and contain small round cells. There may also occasionally be dispersed macrophages in the dish. These cells are significantly larger in size, and the cell number per grouping is often less than 30.

References

1. Hardy, R. R. and Hayakawa, K. (2001) B cell development pathways. *Annu. Rev. Immunol.* **19,** 595–621.
2. Fleming, H. E. and Paige, C. J. (2002) Cooperation between IL-7 and the pre-B cell receptor: a key to B cell selection. *Semin. Immunol.* **14,** 423–430.
3. Reichman-Fried, M., Bosma, M. J., and Hardy, R. R. (1993) B-lineage cells in mu-transgenic scid mice proliferate in response to IL-7 but fail to show evidence of immunoglobulin light chain gene rearrangement. *Int. Immunol.* **5,** 303–310.
4. Hayashi, S., Kunisada, T., Ogawa, M., et al. (1990) Stepwise progression of B lineage differentiation supported by interleukin 7 and other stromal cell molecules. *J. Exp. Med.* **171,** 1683–1695.
5. Caldwell, R. G., Wilson, J. B., Anderson, S. J., and Longnecker, R. (1998) Epstein-Barr virus LMP2A drives B cell development and survival in the absence of normal B cell receptor signals. *Immunity* **9,** 405–411.
6. Caldwell, R. G., Brown, R. C., and Longnecker, R. (2000) Epstein-Barr virus LMP2A-induced B-cell survival in two unique classes of EmuLMP2A transgenic mice. *J. Virol.* **74,** 1101–1113.
7. Cooper, L. and Longnecker, R. (2002) Inhibition of host kinase activity altered by the LMP2A signalosome—a therapeutic target for Epstein-Barr virus latency and associated disease. *Antiviral Res.* **56,** 219–231.

20

Luciferase Real-Time Bioluminescence Imaging for the Study of Viral Pathogenesis

Gary D. Luker and David A. Leib

Summary

Herpes simplex virus 1 (HSV-1) is a common and significant neurotropic human pathogen that infects 80% of all persons by adulthood. During acute HSV-1 infection, virus replicates peripherally in epithelia, enters axonal terminals, and is transported retrogradely to sensory nerve ganglia, where HSV-1 may establish latency or progress to life-threatening infection of the central nervous system. Studies of viral and host factors that influence pathogenesis have largely used experimental mouse models that rely on sacrifice of infected mice to determine distribution and titer of virus. Although this experimental paradigm has provided important data, it precludes real-time investigations of the same animal over the entire course of disease progression. This limits potentially significant insights from animal-to-animal variations in host–pathogen relationships. Unexpected sites of infection also may be missed because appropriate tissues are not analyzed for virus. To improve investigations of viral and host factors that determine HSV-1 pathogenesis, we have validated bioluminescence imaging (BLI) as a technique to monitor infection with a recombinant strain KOS HSV-1 virus that expresses firefly luciferase (FL). This imaging technique allows repetitive, noninvasive monitoring of HSV-1 in living mice. In this chapter, we describe the protocols that we use for in vivo BLI of HSV-1 infection.

Key Words: Herpes simplex virus; pathogenesis; real-time imaging; bioluminescence.

1. Introduction

Studies of pathogens in small animal models, most often in mice, usually depend on observations of clinical symptoms, as well as sacrifice and harvesting of tissues for use in histopathology or in assays for the acutely replicating pathogen. Such methods have been standard in the field for many years and have yielded a great deal of useful information. The major drawback of such approaches, however, is that the sequential sacrifice of mice precludes the subsequent observation of microbiological, clinical, behavioral, and other out-

From: *Methods in Molecular Biology, vol. 292: DNA Viruses: Methods and Protocols*
Edited by: P. M. Lieberman © Humana Press Inc., Totowa, NJ

comes in the mice being sampled. Significant insights, therefore, from animal-to-animal variations in host–pathogen interactions and therapeutics are therefore likely to be missed. Furthermore, the ability to perform serial studies on the same animal would allow the experimenter to better interpret such animal-to-animal variation.

New imaging technology now allows real-time in vivo imaging of luciferase reporter genes in living mice using a cooled charge-coupled device (CCD) camera *(1,2)*. Because bioluminescence imaging (BLI) has minimal background, the technique is very sensitive for detecting sites of luciferase activity *(3)*. The substrate for firefly luciferase, D-luciferin, crosses cell membranes and the intact blood–brain barrier after intraperitoneal injection, thereby enabling luciferase activity to be detected in any anatomic site in the living mouse. In addition, the low immunogenicity and negligible toxicity of luciferin allows repetitive imaging of the same mouse. Relative amounts of bioluminescence produced in vivo can be quantified by region-of-interest (ROI) analysis, which facilitates comparisons among mice in various experimental groups *(2,4)*. The sensitivity for detecting bioluminescence with BLI is determined by the combination of signal strength and anatomic location. Because hair and overlying tissues scatter and absorb light, luciferase activity is detected more readily in superficial than internal sites in an animal. Darkly pigmented organs and tissues, such as liver, spleen, and skin of some mouse strains, also attenuate light transmission. Nevertheless, BLI can readily detect luciferase activity in internal organs such as brain, liver, lung, and spleen *(5,6)*. Other limitations of the technique include the 2–3-mm spatial resolution of BLI and the 2D imaging capability of first-generation instruments. Nevertheless, BLI provides certain advantages over use of position emission tomography (PET) or magnetic resonance imaging (MRI) in terms of cost, feasibility, throughput, and simplicity for real-time imaging of infectious diseases.

A large number of bacteria and viruses have now been studied using BLI *(7–11)*. Any recombinant microbe to be used for in vivo pathogenesis studies with this methodology ideally should contain a luciferase transgene that is expressed at the highest possible levels in all infected cells. Importantly, the transgene should have minimal, if any, effects on natural pathogenesis of the microbe. We describe here the use of a herpes simplex virus type 1 (HSV-1) recombinant that has been used very successfully in combination with BLI to elucidate the spread and tropism of HSV-1 in vivo in both wild-type and knock-out mice *(10–12)*. The method described here elucidates the background information of the generation of the particular virus we have used in our study. Obviously such a description will need to be adapted to the particular needs of the investigator, but it should give some useful general points for the design and generation of other viruses. We anticipate that this methodology will prove useful to many investigators for the study of the pathogenesis of HSV-1 and other viruses.

2. Materials
2.1. Virus Infection Materials

1. High titer ($\geq 4 \times 10^8$ PFU/mL) stock of KOSDlux/oriL (*see* **Note 1**).
2. Specific pathogen-free 6–8-wk-old mice (*see* **Note 2**).
3. Ketamine (Ketaset, Fort Dodge Animal Health).
4. Xylazine (Boehringer Ingelheim Vetmedica).
5. 25-Gage 5/8-inch needles (Becton Dickinson).
6. 28-Gage 1/2-inch U-100 tuberculin syringes (Becton Dickinson).
7. Eppendorf P10 pipetor and tips.

2.2. BLI

1. CCD camera and computer analysis software (IVIS, Xenogen) (*see* **Note 3**).
2. D-Luciferin (Xenogen, Alameda, CA), prepared as a 15 mg/mL stock solution in phosphate-buffered saline (PBS). The solution is sterile-filtered and stored in aliquots at –20°C. The compound is light-sensitive, so the reagent and solution should be protected from light as much as possible.
3. Isoflurane anesthesia delivery system (Harvard) with nose cones and tank.
4. Black paper to use on the imaging shelf of the CCD camera. We use Art Again black paper from Strathmore, which does not produce any background light.
5. 28-Gage 0.3- and 0.5-mL syringes.
6. Balance.
7. 70% Ethanol solution for decontaminating imaging equipment.

3. Methods
3.1. Infection of Mice

1. Anesthetize mice with 1.75 mg of ketamine and 0.26 mg of xylazine by intraperitoneal (ip) injection. The anesthesia will begin after approx 5 min and last for 15 min.
2. For corneal infection, lightly abrade the cornea with eight interlocking strokes of a 25-gage needle. Drop HSV-1 in a 5 µL vol onto both eyes at a maximum concentration of 2×10^6 PFU per eye. Gently massage the eyes with the lids closed for 10 s.
3. For footpad injection, using a 0.5-mL tuberculin syringe, inject up to 20 µL of virus subcutaneously into the footpad.

3.2. Animal Preparation for Imaging

1. To shave mice, anesthetize animals with 2% isoflurane or other approved anesthetic, and use electric clippers to remove hair from areas of interest. Shaving is repeated as needed over the course of an experiment, typically every 4–5 d (*see* **Note 4**).
2. Label individual mice with a permanent identification marker, such as an ear punch. If mice were anesthetized for shaving, this is a convenient time to perform

the ear punch. To allow rapid identification of mice, use a permanent marker to label the tail of each mouse with the appropriate number.

3. Weigh each mouse in a tared plastic beaker. Anesthesia is not necessary to obtain reliable weights.

3.3. CCD Camera Setup

1. Select the smallest field of view (FOV) that will accommodate the desired number of mice to be imaged at one time (*see* **Note 5**). For all our pathogenesis studies, we use the smallest FOV (10 cm) that allows two normal-sized adult mice to be imaged simultaneously.
2. Select the size of image matrix to optimize sensitivity vs spatial resolution. We use a 128×128 matrix to enhance camera sensitivity for studies with KOSDlux/oriL.
3. Define image acquisition time. We begin with a 1-min image and subsequently adjust acquisition time based on amounts of bioluminescence.

3.4. Imaging

1. Inject D-luciferin ip at a dose of 15 mg/kg from the 15 mg/mL stock solution, using a 28-gage, 0.3- or 0.5-mL syringe (*see* **note 6**).
2. Return animals to cages for 5 min to allow distribution of luciferin throughout tissues.
3. Anesthetize mice with 2% isoflurane, which typically requires 2–3 min.
4. Transfer mice to the stage of the CCD camera, placing anatomic sites of interest closer to the camera. Anesthesia is maintained with 2% isoflurane delivered via nose cones.
5. Begin imaging 10 min after injection of luciferin. Initially, the system acquires a gray-scale photograph of mice. Check this gray-scale image to determine that animals are centered properly under the camera. If animals need to be repositioned, stop the image acquisition and move mice as needed.
6. The system proceeds automatically to the bioluminescent photograph, using the selected imaging parameters.
7. Based on the initial image, adjust matrix size and/or acquisition time as needed to detect bioluminescence without saturating the CCD camera.
8. Obtain images from as many positions as desired. Bioluminescence is automatically presented as a pseudocolor representation superimposed on the gray-scale photograph of each mouse (**Fig. 1**).
9. Remove mice from the CCD camera and return to cages. Animals typically recover within 5 min.

3.5. Image Analysis

1. Define a uniform minimum threshold for pseudocolor display of relative amounts of light on images (*see* **Note 7**).
2. Manually define ROIs for all desired anatomic sites for each mouse (**Fig. 1**). Use the saturation map feature of the processing software to avoid quantifying light emission from pixels that saturate the CCD camera.

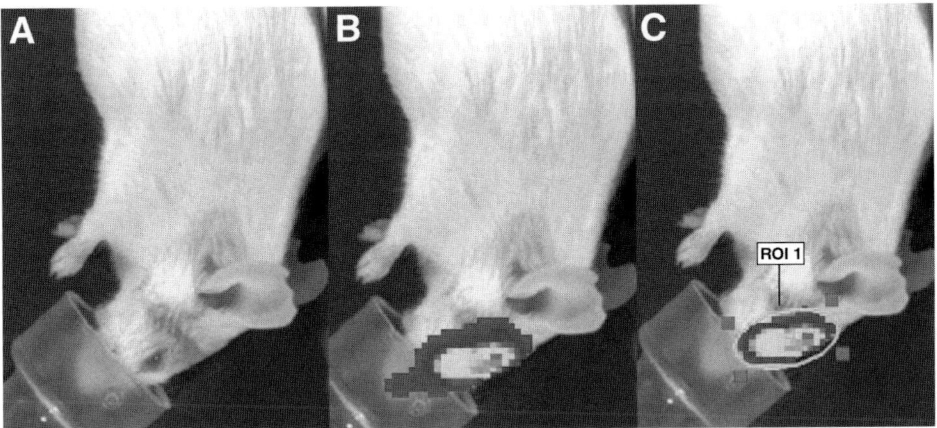

Fig. 1. Bioluminescence imaging 4 d after corneal infection with KOSDlux/oriL. (**A**) Gray-scale photograph of mouse. (**B**) Bioluminescence superimposed on gray-scale photograph of mouse. Relative levels of bioluminescence are depicted as a pseudocolor display, with red and blue representing the highest and lowest amounts of photon flux. (**C**) After applying a minimum threshold value to the image, a ROI is manually defined around the light projected from the eye and periocular tissues.

3. Determine background bioluminescence from an ROI of the same size. This may be obtained from a mock-infected mouse that was injected with luciferin.
4. Measure bioluminescence units of photon flux (photons/s) in each ROI to correct for differences in image acquisition time.
5. Subtract background bioluminescence from photon flux in each ROI of interest to quantify relative luciferase activity as a measure of amounts of virus.
6. If desired, summarize ROI data over the entire experiment by area-under-the-curve (AUC) analysis, using Kaleidograph (Synergy Software) or other appropriate data analysis software.

3.6. Conclusion

The methods presented in this chapter describe how BLI can be applied to studies of virus–host pathogenesis in living mice. As demonstrated by our research and others, BLI has significant advantages over conventional methods used to investigate viruses in mouse models. Noninvasive imaging allows the same group of animals to be studied over time, which reduces the number of animals used in experiments. By allowing serial studies on the same animal, this imaging technique can overcome animal-to-animal variations in experimental data. Moreover, BLI allows the entire mouse to be monitored for viral replication. Using conventional methods that require sacrifice of mice for determinations of viral titers, sites of infection may potentially be missed because the appropriate tissues were not analyzed. As described above, limitations of spatial resolution may hinder selected applications of this imaging technique. In addition, the success of the technique depends on engineering a recombinant reporter virus that is not attenuated relative to the appropriate wild-type strain. With knowledge of the strengths and limitations of the technology, we anticipate that BLI will provide new insights into pathogenesis of HSV-1 and other viruses in mouse models of disease.

4. Notes

1. *Generation of luciferase-expressing viruses.* The virus used in this study, KOSDlux/oriL, has two critical properties that make it successful for these studies. First, the luciferase cassette was inserted into a locus that only minimally affects HSV-1 growth, tropism, and pathogenesis. This locus, between UL49 and UL50, is marked in the wild-type genome by a *Bgl*II site at map position 106,750. Insertion of reporter cassettes into this locus actually truncates the carboxy-terminal 11 residues of the *UL49.5* gene but results in no alteration in ability of the virus to grow in cell culture, and no more than a 10-fold reduction in replication in any tissue tested in vivo at any time-point. Second, the placement of the firefly luciferase cassette under control of the ICP8 (single-stranded DNA binding protein) promoter resulted in expression of very high levels of luciferase. To be able to detect luciferase activity from anything other than very superficial tissues, it is critical that the virus express as much luciferase as possible. In addition, although KOSDluxoriL also expresses renilla luciferase, we have demonstrated that bioavailability of coelenterazine, the substrate for renilla luciferase, is poor. Therefore, the recombinant should contain the firefly luciferase gene, and renilla luciferase is not recommended for use in this type of system. We employed firefly luciferase from the plasmid pGL3 (Promega), but it is likely that other versions of firefly luciferase would be satisfactory. The KOSDlux/oriL recombinant was generated by a cotransfection of HSV-1 KOS infectious DNA with a plasmid containing the luciferase cassette cloned into the *Bgl*II site between UL49 and UL50. Progeny virus plaques were screened very simply for their ability to express luciferase in a 48-well assay. Plaques with recombinant viruses could also potentially be detected by BLI of tissue culture plates. Plaques of interest were followed up with Southern blotting and further rounds of plaque purification.
2. *Mouse strains.* Effects of hair and skin pigmentation are important considerations in selecting a mouse model for studying HSV-1 infection with BLI. If possible, mice with white hair, including Balb/c and CD-1 strains, should be used for imaging examinations because white hair decreases light transmission by only 18% *(2)*. By comparison, strains with dark hair and pigmentation, such as C57Bl/6, reduce transmitted light by as much as 10-fold relative to hairless mice. One alternative is to use albino C57BL/6-Tyr^{c-2J} mice (Jackson Laboratories), although additional breeding is necessary to apply this strategy in genetically engineered mice. For studies with pigmented mouse strains, we find that shaving hair with electric clippers is a rapid, effective strategy for improving sensitivity of BLI studies. We avoid

depilatory agents because these compounds may irritate the skin of mice. Mice are anesthetized with 2% isoflurane delivered via nose cone and then shaved prior to the first day of imaging. We shave the animals again as needed during the course of infection, typically every 4–6 d.

3. *General limitations of BLI.* Although BLI provides unique advantages of repetitive, noninvasive monitoring of acute HSV-1 infection in living mice, users of this technology should be aware of limitations to this method. Light transmission is attenuated by hair, overlying soft tissues, and pigmentation of organs such as liver or spleen. Therefore, sources of light that are closer to the animal surface and CCD camera produce relatively greater photon flux. At a defined anatomic site, amounts of bioluminescence correspond directly to viral titers. However, photon flux from a superficial infection, such as periocular skin, will be greater than the same titer of virus within an internal organ, such as brain.

   The current first-generation technology for BLI provides 2D images, so bioluminescence detected by the CCD camera may arise from overlapping sites of viral infection. As described above, imaging in multiple projections can improve spatial localization of virus within infected tissues. This limitation will be overcome in the future by 3D imaging systems that are currently under development.

4. *Anesthesia.* An inherent limitation of the system is the use of volatile anesthetics for sedation. Volatile anesthetics bind to luciferase and inhibit the reaction of luciferase with luciferin, and the potency of anesthetics in vivo correlates directly with suppression of luciferase activity *(13,14)*. This competitive inhibition has not prevented us from detecting KOS/dlux at a variety of anatomic sites in living mice, including brain and liver. However, it does emphasize the need to maintain a constant percent of anesthetic and duration of anesthesia for all mice throughout an experiment.

5. *CCD camera parameters.* The smallest FOV possible should be used in each experiment to optimize spatial resolution and sensitivity for bioluminescence. Therefore, we use the 10-cm FOV for essentially all our pathogenesis studies. This FOV readily accommodates two normal-sized adult mice without overlap of bioluminescence from one mouse to another. If an experiment requires imaging more than two mice at a time, then a larger FOV should be selected.

   Image resolution is determined by the number of picture elements (pixels) that are used to generate each image. The highest spatial resolution image, which consists of a 1024×1024 matrix, decreases sensitivity for light emission relative to matrices with fewer pixels. Because BLI has a spatial resolution of 2–3 mm *(17)*, distinguishing between two adjacent anatomic sites may not be feasible regardless of image matrix. Therefore, we usually select a 128×128 matrix to enhance sensitivity for bioluminescence.

   Image acquisition time is empirically determined based on anticipated amounts of luciferase activity. The CCD camera can detect a maximum of 65,535 counts per pixel. Greater amounts of emitted light will saturate the camera, so the true bioluminescence value in saturated pixels will exceed the recorded measurement by an unknown amount. If luciferase activity saturates the camera, a warning label

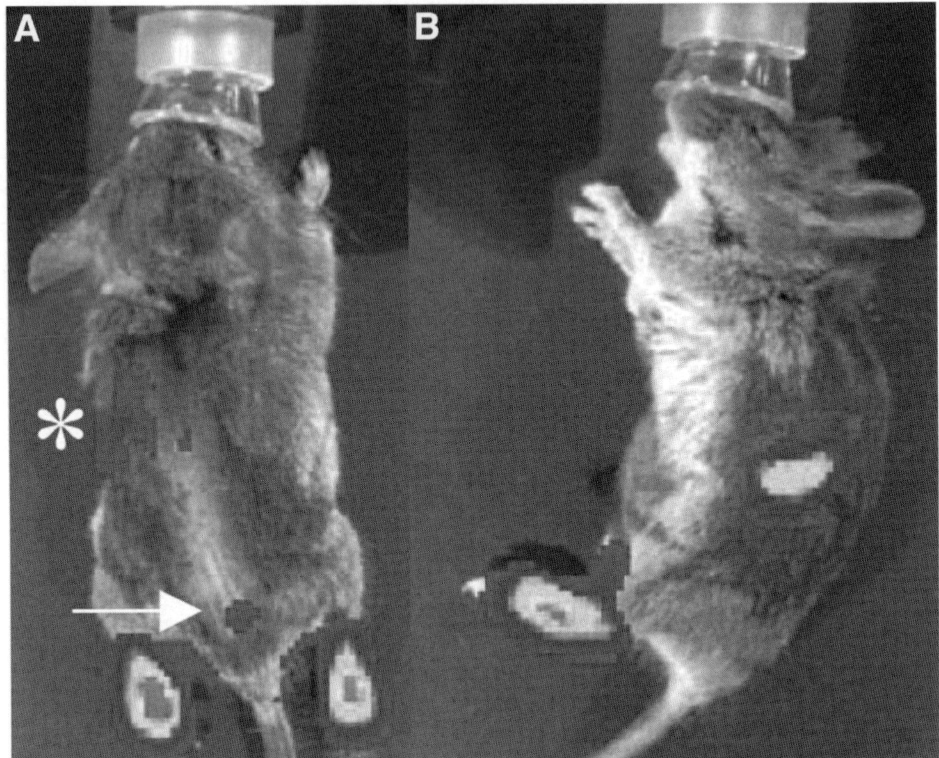

Fig. 2. Imaging in multiple projections 3 d after bilateral footpad infection with KOSDlux/oriL. **(A)** Dorsal view shows luciferase activity in bilateral footpads and the sacral spine (arrow). Additional activity is seen over the left side of the abdomen (asterisk). **(B)** Left lateral view shows bioluminescence arising from the spleen.

appears with the image. Reducing acquisition time or increasing the number of pixels in each image can eliminate image saturation. For very low amounts of bioluminescence, imaging times of 5–10 min may be needed to detect luciferase activity above background levels of the system. We typically begin experiments with a 1-min acquisition and then adjust imaging time or resolution as needed to optimize the signal-to-noise ratio or eliminate image saturation.

6. *Imaging.* Bioluminescent images are 2D, and light is attenuated approx 10-fold per cm of tissue *(16)*. Therefore, additional images may be needed to position different parts of mice closer to the CCD camera **(Fig. 2)**. Obtaining multiple projections can enhance detection of luciferase activity and improve localization of bioluminescence produced from organs and tissue that overlap on a single view. Again, exposure parameters may need to be optimized for amounts of bioluminescence produced by virus at various anatomic sites.

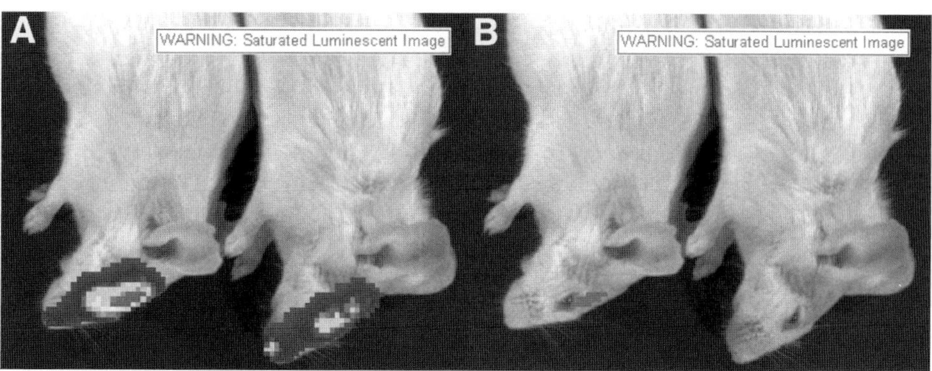

Fig. 3. Image with saturated pixels 4 d after corneal infection. (**A**) One-minute image shows saturation warning label. (**B**) Saturation map shows saturated pixels on the image of the mouse on the left. Photon flux cannot be quantified accurately from this site, although this image can be used to measure luciferase activity from KOSDlux/oriL in the right mouse.

We routinely monitor progression of acute HSV-1 infection with BLI studies performed approximately every 24 h. However, BLI may be repeated more frequently if necessary. Bioluminescence falls to background levels approx 6 h after ip injection of luciferin. Therefore, imaging may be performed every 6 h without accounting for residual bioluminescence. If more frequent examinations are indicated, then an image should be obtained prior to injection of additional luciferin, thereby allowing correction for residual bioluminescence at defined sites of infection.

7. *Data analysis.* Bioluminescence from an anatomic site of infection is quantified by ROI analysis, using the LivingImage software on the imaging system. For all studies within an acute infection experiment, we define a minimum threshold value for pseudocolor display of light emission. This threshold is applied to each image, and then the manually defined ROI is drawn to encompass all light bioluminescence at a site (**Fig. 1**). As luciferase activity increases above the threshold value, relative amounts of bioluminescence become detectable on the pseudocolor scale and are included in the ROI. ROIs are drawn at each site of interest for all animals and quantified as photon flux (photons/s). By measuring light emission as photon flux, data are normalized for differences in image acquisition time and distance of mice from the CCD camera.

Data can be quantified accurately from any site that did not saturate the CCD camera. If any area of an image saturates the recording capability of the CCD camera, a warning label is displayed automatically on the image. Areas of saturation may be identified by using the saturation map feature of the software analysis program (**Fig. 3**). Because saturated sites have exceeded the capacity of the CCD camera to detect individual photons accurately, photon flux from these pixels will under-report actual luciferase activity. An image without saturation, such as images with shorter exposure time or higher binning, should be selected for analysis of these sites of infection.

To summarize cumulative bioluminescence over the course of an experiment, we commonly use AUC analysis of photon flux. AUC allows comparisons of photon flux in defined regions over the time-course of acute HSV-1 infection. Using AUC may show significant overall differences in photon flux and viral replication that are not identified when data are analyzed on individual days of infection. We calculate AUC with Kaleidagraph (Synergy Software, Reading, PA), although other statistical programs also may be used.

Acknowledgments

The authors thank David Piwnica-Worms for encouragement, Kathryn Luker for helpful discussions, and Matthew Smith for technical assistance. This study was funded by NIH grants RO1 EY09083, P50 CA94056, the McDonnell Center for Cellular and Molecular Neurobiology, and P30-EY02687 to the Department of Ophthalmology and Visual Sciences. Support from Research to Prevent Blindness to the Department of Ophthalmology and Visual Sciences and a Lew Wasserman Scholarship to David A. Leib are gratefully acknowledged.

References

1. Contag, C., Spilman, S., Contag, P., et al. (1997) Visualizing gene expression in living mammals using a bioluminescent reporter. *Photochem. Photobiol.* **66,** 523–531.
2. Wu, J., Sundaresan, G., Iyer, M., and Gambhir, S. (2001) Noninvasive optical imaging of firefly luciferase reporter gene expression in skeletal muscles of living mice. *Mol. Ther.* **4,** 297–306.
3. Lipshutz, G., Gruber, C., Cao, Y., Hardy, J., Contag, C., and Gaensler, K. (2001) In utero delivery of adeno-associated viral vectors: intraperitoneal gene transfer produces long-term expression. *Mol. Ther.* **3,** 284–292.
4. Sadikot, R., Jansen, E., Blackwell, T., et al. (2001) High-dose dexamethasone accentuates nuclear factor-$\kappa\beta$ activation in endotoxin-treated mice. *Am. J. Respir. Crit. Care Med.* **164,** 873–878.
5. Contag, C. and Ross, B. (2002) It's not just about anatomy: in vivo bioluminescence imaging as an eyepiece into biology. *J. Magn. Reson. Imaging* **16,** 378–387.
6. Benaron, D., Contag, P., and Contag, C. (1997) Imaging brain structure and function, infection and gene expression in the body using light. *Philos. Trans. R. Soc. Lond. B Biol. Sci.* **352,** 755–761.
7. Francis, K., Joh, D., Bellinger-Kawahara, C., Hawkinson, M., Purchio, T., and Contag, P. (2000) Monitoring bioluminescent *Staphylococcus aureus* infections in live mice using a novel luxABCDE construct. *Infect. Immun.* **68,** 3594–3600.
8. Francis, K., Yu, J., Bellinger-Kawahara, C., et al. (2001) Visualizing pneumococcal infections in the lungs of live mice using bioluminescent *Streptococcus pneumonia* transformed with a novel gram-positive *lux* transposon. *Infect. Immun.* **69,** 3350–3358.

9. Cook, S. and Griffin, D. (2003) Luciferase imaging of a neurotropic viral infection in intact animals. *J. Virol.* **77,** 5333–5338.
10. Luker, G. D., Bardill, J. P., Prior, J. L., Pica, C. M., Piwnica-Worms, D., and Leib, D. A. (2002) Noninvasive bioluminescence imaging of herpes simplex virus type 1 infection and therapy in living mice. *J. Virol.* **76,** 12149–12161.
11. Luker, G., Prior, J., Song, J., Pica, C., and Leib, D. (2003) Bioluminescence imaging reveals systemic dissemination of HSV-1 in the absence of interferon receptors. *J. Virol.* **77,** 11,082–11,093.
12. Summers, B. C. and Leib, D. A. (2002) Herpes simplex virus type 1 origins of DNA replication play no role in the regulation of flanking promoters. *J. Virol.* **76,** 7020–7029.
13. Eckenhoff, R., Tanner, J., and Liebman, P. (2001) Cooperative binding of inhaled anesthetics and ATP to firefly luciferase. *Proteins* **42,** 436–441.
14. Zhang, Y., Stabernack, C., Dutton, R., et al. (2001) Luciferase as a model for the site of inhaled anesthetic action. *Anesth. Analg.* **93,** 1246–1252.
15. Weissleder, R. (2002) Scaling down imaging: molecular mapping of cancer in mice. *Nat. Rev. Cancer* **2,** 1–8.
16. Contag, C. H. and Bachmann, M. H. (2002) Advances in in vivo bioluminescence imaging of gene expression. *Annu. Rev. Biomed. Eng.* **4,** 235–260.

VII

COMPLEX CELL SYSTEMS

21

Culturing Primary and Transformed Neuronal Cells for Studying Pseudorabies Virus Infection

Toh Hean Ch'ng, E. Alexander Flood, and Lynn William Enquist

Summary

This chapter discusses the culture of primary sympathetic neurons (superior cervical ganglia) from rat embryos and PC12 cells differentiated into neurons for use in viral infection experiments. Methods are described for the use of a neurotropic herpesvirus, pseudorabies virus (PRV), to analyze the assembly, egress, and transport of viral antigens in neurons.

Key Words: Cultured neuron; herpesvirus; immunofluorescence; infection; laminin; nerve growth factor; neuron; ornithine; PC12; pheochromocytoma; pseudorabies virus; superior cervical ganglia; sympathetic neurons.

1. Introduction

This chapter focuses on the culture of primary sympathetic neurons (superior cervical ganglia) from rat embryos and PC12 cells differentiated into neurons for use in viral infection experiments. Although methods for culturing these types of neuronal cells are well established (*1,2*), it is often necessary to modify the cell culture method to suit the requirements of a particular experiment. We use these neuronal cultures to understand how the neurotropic herpesviruses traverse the long distances of axons between nerve cells. An important technical advance was to develop a protocol for analyzing infected cultures by immunofluorescence assays. This assay is challenging because the cytopathic effects of viral replication weaken the attachment of the neurons to their substrate, and the neurons are easily washed away during changes of medium. The following methods were developed for use of the neurotropic herpesvirus pseudorabies virus (PRV) to analyze the assembly, egress, and transport of viral antigens in neurons (*3*). PRV is a useful model to study herpesvirus infection because it is closely related to the human herpesviruses, herpes simplex and

From: *Methods in Molecular Biology, vol. 292: DNA Viruses: Methods and Protocols*
Edited by: P. M. Lieberman © Humana Press Inc., Totowa, NJ

varicella-zoster. Additionally, PRV infects a wide range of mammals, except for higher primates, providing flexibility in cell culture use and relative safety for laboratory workers.

2. Materials

2.1. Preparation of Serum-Free Culture Media and Culture Dishes

2.1.1. Serum-Free Neuronal Culture Media

1. 1X F12 nutrient mixture (Ham's) liquid (Invitrogen).
2. 1X Dulbecco's modified Eagle's medium (DMEM), liquid, high glucose (Cellgro).
3. Insulin, bovine pancreas, powder (Sigma). Dissolve insulin in tissue culture-grade water (Sigma), and filter-sterilize. Working solution is stable at 4°C for 1 wk.
4. Human holo-transferrin, powder, cell culture grade (Sigma). Dissolve transferrin in 0.9% saline solution, filter-sterilize, and store in –20°C freezer. Do not freeze-thaw. Keep thawed aliquots at 4°C. Discard after 2 wk.
5. Putrescine (tetramethylenediamine dihydrochloride; Sigma). Dissolve putrescine in 0.9% saline solution, filter-sterilize, and store in –20°C freezer. Do not freeze-thaw. Keep thawed aliquot at 4°C. Discard after 2 wk.
6. Progesterone, cell culture (Sigma). Dissolve progesterone in tissue culture-grade water, filter-sterilize, and store in –20°C freezer. Do not freeze-thaw. Keep thawed aliquot at 4°C. Discard working solution after 1 mo.
7. Selenium (in sodium selenite form; Sigma). Dissolve sodium selenite in tissue culture grade water, filter-sterilize, and store stock at –20°C freezer. Prepare and store working solution in 4°C. Discard stock solution after 1 yr.
8. Glucose. Dissolve glucose in tissue culture-grade water, filter-sterilize, and store at 4°C.
9. 200 m*M* L-Glutamine (Invitrogen). Keep frozen aliquots of L-glutamine at –80°C. Discard remainder of aliquot after use.
10. Bovine serum albumin (BSA), fraction V via cold alcohol precipitation (Sigma).
11. Penicillin/streptomycin, liquid (Invitrogen).
12. Nerve growth factor, 2.5S (Invitrogen; *see* **Note 1**).
13. Luer-Lok sterile syringe with different volumes (Becton Dickinson).
14. 0.22-μm Syringe-driven filter units (Millipore).
15. 50-mL Disposable screw-cap conical tubes (Sarstedt).

2.1.2. Preparing Neuron Culture Dishes

1. Poly-DL-ornithine hydrobromide 3000–15,000 MW (Sigma). Dissolve poly-DL-ornithine in 0.1 *M* borate buffer, pH 8.3.
2. Polystyrene tissue culture dishes (35 × 10 mm; Becton Dickinson).
3. Natural mouse laminin (Invitrogen).
4. Saline G (Sal G) buffer, pH 7.4: 6.1 m*M* glucose, 5.3 m*M* KCl, 136.9 m*M* NaCl, 1.1 m*M* KH_2PO_4, 1.08 m*M* Na_2HPO_4, 0.11 m*M* $CaCl_2$ (anhydrous), 0.63 m*M* $MgSO_4$ (*see* **Note 2**).

5. Ca^{2+}/Mg^{2+}-free (CMF) saline G buffer, pH 7.4: 6.1 mM glucose, 5.3 mM KCl, 136.9 mM NaCl, 1.1 mM KH$_2$PO$_4$, 1.08 mM Na$_2$HPO$_4$ (*see* **Note 2**).
6. Tissue culture-grade water (Sigma).

2.2. Culturing Dissociated Rat Embryonic Sympathetic Ganglia

2.2.1. Dissection of Embryonic Rat Sympathetic Ganglia

1. 6-in Tissue forceps (1 × 2 teeth; Roboz Surgical).
2. 5.5-in Straight, sharp-blunt operating scissors (Roboz Surgical).
3. Dumostar no. 5 microdissecting tweezers (Roboz Surgical).
4. Two 125-mm no. 2 microdissecting knife (blade 1.7 × 27 mm; Roboz Surgical).
5. Two 100-mm Pyrex glass Petri dishes bottoms (Corning).
6. Two 100-mm bacterial polystyrene Petri dishes (Fisher).
7. Saline G buffer.
8. Glass Pasteur pipets with flame-smoothened tips.
9. Styrofoam surgical stage layered with dissection diapers.
10. Pregnant Sprague-Dawley female rat at stage E15.5 (Hilltop).
11. 70% Ethanol.

2.2.2. Plating the Dissociated Sympathetic Neuron Culture

1. CMF saline G buffer (*see* **Subheading 2.1.2., item 5**).
2. Glass Pasteur pipets with flame-smoothened tips.
3. Poly-DL-ornithine and laminin-treated tissue culture dishes.
4. Trypsin (Worthington Biochemicals). Dissolve trypsin in CMF saline G buffer, and filter-sterilize.
5. Trypsin inhibitor, type I-S from soybean (Sigma). Dissolve trypsin inhibitor in CMF saline G buffer, and filter-sterilize.
6. Serum-free neuronal culture media.

2.2.3. Maintaining the Dissociated Sympathetic Neuron Culture

1. Serum-free neuronal culture media.
2. Cytosine β-D-arabinofuranoside (Ara-C; Sigma).

2.3. PC12 Cell Culture and Differentiation

1. Tissue culture dishware (Becton Dickinson, Falcon).
2. PC12 cells (ATCC).
3. Horse serum (Invitrogen).
4. Fetal calf serum (FCS; Invitrogen).
5. Nerve growth factor 2.5S subunit, murine (Invitrogen; *see* **Note 1**).
6. Growth medium: 85% RPMI, 10% horse serum, 5% FCS.
7. Differentiation medium: 99% RPMI, 1% horse serum, (50 µg/mL) nerve growth factor (NGF).
8. Freezing medium: 80% RPMI, 10% horse serum, 10% dimethylsulfitide (DMSO).
9. 0.1 M Borate buffer, pH 8.3, filter-sterilized.

10. 10X HEPES-buffered saline solution (HBSS): 1.3 M NaCl, 100 mM HEPES, 54 mM KCl, 42 mM NaHCO$_3$, 4.5 mM KH$_2$PO$_4$, 3.4 mM Na$_2$HPO$_4$ Adjust pH to 7.2 with NaOH, filter-sterilize, and store 45-mL aliquots at –20°C.
11. Natural mouse laminin. Store 20-µL aliquots at –80°C. (Invitrogen).
12. Poly-DL-ornithine hydrobromide 3000–15,000 MW (Sigma).
13. Rat tail collagen, type 1. Store at 4°C, up to 3 mo (Becton Dickinson Biosciences).

2.4. Infection and Immunofluorescence of Neuron Cultures

1. High titer viral stock.
2. 1X DMEM, liquid, high glucose, (Cellgro).
3. 16% Paraformaldehyde ampules (Electron Microscopy Sciences).
4. 3% BSA (Roche) in phosphate-buffered saline (PBS; Cellgro).
5. 3% BSA (Roche) and 1% saponin in PBS (Cellgro).
6. Infection medium: 99% RPMI, 1% horse serum.
7. Appropriate primary and secondary antibodies.
8. Vectashield mounting medium for fluorescence samples (Vector Laboratories).
9. Cover slips, no. 1, 22 mm^2 (Corning).
10. Nail varnish.
11. Aqua Poly/Mount (Polysciences).
12. 18-mm Round glass no. 1 cover slips (Fisher Scientific, cat. no. 1254584-18cir1D) (*see* **Note 3**).

3. Methods

3.1. Preparation of Serum-Free Neuron Culture Media and Culture Dishes

Our laboratory uses a serum-free and chemically defined medium to culture the sympathetic neurons. By using serum-free media, we eliminate the variability arising from different batches of serum. Typically, a single dissection of 10 embryos (20 superior cervical ganglia) can yield sufficient dissociated neurons for twenty 35-mm tissue culture dishes. The density of each culture can vary depending on the number of ganglia dissected or the number of tissue culture dishes used for plating. Since each dish requires 2 mL of neuron culture medium, prepare 40 mL of the medium prior to dissection. To avoid any contamination, it is important to filter-sterilize each individual stock solution.

3.1.1. Serum-Free Neuronal Culture Media

1. Pipet 10 mL of Ham's F12 nutrient mix into a 50-mL disposable conical tube.
2. Add 400 mg of BSA into the tube, and mix well by vigorous pipeting. Bring volume up to 40 mL by adding 10 mL of Ham's F12 and 20 mL of DMEM.
3. Filter-sterilize basal medium using a 0.22-µm syringe-driven filter unit.
4. Add each supplement listed in **Table 1** to the basal medium (*see* **note 4**). Mix well.
5. Maintain temperature and pH of the serum-free medium by placing in a 37°C, CO$_2$ tissue culture incubator (*see* **Note 5**).

Table 1
List of Supplements in a Serum-Free Neuronal Culture Media

Supplement	Stock concentration	Volume (μL)	Final concentration
Glucose	460 mg/mL	400	4.6 mg/mL
Transferrin	10 mg/mL	400	100 μg/mL
Putrescine	1.6 mg/mL	400	16 μg/mL
Insulin	1 mg/mL	400	10 μg/mL
Glutamine	200 mM	400	2 mM
Penicillin/streptomycin	10,000 μg/mL or U/mL	200	50 μg/mL or U/mL
Selenium	30 μM	40	30 nM
Progesterone	66.7 μM	12	20 nM
Nerve growth factor	100 μg/mL	40	100 ng/mL

3.1.2. Preparing the Tissue Culture Dishes for Plating Neurons

Different tissue culture dishes can be used to grow sympathetic neurons. Some protocols require glass cover slips, and others use Aclar dishes. Our protocol uses the 35-mm standard plastic tissue culture dishes from Falcon. The 60-mm, 6-well or multiwell tissue culture dishes can also be used in place of the 35-mm dish.

1. Dissolve poly-DL-ornithine in borate buffer at a concentration of 500 μg/mL. Filter-sterilize, and store aliquots at –20°C.
2. One day prior to dissection, add 1 mL of poly-DL-ornithine into each 35-mm tissue culture dish. Place dishes in a 37°C CO_2 incubator overnight (*see* **Note 6**).
3. Four to 5 h prior to dissection, remove poly-DL-ornithine from all the dishes, and wash each dish three times with 1 mL of tissue culture-grade water. After the final wash, add 1 mL of laminin (10 μg/mL diluted in CMF saline G buffer) to each dish, and place them back in the incubator. Allow laminin to coat the surface of the dishes for 4–5 h before starting the dissection.

3.2. Culturing Dissociated Rat Embryonic Sympathetic Ganglia

All subsequent protocols including the dissection are carried out in a laminar flow hood to minimize contamination.

3.2.1. Dissection of Embryonic Rat Sympathetic Ganglia

1. Sterilize dissection tools by autoclaving and then immersing them in 70% ethanol.
2. Prepare a pregnant Sprague-Dawley female rat (E15.5) for dissection. Euthanize the rat by placing it in a CO_2-filled chamber for 5 min. Pinch the rat's hind footpad and observe for reflex to verify that the rat has expired.
3. Place the rat on a styrofoam surgical stage layered with absorbent diapers, with her back facing down, and then immobilize the rat by pinning the hind legs down to the

Fig. 1. Location of the superior cervical ganglia (SCG) within the thin slice of the lower mandible.

    surgical stage with pins. Next, sterilize the fur around her underbelly by dousing the fur with 70% ethanol.

4. Using the tissue forceps, lift skin and make an incision 2–3 cm ventral to the urogenital opening with the operating scissors. Make a ventral midline incision all the way to the diaphragm. Do not cut through the muscle layer yet; just separate the skin from the muscle tissue.

5. Next, make a second incision at the muscle tissue at the location of the first incision near the urogenital opening. As described in **step 3,** open the peritoneal cavity by cutting through the muscle tissue from the urogenital opening to diaphragm. Avoid lacerating any organs within the peritoneal cavity.

6. Using a sterile tissue forceps, gently lift the embryos from each horn of the uterus. Detach the embryos from the tissue by cutting off the fallopian tube. Transfer the string of embryos to a tissue culture dish.

7. Using a pair of sterile microdissecting tweezers, carefully remove each embryo from the uterine sac. Avoid decapitating the embryos during the extraction, as they are fragile. Make sure the placenta, as well as the amniotic membrane, has been removed from each embryo. Transfer the embryos to a tissue culture dish filled with Sal G buffer. Wash the embryos gently by swirling the dish and buffer around. Repeat wash one more time by transferring the embryos to a new dish. Do not discard the buffer from the first embryo wash (*see* **Note 7**).

8. After both washes, transfer the embryos to a glass Petri dish bottom containing Sal G buffer. With the aid of a dissection microscope, decapitate the embryos at the neckline using a pair of sterile microdissecting knives **(Fig. 1)**. Next, make an incision just below the ear to obtain a thin slice of the lower mandible (dotted lines in **Fig. 1;** *see* **Note 8**). Repeat for all embryos.

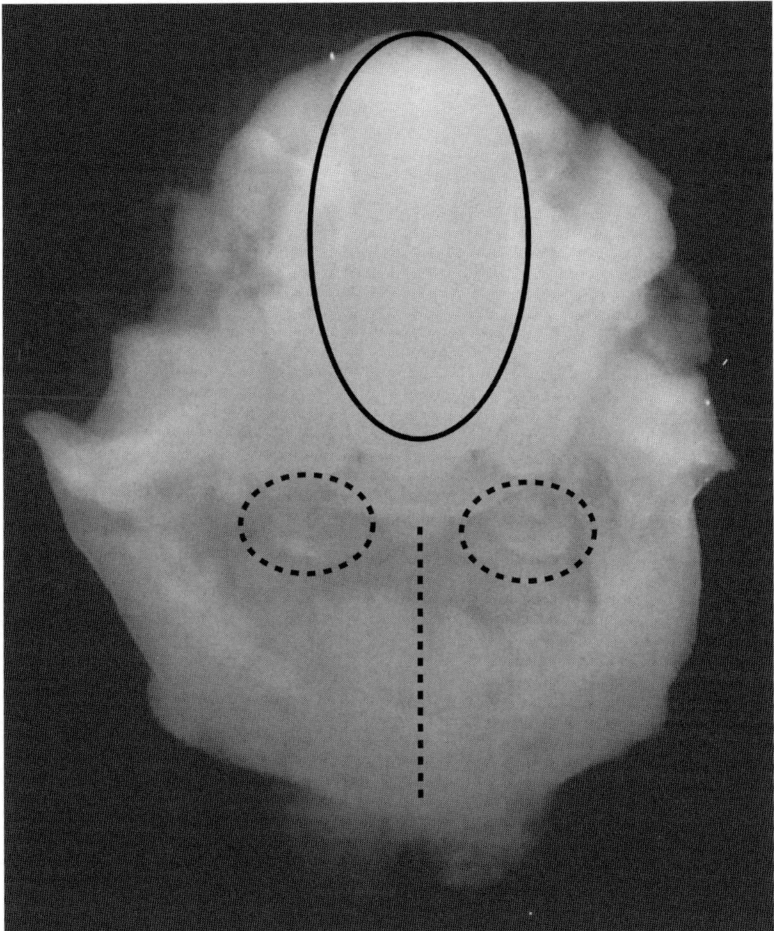

Fig. 2. A magnified thin-slice image of the lower mandible. The solid ellipse outlines the tongue, and the dotted circles denote the location of the SCG. The dotted midline incision from the larynx to the spinal column separates the lower half of the thin slice for easier access to the SCG.

9. Transfer all the thinly sliced lower mandibles to a new glass Petri dish bottom containing Sal G buffer. Before starting the microdissection, align each slice so that the larynx and tongue are facing up and on top, and the spinal column is at the bottom (**Fig. 2**). With the microdissecting knives, first remove the spinal column. Next, make an incision in the center of each slice from the larynx to where the spinal column was located (dotted line in **Fig. 2**). This incision separates the bottom half of the slice, revealing the ganglia. Finally, gently tease apart the surrounding tissue to extract the ganglia from both sides of the slice (circles in **Fig. 2**; *see* **Note 9**).

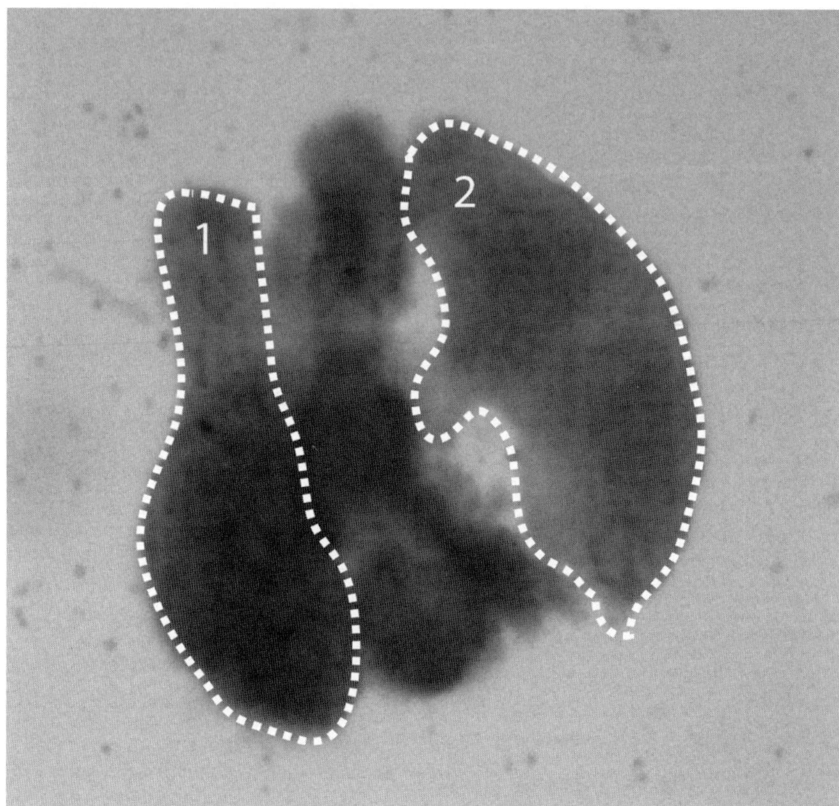

Fig. 3. A magnified image of the superior cervical ganglion (2) that is still attached to the carotid artery. The nodose ganglion (1) is usually adjacent to the SCG.

10. Each superior cervical ganglion (SCG) is attached to the carotid artery and is tightly associated with the neighboring nodose ganglion **(Fig. 3)**. Since they are adjacent to each other, problems may arise when attempting to differentiate the two ganglia. The SCG is crescent-shaped, and the carotid artery is often still attached to the ganglion when first extracted. The nodose ganglion, which contains sensory neurons and is attached to the vagus nerve, looks like a chicken drumstick and has a short stem.

11. Dissect the remaining slices, and extract all the ganglia. A typical dissection should yield 10 embryos with 20 ganglia. Sequester all the dissected ganglia in one corner of the dish free of any discarded tissue debris. This action will help maintain a cleaner and more homogenous culture.

3.2.2. Plating the Dissociated Sympathetic Neuron Culture

1. Obtain a glass Pasteur pipet with a flame-smoothened tip, and coat the inner surface of the pipet with the first embryo wash buffer (*see* **Subheading 3.2.1., step 7**).

This solution prevents the dissected ganglia from adhering to the surface of the glass pipet (*see* **Note 10**).

2. Using the glass Pasteur pipet, gently transfer the dissected ganglia from the glass Petri dish bottom into a 15-mL screw-cap conical tube. Avoid pipeting too much liquid into the tube. Briefly centrifuge the suspension in a clinical centrifuge to pellet the ganglia.

3. Remove the supernatant from the conical tube with a micropipet, and add 1 mL of CMF saline G buffer. Allow the tube to sit in the flow hood for 5 min before pelleting the ganglia in a clinical centrifuge.

4. Carefully remove the CMF saline G buffer with a micropipet, and add 1 mL of trypsin (250 µg/mL). Make sure that all the ganglia are covered in trypsin and they remain settled at the bottom of the tube.

5. Place the tube in a 37°C water bath and incubate for 15 min.

6. During this incubation period, wash the laminin-coated tissue culture dishes twice with CMF saline G buffer. After the final wash, remove all excess liquid from the dishes, and allow the dishes to dry in the flow hood (*see* **Note 11**).

7. After trypsinization, remove the tube from the water bath, and spin for 1 min in a clinical centrifuge to pellet the ganglia. At this stage, the trypsinized ganglia should be "sticky" and clumped together. Carefully remove the trypsin using a micropipet and add 1 mL of trypsin inhibitor (1 mg/mL). Place the tube in a 37°C water bath for another 5 min.

8. Pellet the ganglia by spinning for 1 min in a clinical centrifuge. Remove the trypsin inhibitor with a micropipet and add 2 mL of the neuron culture medium.

9. Using a glass Pasteur pipet prepared as described in **step 1,** gently triturate the trypsinized ganglia by repeatedly pipeting up and down. The ganglia will slowly shrink and dissociate into individual neurons. Triturate until most of the ganglia are no longer visible by naked eye (*see* **Note 12**).

10. Place the tube in the flow hood for 1 min to allow the undissociated ganglia to settle at the bottom of the tube.

11. Using a micropipet, pipet 100 µL of the dissociated neuron culture onto a laminin-coated tissue culture dish. Before plating, make sure the dish is dry. Place the drop of culture in the center of the dish. The surface tension should hold the drop together and prevent the neurons from flowing to the sides of the dish. Repeat this procedure for all 20 tissue culture dishes.

12. Add 1.9 mL of the neuron culture medium to each dish by dribbling the medium down the sides of the dish and placing the dishes back in the 37°C CO$_2$ incubator (*see* **Note 13**).

3.2.3. Maintaining the Dissociated Sympathetic Neuron Culture

1. Two days after dissection, the neuron cell bodies will be attached to the substratum at the center of the dish. Some of the cell bodies will already have neurites. To eliminate the dividing nonneuronal cells, add 1 µ*M;* Ara-C. Remove Ara-C treatment the next day by replacing with fresh neuron culture medium. Repeat Ara-C treatment, if necessary, 4 or 5 d after the dissection.

2. Typically, neuron culture medium is replaced once every 3 d. Allow the fresh medium to warm up and equilibrate in a 37°C CO_2 incubator for at least 30 min prior to replacing the medium (*see* **Note 14**).
3. Neurons are generally cultured for 2–3 wk to allow for terminal differentiation before they are ready for viral infection.

3.3. Growth and Differentiation of PC12 Cells With Nerve Growth Factor

PC12 cells are derived from a rat pheochromocytoma cell line that differentiates into sympathetic-like neurons in response to NGF *(4)*. These cells have a doubling time of approx 48 h and do not adhere well to either plastic or glass substrates, which necessitates coating dishes with either a positively charged molecule or an extracellular matrix component. The major advantages of using a cell line, compared with primary tissue, is that variability from experiment to experiment is reduced, large amounts of cells can be grown, and animals are not needed. To maximize this benefit, we routinely differentiate a large number of PC12 cells and store them "primed" ready for use *(5)*. For a comprehensive discussion of the methods for routine culture and other experimental protocols with PC12 cells, see **ref. 6**.

We routinely grow PC12 cells on plastic dishes coated with collagen, according to the original method of culturing these cells, although poly-D-lysine appears to work equally well for attachment. However, growing PC12 cells on a glass surface requires coating the glass with a combination of poly-ornithine and laminin. The different substrates are described just below.

3.3.1. Collagen Substrate

Coat the inside surface of plastic dishes with collagen at 5 µg/cm² in 0.02 *N* acetic acid as follows (*see* **Note 15**):
1. Aseptically apply the appropriate volume of collagen to each dish for 1 h at room temperature (**Table 2**) (*see* **Note 16**).
2. Aspirate unbound collagen and rinse dish three times with appropriate volume of sterile water.
3. Dishes are ready for use or they can be stored at 4°C for 1 wk.

3.3.2. Poly-DL-Ornithine Substrate (see **Note 17**)

1. Dissolve entire 100 mg bottle of poly-DL-ornithine in 200 mL of 0.1 *M* borate buffer, and filter-sterilize.
2. Store 10-mL aliquots at –20°C.
3. Aseptically coat dishes using 0.5 mg/mL of poly-DL-ornithine using volumes listed in **Table 2.**
4. Incubate dishes overnight in a humidified incubator at 37°C.
5. Rinse dishes five times with sterile water using volumes in **Table 2.**

Table 2
Volumes of Substrates and Washes for Preparing Coated Dishes

Dish diameter (mm)	Volume of substrate applied (mL; *see* **Note 16**)	dH$_2$0 wash (mL)
35	1	2–3
60	2	4–6
100	5	10–15
150	15	30–45

6. Dishes are now ready for use and may be coated with laminin or stored for 1 wk at 4°C.

3.3.3. Laminin Substrate

Laminin is applied after applying the poly-DL-ornithine. Accordingly, coating glass surfaces with these materials requires two overnight incubations.

1. Thaw laminin on ice to prevent it from gelling.
2. Dilute laminin in 1X HBSS to 10 µg/mL.
3. Aseptically coat dishes according to volumes listed in **Table 2.**
4. Incubate dishes overnight in a humidified incubator at 37°C.
5. Rinse dishes three times with HBSS.
6. Rinse dishes once with differentiation medium.
7. Dishes are ready for use or may be stored for 1 wk at 4°C.

3.4. Preparing Stocks of NGF-Primed PC12 Cells (see Note 18)

1. Coat twelve 15-cm dishes with collagen as described in **Subheading 3.3.1.**
2. Wash each dish three times with sterile water (~3 mL per dish), and aspirate completely between washes. Rest the dishes at an angle to facilitate collecting the water.
3. Plate undifferentiated PC12 cells into four 15-cm dishes (a 1:3 split ratio), and allow them to grow for 4 d in complete medium.
4. Collect the cells by trituration, and divide them equally among the eight remaining 15-cm dishes. Incubate overnight in growth medium. Cell clumps can be dissociated using either Pasteur pipets or a 23-gage needle. Trypsin cannot be used to dissociate PC12 cells, as even a brief exposure to trypsin is lethal.
5. The next day replace the growth medium with differentiation medium (*see* **Note 19**).
6. Replace the differentiation medium every other day for 9 d (total of four changes of medium, exposing cells to NGF for 10 d; *see* **Note 20**). After 5 d an extensive network of neurites should be visible (**Fig. 4**).
7. On the 10th day, collect the cells in a volume of 10 mL freezing medium per 15-cm dish.
8. Freeze 1-mL aliquots in a styrofoam box overnight at –80°C.

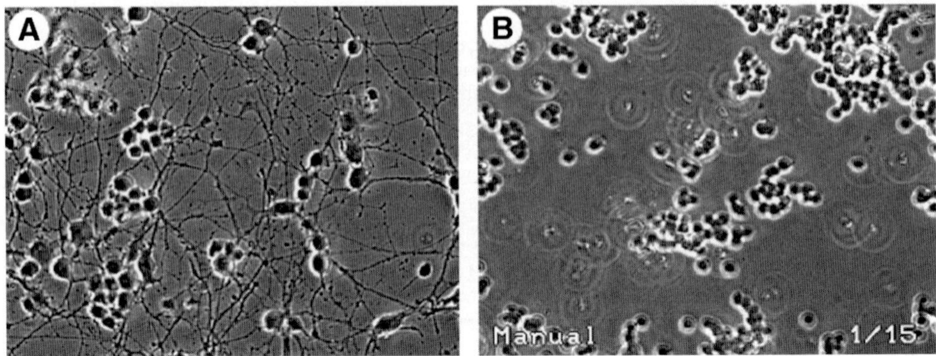

Fig. 4. Neurite outgrowth of PC12 cells grown on dishes treated with poly-ornithine and laminin. PC12 cells after 5 d in differentiation medium (**A**) or in growth medium (**B**).

9. Move the cells to liquid nitrogen storage.
10. Test differentiation of cells by thawing one aliquot and plating various dilutions of cells in a collagen-coated 24-well plate in differentiation medium. Within 24 h, the primed cells will extend visible neurites (*see* **Note 21**).

3.5. Infection and Immuneofluorescence of Dissociated Sympathetic Neurons and NGF-Differentiated PC12 Cells (see Note 22)

Although both sympathetic neurons and PC12 cells attach much better to plastic substrates, some microscopy techniques require that the cells be grown on glass surfaces. The following protocols are routinely used for plating sympathetic neurons on plastic substrates (**Subheading 3.5.1.**) and differentiated PC12 cells on glass (**Subheading 3.5.2.**).

3.5.1. Primary Neurons on Plastic Surfaces

1. For a typical PRV infection, prepare the viral inoculum by diluting a high-titer viral stock (1×10^8 PFU/mL) 1:10 with DMEM. Gently remove, and save the neuron culture medium from the dish. Replace the medium with 600 µL of the viral inoculum. Make sure that the inoculum covers the entire surface of the dish. Place the dish and the neuron culture medium back in the incubator for 1 h. Do not rock or tilt the dish during incubation (*see* **Note 23**). Viral particles will attach to and enter the neurons during this 1-h absorption period.
2. After 1 h, remove viral inoculum from the dish, and replace with the original neuron culture medium saved prior to infection. Place the dish back in the 37°C incubator.
3. At the appropriate time after infection, remove dishes from the incubator, and wash each dish twice with PBS. Gently flow buffer down the sides of the dish. After

infection, the network of neurons is no longer as firmly attached to the substratum and can be easily dislodged.

4. After the final wash, add 1 mL of 3.2% paraformaldehyde, diluted in PBS. Allow the paraformaldehyde to fix the cells for 10 min before washing the dish three times with PBS.

5. After the final wash, add 1 mL of PBS in 3% BSA to the dish, and store the dish at 4°C, overnight. The BSA in the PBS wash buffer acts as a blocking agent and helps stabilize the cells.

6. The next day, incubate the dish in PBS containing 3% BSA and 1% saponin for 10 min (*see* **Note 24**).

7. Remove the buffer, and add 600 µL of the primary antibody diluted in the PBS/BSA/saponin buffer. Incubate for 1 h in a 37°C humidified incubator.

8. Remove the primary antibody, and wash three times with PBS/BSA/saponin buffer. Add 600 µL of the appropriate secondary antibody. Incubate for 1 h as before.

9. Remove the secondary antibody, and wash twice with PBS/BSA/saponin buffer. Rinse the dish one final time with distilled, tissue culture-grade water. Carefully add a drop of munting medium (Vectashield or Aqua Poly/Mount) in the center of the dish. After 30 s, place a cover slip on top of the neuron culture. Or, if cells were grown on cover slips, place a drop of mounting medium on a glass slide, and lay the cover slip on top of the droplet, cell side down.

10. Using a glass Pasteur pipet connected to a vacuum filter, remove excess liquid around the tissue culture dish and around the edges of the coverslip (*see* **Note 25**). Be careful not to remove the mounting medium underneath the cover slip.

11. If you use Vectashield, seal the edges of the cover slip with nail varnish. Make sure the nail varnish dries completely before doing any microscopy work. If you use Aqua Poly/Mount, store the dish at 4°C to allow the medium to polymerize overnight (*see* **Note 26**).

3.5.2. Primed PC12 Cells on Glass Surfaces

1. Place 18-mm glass cover slips (*see* **Subheading 2.4., item 16**) in 12-well tissue culture dishes. Sterilize the cover slip by dipping it in 95% ethanol, and then pass it over an open flame.

2. Coat cover slips with poly-ornithine (*see* **Subheading 3.3.2.**), followed by laminin (*see* **Subheading 3.3.3.**).

3. Thaw primed PC12 cells, and seed them at a density previously determined empirically to yield cells sparse enough that they are not overlapping, yet dense enough that they provide sufficient samples per field.

4. PC12 cells are infected by gently replacing the culture medium with 500 µL of differentiation medium containing 2×10^5 PFU of PRV. Incubate the cells for 1 h in a cell culture incubator. After the 1-h period of adsorption of virus to the cells, replace the viral inoculum with 1 mL of prewarmed differentiation medium, and return the cells to their incubator.

5. Immunostaining of differentiated PC12 cells is carried out according to **steps 3–9** (if using a nonpolymerizing mounting medium, *see* **step 11**) in **Subheading 3.5.1.**

4. Notes

1. It is cost-effective to purify NGF from mouse submaxillary glands (available from Pel-Freez Biologicals). We purify NGF using the protocol described by Mobley et al. *(7)*.
2. Use tissue culture-grade water to prepare buffer solution. Filter-sterilize the buffer with a 0.22-μm filter apparatus.
3. PC12 cells appear to attach more strongly to these cover slips of German origin. Substrate coating is still necessary.
4. It is important to maintain a sterile and contamination-free stock of supplements. Remember to filter-sterilize all the supplements prior to aliquoting the stocks.
5. Loosen the cap on the conical tube to allow proper equilibration of CO_2 inside the tube with the incubator. The CO_2 buffers the pH of the solution. The levels of CO_2 in the incubator should be within the range of 5.5–6.0%, and this level remains unchanged for any subsequent use of the CO_2 incubator in this protocol.
6. Place the 35-mm dishes in a larger 15-cm Petri dish to provide a stable secondary container when transporting the dishes.
7. The Sal G buffer used to wash the embryos must be sterile because the wash buffer will be used to coat the flame-smoothened glass Pasteur pipets.
8. It is easier to immobilize the isolated head by inserting one blade in the eye socket. Next, align the other blade at the jaw. Once the blade is fully anchored, remove the first blade from the eye socket, and place it close to the neck at the opposing end of the jaw. The two blades should now be crossed to form an X. Pull the two blades toward each other in one swift stoke to obtain a clean and thin slice of the lower mandible.
9. It is critical to obtain a thin slice of the lower mandible because the thinner the slice, the easier it is to determine the location of the ganglia.
10. Use glass Pasteur pipets with flame-smoothened tips to avoid shearing or lysing the neurons during pipeting.
11. To accelerate the drying process, tilt the 35-mm tissue culture dish slightly by placing the lid at the bottom of the dish to prop it up. Once the dish is dried, the glossy and shiny surface at the bottom of the dish will be replaced by a thin layer of laminin that looks dull and nonreflective.
12. Trituration requires a steady hand. Work fast, but be gentle. Take great care to avoid making too many bubbles in the medium, as this will lower the yield. Do not expect all the ganglia to dissociate fully into neurons. Let the undissociated "chunks" settle to the bottom of the tube.
13. Use the 2-mL wide-bore disposable pipets when adding the serum-free medium to the dish. Set the discharge flow rate of the pipet-aid to "slow," and allow the medium to dribble down the sides of the dish. This procedure should be followed every time any liquid is exchanged to and from the dish. Never aspirate any media from the dish because the neurons are easily dislodged.
14. Older and more mature cultures are able to survive less frequent neuron culture medium replacements.
15. Collagen applied in this manner is not appropriate for glass surfaces. Store collagen at 4°C for up to 3 mo.

16. Do not use less than the amounts listed in **Table 2,** as smaller dishes require disproportionately larger volumes than the larger dishes. If you notice that the cells are attached to the dish in a ring pattern (i.e., along the outer edge of the dish and relatively sparse in the center), increase the volume of substrate applied and/or the volume of media in which the cells are plated.
17. The D-isomer may allow cells to remain attached for a longer period, since it is not susceptible to host enzymes.
18. Differentiating PC12 cells into neuron-like cells requires 10 d, replacing media every other day. Preparing a large batch at once ensures homogeneity and simpler scheduling demands. We have stored the cells up to a year without noticing any adverse effect.
19. Do not thaw and freeze NGF repeatedly. It is important to store it as small aliquots. We prepare a large amount of RPMI with 1% horse serum and add NGF only to the amount needed for a given day.
20. During differentiation, many cells will detach from the dish. Under these conditions, only the cells that respond to NGF will survive. Presumably the cells not responding to NGF are dying and consequently lift off the dish *(8)*.
21. When the cells are collected by trituration for storage, the neurites are sheared off. This action does not affect their differentiation. In fact, visibly apparent neurites are not a requirement for PC12 cells to exhibit their neuronal character in response to NGF *(9)*. Additionally, the length of the neurites varies with the cell density: too many cells and neurites will not extend. However, when making stocks of primed PC12 cells, we plate the cells at high density to obtain the most primed cells per unit of surface area and because approximately half of the cells will detach over the 10-d differentiation process. Once primed, differentiated PC12 cells will extend neurites rapidly when replated on collagen-coated or laminin-coated substrates and will retain their neuronal properties **(Fig. 5)**. Upon thawing, wash cells with 20 volume equivalents of RPMI, to remove as much FCS as possible.
22. Approximately 1000-fold fewer plaque-forming units are released from NGF-treated cells vs undifferentiated PC12 cells (data not shown). Furthermore, prolonged growth of PC12 cells with NGF (>21 d) may render the cells refractory to lytic infection *(10)*.
23. The calculated multiplicity of infection (MOI) is around 2000 PFU per cell plated. However, because the virions adhere avidly to the substratum on the surface of the dish and the cell density is quite low, the effective MOI for the infection is much lower and cannot be accurately determined. The least amount of virus needed to infect 95% of all cells should be determined empirically for each virus and/or type of neuron.
24. These protocols can also be adapted for nonpermeabilized, surface staining of neurons. To surface stain omit saponin from all subsequent washes.
25. The tissue culture dish can also be air-dried for 30 min in the dark before sealing the cover slip with nail varnish.
26. The solvents in nail varnish diffuse into the mounting medium and quench the fluorescence along the edges. This is not a problem for samples on a large surface, such

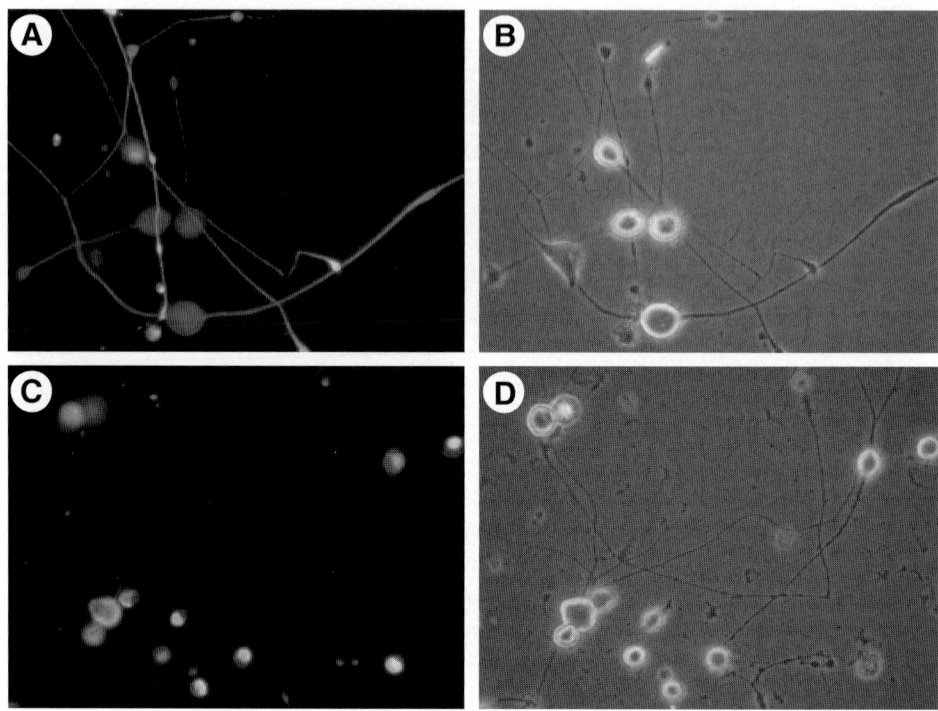

Fig. 5. Polarized protein sorting in NGF-differentiated PC12 cells. An aliquot of primed PC12 cells (**Subheading 3.4.**) was thawed, plated for 2 d in differentiation medium on glass slides, coated with poly-ornithine and laminin (**Subheading 3.3.**), and analyzed by immunofluorescence microscopy (**Subheading 3.5.2.**). Cells were stained with an antibody against the phosphorylated form of neurofilament H, an axonal marker (**A** and **B**); antibody SMI-31, Sternberg Monoclonals) or with an antibody against dephosphorylated neurofilament H, a somatodendritic marker (**C** and **D**); SMI-32, Sternberg Monoclonals). Epifluorescence (A and C) and phase contrast (B and D). Secondary antibody is goat antimouse Alexa-488 (Molecular Probes). Nuclei are stained with DAPI (Polysciences).

as a 35-mm dish because the solvents generally do not diffuse into the large interior. However, smaller areas, such as a 12- or 18-mm cover slip, can be significantly affected, depending on how much nail varnish is applied. A polymerizing mounting medium, such as Aqua Poly/Mount, obviates the need to seal the cover slip.

Acknowledgments

The authors would like to thank Dr. Mark Tomishima for his expertise in culturing rat sympathetic neurons. This work was supported by the National Science Foundation for Behavioral Neuroscience (IBN9876754) and the

National Institute of Health (NS33506) E. A. Flood was supported by an NIH NRSA Fellowship (AI10653).

References

1. Banker, G. and Goslin, K. (eds.) (1998) *Culturing Nerve Cells,* 2nd ed. MIT Press, Cambridge, MA.
2. Fedoroff, S. and Richardson, A. (eds.) (2001) *Protocols for Neural Cell Culture,* 3rd ed. Humana, Totowa, NJ.
3. Tomishima, M. J., Smith, G. A., and Enquist, L. W. (2001) Sorting and transport of alpha herpesviruses in axons. *Traffic* **2,** 429–436.
4. Tischler, A. S. and Greene, L. A. (1975) Nerve growth factor-induced process formation by cultured rat pheochromocytoma cells. *Nature* **258,** 341–342.
5. Greene, L. A., Burstein, D. E., and Black, M. M. (1982) The role of transcription-dependent priming in nerve growth factor promoted neurite outgrowth. *Dev. Biol.* **91,** 305–316.
6. Greene, L. A., Farinelli, S. E., Cunningham, M. E., and Park, D. S. (1998) Culture and experimental use of the PC12 rat pheochromocytoma cell line, in *Culturing Nerve Cells,* 2nd ed. (Banker, G. and Goslin, K., eds.), MIT Press, Cambridge, MA, pp. 161–188.
7. Mobley, W. C., Schenker, A., and Shooter, E. M. (1976) Characterization and isolation of proteolytically modified nerve growth factor. *Biochemistry* **15,** 5543–5552.
8. Greene, L. A. (1978) Nerve growth factor prevents the death and stimulates the neuronal differentiation of PC12 pheochromocytoma cells in serum-free medium. *J. Cell. Biol.* **78,** 747–755.
9. Dichter, M. A., Tischler, A. S., and Greene, L. A. (1977) Nerve growth factor-induced increase in electrical excitability and acetylcholine sensitivity of a rat pheochromocytoma cell line. *Nature* **268,** 501–504.
10. Su, Y. H., Moxley, M., Kejariwal, R., Mehta, A., Fraser, N. W., and Block, T. M. (2000) The HSV-1 genome in quiescently infected NGF differentiated PC12 cells cannot be stimulated by HSV superinfection. *J. Neurovirol.* **6,** 341–349.

22

Human Papillomavirus Type 31 Life Cycle

Methods for Study Using Tissue Culture Models

Frauke Fehrmann and Laimonis A. Laimins

Summary

The life cycle of human papillomaviruses (HPVs) has been difficult to study in tissue culture owing to its dependence on epithelial differentiation. In this chapter several methods are described to imitate the important steps in the HPV life cycle. Normal human keratinocytes (NHKs) harvested from neonatal foreskins were transfected with HPV type 31 genomes in order to generate stable cell lines containing episomal copies of HPV genomes. HPV-positive keratinocyte cultures were maintained in E medium in the presence of mitomycin C-treated J2 3T3 fibroblast feeders. Finally, the keratinocytes were induced to undergo epithelial differentiation in semisolid medium to provoke viral late functions like genomic amplification and late transcription.

Key Words: HPV; keratinocytes; transfection; E medium; epithelial differentiation; methylcellulose; viral late functions.

1. Introduction

The productive life cycle of human papillomaviruses (HPVs) is directly linked to epithelial cell differentiation *(1)*. Infection by papillomaviruses is believed to occur through microtraumas in the epithelium, exposing the basal cells to entry by viruses. Following the infection of keratinocytes in the basal layer, HPV genomes are established as episomes at approx 50 copies per cell, which replicate in synchrony with cellular DNA replication *(1,2)*. The establishment and maintenance of HPV genomes is associated with expression of early HPV proteins. Following cell division, infected daughter cells leave the basal layer, migrate toward the suprabasal regions, and begin to differentiate. Epithelial differentiation of HPV-infected cells results in amplification of the viral genomes to thousands of copies per cell, synthesis of late proteins, and the

From: *Methods in Molecular Biology, vol. 292: DNA Viruses: Methods and Protocols*
Edited by: P. M. Lieberman © Humana Press Inc., Totowa, NJ

assembly of infectious virions *(3)*. Subsequently, virions are released into the environment as the upper layer of the epithelium is shed *(4)*.

Study of HPV in tissue culture has been problematic because of the difficulty in faithfully recreating the differentiated state of the epithelium to induce the full life cycle. However, recently methods have been successfully developed for a genetic analysis of HPV functions in the context of the productive life cycle *(5–7)*. In these methods cloned HPV genomes are first excised from bacterial plasmid sequences, religated, and then transfected into primary keratinocytes isolated from neonatal human foreskins (Fig. 1). This powerful method allows genetic manipulation of the viral genome and analysis of effects on the HPV life cycle at physiologically relevant concentrations of viral factors. In order to recapitulate the HPV life cycle in tissue culture, it is first necessary to isolate cell lines that stably maintain viral genomes as episomes. It is essential for cells to maintain episomes for productive viral replication to occur upon differentiation. Differentiation of keratinocytes can be induced in tissue culture through the use of organotypic raft cultures *(8)* or by suspension in methylcellulose *(9–12)*. Upon differentiation, these cells induce viral late functions, including genome amplification and activation of late gene expression *(13)*.

2. Materials

1. Transport medium for foreskins: 355.2 mL Hanks' balanced salt mixture (HBSS) with phenol red, without Mg^{2+} and Ca^{2+} (cat. no. 14170-112, Hyclone, Logan, UT), 40 mL fetal bovine serum (FBS; cat. no. SH30070.03, Hyclone), 4 mL penicillin/streptomycin (cat. no. 15140-122, Gibco Invitrogen, Carlsbad, CA), 0.8 mL fungizone (cat. no. 15295-017, Gibco Invitrogen), 7-mL aliquots in 15-mL Falcon tube/foreskin.
2. Dulbecco's phosphate-buffered saline (PBS; cat. no. 14190-136, Gibco Invitrogen).
3. Dispase II (cat. no. 295 825, Roche, Indianapolis, IN).
4. 0.25% Trypsin/1 m*M* EDTA (cat. no. 25200-056, Gibco Invitrogen).
5. FBS (Hyclone).
6. Keratinocyte growth medium (KGM; cat. no. CC-3001, Clonetics BioWhittaker, Walkersville, MD).
7. 37°C 5% CO_2 Humidified incubator.
8. 0.05% Trypsin/0.53 m*M* EDTA (cat. no. 25300-054, Gibco Invitrogen).
9. Dimethylsulfoxide (DMSD; cat. no. D-8779, Sigma, Saint Louis, MO).
10. Restriction enzymes.
11. T4 DNA ligase (400,000 U/mL, cat. no. M0202S, NEB, Beverly, MA).
12. 5X Ligase buffer (cat. no. 46300-018, Gibco Invitrogen).
13. Fugene6 Transfection Reagent (cat. no. 1814443, Roche).
14. G418 (Geneticin, cat. no. G-9516, Sigma).
15. E medium (material and preparation described in **Subheadings 3.4.1.–3.4.2.**).
16. 3T3 J2 fibroblasts.

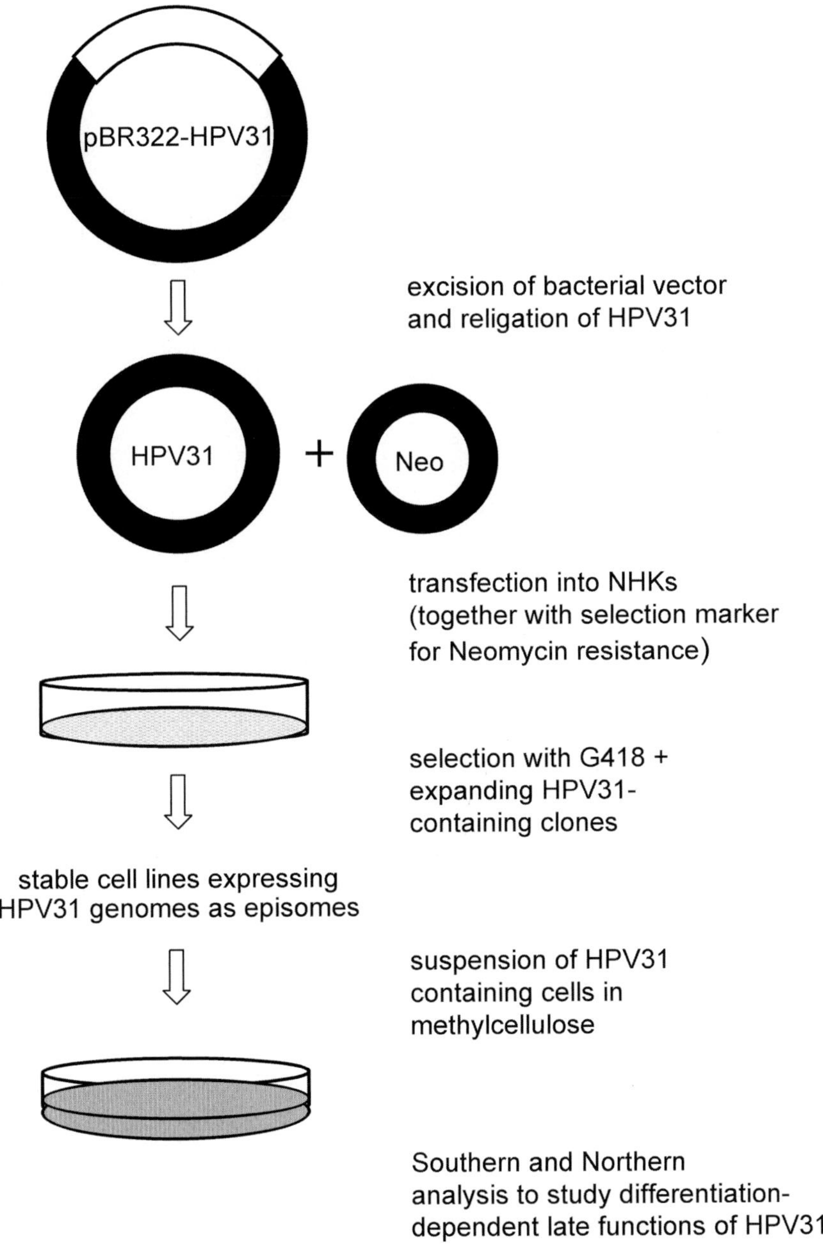

Fig. 1. Experimental design of simulating the human papillomavirus virus (HPV) life cycle in tissue culture. In a first step, stable cell lines harboring HPV31 genomes episomally were produced. The second step includes induction of cell differentiation in semisolid medium to activate viral late functions. NHK, normal human keratinocyte.

17. Dulbecco's modified Eagle's medium (DMEM; cat. no. 11956-092, Gibco Invitrogen).
18. Calf bovine serum (cat. no. 26170-035, Gibco Invitrogen).
19. Gentamycin Reagent Solution (cat. no. 15710-072, Gibco Invitrogen).
20. Mitomycin C (cat. no. 107 409, Roche).
21. Versene: 1 mL 0.5 M sterile EDTA, pH8.0 in 1 L of PBS.
22. Glycerol (cat. no. 4043-00, J. T. Baker, Philipsburg, NJ).
23. Methylcellulose (400 cps, cat. no. M-0512, Sigma).

3. Methods

The methods described below outline (1) the isolation and (2) culture of normal human keratinocytes (NHKs), (3) their stable transfection with HPV genomes, (4) culturing of HPV-containing cell lines, and (5) induction of epithelial differentiation in keratinocytes.

3.1. Isolation of NHKs From Neonatal Foreskins

Each foreskin should be kept in transport medium at 4°C and used to isolate NHKs within 24–72 h after circumcision.

3.1.1. Day 1

1. Rinse a foreskin twice with PBS.
2. Using sterile tweezers, transfer foreskin to a 10-cm dish containing 5 mL PBS.
3. With sterile tweezers and scissors, trim off fat and vessels from the dermal side. Cut into several (four to six) small pieces.
4. Place 4 mL dispase solution into a 6-cm dish. Add the foreskin pieces, epidermal side up. Incubate overnight at 4°C.

3.1.2. Day 2

1. Using two sterile tweezers, peel off the epidermis from each piece and transfer to a new 6-cm dish containing 0.25% trypsin/1 mM EDTA. Incubate for 10–15 min at 37°C.
2. To inactivate the trypsin, add 0.5 mL bovine serum to the dish.
3. With sterile tweezers grasp each piece of tissue and vigorously rub and scrape the tissue with a second tweezers against the bottom of dish to release the keratinocytes. The tissue should be totally dispersed, and the microscopic analysis should reveal a dense suspension of cells.
4. Using a pipet, transfer the cell suspension along with debris to a 15-mL Falcon tube. Rinse the dish once with PBS, and add it to the cell suspension.
5. Centrifuge at 250 g for 5 min at room temperature.
6. Discard supernatant, tap tube to disrupt cell pellet, and resuspend pellet in KGM. Plate keratinocytes onto a 10-cm tissue culture dish containing 10 mL KGM. Grow cells in a 37°C 5% CO_2 humidified incubator.

3.1.3. Day 3

1. Aspirate the medium, debris, and nonadherent cells. Replace with fresh KGM.

3.2. Culture of NHKs

1. Maintain NHKs as monolayers with changes of medium every other day. Passage the cells when they reach about 80% confluence.
2. To passage cells, aspirate medium and wash plate once with PBS.
3. Aspirate PBS and add 2 mL 0.05% trypsin/0.53 mM EDTA per 10-cm dish. Dispense the trypsin all over the surface, and place plate in 37°C incubator for 3–10 min (time varies considerably for different isolates) in order to release the cells from the dish.
4. Periodically check the plate under the microscope to see whether cells have been released from the surface. Tap on the sides of the plate to facilitate the process.
5. When most of the cells have been released, transfer the cell suspension into a 15-mL Falcon tube, which is already provided with an appropriate volume of any serum-containing medium to inactivate the trypsin. Rinse the dish once with serum-containing medium, and add it to the cell suspension.
6. Centrifuge at 250 g for 5 min at room temperature.
7. Aspirate the supernatant, tap the tube to disrupt the cell pellet, and resuspend the pellet in KGM. Plate keratinocytes onto 10-cm tissue culture dishes containing 10 mL KGM. Usually one 80% confluent plate with primary keratinocytes is split at a 1:5 ratio.

3.2.1. Removal of Fibroblasts by Short Trypsin Treatment

1. If the primary keratinocyte culture contains too many fibroblasts, they have to be removed so as not to interfere with the growth of keratinocytes. Fibroblasts are much more sensitive to trypsin than epithelial cells and can therefore be selectively removed by a short trypsin treatment.
2. Aspirate medium, and wash the plate once with PBS.
3. Aspirate PBS and add 2 mL 0.05% trypsin/0.53 mM EDTA per 10-cm dish. Dispense the trypsin all over the surface, and place the plate in a 37°C incubator for about 1 min.
4. Remove fibroblasts by aspirating off the trypsin. Inactivate remaining trypsin immediately by adding 10 ml of any serum-containing medium to the plate.
5. Wash the plate twice with PBS, and add 10 mL KGM.

3.2.2. Freezing of NHKs

1. Prepare freezing medium: for 100 mL, mix 80 mL of KGM with 10 mL FBS, and add 10 mL of sterile DMSO. Mix thoroughly and filter-sterilize. Store at 4°C for 4–6 wk.
2. Trypsinize NHKs when they reach approx 80% confluency, and spin them down in serum-containing medium. Resuspend in ice-cold freezing medium, and slow-cool

using a freezing chamber in –80°C freezer overnight prior to transfer to liquid nitrogen.

3. To thaw cells frozen in glycerol, thaw cryovial of cells in a 37°C water bath. Remove cells from cryovial into 15 mL fresh KGM in a 10-cm dish, and swirl gently. Change the medium the next day.

3.3. Transfection of NHK With HPV Genomes

3.3.1. Preparation of HPV Genomes for Fugene6 Transfection

1. Digest appropriate amount of plasmid containing the HPV genome (e.g., 10 μg pBR322-HPV31 DNA) to release the HPV DNA from the vector background in a volume of 52.5 μL.
2. Check 2.5 μL of solution on agarose gel for complete digestion.
3. Heat-inactivate enzyme by placing solution at 65°C for 10 min.
4. Bring volume to 900 μL for religation of the HPV genome: 50 μL DNA digestion mix, 180 μL 5X ligation buffer, 1 μL T4 DNA ligase (400 U), 669 μL H_2O; incubate overnight at 16°C.
5. Perform isopropyl alcohol precipitation with NaCl (do not use acetate because salt changes pH, which can affect lipid formation): 900 μL ligation, 600 μL isopropyl ethanol, 180 μL 5 *M* NaCl. Precipitate overnight at –20°C.
6. Spin for 30 min at 4 °C in microcentrifuge.
7. Wash with 70% EtOH
8. Resuspend pellet in 15 μL TE.
9. Load 1 μL on gel to check for successful ligation.

3.3.2. Transfection With Fugene6 and Selection of NHKs (see **Note 1**)

1. Seed an early passage of NHKs in 6-cm tissue culture plates, and let them grow to 50–60% confluency
2. For each transfection, set up: 94 μL KGM + 6 μL Fugene. Add to each aliquot: 1 μg HPV DNA + 1 μg pSV2neo, and tap gently.
3. Incubate at room temperature for 45 min.
4. Change medium in NHK plates for 4 mL fresh KGM.
5. Add the Fugene–DNA mixture dropwise, and swirl gently.

3.3.3. G418 Selection

1. Next day (Day 1): plate transfected keratinocytes on mitomycin C-treated feeders (*see* **Subheading 3.4.3.**) in E media + 5 ng epidermal growth factor (EGF)/mL (*see* **Subheading 3.4.2.**) in 10-cm tissue culture plates.
2. Days 2 and 4: replace media with E media/EGF and 200 μg/mL G418.
3. Days 6 and 8: replace media with E media/EGF and 100 μg/mL G418.
4. Days 3, 5, and 7: add mitomycin-treated feeders.
5. Day 9: stop selection, replace media with E media/EGF, and add mitomycin-treated feeders if necessary.

It takes about 5 d to see the effects of G418 selection (*see* **Note 2**). About 2–3 wk after finishing selection, numerous colonies of keratinocytes should be grown and visible even without microscope. During the whole time, fresh mitomycin-treated feeders have to be provided every few days.

3.4. Culture of HPV-Positive Keratinocytes

Keratinocyte cell lines containing HPV genomes are maintained as monolayers in E medium supplemented with 5 ng EGF/mL in the presence of mitomycin C-treated J2 3T3 fibroblast feeders. Cells are grown in a 37°C 5% CO_2 humidified incubator and passaged when they reach 80–90% confluence.

3.4.1. E Medium: Materials

1. DMEM (cat. no. 1200-061, Gibco Invitrogen).
2. Ham's F-12 (cat. no. 21700-075, Gibco Invitrogen).
3. $NaHCO_3$ (cat. no. 11810-033, Gibco Invitrogen).
4. Penicillin/streptomycin (Gibco Invitrogen).
5. Hydrocortisone (cat. no. H-0888, Sigma).
6. Cholera enterotoxin (cat. no. 856011, ICN Biomedicals, Aurora, OH).
7. HCl (cat. no. 9535-05, J. T. Baker).
8. Defined FBS (cat. no. SH30070.03, Hyclone).
9. Adenine (cat. no. A-2786, Sigma).
10. Insulin (cat. no. I-6634, Sigma).
11. Transferrin (cat. no. T-1147, Sigma).
12. 3,3′,5-Triiodo-L-thyronine (T_3; cat. no. T-6397, Sigma).
13. HEPES (cat. no. 391338, Calbiochem-Novabiochem, San Diego, CA).
14. Mouse EGF (cat. no. 354010, Collaborative Research/Biomedical, Bedford, MA).
15. 0.2 μm Mediakap Hollow Fiber Media Filter (cat. no. F348-10, MG Scientific, Pleasant Prairie, WI).

3.4.2. Preparing E Medium

For 40 L, combine the following in a 40-L carboy:
1. 30 L Distilled, deionized water (ddH_2O).
2. Three (10-L) packets of DMEM powder, dissolved in 3 L ddH_2O.
3. Ten (1-L) packets of Ham's F-12 powder, dissolved in 1 L ddH_2O.
4. 122.74 g Tissue culture-grade $NaHCO_3$, dissolved in 1 L ddH_2O.
5. 400 mL 100X cocktail (*see* item 1 under supplements, just below).
6. 400 mL Penicillin/streptomycin.
7. 40 mL 1000X Hydrocortisone.
8. 40 mL 1000X Cholera toxin.
9. Shake briefly to mix.
10. Add 12.5 mL concentrated HCl (dilute in ddH_2O first), shake, and remove a small quantity to check that pH is approx 7.1.
11. For 5% FBS, add 2 L of FBS.

12. Make up to 40 L with deionized ddH_2O water, and shake.
13. Filter-sterilize using a low-protein-binding 0.2-μm filter, and aliquot aseptically. Store at 4°C in the dark.

The supplements are as follows:

1. 100X Cocktail: combine 200 mL 1.8×10^{-1} adenine, 200 mL 5 mg/mL insulin, 200 mL 5 mg/mL transferrin, 200 mL 2×10^{-8} M T_3, and 1200 mL sterile PBS; mix well, and filter-sterilize. Store in 400-mL aliquots at –20°C.
   a. Adenine: add 4.86 g to 150 mL ddH_2O. Then add (while stirring) about 6 mL of 10 N HCl slowly, until all powder is dissolved. Do not add any more HCl. Make up to 200 mL with ddH_2O.
   b. Insulin: dissolve 1 g in 200 mL of 0.1 N HCl.
   c. Transferrin: dissolve 1 g in 200 mL sterile PBS.
   d. T_3: weigh 13.6 mg of T_3, and dissolve in 100 mL 0.02 N NaOH (2×10^{-4} M T_3). Take 0.1 mL of 2×10^{-4} M T_3, and add 9.9 mL sterile PBS (2×10^{-6} M T_3). Take 2 mL of 2×10^{-6} M T_3, and add 198 mL sterile PBS (2×10^{-8} M $T3_3$).
2. HEPES buffer: Make A 1 M stock solution by dissolving 71.49 g into 250 mL ddH_2O. Use 10 N NaOH to adjust the pH to 7.0. Bring the volume to 300 mL, and filter-sterilize.
3. 1000X Hydrocortisone: dissolve 100 mg in 20 mL 100% EtOH to make a 5 mg/mL stock. Take 19.2 mL of the stock, and add 220.8 mL HEPES buffer. Make 6 × 40 mL aliquots (0.4 μg/mL), and store at –20°C.
4. 1000X Cholera enterotoxin: reconstitute a 1-mg vial with 1 mL ddH_2O to make a 1×10^{-5} M stock solution. Further dilute with 99 mL ddH_2O to obtain a 1000X stock. Store at 4°C in the dark.
5. 200X EGF: combine 100 μg EGF and 10 mg BSA, each dissolved in 10 mL ddH_2O, and bring up to 100 mL with ddH_2O. Filter-sterilize, aliquot, and store at –20°C. Supplement E medium with 5 ng/mL EGF immediately before use.

3.4.3. Coculture of 3T3 J2 Fibroblasts

1. Maintain 3T3 J2 fibroblasts as monolayers in DMEM, supplemented with 10% calf serum and Gentamycin Reagent Solution (0.05 mg/mL). Grow cells in a 37°C 5% CO_2 humidified incubator and passage until they reach 80% confluency. Do not allow fibroblasts to grow too confluent. They can be used until passages 20–25 following thawing of an early passage isolate or until spontaneously transformed foci are observed. Passaging of cells should be performed according to the standard procedure as described in **Subheading 3.4.4.** J2 3T3 fibroblasts are usually split 1:5 or 1:10 but can be split up to 1:20 if desired.
2. J2 3T3 fibroblasts must be treated with mitomycin C for use as feeder layers to grow keratinocytes. To treat cells, add 100 μL of 0.4 mg/mL mitomycin C solution to 5 mL medium on a preconfluent plate of fibroblasts, swirl gently, and incubate for 2–4 h at 37°C in a 5% CO_2 humidified incubator.
3. Aspirate off medium containing mitomycin C, and wash three times with PBS.

4. If cells are to be used right away, trypsinize cells from the dish, and add to keratinocytes. One preconfluent plate will feed two to four 10-cm dishes of keratinocytes.
5. Mitomycin C-treated fibroblasts can be maintained for up to 48 h in DMEM with 10% calf serum and Gentamycin Reagent Solution at 37°C in 5% CO_2 prior to addition to keratinocytes.

3.4.4. Passaging of HPV-Positive Keratinocytes

1. For passaging of HPV-positive keratinocyte lines, remove the medium, and wash the plate once with PBS.
2. Aspirate the PBS, and add 2 mL 0.05% trypsin/0.53 m*M* EDTA per 10-cm dish. Dispense the trypsin all over the surface, and place the plate in a 37°C incubator for 3–10 min (time varies considerably for different cell lines) to release the cells from the dish.
3. Periodically check the plate under the microscope to see whether cells have been released from the surface. Tap on the sides of the plate to facilitate the process.
4. When most of the cells have been released, aspirate off the cell suspension, and rinse the surface of the dish several times to get all the cells.
5. Dispense aliquots of the cell suspension into dishes containing fresh E medium and mitomycin C-treated J2 3T3 fibroblast feeders. Keratinocytes should usually be passaged 1:5 to 1:10.
6. Refeed cells every third day until the next passage.

3.4.5. Versene Treatment

1. For collecting DNA, RNA, or proteins from keratinocytes, it is necessary to remove fibroblast feeders first, through the use of Versene. Versene selectively removes fibroblasts, leaving epithelial cells attached.
2. First remove the medium from keratinocytes growing in the presence of fibroblast feeders, and wash the plate once with PBS.
3. Aspirate the PBS, and add 5 mL Versene per 10-cm dish. After 2 min remove the fibroblasts by using a 5-mL pipet, and spray the whole plate vigorously several times with the Versene.
4. Monitor the detachment of fibroblasts by the microscope, observing whether most fibroblasts have been eliminated. Repeat the Versene procedure if necessary (*see* **Note 3**).

3.4.6. Freezing Keratinocytes

1. Prepare freezing medium: for 100 mL, mix 70 mL of E medium (which already contains 5% FBS) with 10 mL FBS, and add 20 mL of sterile glycerol. Mix thoroughly and filter-sterilize. Store at 4°C for 4–6 wk.
2. Trypsinize keratinocytes when they reach approx 80% confluency, and spin them down in serum-containing medium. Resuspend in ice-cold freezing medium, and slow-cool using the freezing chamber in a –80°C freezer overnight prior to transfer to liquid nitrogen.

3. To thaw cells frozen in glycerol, thaw cryovial of cells in a 37°C water bath. Remove cells from the cryovial, place into 15 mL fresh E medium in a 10-cm dish containing mitomycin-treated feeders, and swirl gently. Change medium the next day.

3.5. Induction of Epithelial Differentiation in Keratinocytes Using Semisolid Medium

3.5.1. Prepare E Medium Supplemented With 1.5% Methylcellulose

1. Autoclave dry methylcellulose with a stir bar in a bottle.
2. Add 1/2 of E medium to autoclaved methylcellulose, and heat in a water bath to 60°C for 20 min. (No stirring is needed.)
3. Add remaining 1/2 of E medium except leave out 5% volume to add FBS later.
4. Stir at 4°C for about 3 h (or longer)—solution should clear.
5. Add an additional 5% FBS, and mix.
6. Store at 4°C.

3.5.2. Suspend Keratinocytes

1. Add approx 15 mL methylcellulose medium to 6-cm Petri dishes. (If tissue culture plates are used, they need to be coated with poly-HEMA to prevent cells from adhering to the plastic.)
2. Treat plates with keratinocytes with Versene to get rid of fibroblast feeder cells.
3. Trypsinize keratinocytes, and spin down (for 5 min at 275 g).
4. Resuspend pelleted cells in 0.5 mL E medium/dish, and add solution dropwise to one plate filled with methylcellulose, distributing it around the entire plate. Take a pipet, and stir in cells until fairly homogeneous.

3.5.3. Harvesting Cells From Methylcellulose

Typically HPV late function and full cellular differentiation occur after 24 and 48 h in suspension in methylcellulose *(13,14)* (**Fig. 2**).

1. Scrape cells in methylcellulose from each 6-cm Petri dish into two 50-mL conical tubes. Wash the Petri dish with PBS several times, and add to the 50-mL conical tubes. Fill up to 50 mL with PBS.
2. Spin cells down at 275 g for 10 min in a table-top centrifuge (Beckman, Fullerton, CA) Combine pellets into one 50-mL tube, and fill up to 50 mL with PBS.
3. Spin down. Resuspend pellet in 14 mL PBS, transfer to a 15-mL tube, and spin again.
4. Use pellets for protein, DNA, RNA, and so on (*see* **Note 4**).

4. Notes

1. Successfully transfected cell lines carrying HPV genomes episomally were also achieved by using LipofectAce *(7,15)* and LipofectAmine *(16,17)* lipofection as described by the manufacturer (Gibco Invitrogen).

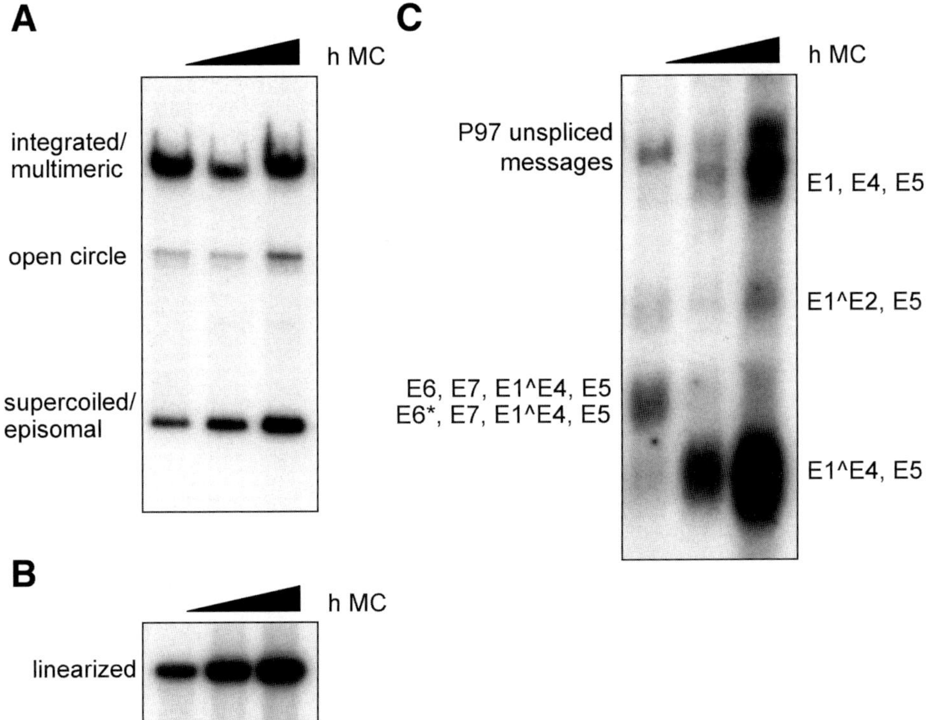

Fig. 2. Differentiation-dependent late functions of HPV31 following suspension in methylcellulose. **(A)** Autoradiogram of Southern analysis of NHKs stably transfected with HPV31 DNA. Total genomic DNA was harvested from monolayer culture, and cells were cultured in methylcellulose (MC) for 24 and 48 h and prepared as described before *(14)*. The Southern blot was hybridized with a probe corresponding to the complete HPV31 genome. *(B)* Equal amounts of total genomic DNA were digested with a single cutter *(XbaI)* to linearize the HPV31 genomes and facilitate copy number analysis. *(C)* Northern blot analysis of differentiation-dependent transcripts in an HPV31-containing cell line following suspension in methylcellulose. Total RNA was isolated from monolayer culture, and cells were cultured in methylcellulose for 24 and 48 h. Bands of early transcripts are labeled on the left and bands of late transcripts on the right. The Northern blot was hybridized with a probe corresponding to the complete HPV31 genome.

2. Since the G418 effect on nonresistant cells takes several days, it might be necessary to split the keratinocytes during selection. In this case split them 1:2 or 1:3 onto mitomycin-treated feeders in E medium/EGF without G418, and continue with selection the next day.

3. Repeated Versene treatment can select for resistant fibroblasts that can no longer be removed by this treatment. An alternative, but less selective, method to get rid of fibroblasts is the short trypsin treatment, described in **Subheading 3.2.1.**

4. Poor differentiation of keratinocytes in methylcellulose can be caused by already overgrown monolayer cells. It is necessary to harvest keratinocytes for suspension in methylcellulose when they are still growing exponentially and are no more confluent than 80%.

References

1. Howley, P. M. and Lowy, D. R. (2001) Papillomaviruses and their replication, in *Virology,* vol. 2 (Fields, B. N., Knipe, J. B., and Howley, P. M., eds.), Lippincott Williams & Wilkins, Philadelphia, pp. 2197–2229.

2. Stubenrauch, F. and Laimins, L. A. (1999) Human papillomavirus life cycle: active and latent phases. *Semin. Cancer Biol.* **9,** 379–386.

3. Lambert, P. F. (1991) Papillomavirus DNA replication. *J. Virol.* **65,** 3417–3420.

4. Laimins, L. A. (1996) Human papillomaviruses target differentiating epithelia for virion production and malignant conversion. *Semin. Virol.* **7,** 305–313.

5. Meyers, C., Mayer, T. J., and Ozbun, M. A. (1997) Synthesis of infectious human papillomavirus type 18 in differentiating epithelium transfected with viral DNA. *J. Virol.* **71,** 7381–7386.

6. Frattini, M. G., Lim, H. B., Doorbar, J., and Laimins, L. A. (1997) Induction of human papillomavirus type 18 late gene expression and genomic amplification in organotypic cultures from transfected DNA templates. *J. Virol.* **71,** 7068–7072.

7. Frattini, M. G., Lim, H. B., and Laimins, L. A. (1996) In vitro synthesis of oncogenic human papillomaviruses requires episomal genomes for differentiation-dependent late expression. *Proc. Natl. Acad. Sci. USA* **93,** 3062–3067.

8. McCance, D. J., Kopan, R., Fuchs, E., and Laimins, L. A. (1988) Human papillomavirus type 16 alters human epithelial cell differentiation in vitro. *Proc. Natl. Acad. Sci. USA* **85,** 7169–7173.

9. Meyers, C. and Laimins, L. A. (1994) In vitro systems for the study and propagation of human papillomaviruses. *Curr. Top. Microbiol. Immunol.* **186,** 199–215.

10. Flores, E. R. and Lambert, P. F. (1997) Evidence for a switch in the mode of human papillomavirus type 16 DNA replication during the viral life cycle. *J. Virol.* **71,** 7167–7179.

11. Green, H. (1977) Terminal differentiation of cultured human epidermal cells. *Cell* **11,** 405–416.

12. Ruesch, M. N. and Laimins, L. A. (1998) Human papillomavirus oncoproteins alter differentiation-dependent cell cycle exit on suspension in semisolid medium. *Virology* **250,** 19–29.

13. Ruesch, M. N., Stubenrauch, F. and Laimins, L. A. (1998) Activation of papillomavirus late gene transcription and genome amplification upon differentiation in semisolid medium is coincident with expression of involucrin and transglutaminase but not keratin-10. *J. Virol.* **72,** 5016–5024.

14. Fehrmann, F., Klumpp, D. J. and Laimins, L. A. (2003) Human papillomavirus type 31 E5 protein supports cell cycle progression and activates late viral functions upon epithelial differentiation. *J. Virol.* **77,** 2819–2831.
15. Thomas, J. T., Oh, S. T., Terhune, S. S., and Laimins, L. A. (2001) Cellular changes induced by low-risk human papillomavirus type 11 in keratinocytes that stably maintain viral episomes. *J. Virol.* **75,** 7564–7571.
16. Terhune, S. S., Hubert, W. G., Thomas, J. T., and Laimins, L. A. (2001) Early polyadenylation signals of human papillomavirus type 31 negatively regulate capsid gene expression. *J. Virol.* **75,** 8147–8157.
17. Hubert, W. G., Kanaya, T., and Laimins, L. A. (1999) DNA replication of human papillomavirus type 31 is modulated by elements of the upstream regulatory region that lie 5′ of the minimal origin. *J. Virol.* **73,** 1835–1845.

VIII

RECOMBINANT GENETICS

23

Molecular Genetics of Herpesviruses

A Recombinant Technology Approach

Jason S. Knight, Subhash C. Verma, Ke Lan, and Erle S. Robertson

Summary

Herpesvirus genetics have long been hindered by the large size of the typical herpesvirus genome and the consequent recalcitrance of these genomes to manipulation by standard molecular genetics techniques. However, two primary strategies have emerged that allow for the generation of targeted viral mutants. With these mutants, investigators can pursue critical questions regarding the relationship between specific viral genetic elements and the viral life cycle. The first strategy of viral genome manipulation utilizes the mammalian homologous recombination machinery to introduce specific changes into the native viral genome. This approach involves construction of a targeting vector containing both the desired mutation and a significant flanking viral sequence to permit efficient recombination. The targeting vector is then introduced into mammalian cells, along with viral DNA, and recombinant virus is subsequently selected, harvested, and purified. The second, and more recent, approach utilizes bacterial artificial chromosome (BAC) technology to reconstitute complete herpesvirus genomes in the context of a prokaryote, *E. coli*. This artificial genome is then manipulated in, and purified from, *E. coli* before introduction into a mammalian background in which viral phenotypes can be assessed. Both strategies are discussed in this review, with particular emphasis on the homologous recombination strategies that have continued to be a powerful genetic tool in many herpesvirus systems.

Key Words: Molecular genetics; herpesvirus; herpes simplex; Epstein-Barr virus; homologous recombination; bacterial artificial chromosome; BAC.

1. Introduction

Understanding the role of specific herpesvirus gene products in the context of intact virus has long been hampered by the large size of the herpesvirus genome. These genomes have been difficult to manipulate experimentally in vitro, limiting the ability to functionally characterize and create specifically targeted changes in the genetic material. The background mutation rate for DNA

From: *Methods in Molecular Biology, vol. 292: DNA Viruses: Methods and Protocols*
Edited by: P. M. Lieberman © Humana Press Inc., Totowa, NJ

viruses is generally low; therefore, early characterization of herpesvirus genomes employed random mutagenesis to identify genomic elements that imparted either sensitivity or resistance to a variety of physical, chemical, and biological conditions; such selective conditions included host range properties, plaque morphology, drug resistance, and temperature sensitivity *(1–4)*. These forward genetic approaches required eventual mapping and comparison of mutations by a variety of cumbersome techniques including cross-complementation, recombination analysis, and marker rescue assays.

With the advent of current molecular cloning technologies, progress in understanding small mammalian DNA tumor viruses advanced rapidly; however, progress in applying such technologies to more complex herpesvirus genomes has been a slower, stepwise process. Today, targeted manipulation of the herpesvirus genome generally employs one of three broad strategies. The first specifically introduces a selectable marker into the genome by co-expression of an intact wild-type viral genome along with a targeting vector containing the marker flanked by the herpesvirus sequence to be disrupted. The recombination machinery of the host cell then generates a small population of the desired mutant virus. This method has been particularly successful for herpes simplex virus (HSV), which propagates robustly in tissue culture *(5–7)*. In contrast, this technology has been less useful for viruses such as cytomegalovirus (CMV) that are strongly cell-associated, that replicate slowly in tissue culture, or that enter the lytic cycle only under specific and limited culture conditions *(8–11)*. Another disadvantage of this approach is that wild-type virus is invariably vastly overrepresented, and consequently selection and/or extensive purification of mutant virus are required.

A second genetic approach for generating viral mutants also utilizes the eukaryotic recombination machinery. A complete viral genome is generated in cells by cotransfection of a complete set of overlapping cosmid clones in the absence of viral infection. This method has several advantages including isolation of mutant virus in the absence of wild-type and abrogation of the need for a selectable marker *(12)*. As this approach essentially utilizes the same technologies as for the aforementioned selective marker approaches, it is only discussed briefly here.

A third and more recent approach to herpesvirus genetics employs bacterial artificial chromosomes (BACs) to maintain and manipulate complete herpesvirus genomes in *E. coli*. This strategy has the obvious and powerful advantage of employing the recombinogenic activities of bacteria *(E. coli)* for manipulation of large viral genomes. However, the complex and highly repetitive nature of the herpesvirus genome has in some cases called into question the stability or integrity of the *E. coli*-manipulated viral genome. In this chapter we provide a detailed description of the traditional generation of mutant viruses utilizing homologous recombination and an overview of the more recent employment of BAC technologies for herpesvirus genetic manipulation.

2. Materials

2.1. Homologous Recombination

1. Viral cosmid DNA for targeting vector construction (frequently available from commercial sources).
2. Restriction enzymes for manipulation of cosmid DNA (e.g., New England Biolabs).
3. Plasmid DNA containing the green fluorescent protein (GFP) gene (e.g., EGFP-C1, BD Biosciences Clontech).
4. Cell line permissive for viral infection and propagation (e.g., BHK cells for herpes simplex, purified primary B cells for Epstein-Barr virus [EBV], and so on).
5. Tissue culture medium (e.g., Gibco).
6. Reagents for standard calcium phosphate transfection.
7. Electroporation unit (e.g., Bio-Rad Gene Pulser II).
8. Fluorescence microscope for visualization of GFP virus (e.g., Olympus X170).

2.2. BAC

1. Virus and cells: viral cosmid DNA or virally infected cells for producing viral DNA.
2. pBAC vector for generation of shuttle vector (commercially available from various molecular biology companies).
3. Restriction endonucleases for the manipulation of shuttle vectors and viral DNA (e.g., New England Biolabs and Invitrogen).
4. Cell line permissive for viral latent and lytic replication (e.g., HEK293 and BJAB cells).
5. *E. coli* DH10B for transformation of pBAC containing viral DNA.
6. Reagents and equipment for standard transformation and electroporation into *E. coli*.
7. Antibiotics, available commercially (e.g., Sigma).
8. 12-*O*-Tetradecanoyl phorbol-13-acetate (TPA, e.g., Sigma) for the induction of lytic replication

3. Methods

3.1. Homologous Recombination Approaches to Herpesvirus Genetics

3.1.1. Design and Generation of Targeting Vector

Here we describe the preparation of a targeting vector for production of a general GFP herpesvirus that can facilitate the identification of herpesvirus-infected cells or trace dissemination of virus from a peripheral site of inoculation to different tissues.

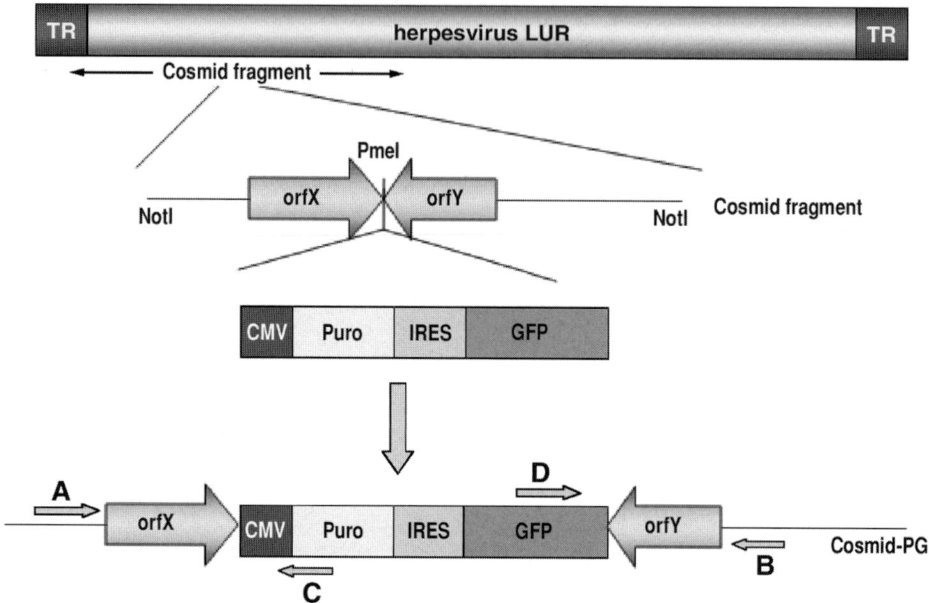

Fig. 1. Schematic representation of targeting vector for homologous recombination. A cassette containing the puromycin and green fluorescent protein (GFP) genes under the control of the CMV promoter is inserted into a unique *Pme*I restriction site in a commercially available herpesvirus cosmid. a, b, c, and d represent primers for differential determination of wild-type and recombinant virus. ab amplifies wild-type virus. ac and db amplify recombinant virus. IRES, internal ribosome entry; LUR, long unique region; TR, tandem repeat.

3.1.2. Construction of GFP Expression Cassette

This cassette potentially contains several elements including the *GFP* gene, a eukaryotic promoter to drive *GFP* gene expression, a drug resistance gene (such as puromycin or neomycin), and a eukaryotic promoter to drive expression of the drug-resistance gene. In some cases a polycistronic gene may be constructed using an internal ribosome entry site ([IRES] such as that of the hepatitis C virus [HCV]) to facilitate expression of GFP and drug resistance from a common promoter. This cassette should be constructed in a vector background with unique flanking restriction sites such that the cassette can be excised for insertion into the targeting vector (**Fig. 1**).

For viruses such as herpes simplex, with a tissue culture system that robustly supports lytic infection, the insertion of a drug resistance gene into the aforementioned GFP cassette may be unnecessary. However, for viruses such as EBV and Kaposi's sarcoma-associated herpesvirus (KSHV), which are primarily maintained in tissue culture as latently infecting genomes, drug selection may allow the establishment of latently infected producer cell lines that can be used to generate free virus by induction of lytic replication with chemical inducers or viral transactivating genes. This distinction will be discussed further below.

3.1.3. Construction of Targeting Vector for Generation of Recombinant Virus

Ideally the GFP cassette should be inserted into a unique restriction site in a region of the genome that does not encode a gene product, eliminating the possibility that the cassette will disrupt a viral open reading frame (ORF) and minimizing the potential for disruption of viral gene transcription. Careful consideration of this site is important, as the end goal of this strategy is to generate a recombinant GFP herpesvirus that mimics the functionality of the wild-type virus. Ideally, the aforementioned restriction site should be chosen with careful consideration of available cosmid sequences such that digestion with the appropriate enzyme results in a single cut and linearization of the cosmid. The GFP/drug resistance cassette can then be excised from its plasmid and inserted into the digested cosmid by a standard molecular ligation reaction. A restriction site can be chosen such that the insert is flanked by at least 3 kb of viral DNA sequence on the 5′ and 3′ termini. Again, flanking sequence should be chosen to minimize the dysregulation of the gene expression pattern of viral genes in the adjacent areas.

An example of such a strategy utilizes the *Pme*I site that lies downstream of ORF 18 and ORF 19 of the KSHV genome *(13)*. Because ORF 18 is transcribed to the right and ORF 19 to the left, this site is downstream of both transcripts, minimizing the likelihood that insertion of the GFP cassette will disrupt viral gene expression. Importantly, this is the only *Pme*I restriction site in the available cosmid fragment, permitting unique insertion of the GFP cassette. This site has previously been used in the construction of a GFP KSHV BAC *(13)*, but the same advantages apply in the construction of a targeting vector for homologous recombination.

3.1.4. Mutagenesis Strategies

The desired recombinant virus may possess either a manipulated or disrupted viral gene, in contrast to the general introduction of a foreign gene, *GFP,* as described just above. This usually depends on the specific application or genetic question. We will briefly outline approaches for introducing such mutations.

As mentioned above, *GFP* provides a signal that marks *GFP* recombinant virus-infected cells. However, for addressing specific gene functions in the context of the whole virus, genes can be knocked out and the resultant virus can be assayed for viral replication, packaging, infection, and so on. Based on the *GFP* recombinant virus backbone, another cassette containing a second selectable marker can be introduced into a specific viral ORF by homologous recombination, disrupting the coding region of the targeted gene. The advantage of introduction of another selection marker into the viral genome is to facilitate double selection or selection of cloned infected populations, which include the *GFP* (initial selection marker) "wild-type" recombinants with a green color in infected cells and those that can be positively selected by antibiotics as "mutant" recombinants. This strategy might also employ one of the many fluorescent color proteins available as the second selection marker.

The strategy for construction of this second targeting vector is very similar to that described above for *GFP*. The primary difference is that the cassette containing the second selection marker should be introduced into a unique restriction site within the coding region of the viral gene of interest. A cassette encoding neomycin/puromycin/red fluorescent protein (RFP) under control of the CMV promoter can be digested from a commercially available construct. For generation of mutant virus, this cassette should be introduced by restriction digest into a cosmid vector that contains a large fragment of viral genome including the target gene of interest. Cells stably infected with *GFP* recombinant virus can then be transfected with the mutant targeting vector. With induction of lytic replication, the targeting vector can recombine with *GFP* recombinant virus. To get a pure population of mutant virus, the supernatant containing *GFP* "wild-type" virus and *GFP* "mutant" virus can be used to infect permissive cells. By serial dilution of supernatant, and by employing the selection principle described above, pure mutant virus may be obtained.

3.1.5. Transfection and Induction of Lytic Replication

Cell type and transfection method are clearly dependent on the specific herpesvirus being manipulated. For HSV, BHK cells are generally cotransfected with intact viral DNA and the targeting vector by a standard calcium phosphate transfection protocol. A similar approach is taken for varicella-zoster virus (VZV) and CMV with melanoma and fibroblast cell lines, respectively.

As no robust system for EBV or KSHV lytic infection exists in vitro, the targeting vector is generally transfected into a virus-infected cell line (typically a B cell) by electroporation. Successful electroporation of B cells can be performed with the Bio-Rad Gene Pulser at 200 V and 960 µF (*see* **Note 1**). Lytic infection is induced either by cotransfection of a lytic transactivator (BZLF1 for EBV and Rta for KSHV) or by treatment with chemical agents such as phorbol esters or butyric acid *(14,15)*. Although treatment with chemical agents produces a more robust overall lytic response, cotransfection of the transactivator has the advantage of enriching for lytic replication only in the transfected population of cells, potentially reducing the background of wild-type virus with no potential to recombine. Similarly, cotransfection of lytic gene transactivators has proved useful for other viruses such as the use of VZV ORF62 in the generation of intact virus from overlapping cosmid clones *(16)*.

The efficiency of lytic virus induction can be monitored by immunofluorescence of a lytic gene (e.g., gB) and also by specific polymerase chain reaction (PCR) analysis of the cell supernatant for viral DNA (*see* **Note 2**). PCR analysis may differentiate wild-type from recombinant virus by specific PCR primers that will detect and distinguish the wild-type from recombinant virus DNA. It should, however, be noted that these methods only indicate the production of viral progeny and not the infectious nature of these progeny virions. Returning to the example of the unique *Pme*I site in the viral cosmid construct, PCR primers for distinguishing wild-type and recombinant virus might be designed as indicated in **Fig. 1.**

3.1.6. Selection and Isolation of Recombinant Virus

For herpesviruses such as HSV, which lytically replicate to high titer and form robust plaques in tissue culture, permissive cells are simply transfected with targeting vector and virion DNA, and plaques of *GFP*-expressing virus are purified from wild-type by limiting dilution strategies. For lytic viruses this strategy has proved highly efficient in genetic analysis of virus-encoded genes with mutation of these ORFs determining functional domains required for infection and replication *(17)*.

For viruses with tissue culture systems that less robustly support lytic infection, induction of latently infected cells may be necessary, as discussed previously. Ideally, upon induction recombinant virus will be produced and will egress from the cell into the supernatant, where it can be collected for further purification and study. In some cases, recovery of recombinant virus may be enhanced by transfecting latently infected cells with the targeting vector followed by drug selection. Selection will enrich for cells transfected with the targeting vector and should increase the ratio of recombinant to wild-type virus in the cell supernatant. To obtain a pure recombinant virus population, the supernatant can be diluted and used to infect permissive cells at varying dilutions. The infected cells are again drug-selected, and, under continuous selection, stable recombinant virus-infected cells can be obtained. These latently infected cell lines can then be induced to generate recombinant virus. After several cycles of limiting supernatant dilution, cell infection, and selection, a pure recombinant virus-producer cell line can be obtained (*see* **Note 3**).

3.1.7. Characterization

To determine whether the desired recombinant herpesvirus was successfully generated, it is necessary to complete genetic and phenotypic characterization of the virus (*see* **Notes 4** and **5**). The first line of characterization involves restriction analysis using enzymes that cut with moderate frequency in herpesvirus genomes such as *Bam*H1, *Eco*R1, *Not*1, and *Sal*1. It is recommended that analysis of at least three restriction profiles be determined. Any discord between wild-type and recombinant profiles is probably attributable to the desired recombination event. The identity of bands modified by the recombination event should then be confirmed by Southern hybridization with probe/probes specific to the sequence immediately flanking the recombination event as well as any genetic material inserted in the site (selectable marker). Once the alteration of the genetic structure of the genome has been confirmed, the integrity of the recombination event should be confirmed by PCR amplification and sequencing of any junctions between the viral and foreign DNA (*see* **Note 6**).

Phenotypically, the recombinant virus can be characterized by gene expression profiling of genes immediately flanking the recombination event. Such analysis can be done by standard methods such as Northern blotting or real-time PCR. For viruses with a cell culture system that supports lytic infection, a one-step growth curve can be performed to compare the in vitro growth kinetics of the recombinant virus with that of the wild-type virus. For viruses lacking a system for characterizing lytic infection, other phenotypic properties should be considered, such as the efficiency of recombinant EBV to immortalize B lymphocytes in vitro in comparison with the efficiency of wild-type EBV. A final, and costly, means of characterization involves putting the recombinant virus into an animal system for consideration of pathogenicity. Such characterization is undoubtedly warranted when the recombinant virus, or subsequent derivatives, will ultimately be employed in animal pathogenesis experiments.

3.2. BAC Approaches to Herpesvirus Genetics

3.2.1. Background

A BAC is a DNA molecule of plasmid origin into which another large DNA fragment (100–300-kb insert size; average, 150 kb) can be integrated without loss of the plasmid's capacity for self-replication. Yeast artificial chromosomes (YACs) were the first vectors used for cloning large genomic DNA fragments; however, YACs have numerous problems including difficulty in purifying cloned DNA from contaminating yeast DNA and frequent spontaneous rearrangements of the foreign DNA *(18,19)*. In contrast, BAC clones show surprising stability of the foreign DNA *(20)* and can accommodate DNAs larger than 300 kb. One of the first technologies employed by Shizuya et al. in 1992 *(21)* to clone large DNA fragments was based on the *E. coli* F factor. Strict control of the F-factor replicon maintains a single copy of the plasmid DNA per cell, reducing the risk of recombination events via repetitive DNA elements present in the foreign DNA. Interestingly, cosmid clones have also been shown to maintain foreign DNA more stably under the control of F-derived replicons than the common origin of plasmid replication *(21)*.

3.2.2. Cloning of Herpesvirus Genomes Into BACs

The cloning limit of BACs exceeds the size of the largest known herpesviruses (hCMV); therefore, BACs can theoretically support and maintain complete herpesvirus genomes. In 1997, the first infectious herpesvirus BAC, the murine CMV, was constructed and was subsequently followed by other viruses *(22–28)*. The large size and the recalcitrance to in vitro manipulation of the herpesvirus genome prevented simple ligation of the viral genome into a BAC by standard molecular cloning procedures. Efficient cloning required a preliminary step in which the BAC plasmid cassette is itself introduced into the viral genome, essentially through the homologous recombination strategy described above. Insertion of the BAC sequence is followed by isolation of a BAC recombinant virus **(Fig. 2)**. Frequently, a selection marker (antibiotic resistance) or reporter gene such as *GFP/lacZ* is employed to facilitate recovery of recombinant virus.

3.2.3. Purification and Recovery of BAC Recombinant Virus

Before genetic manipulation, the BAC viral recombinant must be purified, followed by transformation into *E. coli* competent cells. For successful transformation, an intact, circularized form of the recombinant virus must be isolated prior to transformation into *E. coli*. Two approaches are available for isolation and purification of the viral genome. The most common approach takes advantage of the herpesvirus replication cycle, which results in circular replicative intermediates of viral DNA prior to cleavage into unit-length linearized forms for packaging in the virus particle. DNA harvested by *Hirt* extraction early after transfection gives a significant amount of circular replicative intermediate *(29)*. Alternatively, linear DNA can be harvested and ligated in vitro **(Fig. 3)**. This procedure can potentially give both circular and concatamaric version of the genome, but only circular versions will propagate.

3.2.4. BAC Stability in E. coli

Herpesvirus genomes contain highly repetitive sequences that are potential targets for recombination. Therefore, a herpesvirus BAC can be maintained in a bacterial strain of *E. coli* that is devoid of the *recABCD* gene *(20)*. However, for mutagenesis and gene manipulation, the recombination system may need to be re-expressed. This can be achieved by shuttling the BAC recombinant virus

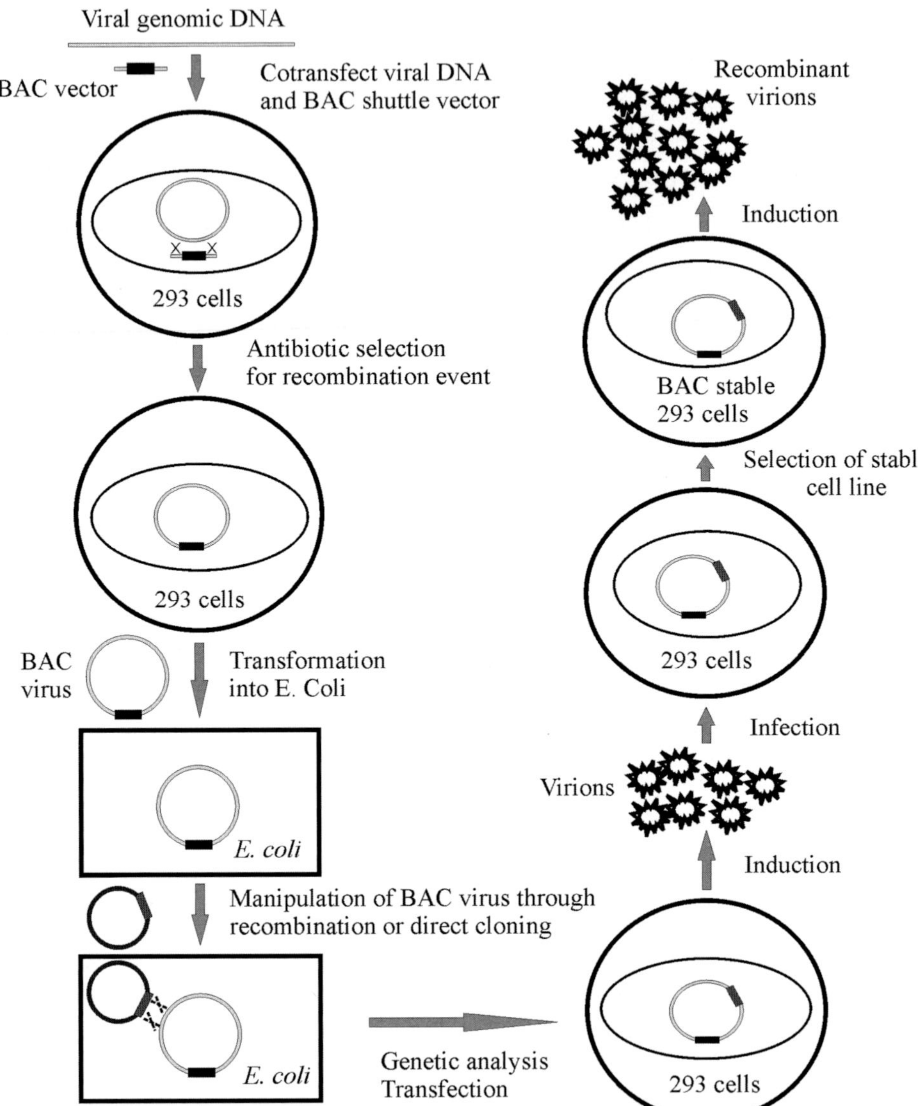

Fig. 2. Construction of bacterial artificial chromosome (BAC) herpesvirus by homologous recombination. Efficient cloning of the BAC herpesvirus requires a preliminary step in which the BAC plasmid cassette is introduced into the viral genome through homologous recombination. Insertion of the BAC sequence is followed by isolation of a BAC recombinant virus. A selection marker (antibiotic resistance) may be employed to facilitate efficient recovery of recombinant virus.

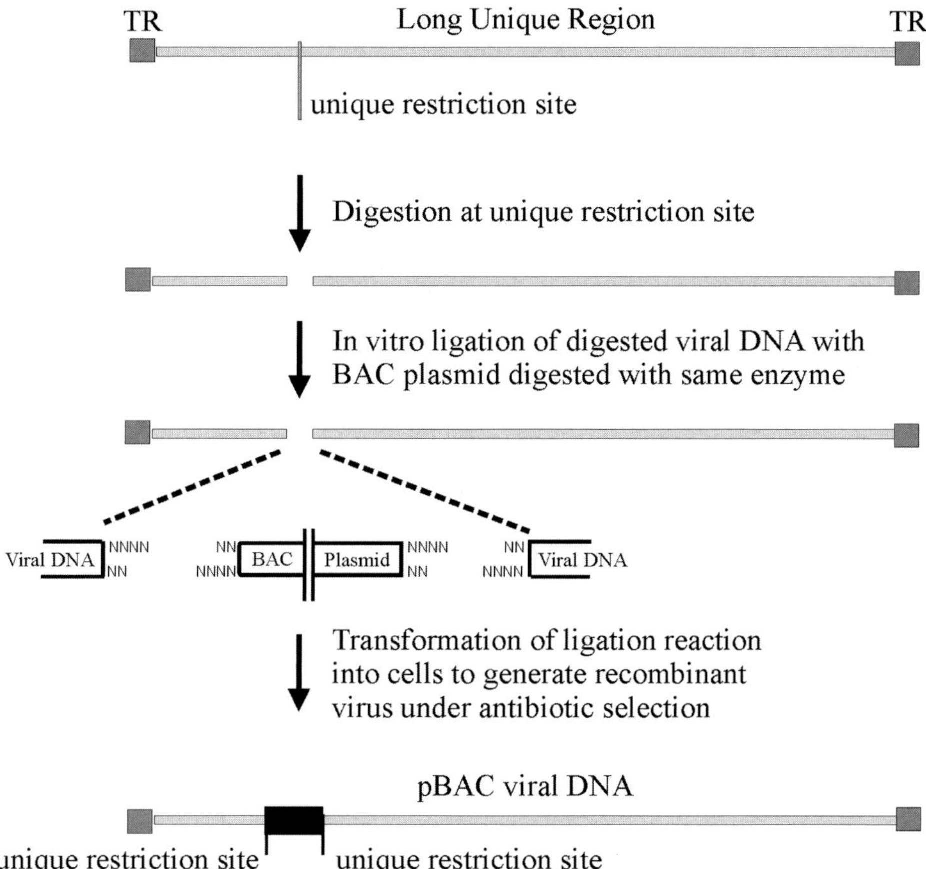

Fig. 3. Alternative construction of bacterial artificial chromosome (BAC) herpesvirus by in vitro ligation. A unique restriction site within the viral genome may permit direct ligation of the BAC plasmid into the viral genome.

from a *recA⁻* to a *recA⁺* strain *(22,30,31)* or by using a bacterial strain in which expression is controlled by an inducible promoter *(22,32)*.

3.2.5. Alternatives to Homologous Recombination for Generation of BAC Clones

Although BAC cloning typically requires homologous recombination between viral sequence flanking the BAC plasmid and viral DNA, this is not always the case. If unique restriction sites are present in the viral genome, BAC vector sequence can potentially be ligated in vitro. An example of ligation of the BAC plasmid is shown in **Fig. 3.** The viral DNA is digested at unique restriction sites and ligated with the BAC plasmid similarly digested with the same restriction enzyme. In one example, transfection of the ligation mix generated a recombinant BAC-containing virus. Circular viral DNA was isolated and used to transform into *E. coli.* Transformed colonies screened for the BAC contained full-length viral genomes *(33).*

3.2.6. Manipulation of the Herpesvirus–BAC Genome

Once the viral genome is cloned into a BAC system, a variety of techniques are available for the manipulation of viral genome following transformation into *E. coli.* Some commonly used methods are described here.

3.2.6.1. SHUTTLE PLASMIDS

Shuttle plasmids are used for the mutagenesis of the herpesvirus genome in *E. coli* via homologous recombination. This method is believed to be one of the simplest methods for the introduction of various kinds of mutations (e.g., point mutation, deletions, insertions, or sequence replacement) into viral BAC. A schematic of this strategy is shown in **Fig. 4,** in which the desired mutations are cloned into a suicide plasmid (unable to replicate in *E. coli*) that contains a viral sequence homologous to the target site in the viral BAC. The resulting shuttle vector is then cotransformed along with the viral BAC into an *E. coli* strain that conditionally expresses RecA. Homologous recombination via RecA activity potentially leads to the insertion of a mutated copy of the gene into BAC viral DNA.

3.2.6.2. LINEAR DNA

In this method the mutations are introduced into linear DNA with the use of *recET* from prophage Rac or redαβ from bacteriophage λ. A linear DNA fragment containing a selectable marker and a homologous sequence flanking the target mutated sequence is transferred into a recombination-positive *E. coli* strain. By a double crossover event, the target gene (with selectable marker) is introduced into the BAC vector **(Fig. 5)**. This method has advantages over shuttle plasmid-based recombination because the *RecET* recombination is more efficient and requires homologies of only 25–50 nucleotides for crossover. These homologous sequences can be provided by the synthetic oligonucleotide primers used to amplify the liner target DNA sequence.

3.2.6.3. RANDOM TRANSPOSON MUTAGENSIS

Transposons (Tn) are mobile genetic elements that insert themselves into DNA at random sequences. Transformation of a Tn donor plasmid into *E. coli*-containing viral BAC can lead to mutation in the BAC sequence, as the transposon element has the ability to integrate randomly on DNA **(Fig. 6)**. By using

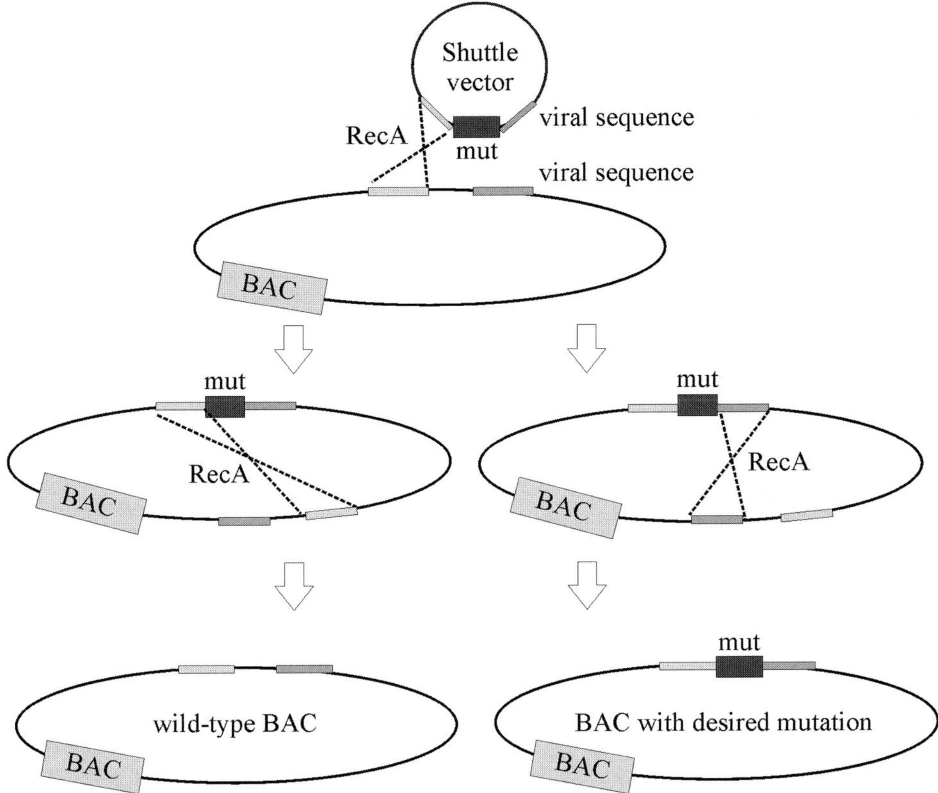

Fig. 4. Introduction of mutations into the viral genome by a shuttle vector and recombinant A (recA)-dependent recombination in *E. coli*. BAC, bacterial artificial chromosome; ORF, open reading frame.

a suicide origin of replication on the Tn donor vector, propagation of the vector can be eliminated. The Tn insertion site can be easily mapped and sequenced from primer sites within the Tn element; this is a major advantage over chemical mutagenesis when one is screening large BAC genomes.

3.2.7. Limitations of BAC Herpesviruses

One major concern of BAC-derived technology is the long-term stability of eukaryotic DNA in *E. coli*. To date, even complex sequences such as inverted and direct repeats appear to be extremely stable in *E. coli*. However, Smith and Enquist *(30)* reported that F-plasmid sequences were unstable following passage of virus in experimental animals. Additionally, it is not well understood how many heterologous genes can be packaged into the BAC-derived virus

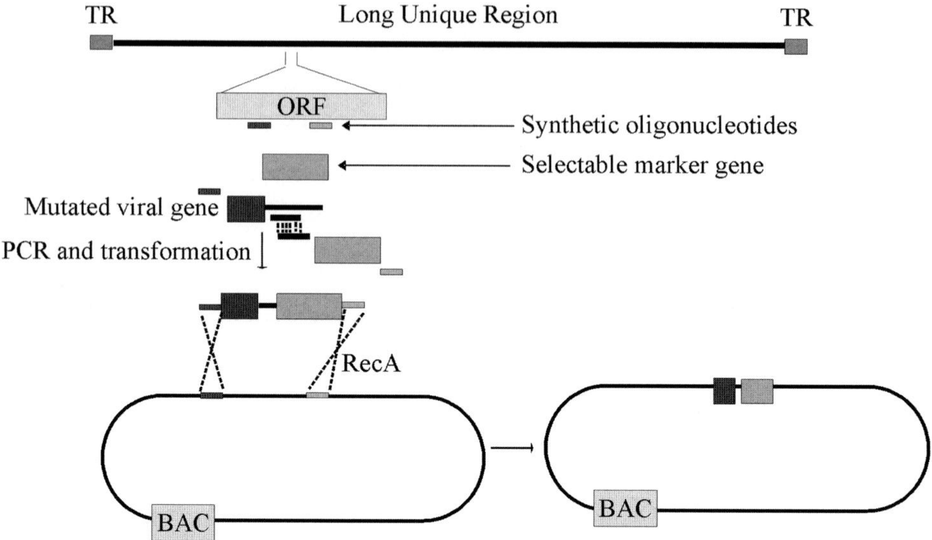

Fig. 5. Mutations can be introduced into linear DNA with the use of recET from prophage or redαβ from bacteriophage λ. A linear DNA fragment containing selectable marker and homologous sequence flanking the target mutated sequence is transferred into a recombination-positive *E. coli* strain. By a double crossover event, the target gene (with selectable marker) is introduced into the bacterial artificial chromosome (BAC) vector. Recombination requires only 25–50 nucleotides for crossover, and these homologous sequences can be provided by the synthetic oligonucleotide primers used to amplify the linear target DNA sequence. ORF, open reading frame; RecA, recombinant A; TR, tandem repeat.

before size constraints becomes problematic. For HSV, it has been reported that up to 30 kb in addition to the viral genome can be packaged. Other viruses remain to be more thoroughly explored by this approach.

The second major problem with BAC-based technology is in developing homologous recombination strategies for shuttle insertion of novel genes into BAC viruses. This strategy is limited because of the lack of unique and suitable restriction sites in most viral genomes. However, this drawback can be partially addressed by the diverse mutagenesis techniques available in *E. coli*. Transposon mutagenesis has been most useful to date in allowing rapid generation of herpesvirus BAC mutants by insertion of Tn5 across the viral genome.

3.2.8. Further Application of Herpesvirus–BAC Technology

BAC technology has significantly advanced the analysis and molecular genetics of the herpesvirus genomes. BAC is probably the most flexible and efficient technology for generating mutant viruses, especially when the specif-

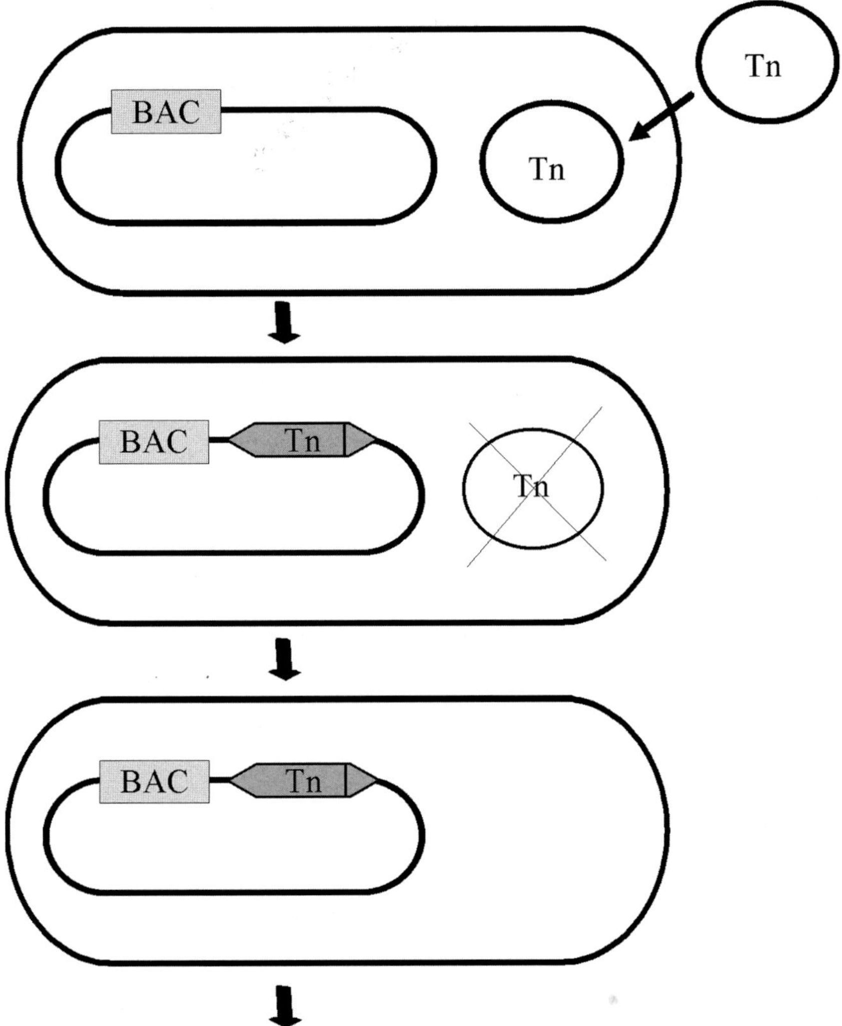

Isolation of BAC and sequence analysis.
Functional analysis by transfection.

Fig. 6. Random mutagenesis of the bacterial artificial chromosome (BAC) herpesvirus genome by insertion of a transposable element. Transformation of a transposon (Tn) donor plasmid into *E. coli* containing viral BAC can lead to mutation in the BAC sequence as the transposon element has the ability to integrate randomly on DNA.

ic herpesvirus BAC of interest has already been developed. This has been particularly true for α-herpesviruses, as, to date, such viruses have aided greatly in defining the roles of specific viral gene products in pathogenesis. Random mutagenesis strategies have been developed for insertional, deletional, or transposon-mediated approaches that can quickly establish the roles of specific genes in the viral life cycle *(26,34–36)*. BAC technology may also prove to be a powerful tool for generating recombinant viruses expressing heterologous therapeutic genes utilized as herpesvirus gene therapy vectors *(37)*. Additionally, herpesvirus BACs can function as helper-virus-free reagents in the propagation of gene therapy vectors such as HSV amplicons and adeno-associated virus vectors. Moreover, BACs can also be useful in vaccine design, by not only providing a substrate to generate safe live attenuated vaccine but also as immunogens for DNA vaccines.

The use of BAC technology for generating γ-herpesvirus recombinants is still somewhat in its infancy. The stability and successful generation of mutants in the γ-herpesviruses EBV and KSHV will further enhance the capability of this technology in analysis of the large DNA virus genomes. However, studies utilizing this technology for EBV and KSHV have not produced consistent results in targeting specific viral genes. Success in generation of specific mutations in a reproducible and stable fashion will further increase the utility of this technology.

4. Notes

1. For transfection of cells to introduce the recombinant plasmids, one must be sure that the cells are greater than 98% viable to ensure effective transfection and sufficient recombination efficiency. Cells should be fed 24 h prior to transfection and growing exponentially. DNA should be mixed thoroughly in buffer to ensure proper solubility before transfection. Usually DNA stocks for transfection should be kept at 4°C and at a concentration of no greater than 0.5 µg/mL. It is important to note that your DNA preparation should be of the highest purity, and we recommend using CsCl-purified DNA, prepared with two spins before collecting for use. Once these DNA preparations are prepared, we have found they can be stored and used indefinitely with little loss in activity.

2. Transfected cells can be monitored for production of virion particles. Samples of the supernatant can be collected and spun at maximum speed in a microcentrifuge for 20 min to collect possible virus particles. The resulting pellet can then be heated to 95°C for 15 min, followed by treatment with proteinase K (followed by killing of the protease). This crude lysate can then be used for PCR analysis of the viral DNA or a specific sequence introduced into the virus. PCR analysis should always include the appropriate positive and negative controls as transfected DNA can sometimes be stable and carried forward in the supernatant.

3. The use of *GFP* as a means of tracking recombinant virus production should be carefully monitored. In some cases we have seen *GFP* transferred to virion particles and introduced into infected cells without recombinant virus. This false transduction can be misleading and lead to incorrect conclusions as to the presence of a recombinant virus.

4. The production of a recombinant virus is very time-consuming, and viruses typically have a tendency to acquire their most stable genomic state. Therefore, if there are small amounts of wild-type virus in the viral preparation, over time this can become the dominant genome if the selective pressures are absent. This potential problem has to be carefully considered, as continued passage in culture can quickly lead to loss of your recombinant virus.

5. Another common pitfall is the assumption that the presence of either *GFP* DNA or selective marker plasmid DNA confirms the presence of the recombinant virus. This should always be checked with an induction and passage of the virus into fresh uninfected cells. Additionally, the resulting virus genome should be characterized fully to be sure that deletions and rearrangements have not occurred that might impair viral viability. Herpesviruses have a strong propensity for recombination and rearrangements. There have been many occasions on which an investigator has identified a recombinant virus by *GFP* or selective marker analysis, but the remaining viral genome was deleted or rearranged. This rearranged or deleted virus then becomes the predominant genome, as it is usually selected for by the investigator.

6. One highly efficient approach to determination of a stable recombinant virus genome is to select a number of unique sites spanning the genome that can be amplified by PCR analysis. This can be performed with minimal effort and confirms the presence of a relatively complete genome. The presence of specific mutations can also be determined for stability over multiple passages by this method.

Acknowledgments

J. S. K. is supported by the Lady Tata Memorial Trust. E. S. R. is funded by NIH grants NCI CA72150-07, NCI CA91792-01, and DCR DE14136-01 and is also a scholar of the Leukemia and Lymphoma Society of America.

References

1. Aurelian, L. and Roizman, B. (1964) The host range of herpes simplex virus. *Virology* **22,** 452–461.

2. Ejercito, P. M., Kieff, E. D., and Roizman, B. (1968) Characterization of herpes simplex virus strains differing in their effects on social behaviour of infected cells. *J. Gen. Virol.* **2,** 357–364.

3. Kit, S. and Dubbs, D. R. (1963) Nonfunctional thymidine kinase cistron in bromodeoxyuridine resistant strains of herpes simplex virus. *Biochem. Biophys. Res. Commun.* **13,** 500–504.

4. Schaffer, P., Vonka, V., Lewis, R., and Benyesh-Melnick, M. (1970) Termperature-sensitive mutants of herpes simplex virus. *Virology* **42,** 1144–1146.

5. Goldstein, D. J. and Weller, S. K. (1988) An ICP6::lacZ insertional mutagen is used to demonstrate that the UL52 gene of herpes simplex virus type 1 is required for virus growth and DNA synthesis. *J. Virol.* **62,** 2970–2977.

6. Neidhardt, H., Schroder, C. H., and Kaerner, H. C. (1987) Herpes simplex virus type 1 glycoprotein E is not indispensable for viral infectivity. *J. Virol.* **61,** 600–603.

7. Post, L. E., Mackem, S., and Roizman, B. (1981) Regulation of alpha genes of herpes simplex virus: expression of chimeric genes produced by fusion of thymidine kinase with alpha gene promoters. *Cell* **24,** 555–565.

8. Jones, T. R., Muzithras, V. P., and Gluzman, Y. (1991) Replacement mutagenesis of the human cytomegalovirus genome: US10 and US11 gene products are nonessential. *J. Virol.* **65,** 5860–5872.

9. Kemble, G. W. and Mocarski, E. S. (1989) A host cell protein binds to a highly conserved sequence element (pac-2) within the cytomegalovirus A sequence. *J. Virol.* **63,** 4715–4728.

10. Spaete, R. R. and Mocarski, E. S. (1987) Insertion and deletion mutagenesis of the human cytomegalovirus genome. *Proc. Natl. Acad. Sci. USA* **84,** 7213–7217.

11. Vieira, J., Farrell, H. E., Rawlinson, W. D., and Mocarski, E. S. (1994) Genes in the *Hind*III J fragment of the murine cytomegalovirus genome are dispensable for growth in cultured cells: insertion mutagenesis with a lacZ/gpt cassette. *J. Virol.* **68,** 4837–4846

12. Wagner, M., Ruzsics, Z., and Koszinowski, U. H. (2002) Herpesvirus genetics has come of age. *Trends Microbiol.* **10,** 318–324.

13. Zhou, F. C., Zhang, Y. J., Deng, J. H., Wang, X. P., Pan, H. Y., Hettler, E., and Gao, S. J. (2002) Efficient infection by a recombinant Kaposi's sarcoma-associated herpesvirus cloned in a bacterial artificial chromosome: application for genetic analysis. *J. Virol.* **76,** 6185–6196.

14. Nakamura, H., Lu, M., Gwack, Y., Souvlis, J., Zeichner, S. L., and Jung, J. U. (2003) Global changes in Kaposi's sarcoma-associated virus gene expression patterns following expression of a tetracycline-inducible Rta transactivator. *J. Virol.* **77,** 4205–4220.

15. Tomkinson, B., Robertson, E., and Kieff, E. (1993) Epstein-Barr virus nuclear proteins EBNA-3A and EBNA-3C are essential for B-lymphocyte growth transformation. *J. Virol.* **67,** 2014–2025.

16. Cohen, J. I. and Seidel, K. E. (1993) Generation of varicella-zoster virus (VZV) and viral mutants from cosmid DNAs: VZV thymidylate synthetase is not essential for replication in vitro. *Proc. Natl. Acad. Sci. USA* **90,** 7376–7380.

17. Roizman, B. and Knipe, D. M. (2002) Herpes simplex viruses and their replication, in *Fields Virology,* vol. 2 (Knipe, D. and Howley, P., eds.), Lippincott Williams & Wilkins, pp. 2399–2459.

18. Ramsay, M. (1994) Yeast artificial chromosome cloning. *Mol. Biotechnol.* **1,** 181–201.

19. Schalkwyk, L. C., Francis, F., and Lehrach, H. (1995) Techniques in mammalian genome mapping. *Curr. Opin. Biotechnol.* **6,** 37–43.

20. Shizuya, H., Birren, B., Kim, U. J., Mancino, V., Slepak, T., Tachiiri, Y., and Simon, M. (1992) Cloning and stable maintenance of 300-kilobase-pair fragments of

human DNA in *Escherichia coli* using an F-factor-based vector. *Proc. Natl. Acad. Sci. USA* **89,** 8794–8797.

21. Kim, U. J., Shizuya, H., de Jong, P. J., Birren, B., and Simon, M. I. (1992) Stable propagation of cosmid sized human DNA inserts in an F factor based vector. *Nucleic Acids Res.* **20,** 1083–1085.

22. Messerle, M., Crnkovic, I., Hammerschmidt, W., Ziegler, H., and Koszinowski, U. H. (1997) Cloning and mutagenesis of a herpesvirus genome as an infectious bacterial artificial chromosome. *Proc. Natl. Acad. Sci. USA* **94,** 14759–14763.

23. Stavropoulos, T. A. and Strathdee, C. A. (1998) An enhanced packaging system for helper-dependent herpes simplex virus vectors. *J. Virol.* **72,** 7137–7143.

24. Delecluse, H. J., Hilsendegen, T., Pich, D., Zeidler, R., and Hammerschmidt, W. (1998) Propagation and recovery of intact, infectious Epstein-Barr virus from prokaryotic to human cells. *Proc. Natl. Acad. Sci. USA* **95,** 8245–8250.

25. Saeki, Y., Ichikawa, T., Saeki, A., et al. (1998) Herpes simplex virus type 1 DNA amplified as bacterial artificial chromosome in *Escherichia coli:* rescue of replication-competent virus progeny and packaging of amplicon vectors. *Hum. Gene Ther.* **9,** 2787–2794.

26. Wagner, M., Jonjic, S., Koszinowski, U. H., and Messerle, M. (1999) Systematic excision of vector sequences from the BAC-cloned herpesvirus genome during virus reconstitution. *J. Virol.* **73,** 7056–7060.

27. Adler, H., Messerle, M., and Koszinowski, U. H. (2003) Cloning of herpesviral genomes as bacterial artificial chromosomes. *Rev. Med. Virol.* **13,** 111–121.

28. Delecluse, H. J., Kost, M., Feederle, R., Wilson, L., and Hammerschmidt, W. (2001) Spontaneous activation of the lytic cycle in cells infected with a recombinant Kaposi's sarcoma-associated virus. *J. Virol.* **75,** 2921–2928.

29. Hirt, B. (1967) Selective extraction of polyoma DNA from infected mouse cell cultures. *J. Mol. Biol.* **26,** 365–369.

30. Smith, G. A. and Enquist, L. W. (1999) Construction and transposon mutagenesis in *Escherichia coli* of a full-length infectious clone of pseudorabies virus, an alphaherpesvirus. *J. Virol.* **73,** 6405–6414.

31. Horsburgh, B. C., Hubinette, M. M., Qiang, D., MacDonald, M. L., and Tufaro, F. (1999) Allele replacement: an application that permits rapid manipulation of herpes simplex virus type 1 genomes. *Gene Ther.* **6,** 922–930.

32. Kempkes, B., Pich, D., Zeidler, R., and Hammerschmidt, W. (1995) Immortalization of human primary B lymphocytes in vitro with DNA. *Proc. Natl. Acad. Sci. USA* **92,** 5875–5879.

33. McGregor, A. and Schleiss, M. R. (2001) Molecular cloning of the guinea pig cytomegalovirus (GPCMV) genome as an infectious bacterial artificial chromosome (BAC) in *Escherichia coli. Mol. Genet. Metab.* **72,** 15–26.

34. Brune, W., Messerle, M., and Koszinowski, U. H. (2000) Forward with BACs: new tools for herpesvirus genomics. *Trends Genet.* **16,** 254–259.

35. Muyrers, J. P., Zhang, Y., Testa, G., and Stewart, A. F. (1999) Rapid modification of bacterial artificial chromosomes by ET-recombination. *Nucleic Acids Res.* **27,** 1555–1557.

36. Brune, W., Menard, C., Hobom, U., Odenbreit, S., Messerle, M., and Koszinowski, U. H. (1999) Rapid identification of essential and nonessential herpesvirus genes by direct transposon mutagenesis. *Nat. Biotechnol.* **17,** 360–364.
37. Borst, E., and Messerle, M. (2000) Development of a cytomegalovirus vector for somatic gene therapy. *Bone Marrow Transplant* **25 Suppl 2**, S80–82.

24

Molecular Genetics of DNA Viruses

Recombinant Virus Technology

Bernhard Neuhierl and Henri-Jacques Delecluse

Summary

Recombinant viral genomes cloned onto BAC vectors can be subjected to extensive molecular genetic analysis in the context of *E. coli*. Thus, the recombinant virus technology exploits the power of prokaryotic genetics to introduce all kinds of mutations into the recombinant genome. All available techniques are based on homologous recombination between a targeting vector carrying the mutated version of the gene of interest and the recombinant virus. After modification, the mutant viral genome is stably introduced into eukaryotic cells permissive for viral lytic replication. In these cells, mutant viral genomes can be packaged into infectious particles to evaluate the effect of these mutations in the context of the complete genome.

Key Words: Herpesviruses; Epstein-Barr virus; viral recombinants; genetic analysis; viral mutants.

1. Introduction

1.1. Construction of Mutants by Recombination Using Linearized Targeting Vectors

Molecular genetic analysis of viruses offers a direct approach for the analysis of viral gene function and is becoming a standard tool in many laboratories. The scope of introduced modifications ranges from knockout mutations that render viral genes or groups of viral genes nonfunctional to more subtle changes such as point mutations. Insertion of conditional expression systems within the complete virus genome even allows sequential activation and repression of viral gene expression, allowing a more thorough understanding of how viruses interact with their host cells. Small virus genomes (<20 kb) can be modified by standard recombinant DNA technology, i.e., cloning using restriction enzyme cleavage

From: *Methods in Molecular Biology, vol. 292: DNA Viruses: Methods and Protocols*
Edited by: P. M. Lieberman © Humana Press Inc., Totowa, NJ

and ligation *(1)*. With increasing size, virus genomes become more difficult to manipulate and propagate. Herpesviruses, for example, possess large genomes (up to 200 kb) that are not amenable to standard cloning techniques, and modification of these genomes relies on homologous recombination.

Classically, homologous recombination was performed in eukaryotic cells carrying the virus to be modified *(2–4)*. A major difficulty of this kind of approach is that the mutant virus coexists with the wild-type virus in the infected cell and therefore must first be purified. More recently, single-copy F-plasmid replicons (also known as bacterial artificial chromosomes [BACs]) were used to clone up to 300 kb of genomic DNA *(5)*. Introduction of these replicons into large viral DNAs has allowed cloning and manipulation of these genomes in the context of a prokaryotic host *(6–10)*. Genetic analysis using viruses cloned onto BACs is less time-consuming than the classical methods. Sequential introduction of multiple mutations into viral genomes cloned onto BACs is possible within a reasonable amount of time. In this review, we focus on a few techniques commonly performed in our laboratories to construct and propagate viral mutants. Our own experience is strongly focused on γ-herpesviruses, in particular the Epstein-Barr virus (EBV), but all large DNA viruses cloned onto BACs should be amenable to this technology. In fact, many large DNA viruses, including most herpesviruses, have now been cloned onto BACs and are already available for further manipulations *(6–10)*. This review makes no pretence of being exhaustive, and many alternative protocols can be found in the literature available on BACs.

In all methods presented, viral mutants are constructed by homologous recombination between a targeting vector containing the mutated version of the gene of interest and the cloned viral genome. The type of recombinase (*recA* or λ-phage recombinase) and the structure of the targeting vector (linear or circular) vary in the different methods used. Some characteristics of these techniques, all of which exclusively make use of viral recombinant genomes cloned onto BACs, are given in **Table 1.** The first type of method reported here makes use of linearized vectors and a selectable marker gene, which is either left on the virus genome or only partially removed in a further step. This method will therefore not be applicable when it is imperative to avoid the presence of foreign sequences on the viral genome. The targeting vector includes the gene of interest in its mutated version as well as its left and right viral flanking sequences that will provide targets for homologous recombination **(Fig. 1)**. The viral sequences cloned onto the targeting vector are therefore in the same configuration as in the wild-type viral genome with the exception of the gene of interest that carries mutations (or deletions). The minimal size of the flanking regions varies with the recombination system used. If an *E. coli recA*-dependent system is used, recombination frequencies are usually good when targeting vector and viral genomes share more than 1 kb of homologous sequences on

Table 1
Methods for Homologous Recombination on Large Plasmids

Method	Structure of targeting vector	Length of homology arms	Foreign DNA left on virus	Efficiency of the method	Time required
BJ5183[a]	Linear	>1 kb	Yes	High	7–10 d
recA-plasmid	Linear	>1 kb	Yes	Low	4–7 d
λ	Linear	≥40 bp	Yes	High	4–7 d
Chromosomal building	Circular	>1 kb	No	High	2–4 wk

[a] Mutated genomes need to be moved to a recA⁻ strain to avoid unwanted recombination.

each side (**Table 1**). In more recent years, recombination methods using the λ-phage recombinase have been described (reviewed in **ref. *11***).

1.2. Construction of Mutants Using Chromosomal Building

Transformation of a linear targeting vector is a quick and efficient method for the generation of viral recombinants, provided that the introduction of the selection marker (or its remnants after FLP recombination) does not interfere with any viral function. When it is essential to avoid the presence of foreign sequences in the viral genome, chromosomal building is the method of choice. However, this method is more time-consuming and can be quite inefficient in some cases. It makes use of plasmids with a temperature sensitive (ts) origin of replication and has initially been described for *E. coli* (**12**): a mutated gene (along with flanking regions of homology) is cloned onto a plasmid with a ts origin of replication. When cells are shifted to a nonpermissive temperature, the ts plasmid will be eliminated after a few cell divisions unless it integrates into the bacterial chromosome via homologous recombination. Because the introduced ts plasmid is circular (compare with transformation of linear fragments in **Subheading 1.1.**), integration leads to a juxtaposition of the targeted gene (wild-type) with the mutated version of this gene present on the targeting vector (**Fig. 2**). The presence of an antibiotic resistance gene allows selection of bacterial cells carrying an integrated copy of the ts plasmid. When cells are shifted back to a permissive temperature, the integrated ts plasmid can be excised from the chromosome, and the chances are (theoretically) 50% that the excised plasmid will include the stretch of homology that was originally on the chromosome (leading to mutation) and 50% that the excised plasmid will contain its original sequences (leading to reversion to the wild-type genotype) (**Fig. 2**). In our experience, this method is reliable in most cases, although more time-consuming than transformation of linear DNA pieces. In some cases, however, the mutated gene

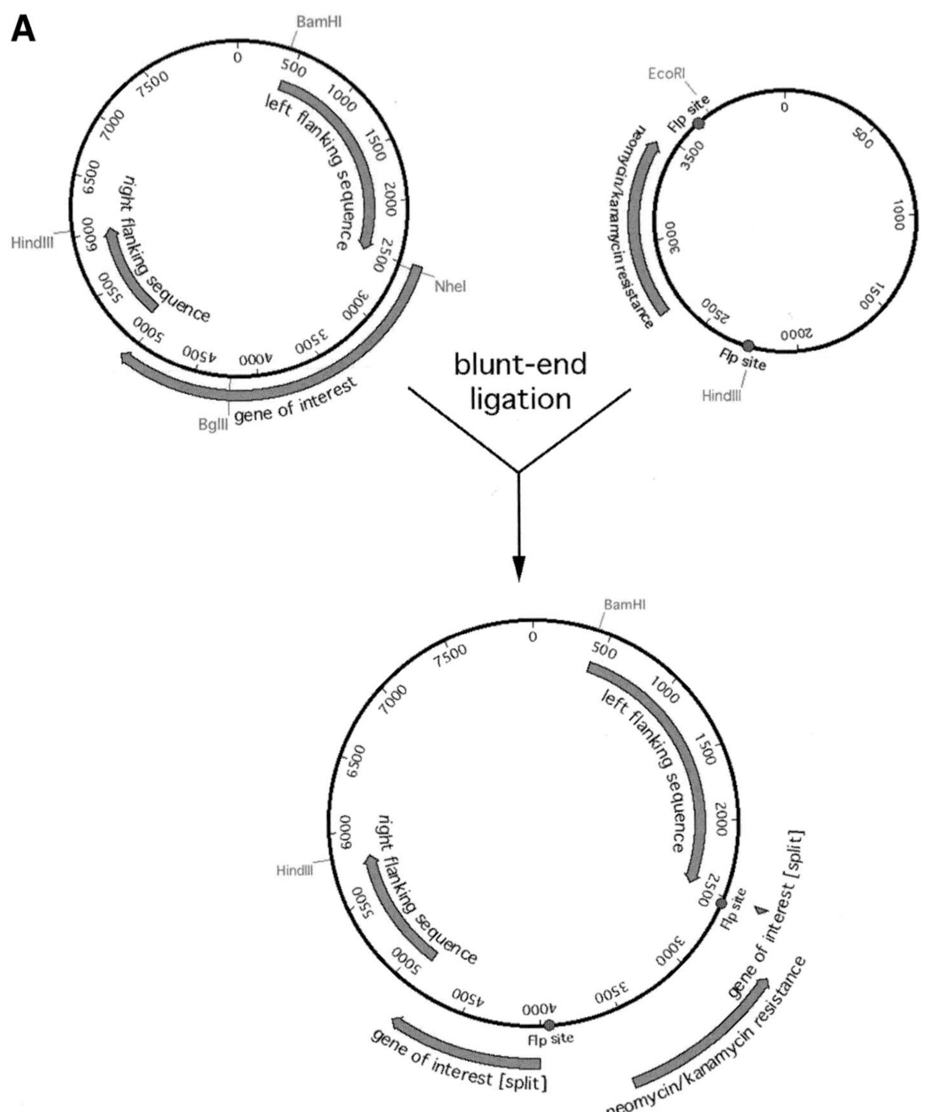

Fig. 1. Schematic representation of targeting vectors to be used for linear transformation. **(A)** Targeting vectors for knockout mutations. This figure is a schematic representation of the strategy followed to knock out the gene of interest. The upper right plasmid (pCP15) shows the kanamycin resistance gene flanked by FLP sites. Digestion of pCP15 with *Eco*RI and *Hin*dIII generates a single fragment containing the kanamycin resistance cassette flanked by FLP sites. The upper left plasmid represents an EBV subclone containing the gene of interest as well as its left and right flanking sequences. The gene of interest contains restriction sites for the *Nhe*I and *Bg*lII restriction enzymes. Digestion with these two enzymes allows excision of two-thirds of the gene of interest from the EBV subclone. This linearized vector and the fragment carrying the kanamycin resistance cassette are then treated with Klenow enzyme, and both

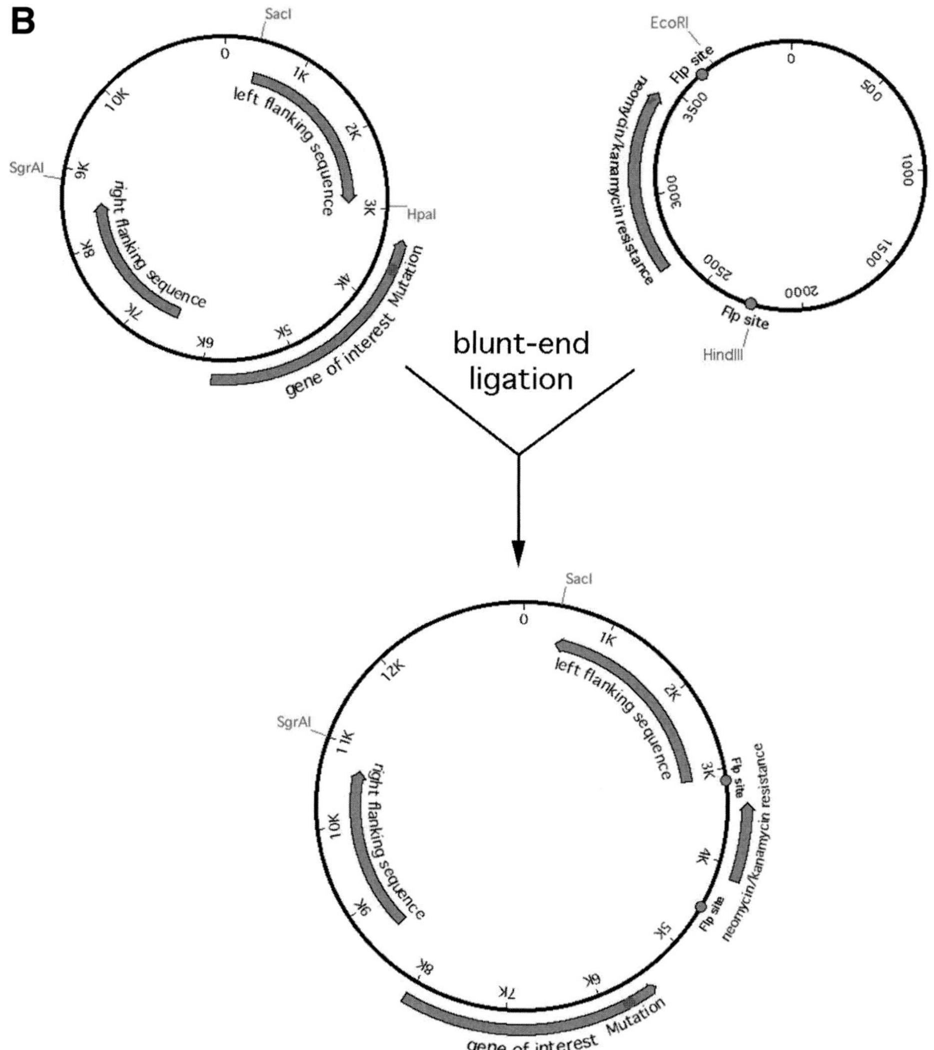

Fig. 1. *(continued)* blunt-ended fragments are ligated. The lower panel shows the targeting vector that consists of the flanking sequences, the kanamycin resistance cassette, and remaining sequences from the gene of interest. However, insertion of this cassette is designed to lead to a frame shift in these remaining sequences, resulting in its complete inactivation. The *Bam*HI and *Hin*dIII restriction sites are unique and located outside the flanking regions and can therefore be used for linearization of the targeting vector. **(B)** Schematic representation of targeting vectors for introduction of point mutations. The upper left plasmid represents an EBV subclone plasmid containing a mutated version of the gene of interest as well as its left and right flanking sequences. There is a unique restriction site (*Hpa*I) between the gene of interest and the left flanking sequences that allows insertion of the kanamycin resistance cassette. The targeting vector possesses two unique sites (*Sac*I and *Sgr*I) that can be used for linearization.

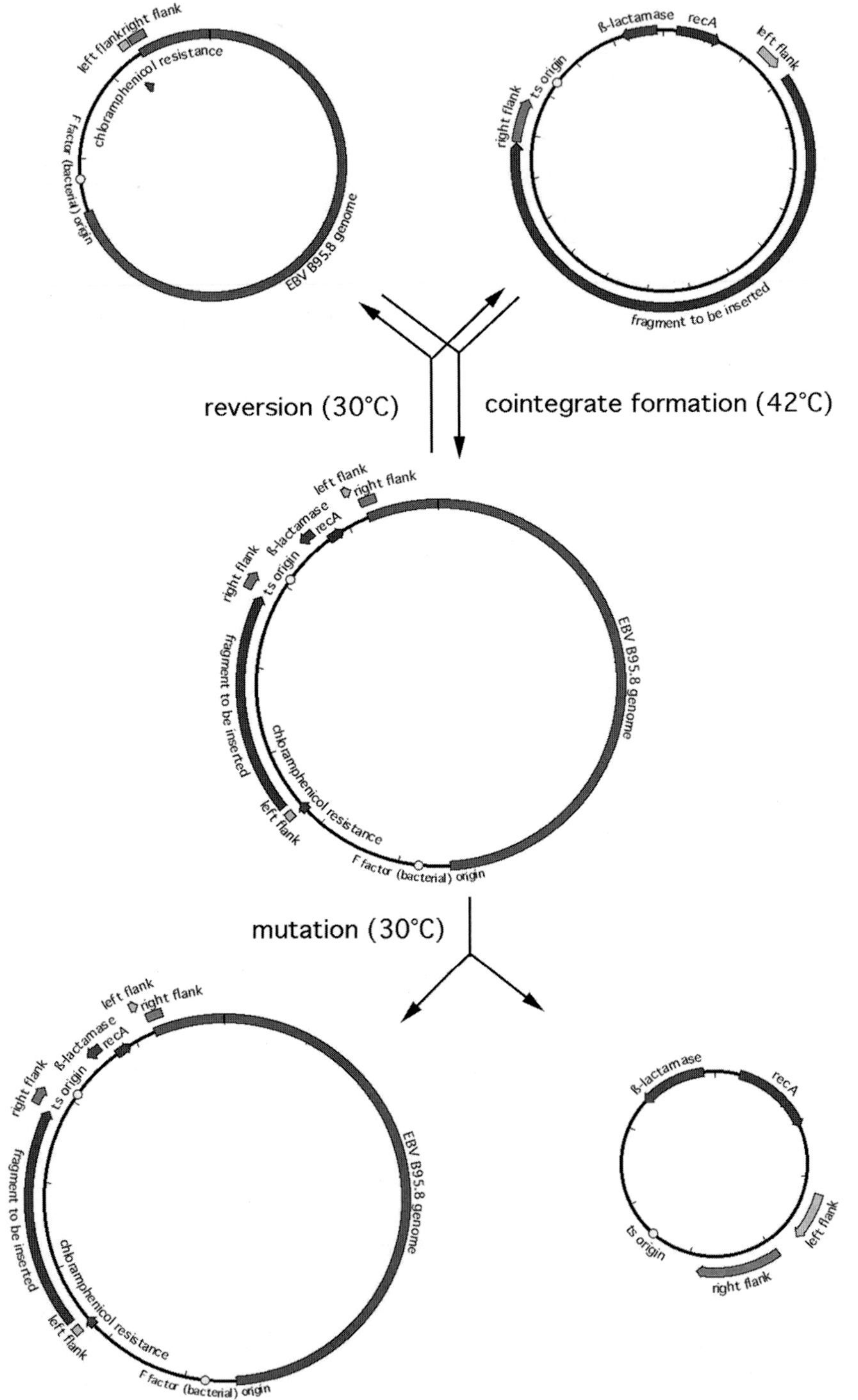
left flank
right flank
chloramphenicol resistance
F factor (bacterial) origin
EBV B95.8 genome
ß-lactamase
recA
left flank
right flank
ts origin
fragment to be inserted
reversion (30°C)
cointegrate formation (42°C)
left flank
right flank
ß-lactamase
recA
right flank
ts origin
fragment to be inserted
chloramphenicol resistance
left flank
F factor (bacterial) origin
EBV B95.8 genome
mutation (30°C)
left flank
right flank
ß-lactamase
recA
right flank
ts origin
fragment to be inserted
chloramphenicol resistance
left flank
F factor (bacterial) origin
EBV B95.8 genome
ß-lactamase
recA
ts origin
right flank
left flank

Fig. 2. The upper left plasmid represents the wild-type EBV genome cloned onto the F factor containing the gene to be modified. The upper right upper plasmid represents the targeting vector that consists of the mutated version of the gene of interest with its flanking sequences cloned onto a plasmid that carries *recA*. At 42°C, the targeting vector becomes integrated into the recombinant EBV genome, leading to the formation of a co-integrate (middle plasmid) that contains both the wild-type and the mutated version of the gene of interest as well as duplicated flanking sequences. When the temperature is switched to 30°C, the inserted targeting vector can be excised using either the duplicated right or left flanking sequences. If *recA*-mediated recombination takes place through the right flanking sequences, excision will lead to a reversion to the wild-type phenotype, whereas if it takes place through the left flanking sequences, it will lead to the formation of a recombinant virus carrying the mutated version of the gene of interest (left lower plasmid).

is preferentially re-excised, and a large number of candidates has to be screened before a recombinant virus carrying the mutation can be identified.

1.3. Encapsidation of the Recombinant Viral Genome Into an Infectious Virus

Large viral recombinants are easily modified in a prokaryotic background, but bacterial cells are obviously not permissive to viral lytic replication and virus production. The cloned viral DNA must therefore be transferred back to eukaryotic cells in which the viral genome can be packaged into infectious particles. In the EBV example, not all eukaryotic cell lines will support the viral lytic cycle, but a few cell lines, including 293, are available. 293 is a kidney epithelial cell line immortalized by transfection of the papillomavirus oncogenes E1a and E1b *(13)*. The viral DNA is introduced into 293 cells by lipotransfection; stably transfected 293 cells can be selected using hygromycin, because the viral recombinant carries the hygromycin resistance gene.

2. Materials
2.1. Construction of Mutants Using Linearized Targeting Vectors

1. Viral genome cloned onto a BAC in *recA*$^+$ *recBCD*$^-$ *E. coli* strain (BJ5183) or *E. coli* strain DH10B.
2. Plasmid carrying *recA*.
3. Targeting vector.
4. Oligonucleotide primers.
5. PCR equipment.
6. Agarose gel equipment.
7. Restriction enzymes.
8. Antibiotics: chloramphenicol, ampicillin, kanamycin.

9. Transformation solutions.
10. 10% w/v L-(+) Arabinose solution.

2.2. Construction of Mutants Using Chromosomal Building

1. *E. coli* strain DH10B containing a plasmid carrying the virus genome to be modified (DH10B/p2089 in this example).
2. Targeting vector on the basis of the temperature-sensitive plasmid p2423.
3. Antibiotics: chloramphenicol, ampicillin, kanamycin.
4. Transformation solutions.
5. Agarose gel equipment.
6. Restriction enzymes.

2.3. Encapsidation of the Recombinant Viral Genome Into an Infectious Virus

2.3.1. Stable Transfection

1. 293 Cell line.
2. Lipofection reagent.
3. Recombinant viral DNA.
4. Hygromycin.

2.3.2. Clone Selection

1. Epifluorescence microscope.
2. BZLF1 expression plasmid.
3. Antibody directed against gp110 or gp350.
4. Acetone.
5. Lipotransfection kit.
6. Target B cell line.

3. Methods

3.1. Construction of Mutants by Recombination Using Linearized Targeting Vectors

3.1.1. Construction of the Targeting Vector

3.1.1.1. TARGETING VECTORS FOR RECOMBINATION WITH RECA

In many cases, the construction of the targeting vector makes use of standard recombinant DNA technology that is widely available *(1)*. We therefore restrict our discussion to some general remarks. Gene deletion is performed by exchanging the gene of interest against an antibiotic resistance cassette (e.g., against kanamycin or tetracyclin). In favorable cases, the gene of interest will contain restriction sites that allow its complete or nearly complete excision in a single step **(Fig. 1A)**. A subclone of the viral genome containing the gene of interest and its flanking sequences can then be digested with these restriction enzymes and ligated to the antibiotic resistance cassette. If it is not possible to delete the gene of interest entirely by this approach, ligation introduction of the antibiotic resistance cassette should be performed in such a way that it leads to a frame shift in the remaining sequences of the deleted gene. In addition, it is essential that the final version of the targeting vector contain one or several unique restriction sites outside the viral sequences to allow its linearization.

Introduction of a mutation in a given viral gene is based on a similar approach. In this case, the antibiotic resistance cassette is introduced within a plasmid containing the mutated version of the gene of interest and its flanking sequences (**Fig. 1B**). Typically the cassette will be cloned directly next to the mutated gene. It is important that the insertion of the cassette does not interrupt any adjacent reading frames or interfere with sequences that govern their gene expression. In some cases, the cloning techniques just described will not be applicable because there are no appropriate restriction sites available within the gene to be modified. In these situations, regions flanking the gene to be deleted first need to be amplified using polymerase chain reaction (PCR) and then assembled in further steps around the selection marker. Sequencing of the PCR products after assembly is recommended to confirm the integrity of the targeting vector.

3.1.1.2. TARGETING VECTOR FOR RECOMBINATION WITH THE λ-PHAGE RECOMBINASE

Targeting constructs can easily be obtained by a single round of PCR that makes use of long primers because this recombinase requires only 40 bp of sequence homology (**Table 1**). The internal part of the primers contains sequences specific for an antibiotic resistance cassette, whereas the external part consists of regions homologous to the target gene locus (**Fig. 3**). The PCR amplification therefore generates a targeting vector construct consisting of the antibiotic resistance cassette flanked with short sequences homologous to sequences around the target gene. After PCR, the amplification product is digested with the restriction endonuclease *Dpn*I, which removes template DNA, precipitated, and dissolved in sterile, double-distilled water at a concentration of 0.5–1 µg/µL.

3.1.2. Construction of the Viral Recombinant

3.1.2.1. RECOMBINATION USING *RECA*

1. Generation of *E. coli* recombination-proficient competent cells containing the wild-type recombinant virus.

    It is first necessary to introduce the recombinant wild-type virus into *recA⁻ E. coli* DH10B cells (*14*) or *recA⁺ E. coli* BJ5183 (*12*) cells using electroporation (*15*).
    a. *E. coli* cells are grown in 400 mL standard medium up to an optical density of 0.5 at 600 nm.
    b. After three successive washes in ice-cold 10% glycerol, bacterial cells are resuspended in 800 µL 10% glycerol.

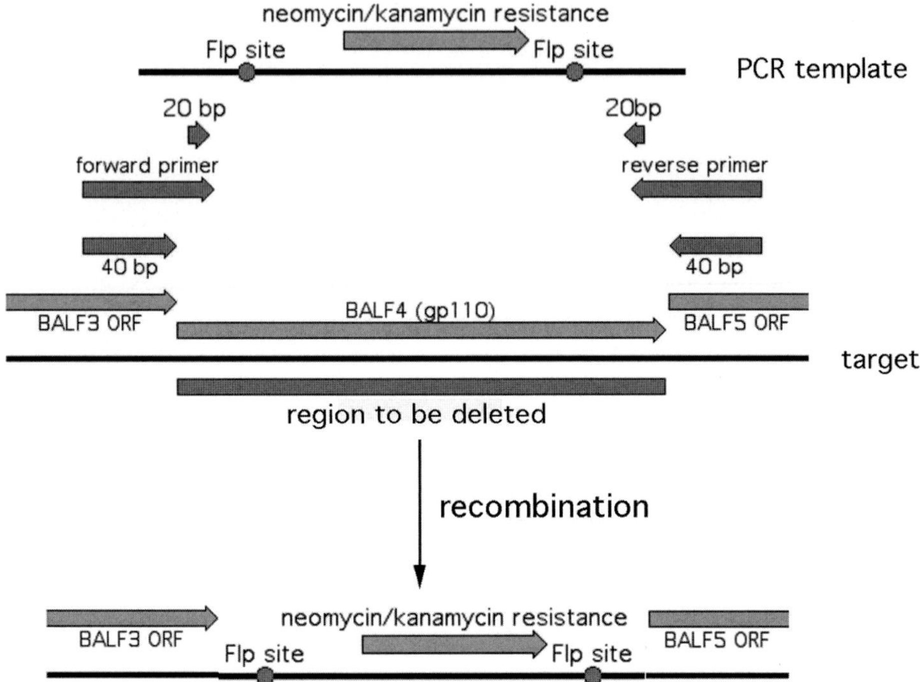

Fig. 3. Schematic representation of λ recombinase-assisted recombination. Long primers are synthesized for use in PCR amplification and subsequent recombination. The internal segments of these primers contain 20 bp of homology with the kanamycin resistance gene and are therefore used to perform PCR-mediated amplification of the kanamycin resistance gene. Because the external segments of the PCR primers share 40 bp of homology with the flanking sequences of the gene of interest, the PCR amplification product consisting of the kanamycin resistance cassette flanked by viral sequences can be used directly for linear recombination.

   c. Then 50 μL of these bacterial cells are mixed with 1 μg DNA of recombinant virus, electroporated (0.1-cm cuvet, 900 V, 100 Ω, 25 μFd), and resuspended in 3 mL of standard medium.

   d. After 1 h at 37°C, bacterial cells are plated onto agar plates containing, e.g., chloramphenicol (our BAC vectors contain the chloramphenicol resistance cassette) at a final concentration of 15 μg/mL.

   e. Alkaline minipreparations of plasmid are performed on a few chloramphenicol-resistant colonies to select bacterial clones that contain intact genomes.

   f. *E. coli* BJ5183 or DH10B cells containing the recombinant wild-type virus are then made competent for use in electroporation (see above) or for standard transformation methods (see, for example, **ref. 1**).

g. Competent cells containing the recombinant virus can either be used fresh or aliquoted and snap frozen.

2. Recombination using *recA*+ cells (BJ5183).

The linearized targeting vector is transformed into competent BJ5183 *E. coli* cells carrying the cloned wild-type viral genome using standard methods. BJ5183 is *recA*+ (a gene that greatly enhances homologous recombination) and *recBC*– (i.e., the bacteria are unable to degrade linear DNA fragments). Bacterial cells that successfully underwent recombination can be selected by adding the antibiotic toward which the cassette included in the targeting vector confers resistance. Resistant bacterial clones are then analyzed (alkaline minipreparation of plasmid) and properly recombined viral BACs transferred into a *recA*– bacterial host such as DH10B to avoid further unwanted recombinations (**Fig. 4**).

3. Recombination using *recA* carrying plasmid.

As an alternative to the previous method, we have constructed a plasmid (p2650) that carries all functions necessary to perform recombination of linear fragments in *recA*– *recBCD*+ *E. coli* strains such as DH10B. p2650 is a plasmid with a temperature-sensitive origin of replication that carries the ampicillin resistance (from pST76-amp [**16**]) gene as well as the *E. coli recA* gene (from plasmid pKY102 [**17**]) along with the redγ function from bacteriophage λ (from pBAD-ETγ [**18**]), which is an inhibitor of RecBCD (exonuclease V). Transient introduction of this plasmid into DH10B cells thus allows recombination of linear fragments to take place and eliminates the need for successive transfer of the viral genome from BJ5183 into DH10B.

a. p2650 is introduced into competent DH10B cells containing the recombinant virus using standard transformation methods (*1*). Bacterial cells are grown in the presence of ampicillin and chloramphenicol at 30°C overnight.

b. Four to eight single resistant colonies are used to produce electrocompetent cells. (DH10B are difficult to transform with linear fragments using standard methods.)

   (i) First, 50 µL of these electrocompetent cells are mixed with 1–3 µg of the targeting vector.

   (ii) After transformation, bacterial cells are grown at 30°C for 3–4 h, spread on plates containing chloramphenicol and the antibiotic corresponding to the resistance cassette present in the targeting vector (must not be ampicillin as this cassette is present on the p2650 plasmid), and grown overnight at 42°C. This will lead to the loss of p2650.

   (iii) Antibiotic-resistant clones are then analyzed by plasmid minipreparation.

   (iv) It is also useful to ensure that the selected clones are sensitive to ampicillin, as this indicates loss of p2650 and thus a *recA*– genotype.

3.1.2.2. RECOMBINATION USING λ RED: TRANSFORMATION OF pKD46 INTO DH10B CARRYING RECOMBINANT VIRUSES

The recombination system based on phage λ recombinase has the advantage that homology stretches of 40 bp are sufficient to achieve decent recombination rates (*19*).

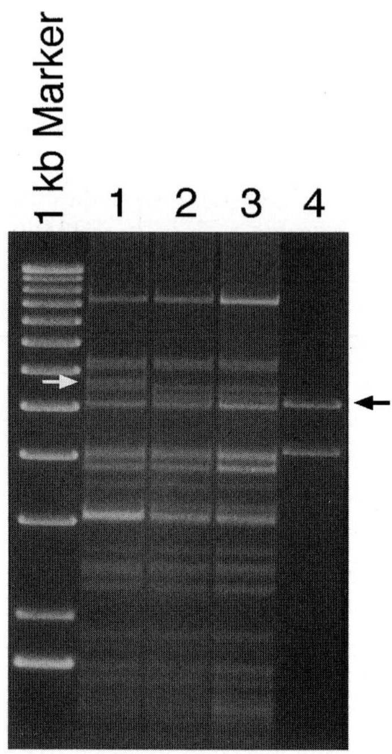

Fig. 4. Generation of a recombinant virus mutant using a linearized targeting vector. This figure shows a restriction analysis of the EBV wild-type genome (lane 1) and a mutant recombinant virus (lane 2) generated by recombination with the linearized targeting vector (lane 4). All DNAs were digested with the *Msc*I restriction enzyme. Recombination of the larger fragment from the linearized targeting vector gives rise to a deletion of the restriction fragment carrying the gene of interest and the appearance of a new restriction fragment roughly similar in size to the targeting vector. After stable transfection of the recombinant vector into 293 cells (*see* **Subheading 3.3.1.**), the recombinant is transferred back into *E. coli* cells. Restriction analysis (lane 3) confirms the integrity of the stably transfected recombinant.

1. pKD46 (plasmid carrying the λ red recombinase under the control of an arabinose-inducible promoter and the ampicillin resistance cassette [*19*]) is transformed into competent DH1OB *E. coli* cells containing the recombinant virus (standard method). Transformed bacterial cells are plated on chloramphenicol and ampicillin plates and grown overnight at 30°C.
2. Electrocompetent bacterial cells containing both pKD46 and the recombinant virus are generated in medium containing L-(+) arabinose at a final concentration of 0.1% as well as chloramphenicol and ampicillin at 30°C according to the previously described protocol (*see* **Subheading 3.1.2.1.**). Then 1–3 μg of the PCR-generated fragment containing short homology sequences is introduced into these cells. After transformation, cells are plated on agar plates containing chloram-

phenicol and the antibiotic corresponding to the resistance cassette on the targeting fragment and grown overnight at 42°C. At this temperature pKD46 is lost.

3. The correct genotype of the clones is verified by restriction enzyme analysis of DNA minipreps. Clones are purified on plates containing the same antibiotics and then tested for ampicillin sensitivity (i.e., absence of pKD46) (*see* **Note 1**).

3.2. Construction of Mutants Using Chromosomal Building

3.2.1. Introduction of a Mutated Viral Gene Flanked by Homology Arms Into Temperature-Sensitive Plasmids

The targeting vector is constructed using standard cloning methods or PCR-based methods. The targeting backbone for the vector includes the mutated/deleted viral gene and 0.5–1 kb of flanking sequences on each side. We currently use plasmid p2423 as a targeting vector, which carries a temperature-sensitive origin of replication *(16)*, an ampicillin resistance gene, and the *E. coli recA* gene *(17)*. This allows recombinations to be performed in *recA⁻* strains such as DH10B.

3.2.2. Recombination

1. Transformation of targeting vector into *E. coli* DH10B carrying the recombinant virus is performed using standard methods. Transformed cells are plated on agar plates containing ampicillin (resistance encoded by p2423) and chloramphenicol (resistance encoded by the recombinant virus) and grown overnight at 30°C. At this temperature p2423 efficiently replicates, and both p2423 and the recombinant virus coexist as plasmids within the bacterial cells.
2. Generation and purification of colonies containing co-integrates.
   a. A few colonies are then streaked on agar plates containing ampicillin and chloramphenicol and grown overnight at 42°C.
   b. Following this temperature shift, only those bacterial cells that contain the temperature-sensitive plasmid integrated into the wild-type genome (co-integrates) will survive antibiotic selection. Therefore only co-integrates should be present in antibiotic-resistant bacterial clones.
   c. A few of these clones (less than 10) are then passaged twice on prewarmed agar plates containing ampicillin and chloramphenicol and grown overnight at 42°C to ensure that all freely replicating copies of the targeting vector are completely lost.
3. Resolution of co-integrates.
   a. To obtain viral mutants, co-integrates must be resolved. To this aim, a few (less than 10) colonies carrying co-integrates are streaked at 30°C overnight on plates containing chloramphenicol only.
   b. To ensure that clones from these plates contain only the viral recombinant, replicates are grown on plates containing ampicillin or chloramphenicol. Clones that have lost p2423 should grow in the presence of chloramphenicol but not in the presence of ampicillin.

c. If no ampicillin-sensitive colonies can be found, streaking at 30°C is repeated until a suitable number of candidates is obtained. Between 10 and 20 of these candidates will be analyzed by minipreparation of plasmid DNA followed by restriction analysis.

d. If none of theses colonies contains the mutant, more clones need to be analyzed (*see* **Note 2**).

3.3. Encapsidation of the Recombinant Viral Genome Into an Infectious Virus

3.3.1. Viral DNA Transfection Into 293 Cells

1. Viral DNA obtained by maxipreparation (using cesium chloride gradients or DNA purification columns) is transfected into 293 cells using lipid micelles. We routinely use Lipofectamine® from Invitrogen, but other lipotransfection methods can also be used.

2. For each transfection, we seed 10^5 293 cells in one well from a 6-well cluster plate the night before transfection.

3. The next day, 1 µg of recombinant DNA is mixed with 6 µL of lipid micelles in 1 mL OptiMEM® solution, incubated for 30 min at room temperature, and applied to semiconfluent 293 cells that were moved to OptiMEM culture medium for 1 h prior to transfection.

4. The transfection mixture is removed 3 h later from the 293 cells and replaced by RPMI, 10% foetal calf serum containing hygromycin (final concentration 100 µg/mL).

5. After 3–6 wk of selection, 293 cell clones stably carrying the viral recombinant become visible and can be further expanded (*see* **Note 3**).

3.3.2. Clone Selection

Efficiency of transfection varies with different DNA transfections, but on average twenty 293 cell clones can be selected. However, in general, not all these clones will be permissive for viral lytic replication, and among those that are permissive, some will produce more virus than others and will be selected for further experiments

1. Clones that survived hygromycin selection are first briefly examined under UV light. Only those clones that produce GFP will be further analyzed. GFP-negative clones either do not carry the recombinant virus or carry a rearranged version of the virus genome and should be discarded.

2. GFP-positive cell clones are transfected with a BZLF1 expression plasmid using exactly the method described for stable transfection of recombinant viral DNA (*see* **Subheading 3.3.1.**). In this case, 0.5 µg of BZLF1 expression vector (this might depend on the expression plasmid used) is introduced into the different clones to be tested, and supernatants are harvested 3 d after transfection.

3. These supernatants are filtered using filters with a 0.45-µm pore diameter, and 10^5 Raji cells are incubated with 0.5 mL of supernatants.

4. Raji is an EBV-positive cell line that can easily be superinfected with EBV super-natants. Because only the recombinant EBV carries GFP, identification of green Raji cells ensures that the supernatants contain infectious virus and that the 293 clones tested are good virus producers. Counting the number of green Raji cells gives an indirect estimation of virus titers and allows selection of the 293 clones with high-est virus production. Infection of 10E5 Raji cells with 100 µL of supernatant from good producer cell lines typically generates 10–15% GFP-positive cells.

4. Notes

1. Construction of mutants by recombination using linearized targeting vectors.
   a. The greatest care must be taken in the design and actual construction of the tar-geting vector. Any flaws at this step will go undetected until the recombinant virus can be functionally tested. It is therefore essential to first construct the tar-geting vector and the recombinant viral mutant using an appropriate computer software. This will guide construction of the recombinant and help early detec-tion of possible mistakes.
   b. The PCR amplification product must include the bacterial promoter for the resistance gene.
   c. Targeting vectors carrying flanking sequences showing more than 1 kb of homology will be suitable.
   d. It is essential to have high-quality electrocompetent cells. Positive controls (e.g., supercoiled DNA of the vector used as a template in the PCR reaction) should yield at least 10E7 transformants/µg DNA.
   e. The use of resistance gene cassettes flanked by frt or loxP sites permits easy removal of the resistance gene. In this case, FLP or LoxP recombinase is tran-siently introduced into bacterial cells containing the mutated recombinant virus, and viral colonies are checked by restriction analysis. Both recombinases work very efficiently.
2. Construction of mutants using chromosomal building.
   a. In principle, the viral DNA can be subcloned directly onto the ts plasmid and manipulated there. However, cloning with such plasmids is not always simple, because yields in DNA preparations are low, and qualities usually are poor. Therefore, if complex modifications are to be performed, it may prove easier to subclone onto a standard vector first and then insert the complete construct into the ts plasmid in a second step.
   b. It is strongly recommended to design mutations in such a way that a distinction between the wild-type and mutated copy is easy, e.g., by introducing or remov-ing a restriction enzyme cleavage site. Screening for recombinants by DNA sequencing can be difficult.
   c. During purification of putative co-integrates, plates preheated at 42°C should be used to ensure complete loss of nonintegrated copies of the ts plasmid. This can be further tested by using PCR with primers that amplify the ts plasmid.
   d. The frequency of generating mutants in the excision step can vary between 2 and 80%.

3. Encapsidation of the recombinant viral genome into an infectious virus.
   a. The permissivity of the stably transfected 293 clones is unpredictable. Some transfection rounds will give rise to a majority of high producer cell lines, whereas others using the same DNA and the same cell lines might deliver only low or nonproducers.
   b. It is strongly recommended to check the integrity of the stably transfected cell line before proceeding with further experiments (e.g., Southern blot analysis to make sure the cell line contains the appropriate viral mutant). It is also sensible to transfer the transfected plasmid back into *E. coli* cells and to reanalyze by restriction enzyme analysis and/or sequencing.

Acknowledgments

This work was supported by grants from Cancer Research UK and the Leukemia Research fund.

References

1. Sambrook, J., Maniatis, T., and Fritsch, E. F. (1989) *Molecular Cloning: A Laboratory Manual.* Cold Spring Harbor Laboratory Press, Cold Spring Harbor, NY.
2. Mocarski, E. S., Post, L. E., and Roizman, B. (1980) Molecular engineering of the herpes simplex virus genome: insertion of a second L-S junction into the genome causes additional genome inversions. *Cell* **22,** 243–255.
3. Smiley, J. R. (1980) Construction in vitro and rescue of a thymidine kinase-deficient deletion mutation of herpes simplex virus. *Nature* **285,** 333–335.
4. Fields, B. N., Knipe, D. M., and Griffin, D. E. (2001) *Fields Virology.* Lippincott-Raven, Philadelphia.
5. O'Connor, M., Peifer, M., and Bender, W. (1989) Construction of large DNA segments in *Escherichia coli. Science* **244,** 1307–1312.
6. Messerle, M., Crnkovic, I., Hammerschmidt, W., Ziegler, H., and Koszinowski, U. H. (1997) Cloning and mutagenesis of a herpesvirus genome as an infectious bacterial artificial chromosome. *Proc. Natl. Acad. Sci. USA* **94,** 14759–14763.
7. Delecluse, H. J., Hilsendegen, T., Pich, D., Zeidler, R., and Hammerschmidt, W. (1998) Propagation and recovery of intact, infectious Epstein-Barr virus from prokaryotic to human cells. *Proc. Natl. Acad. Sci. USA* **95,** 8245–8250.
8. Saeki, Y., Ichikawa, T., Saeki, A., et al. (1998) Herpes simplex virus type 1 DNA amplified as bacterial artificial chromosome in *Escherichia coli:* rescue of replication-competent virus progeny and packaging of amplicon vectors. *Hum. Gene Ther.* **9,** 2787–2794.
9. Delecluse, H. J., Kost, M., Feederle, R., Wilson, L., and Hammerschmidt, W. (2001) Spontaneous activation of the lytic cycle in cells infected with a recombinant Kaposi's sarcoma-associated virus. *J. Virol.* **75,** 2921–2928.
10. Schumacher, D., Tischer, B. K., Fuchs, W., and Osterrieder, N. (2000) Reconstitution of Marek's disease virus serotype 1 (MDV-1) from DNA cloned as

a bacterial artificial chromosome and characterization of a glycoprotein B-negative MDV-1 mutant. *J. Virol.* **74,** 11088–11098.

11. Poteete, A. R. (2001) What makes the bacteriophage lambda Red system useful for genetic engineering: molecular mechanism and biological function. *FEMS Microbiol. Lett.* **201,** 9–14.

12. Hamilton, C. M., Aldea, M., Washburn, B. K., Babitzke, P., and Kushner, S. R. (1989) New method for generating deletions and gene replacements in *Escherichia coli. J. Bacteriol.* **171,** 4617–4622.

13. Graham, F. L., Smiley, J., Russell, W. C., and Nairn, R. (1977) Characteristics of a human cell line transformed by DNA from human adenovirus type 5. *J. Gen. Virol.* **36,** 59–74.

14. Hanahan, D. (1983) Studies on transformation of *Escherichia coli* with plasmids. *J. Mol. Biol.* **166,** 557–580.

15. Sheng, Y., Mancino, V., and Birren, B. (1995) Transformation of *Escherichia coli* with large DNA molecules by electroporation. *Nucleic Acids Res.* **23,** 1990–1996.

16. Posfai, G., Koob, M. D., Kirkpatrick, H. A., and Blattner, F. R. (1997) Versatile insertion plasmids for targeted genome manipulations in bacteria: isolation, deletion, and rescue of the pathogenicity island LEE of the *Escherichia coli* O157:H7 genome. *J. Bacteriol.* **179,** 4426–4428.

17. Ihara, M., Oda, Y., and Yamamoto, K. (1985) Convenient construction of strains useful for transducing recA mutations with bacteriophage P1. *FEMS Microbiol. Lett.* **30,** 33–35.

18. Zhang, Y., Buchholz, F., Muyrers, J. P., and Stewart, A. F. (1998) A new logic for DNA engineering using recombination in *Escherichia coli. Nat. Genet.* **20,** 123–128.

19. Datsenko, K. A. and Wanner, B. L. (2000) One-step inactivation of chromosomal genes in *Escherichia coli* K-12 using PCR products. *Proc. Natl. Acad. Sci. USA* **97,** 6640–6645.

25

Genetic Analysis of Cytomegalovirus by Shuttle Mutagenesis

Manfred Lee and Fenyong Liu

Summary

The genomes of Herpesviridae family members are among the largest of all viruses and therefore present a formidable challenge in understanding the roles of every gene in replication or pathogenesis. For example, murine cytomegalovirus (MCMV) has a genome of 230 kb that encodes more than 170 genes, many of which have unknown functions. Many techniques for the genetic analysis of a herpesvirus have been developed over the past two decades. One such procedure involves the use of a shuttle mutagenesis system, and it has successfully generated a pool of MCMV mutants that contained an engineered Tn3-type transposon inserted within their genome. The process of shuttle mutagenesis involves the construction of a genomic fragment library, transposon mutagenesis of the library, and generation of virus mutants through homologous recombination. This chapter details the methodologies required for implementing a Tn3-based shuttle mutagenesis system for construction of a mutant virus library.

Key Words: Cytomegalovirus; transposon; mutagenesis; herpesvirus; viral pathogenesis.

1. Introduction

Prior to the introduction of systems that allowed for the genetic manipulation of human cytomegalovirus (HCMV), research relied mainly on the use of temporary assays to understand the mechanisms of viral replication and pathogenesis. Based on previous work done with herpes simplex virus 1 (HSV-1) *(1)*, Spaete and Mocarski *(2)* cloned portions of the CMV genome into a plasmid containing the β-galactosidase *(lacZ)* gene *(2)*. The cloned portions were inserted into the plasmid so they would flank either end of the *lacZ* gene. When this construct was introduced into the cell along with HCMV, the cellular recombination factors would act on the homologous flanking regions to catalyze the insertion of the cassette into the viral genome, thereby creating the first HCMV

From: *Methods in Molecular Biology, vol. 292: DNA Viruses: Methods and Protocols*
Edited by: P. M. Lieberman © Humana Press Inc., Totowa, NJ

mutated by an insertion of a gene cassette. The *lacZ* gene within the cassette facilitated the isolation of mutant viruses, as plaques in cell culture appeared blue in the presence of 5-bromo-3-chloro-indolyl-β-D-galactopyransoside (X-gal). This system allowed for the site-specific disruption of any gene in the CMV genome and provided a powerful tool for understanding the genetic basis of CMV pathogenesis and replication. These experiments, however, did not always generate the desired result, as random deletions or rearrangements were sometimes discovered around the site of insertion, suggesting that HCMV could accommodate no more than a 5% addition to its genome *(2)*.

Although the initial attempts to manipulate CMV genetically established a way to identify viruses that had undergone recombination, a method by which researchers could select for viruses that had undergone the rare event of recombination while simultaneously selecting against the wild-type virus was preferable. The first reports of utilizing the mammalian selectable marker *neo* greatly enhanced the ability of scientists to select for recombinants because only those cells infected with recombined virus encoding the *neo* gene could survive in media containing geneticin *(3)*. Similar approaches are still in use; however, they typically depend on resistance to more lethal drugs such as puromycin *(4–6)*. Alternatively, the *Escherichia coli* guanosine phosphoribosyl transferase *(gpt)* gene has been used in the creation of recombinant murine cytomegalovirus (MCMV) and HCMV *(7–8)*. Consequently, mammalian cells infected with recombined virus encoding the *gpt* gene survived better in media containing mycophenolic acid and xanthine.

Recently, other researchers have created novel procedures that allow for a more rapid and efficient production of recombinant CMV. One such system utilized overlapping DNA segments (cosmids) of the avirulent HCMV Towne strain with one fragment from a virulent Toledo strain in human foreskin fibroblasts *(9)*. These overlapping genomic segments were transfected into cells, where they were recombined to generate a hybrid Towne/Toledo strain. The advantage of the cosmid system is that it bypasses the need to enrich for recombinants or purify them by plaque picking; all viruses produced by this technique are reportedly pure recombinants. Taking an entirely different approach, Messerle et al. *(10)* were able to clone the entire MCMV genome as a bacterial artificial chromosome (BAC). This system facilitates the molecular manipulation of the genome in *Escherichia coli* using well-known bacterial site-specific recombination techniques. Subsequently, the mutated BAC can be purified from a clonal bacterial isolate and transfected into cells to yield a clonal, mutated virus, thus bypassing the dependence that all previous systems had on the mammalian host factors to mediate recombination. One of the most recently developed tools for producing recombinant CMV makes use of the green fluorescent protein *(gfp)* gene in conjunction with a *gpt* gene. The fluorescent marker provides the advantage of having a visual indication that the process of recombination has occurred.

In addition to the more common site-directed mutagenesis approaches, the random disruption of the genome by transposon insertion has also been successfully used to identify noncoding regions and open reading frames that are essential for herpesvirus replication or pathogenesis *(11–14)*. Our laboratory has recently developed a shuttle mutagenesis *(15,16)* approach for systematically disrupting the MCMV genome *(17,18)*. To accomplish this, a Tn3-type transposon was engineered to contain the *gpt* gene, which allowed for the selection of recombinant mutants following insertion of the transposon into the viral genome (**Fig. 1**). This chapter describes the reagents and protocols required in the application of a Tn3 shuttle mutagenesis system for creating a library of MCMV mutants.

2. Materials

2.1. Construction of a CMV Genomic Library

1. Full-length, wild-type MCMV (Smith strain, ATCC, Manasaus, VA) genomic DNA from purified virions.
2. *Sau*3AI restriction enzyme (New England Biolabs, Beverly, MA).
3. Agarose (Fisher Scientific, Pittsburgh, PA).
4. Geneclean Kit (Qbiogene, Carlsbad, CA).
5. Klenow fragment (New England Biolabs).
6. 1 m*M* Each of dGTP, dATP, dTTP, and dCTP.
7. *Sal*I restriction enzyme (New England Biolabs).
8. pHSS6-*Sal*I Vector (*see* **Note 1**).
9. Calf intestine alkaline phosphatase (Promega, Madison, WI).
10. Buffer-saturated 25:24:1 phenol/chloroform/isoamyl alcohol (Fisher Scientific).
11. 24:1 Chloroform/isoamyl alcohol (Fisher Scientific).
12. 3 *M* Sodium acetate (Fisher Scientific).
13. 95–100% and 70% Ethanol.
14. T4 DNA Ligase (New England Biolabs).
15. Chemically competent *Escherichia coli* B211-F⁻ *recA1*(Δ*lac-pro*) *rpsE* spectinomycin[r] with plasmid, pLB101 (pACYC184 with *tnpA*, chloramphenicol[r]).
16. Luria-Bertani (LB) broth, Miller and LB agar plates (Fisher Scientific).

2.2. Shuttle Mutagenesis of CMV Genomic Fragments

1. *E. coli* strain XZ95: B224/RDP146 (F⁻ *recA1*[Δ*lac-pro*] *rps,* spectinomycin[r]) with plasmid, pOX38::m-Tn3/*gpt* (pOX38 carrying the Tn3 transposon containing tetracycline[r] and guanosine phophoribosyl transferase).
2. *E. coli* strain 70: NG135 (K12 *recA56 gal-*Δ*S165 strA,* streptomycin[r]) with plasmid, pNG54 (pACYC184 with *tnpR*, choramphenicol[r]).
3. Chemically competent *E. coli* strain DH5α: *endA1 hsdR17 supE44 thi-1 recA1 gyrA relA1* Δ*(lacZYA-argF) U169 deoR* (Φ*80 dlac*Δ*([lacZ]M15)*.

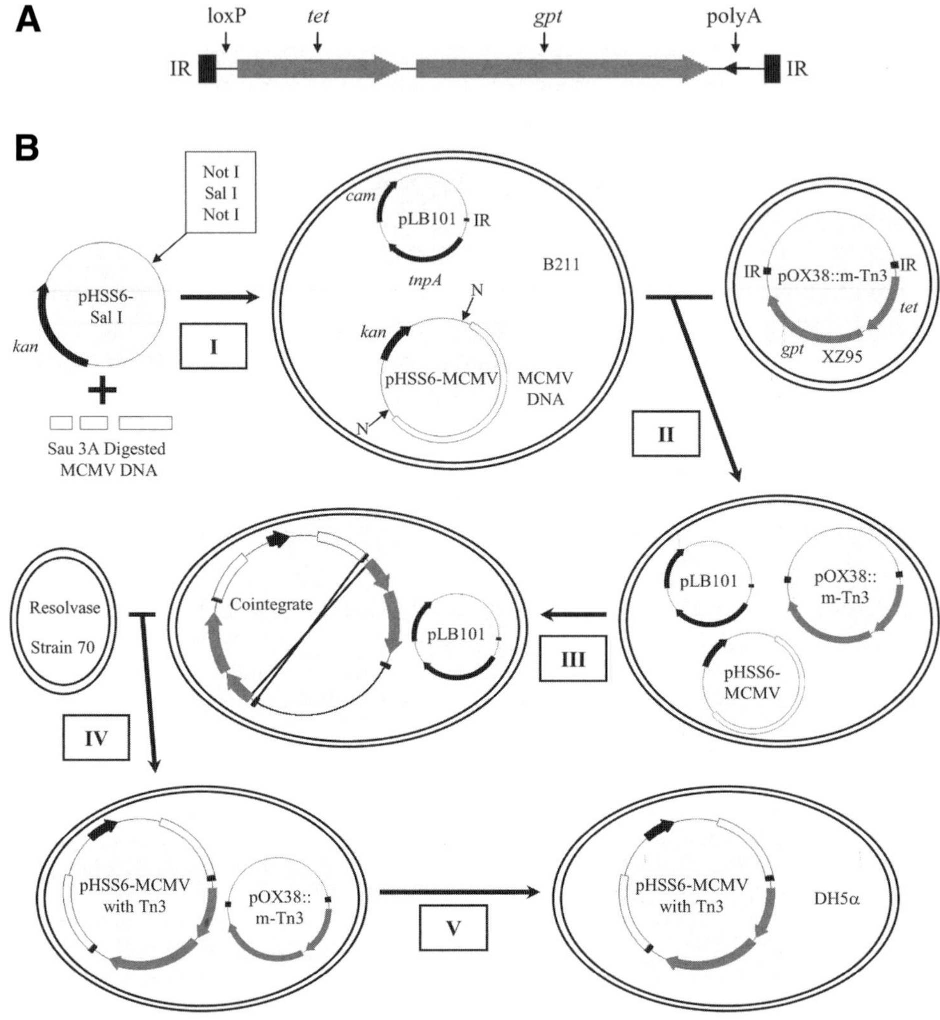

4. Tetracycline (Sigma, St. Louis, MO): 15 mg/mL in H_2O; use at 15 µg/mL.
5. Chloramphenicol (Sigma): 30 mg/mL in ethanol; use at 30 µg/mL.
6. Spectinomycin (Sigma): 100 mg/mL in H_2O; use at 100 µg/mL.
7. Kanamycin (Sigma): 25 mg/mL in H_2O; use at 25 µg/mL.
8. Streptomycin (Sigma): 100 mg/mL in H_2O, use at 100 µg/mL.
9. LB broth and LB agar plates.
10. QIAprep Miniprep kit (Qiagen, Valencia, CA).
11. Sequencing primer: FL110PRIM (5′-GCAGGATCCTATCCATATGAC-3′).

Fig. 1. Elements of the shuttle mutagenesis procedure. **(A)** Schematic representation of the modified Tn3 transposon. It contains six elements: a loxP sequence (loxP), a tetracycline resistance marker *(tet)*, guanosine phosphoribosyltransferase *(gpt)*, an additional poly(A) signal, and two 38-bp terminal inverted repeats (IR). **(B)** Overview of shuttle mutagenesis. I, MCMV DNA that has been partially digested with *Sal*I is cloned into the *Sal*I site of pHSS6-*Sal*I and introduced into B211 containing pLB101, which expresses the transposase gene, *tnpA*. II, Dual plasmid bacteria are selected by resistance to Cm and Kan and are conjugated to XZ95 containing pOX38::m-Tn3/*gpt* to generate a triply resistant bacteria (Tet, Kan, and Cm). III, TnpA catalyzes the insertion of pOX38 into pHSS6 to form a co-integrated plasmid. pLB101 is immune to this process as it carries a 38bp IR. IV, The co-integrate is resolved by conjugation with a bacteria that expresses λ(*cre*+) resolvase, resulting in a pHSS6-MCMV plasmid that now contains a single Tn3 insertion. V, The mutated plasmids are isolated and reintroduced into DH5α bacteria and selected for their double resistance to Tet and Kan. Colonies are screened for the presence of a genomic fragment that also contains a Tn3 insertion. The mutated fragments are sequenced at the Tn-MCMV DNA junction to determine the site of insertion.

2.3. Recombinant Virus Generation

1. *Not*I restriction enzyme (New England Biolabs).
2. Buffer-saturated 25:24:1 phenol/chloroform/isoamyl alcohol (Fisher Scientific).
3. n24:1 Chloroform/isoamyl alcohol (Fisher Scientific).
4. 3 *M* Sodium acetate (Fisher Scientific).
5. 95–100% and 70% Ethanol.
6. Full-length, wild-type murine CMV genomic DNA (Smith Strain, ATCC) from purified virions.
7. Murine NIH/3T3 fibroblasts (ATCC).
8. NIH/3T3 media (Dulbecco's modified Eagle's medium [DMEM] with high glucose, penicillin/streptomycin, nonessential amino acids, essential amino acids [Gibco/Invitrogen, Carlsbad, CA], 10% NuSerum [Becton Dickinson, San Jose, CA]).
9. Calcium phosphate transfection kit (Gibco/Invitrogen).
10. 25 mg/mL Mycophenolic acid in H_2O; use at 25 µg/mL, (Gibco/Invitrogen).
11. 10 mg/mL Xanthine in 1 *M* NaOH; use at 50 µg/mL (Gibco/Invitrogen).
12. Type VII low Gelling agarose, tissue culture grade: 2% w/v in tissue culture grade water (Sigma).
13. 2X NIH/3T3 media (DMEM from powder; Sigma).
14. 10% Nonfat dry milk (NFDM) in tissue culture grade water, autoclaved (Gibco/Invitrogen).

2.4. Recombinant Rescue Virus Generation

The materials needed in this section are identical to those of **Subheading 2.3.** with the addition of:

1. Full-length, mutant murine CMV genomic DNA from purified virions.
2. Murine STO fibroblasts (ATCC).
3. STO media: DMEM with high glucose, penicillin/streptomycin, 10% fetal bovine serum (Gibco/Invitrogen).
4. 25 mg/mL 6-Thioguanosine in 0.2 M NaOH; use at 25 µg/mL (Sigma).

3. Methods

3.1. Construction of a CMV Genomic Library

The first phase follows a standard scheme for the creation of a library in which random MCMV genomic fragments are cloned into the vector, pHSS6-*Sal*I.

1. Digest purified MCMV genomic DNA with *Sau*3AI restriction enzyme using optimal conditions as empirically determined (*see* **Note 2**).
2. Run digested genomic DNA on a 1% agarose gel.
3. Excise fragments in the range of 1.6–4.0 kbp extract DNA with a Geneclean kit following the manufacturer's protocol.
4. Use Klenow fragment (1 U/µg of DNA) with only dGTP (33 µM final concentration) and dATP (33 µM final concentration) in the reaction to fill in the overhanging cytosine and thymidine nucleotides (leaves an overhanging G-A) present on the *Sau*3AI fragments. Phenol-chloroform extract DNA, ethanol-precipitate, and resuspend in molecular-grade water.
5. Prepare pHSS6-*Sal*I vector for ligation by digesting with *Sal*I enzyme.
6. Use Klenow fragment with only dCTP and dTTP (following the conditions described in **step 4**) in the reaction to fill in the overhanging adenosine and guanosine bases (leaves an overhanging T-C) present on the pHSS6-*Sal*I vector. Phenol-chloroform extract DNA, ethanol-precipitate, and resuspend in molecular-grade water.
7. Dephosphorylate pHSS6-*Sal*I vector with CIAP. Phenol-chloroform extract DNA, ethanol-precipitate, and resuspend in molecular-grade water.
8. Ligate genomic fragments into the prepared pHSS6-*Sal*I plasmid.
9. Transform ligated products into chemically competent *E. coli* strain B211 by heat shock, and select for kanamycin and chloramphenicol resistance on LB agar plates. Resulting colonies form the pHSS6-MCMV genomic library.
10. Pick 25–50 random colonies, extract plasmid DNA, and digest with restriction enzymes (*Sal*I or *Not*I) to check for sufficient randomness in the size of the genomic fragments.

3.2. Shuttle Mutagenesis of CMV Genomic Fragments

The second phase generates a library of MCMV genomic fragments that contain a Tn3/*gpt* insertion through a series of conjugations and selection with antibiotics for the appropriate clones (*see* **Fig. 1** for an overview).

3.2.1. Co-Integrate Formation

1. Grow a mixed population of B211 cells containing the genomic library in LB broth (>2 mL) with kanamycin and chloramphenicol overnight at 37°C.

2. In parallel, grow strain XZ95 overnight in LB broth (> 2mL) containing spectino-mycin and tetracycline overnight at 37°C.
3. Dilute each culture to 1:100 in LB broth without antibiotics (> 2 mL), and grow cultures to early log phase density (when cells just start to swirl, ~2 h at 37°C).
4. Mix 100 µL of the XZ95 culture with 200 µL of the B211 culture, and incubate the mixture at 37°C without agitation for 20 min to 1 h to allow for mating (*see* **Notes 3** and **4**).
5. Plate 50-µL aliquots of the mated culture on to LB agar plates containing kanamycin, tetracycline, and chloramphenicol.
6. Grow plates at 30°C for 1–2 d. Resulting colonies contain pHSS6-MCMV co-inte-grated with pOX38::m-Tn3/*gpt*.

3.2.2. Resolution of the Co-Integrate

1. Grow strain 70 in LB broth containing streptomycin and chloramphenicol overnight at 37°C.
2. Collect colonies containing co-integrates (*see* **Subheading 3.2.2.**, **step 6**) by adding 2 mL of LB broth to each plate, gently scraping the colonies with a spread-er, and placing the wash into a tube.
3. Dilute strain 70 to 1:100 in LB broth without antibiotics. Dilute the eluted bacteria to approximately the same starting density as strain 70. Grow each culture to early log phase density (~2 h at 37°C).
4. Mix 100 µL of the co-integrate culture with 200 µL of the strain 70 culture, and incubate the mixture at 37°C without agitation for 20 min to 1 h to allow for mat-ing (*see* **Notes 3** and **4**).
5. Plate 50-µL aliquots of the mated culture onto LB agar plates containing kanamycin, tetracycline, and streptomycin.
6. Grow plates overnight at 37°C.

3.2.3. Recovering a Library of Mutated MCMV Genomic Fragments

1. Collect colonies containing co-integrates (*see* **Subheading 3.2.2.**, **step 5**) by adding 2 mL of LB broth to each plate, gently scraping the colonies with a spread-er, and placing the wash into a tube.
2. Dilute wash to near saturation into LB broth containing kanamycin, tetracycline, and streptomycin, and grow culture at 37°C.
3. Purify DNA from cultures with a QIAPrep miniprep DNA isolation kit following the manufacturer's protocols.
4. Transform DNA into competent DH5α cells, and plate transformants on to LB agar containing tetracycline and kanamycin. Resulting colonies contain pHSS6-MCMV with a Tn3/*gpt* insertion.
5. Screen colonies for insertion of Tn3/*gpt* into the MCMV genomic fragment rather than the pHSS6 vector by restriction digestion of minprep DNA with *Not*I (*see* **Note 5**).
6. Grow large-scale preparations of pHSS6-MCMV DNA that contain a Tn3/*gpt* insertion within the cloned MCMV genomic fragment.

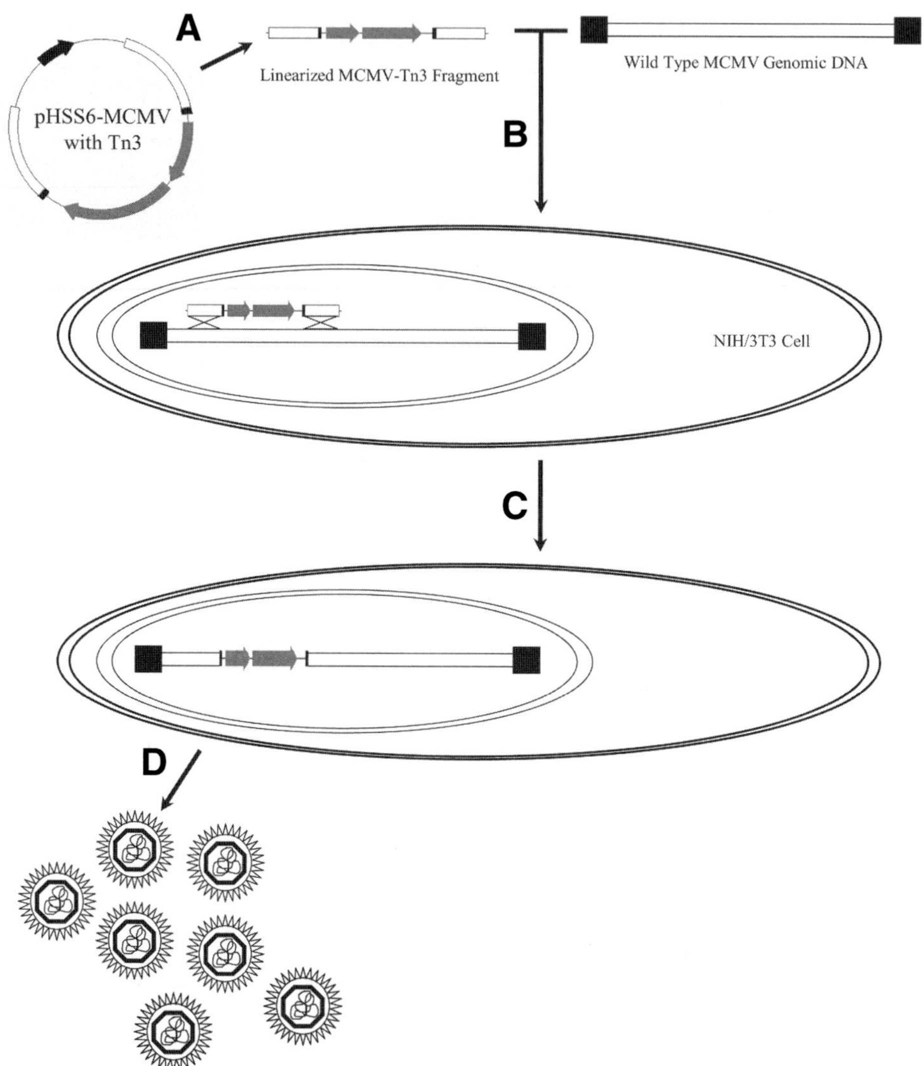

Fig. 2. Generating and purifying recombinant murine cytomegalovirus (MCMV). (**A**) The mutated MCMV fragment is released from pHSS6-*Sal*I by digestion with *Not*I. (**B**) The purified MCMV-Tn3 fragment is cotransfected with wild-type MCMV (Smith strain) DNA into NIH/3T3 cells. (**C**) The cellular machinery within the host mediates homologous recombination between the mutated MCMV fragment and the wild-type genome. (**D**) The transfected cells are grown in selective media containing xanthine and mycophenolic acid to select for cells that harbor viruses that have successfully recombined with the mutated MCMV DNA fragment. Clonal isolates of the recombinant MCMV are purified by serial dilution combined with plaque picking.

7. Large-scale DNA preparations can be sequenced using the FL110 primer in order to identify the site of insertion.

3.3. Recombinant Mutant Virus Generation

The third phase involves the creation of mutant viruses through the process of cotransfection, selection of viruses that have recombined with the mutated MCMV genomic fragment, clonal isolation of the mutant viruses, and molecular verification of the mutation (*see* Fig. 2 for an overview). Mutant recombinant viruses are able to be isolated owing to *gpt* gene present in the Tn3 that has disrupted a particular region of the MCMV genome following recombination. In the presence of mycophenolic acid, the metabolic pathway involved in the synthesis of guanosine nucleotide precursors is blocked (**Fig. 3**) *(19–21)*. However, cells infected with a mutant virus are able to convert the xanthine present in the media into the necessary precursors through a salvage pathway involving the guanosine phosphoribosyl transferase expressed by each mutant virus. Hence, all cells not infected by the mutant virus or infected by the wild-type virus die as a result of selection. The mutants are further purified and individually selected by plaque picking to yield clonal isolates.

3.3.1. Preparation of Mutated Genomic Fragment for Transfection

1. Digest mutated MCMV genomic fragment with *Not*I restriction enzyme by combining the following reagents to a final volume of 100 µL:
   a. 50 µg Mutated MCMV genomic fragment.
   b. 10 µL NEB buffer 3: (1 *M* NaCl, 500 m*M* Tris-HCl, 100 m*M* MgCl$_2$, 10 m*M* dithiothreitol DTT).
   c. 1 µL bovine serum albumin (BSA; 100 mg/mL).
   d. 50 U *Not*I enzyme.
2. After overnight incubation at 37°C, verify digestion on agarose gel, and bring the reaction volume up to 500 µL with water.
3. Extract DNA twice with 1 mL of buffer-saturated 25:24:1 phenol/chloroform/isoamyl alcohol, collecting the aqueous phase each time.
4. Remove residual organic elements with 1 mL of 24:1 chloroform/isoamyl alcohol.
5. Add 0.1 vol of 3 *M* NaOAc to the extracted DNA, mix, and precipitate the DNA with 1 mL of 95–100% ethanol on dry ice for 10 min.
6. Spin down DNA at 10,000*g* for 10 min.
7. Aspirate ethanol, and wash pellet with 250 µL of 70% ethanol. Repeat spin.
8. Aspirate ethanol wash, air-dry DNA for 10 min, and resuspend pellet in a sufficient volume of tissue culture grade (i.e., endotoxin-free) H$_2$O to bring the DNA concentration to 1 µg/µL.

3.3.2. Cotransfection

1. Prior to the day of transfection, seed a 25-cm^2 (T-25) flask with a sufficient number of NIH/3T3 cells to obtain 15–20% confluence.

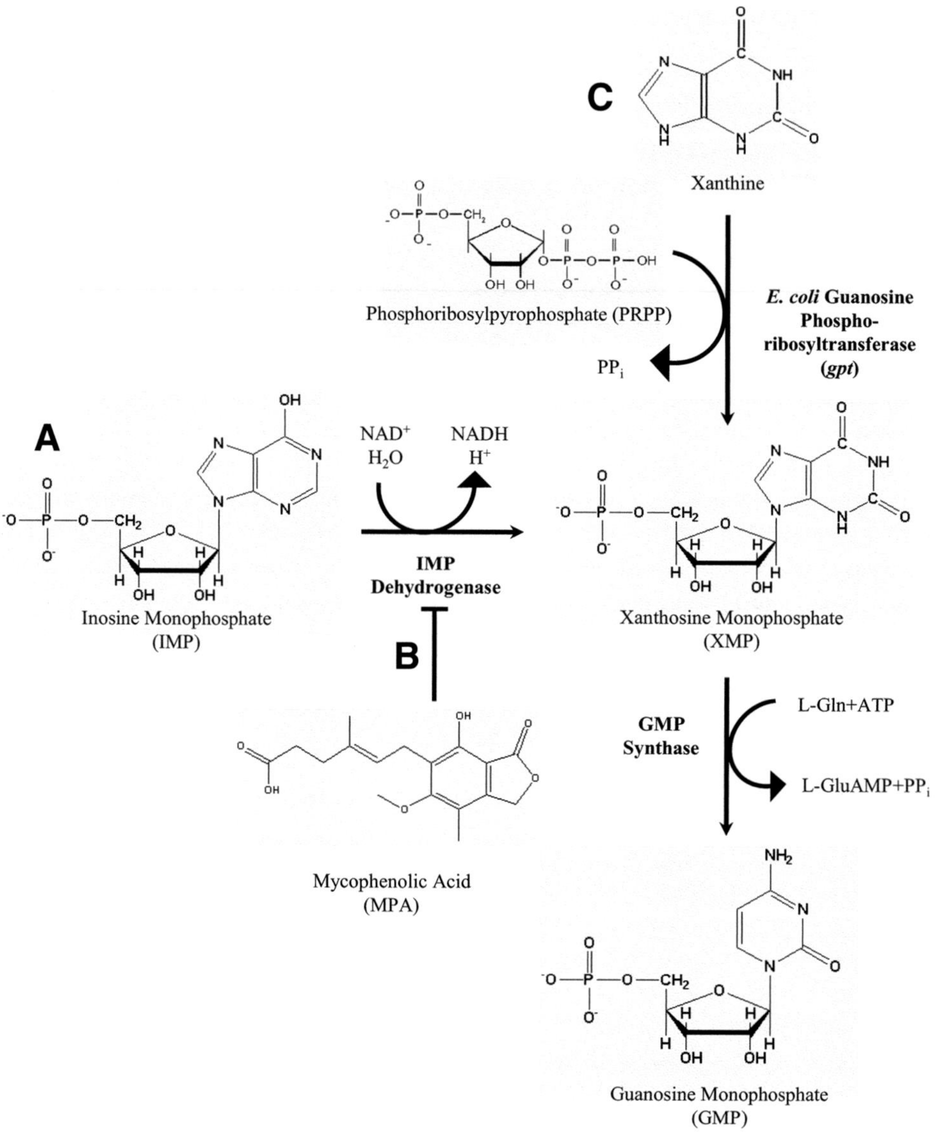

Fig. 3. Biochemical pathways involved in *gpt*-mediated selection. (**A**) Under normal conditions, the cell IMP to GMP in a two-step reaction is mediated by IMP dehydrogenase and GMP synthase. (**B**) When mycophenolic acid is added to the media, a block in *de novo* purine biosynthesis pathway occurs at the step mediated by IMP dehydrogenase, resulting in cell death. (**C**) In the presence of xanthine and the *gpt* selectable marker, a salvage pathway is formed allowing cells that express *gpt* to circumvent inhibition by MPA.

2. Cotransfect purified, full-length, wild-type MCMV genomic DNA and *Not*I-digested pHSS6-MCMV with Tn3/*gpt* into NIH/3T3 cells using the calcium phosphate transfection kit following the manufacturer's protocol (*see* **Note 6**).
3. Allow the resulting viruses to grow until all the cells appear to be infected (~7–10 d), changing the media as necessary.
4. Once 100% cytopathic effect (CPE) is observed, add an equal volume of 10% NFDM, and store flasks cell side down in a –80°C freezer until needed for selection and plaque purification (*see* **Note 7**).

3.3.3. Selection of Recombinant Mutant Viruses

1. Prior to the day on which selection experiments will be performed, seed the wells of a 6-well plate with a sufficient number of NIH/3T3 cells to obtain 50% confluence.
2. Thaw T-25 flasks containing virus (from **step 4, Subheading 3.3.2.**) in a 37°C water bath, and occasionally shake the flask to detach/disrupt the cells. Store thawed virus stock on ice.
3. Sonicate virus stock to complete the disruption of the cell membranes.
4. Perform a 10-fold serial dilution of the virus stock into six tubes containing 1 mL of NIH/3T3 media.
5. Use dilutions to infect each well of a 6-well plate (prepared the day before) for 90 min in a 37°C cell incubator.
6. Wash cells twice with 3 mL of NIH/3T3 media.
7. Add 3 mL of NIH/3T3 media containing xanthine and mycophenolic acid (selection media) to each well, and incubate plates in a 37°C cell incubator.
8. Replace selection media every 3 d.
9. To the well in which viral growth is observed at the highest possible dilution, add 3 mL of 10% NFDM.
10. Scrape the cells, transfer the mixture to a tube, and sonicate the mixture. The sonicated virus can be stored in a –80°C freezer or used immediately for the next round of selection.
11. Repeat **steps 4–10** for at least two additional rounds to complete the selection process. Use the samples collected in the final round of selection for the plaque purification steps.

3.3.4. Plaque Purification of Recombinant Mutant Viruses

1. Prior to the day in which plaque purification of the mutants will be performed, seed the wells of a 6-well plate with a sufficient number of NIH/3T3 cells to obtain 50% confluence.
2. Thaw the virus stocks that have undergone selection in a 37°C water bath. Store thawed virus stock on ice.
3. Perform a 10-fold serial dilution of virus stock that has undergone selection into six tubes containing 1 mL of NIH/3T3 media.
4. Use dilutions to infect each well of a 6-well plate for 90 min in a 37°C cell incubator.
5. Wash cells twice with 3 mL of NIH/3T3 media.

6. Overlay wells with 4 mL of a 1:1 mixture of 2X NIH/3T3 media and 2% low melting, tissue culture grade agarose. Allow DMEM–agarose to gel at room temperature.

7. Incubate plates in a 37°C, 5% CO_2 incubator until individual virus plaques become visible (~3–5 d).

8. Mark plaques from the highest dilution well with a felt-tip pen, and extract plaques and agarose plugs with a P-200 micropipet tip (*see* **Note 8**).

9. Dispense plug and plaque into 2 mL of a 1:1 mixture of NIH/3T3 media and 10% NFDM.

10. Sonicate plug and store at –80°C or use immediately for the next round of plaque purification.

11. Repeat **steps 2–10** for at least two additional rounds of purification.

12. Expand individual plaques to make clonal virus stocks.

13. The identity and purity of the mutant virus stocks can be confirmed by a number of methods including, but not limited to, Southern analysis, restriction fragment analysis, and DNA sequencing with the FL110 primer (*see* **Note 9**).

3.4. Recombinant Rescue Virus Generation

The final phase of generating a rescue virus for each mutant follows a process that is nearly identical to the methods used to make the mutant. Instead of using the wild-type MCMV genomic DNA as the backbone for recombination, the mutant's genomic DNA is used. The full-length, mutated genomic DNA is cotransfected with a DNA fragment containing the intact wild-type open reading frame, and the resulting recombinant viruses are purified by selection against the expression of *gpt* in STO murine fibroblasts in the presence of 6-thioguanosine *(7,20)*. STO murine fibroblasts are cells that do not express hypoxanthine-guanine phosphoribosyl transferase and are therefore unable to convert 6-thioguanine into a toxic nucleotide analog. Since the mutant contains the transposon that expresses *E. coli gpt,* all cells infected with the mutant become sensitive to the 6-thioguanine and are killed. Conversely, any cells in which the mutant has reverted back to wild type will be able to survive longer (until the virus kills the cell), hence enriching the culture for the reversion event.

3.4.1. Preparation of the Rescue DNA Fragment

A number of alternatives are available for the generation of the rescue/wild-type DNA fragment. All that is required is that the fragment includes the region that has been disrupted by the transposon in the mutant. High-fidelity PCR amplification (with proofreading polymerases) of the region is the most direct means of obtaining the rescue fragment and has been the alternative chosen by our laboratory to generate rescue viruses. Ten 100-μL PCR reactions create sufficient DNA for cotransfection. It is highly recommended that the wild-type PCR fragments be subjected to agarose gel purification using any of the commercially available kits.

3.4.2. Cotransfection

This step is accomplished as outlined in **Subheading 3.3.2.,** with the exception that the amplified wild-type fragment is cotransfected with the full-length mutant MCMV genomic DNA into NIH/3T3 cells. As stated in **Subheading 3.3.2.,** the viruses generated by cotransfection are collected and stored in a –80°C freezer until ready for selection.

3.4.3. Selection of Recombinant Rescue Viruses

1. Prior to the day in which selection experiments will be performed, seed the wells of a 6-well plate with a sufficient number of STO cells to obtain 50% confluence.
2. Thaw T-25 flasks containing virus (from **Subheading 3.4.2.**) in a 37°C water bath, and occasionally shake the flask to detach/disrupt the cells. Store thawed virus stock on ice.
3. Sonicate virus stock to complete the disruption of the cell membranes.
4. Perform a 10-fold serial dilution of the virus stock into six tubes containing 1 mL of STO media.
5. Use dilutions to infect each well of a 6-well plate (prepared the day before) for 90 min in a 37°C cell incubator.
6. Wash cells twice with 3 mL of STO media.
7. Add 3 mL of STO media 6-thioguanosine, and incubate plates in a 37°C cell incubator.
8. Replace selection media every 3 d.
9. To the well in which viral growth is observed at the highest possible dilution, add 3 mL of 10% NFDM.
10. Scrape the cells, transfer the mixture to a tube, and sonicate the mixture. The sonicated virus can be stored in a –80°C freezer or used immediately for the next round of selection.
11. Repeat **steps 4–10** for at least two additional rounds to complete the selection process. Use the samples collected in the final round of selection for the plaque purification steps.

3.4.4. Plaque Purification of Recombinant Rescue Viruses

Plaque purification of the recombinant rescue viruses is carried out as described in **Subheading 3.3.4.** on NIH/3T3 cells. The loss of the transposon can be verified by molecular techniques such as PCR with primers specific for the transposon or Southern analysis.

4. Notes

1. pHSS6-*Sal*I is identical to pHSS6[16] (gene accession no. M84115) except that the multiple cloning site between the *Not*I sites has been replaced with a single *Sal*I site.
2. *Sau*3AI digestion of MCMV genomic DNA must be optimized to maximize the diversity of the fragments within the 1.6–4 kbp range. To accomplish this, it is rec-

ommended that one serially dilute the amount of enzyme used in a set of reactions or use a standard enzyme concentration and vary the reaction time. The products from these test experiments are then run on a gel, and those conditions that yield the maximum number of fragments within the 1.6–4 kbp range should be used for the actual creation of a genomic fragment library. From our experience, we found the following reaction (50 µL final volume at 37°C for 30 min) to be a good starting point:

   a. 20 µg of full-length MCMV genomic DNA.

   b. 5 µL of NEB 10X *Sau*3AI buffer: (1 M NaCl, 100 m*M bis* propane-HCl, 100 m*M* MgCl$_2$, 10 m*M* DTT).

   c. 0.5 µL of BSA (1 mg/mL).

   d. 10 U of *Sau*3AI enzyme.

3. It is important to include the appropriate controls throughout the antibiotic selection (i.e., grow the starting strains prior to each mating in the medium/antibiotic in which the resulting mated bacteria will be grown) to ensure that there are no contaminants during the process.

4. Each of the mating procedures can (and are recommended) be scaled up in order to increase the diversity of the resulting mutant libraries.

5. The process of screening to identify clones that contained MCMV genomic fragments with a Tn3 insertion rather than the vector involved digesting miniprep DNA with *Not*I and identifying any clones in which the 2.2-kbp vector sequence had been enlarged by 3.6 kbp to 5.8 kbp owing to the presence of the transposon. This screening procedure is relatively labor-intensive, as our project involved the screening of thousands of colonies and the sequencing of hundreds of clones.

6. Cotransfection of the full-length DNA with the DNA fragment is accomplished by varying the ratios at which they are each transfected. For example, our laboratory uses 5, 10, and 15 µg of full-length genomic DNA combined with 1–10 µg of the DNA fragment for each recombinant virus. This process has been adopted simply to increase the chance of success in obtaining a recombinant virus.

7. After cotransfection, the cells may sometimes reach confluence before the virus has had a chance to replicate. If this is the case, it is recommended that the T-25 flask be divided (with trypsin) into two or three new T-25 flasks to allow the cells additional space to grow.

8. Micropipet tips can be cut to a 0.5-mm diameter to improve agarose plug extraction. Although the protocols only describe the use of a single 6-well plate for selection and plaque purification, it is highly recommended that one increase the number of plates to the maximum amount that the researcher can handle. By increasing the number of plaques that are picked, the chances for a successful generation of a recombinant virus are increased.

9. Restriction digest and Southern analysis of purified mutant viral DNA is absolutely necessary to ensure that the integrity of the genome has been maintained. We have observed rearrangements and spontaneous deletion in some of our mutants. The Tn-MCMV genome junction can also be resequenced using purified genomic DNA (from the mutant) as the template with the fmol sequencing system (Promega) to

confirm that the location of the transposon within the mutant's genome is identical to that seen with the original MCMV genome fragment used to make the recombinant mutant.

Acknowledgments

We would like to thank Dr. Michael Snyder for providing us with the shuttle mutagenesis reagents and protocols. M. L. was a recipient of a dissertation fellowship from the State of California Universitywide AIDS research program. F. L. is a Pew scholar in the biomedical sciences, a Scholar of the Lymphoma and Leukemia Society of America, and a recipient of an Established Investigator Award from American Heart Association. This work was supported by grants from the March of Dimes Birth Defects National Foundation and the National Institutes of Health.

References

1. Mocarski, E., Post, L., and Roizman, B. (1980) Molecular engineering of the herpes simplex virus genome: insertion of a second L-S junction into the genome causes additional genome inversions. *Cell* **22,** 243–255.
2. Spaete, R. R. and Mocarski, E. S. (1987) Insertion and deletion mutagenesis of the human cytomegalovirus genome. *Proc. Natl. Acad. Sci. USA* **84,** 7213–7217.
3. Wolff, D., Jahn, G., and Plachter, B. (1993) Generation and effective enrichment of selectable human cytomegalovirus mutants using site-directed insertion of the neo gene. *Gene* **130,** 167–173.
4. Patterson, C. E. and Shenk, T. (1999) Human cytomegalovirus UL36 protein is dispensable for viral replication in cultured cells. *J. Virol.* **73,** 7126–7131.
5. Abbate, J., Lacayo, J. C., Prichard, M., Pari, G., and McVoy, M. A. (2001) Bifunctional protein conferring enhanced green fluorescence and puromycin resistance. *Biotechniques* **31,** 336–340.
6. Morello, C. S., Cranmer, L. D., and Spector, D. H. (1999) In vivo replication, latency, and immunogenicity of murine cytomegalovirus mutants with deletions in the M83 and M84 genes, the putative homologs of human cytomegalovirus pp65 (UL83). *J. Virol.* **73,** 7678–7693.
7. Greaves, R. F., Brown, J. M., Vieira, J., and Mocarski, E. S. (1995) Selectable insertion and deletion mutagenesis of the human cytomegalovirus genome using the *Escherichia coli* guanosine phosphoribosyl transferase (gpt) gene. *J. Gen. Virol.* **76,** 2151–2160.
8. Vieira, J., Farrell, H. E., Rawlinson, W. D., and Mocarski, E. S. (1994) Genes in the *Hind*III J fragment of the murine cytomegalovirus genome are dispensable for growth in cultured cells: insertion mutagenesis with a lacZ/gpt cassette. *J. Virol.* **68,** 4837–4846.
9. Kemble, G., Duke, G., Winter, R., and Spaete, R. (1996) Defined large-scale alterations of the human cytomegalovirus genome constructed by cotransfection of overlapping cosmids. *J. Virol.* **70,** 2044–8.

10. Messerle, M., Crnkovic, I., Hammerschmidt, W., Ziegler, H., and Koszinowski, U. H. (1997) Cloning and mutagenesis of a herpesvirus genome as an infectious bacterial artificial chromosome. *Proc. Natl. Acad. Sci. USA* **94,** 14759–14763.
11. Weber, P. C., Levine, M., and Glorioso, J. C. (1987) Rapid identification of nonessential genes of herpes simplex virus type 1 by Tn5 mutagenesis. *Science* **236,** 576–579.
12. Jenkins, F. J., Casadaban, M. J., and Roizman, B. (1985) Application of the mini-Mu-phage for target-sequence-specific insertional mutagenesis of the herpes simplex virus genome. *Proc. Natl. Acad. Sci. USA* **82,** 4773–4777.
13. Smith, G. A. and Enquist, L. W. (1999) Construction and transposon mutagenesis in *Escherichia coli* of a full-length infectious clone of pseudorabies virus, an alphaherpesvirus. *J. Virol.* **73,** 6405–6414.
14. Brune, W., Menard, C., Hobom, U., Odenbreit, S., Messerle, M., and Koszinowski, U. H. (1999) Rapid identification of essential and nonessential herpesvirus genes by direct transposon mutagenesis. *Nat. Biotechnol.* **17,** 360–364.
15. Burns, N., Grimwade, B., Ross-Macdonald, P. B., et al. (1994) Large-scale analysis of gene expression, protein localization, and gene disruption in *Saccharomyces cerevisiae. Genes Dev.* **8,** 1087–1105.
16. Seifert, H. S., Chen, E. Y., So, M., and Heffron, F. (1986) Shuttle mutagenesis: a method of transposon mutagenesis for *Saccharomyces cerevisiae. Proc Natl Acad Sci USA* **83,** 735–739.
17. Zhan, X., Lee, M., Abenes, G., et al. (2000) Mutagenesis of murine cytomegalovirus using a Tn3-based transposon. *Virology* **266,** 264–274.
18. Lee, M., Xiao, J., Haghjoo, E., et al. (2000) Murine cytomegalovirus containing a mutation at open reading frame M37 is severely attenuated in growth and virulence in vivo. *J. Virol.* **74,** 11099–11107.
19. Michal, G. (ed.) (2002) *Boehringer Mannheim Biochemical Pathways Chart.* Roche Diagnostics Corporation, Roche Applied Science, Indianapolis, IN.
20. Mulligan, R. C. and Berg, P. (1980) Expression of a bacterial gene in mammalian cells. *Science* **209,** 1422–1427.
21. Mulligan, R. C. and Berg, P. (1981) Selection for animal cells that express the *Escherichia coli* gene coding for xanthine-guanine phosphoribosyltransferase. *Proc. Natl. Acad. Sci. USA* **78,** 2072–2076.

26

Construction of a Gene Inactivation Library for *Bovine herpesvirus 1* Using Infectious Clone Technology

Timothy J. Mahony, Fiona M. McCarthy, Jennifer L. Gravel, and Peter L. Young

Summary

The application of infectious clone technology to herpesvirus biology has revolutionized the study of these viruses. Previously the ability to manipulate these large DNA viruses was limited to methods dependent on homologous recombination in mammalian cells. However, the construction of herpesvirus infectious clones using bacterial artificial chromosome vectors has permitted the application of powerful bacterial genetics for the manipulation of these viruses. A method is described for the construction and characterization of a gene inactivation library of *Bovine herpesvirus 1* using an infectious clone. The method utilizes transposon-mediated gene inactivation, which permits gene inactivation without any prior knowledge of the viral genomic sequence. Furthermore, as the genetic manipulation is performed in bacteria the inactivation of those viral genes that are essential for viral replication is also possible. The method described here can be readily applied to any herpesvirus clone and provides the tools for precise characterization of all the genes contained within a herpesvirus genome.

Key Words: Bovine herpesvirus; homologous recombination; infectious clone; Tn5 transposon; bacterial artificial chromosome; gene deletion; gene insertion.

1. Introduction

The application of infectious clone technology to the study of herpesviruses by propagating these genomes as bacterial artificial chromosomes (BACs) has provided researchers with unprecedented methodologies for studying these viruses. Previously the large sizes of the herpesvirus genomes (100–230 kbp) have prevented the application of standard recombinant DNA techniques for constructing recombinant viruses *(1)*. Standard approaches for investigating gene function in herpesviruses have been dependent on homologous recombination in virus-susceptible cells. In some cases this has limited the study of

From: *Methods in Molecular Biology, vol. 292: DNA Viruses: Methods and Protocols*
Edited by: P. M. Lieberman © Humana Press Inc., Totowa, NJ

gene function, as the host cells may be poorly susceptible to commonly used transfection methodologies.

A significant advantage of constructing recombinant herpesviruses with BAC technology is that recombinants can be made in a manner that is independent of viral replication. Previously this was not possible without some form of complementing technology being used. BAC technology is also particularly useful for studying viruses with poor growth characteristics such as Epstein-Barr virus or Marek's disease virus *(2,3)*. Recombinogenic engineering systems have been developed that permit site-specific modification of BACs *(4–6)*. These methods permit the rapid generation of recombinant viruses although they are dependent on prior knowledge of the genomic sequence in the targeted area.

To demonstrate the power of BAC-based manipulation methods for herpesvirus, a method for generating a complete gene inactivation library is described using the *Bovine herpesvirus 1* (BoHV-1) infectious clone pBAC-BHV37 *(7)*. This method uses a Tn5 transposon system that permits the identification of genes and noncoding regions within the virus genome that could be utilized for transgene insertion. The construction and essential elements of the Tn5 transposon are described in detail. This method could be applied to any herpesvirus infectious clone without any prior knowledge of the genomic sequence to identify genomic regions suitable for transgene insertion. Such studies will lead not only to a better understanding of herpesvirus biology but also to the construction of better vaccine and gene therapy vectors for which these viruses have shown much promise but whose construction has been limited by the availability of poor manipulation technologies.

2. Materials

1. EZ::TN™ pMOD™<MCS> (Epicentre Technologies).
2. pEGFP (Clontech).
3. ElectroMAX DH10B (Invitrogen); store at –70°C. Limit freeze/thaw cycles to one. Electrocompetent *E. coli* strain DH10B are strongly recommended for use with BAC clones.
4. Oligonucleotide primers, as described in the text.
5. Restriction endonucleases: *Eco*47III, *Pin*A1, *Eco*RI, *Hind*III.
6. DNA modifying enzymes: T4 DNA polymerase (Promega) T4 DNA ligase (Promega), shrimp alkaline phosphatase (Roche).
7. SOC medium (Invitrogen); store at 4°C for up to 2 wk after opening.
8. LB (Luria-Bertani) agar containing 25 µg/mL of kanamycin (LBA-Kan); light-sensitive; store at 4°C in the dark for up to 4 wk.
9. Platinum *Taq* DNA polymerase (Invitrogen).
10. 10 m*M* Each of dATP, dCTP, dGTP, and dTTP; store at or below –20°C.
11. Rabbit kidney 13 cells (RK13), stored in liquid nitrogen for long-term storage.

12. Effectene transfection reagent (Qiagen).
13. Phosphate-buffered saline (PBS).
14. Growth media (G media): 1X Dulbecco's modified Eagle's medium (Invitrogen), 1X nonessential amino acids (Invitrogen), 1X glutaMAX (Invitrogen), 1X antibiotic/antimyotic (Invitrogen), 5–20% (v/v) donor calf serum (Invitrogen); light-sensitive; store at 4°C.
15. EZ::TN <Kan-1> Transposon (Epicentre Technologies).
16. pGEM-T (Promega).
17. LB agar plates containing 100 µg/mL ampicillin and 25 µg/mL kanamycin (LBA-Amp/Kan); light-sensitive; store at 4°C in the dark for up to 4 wk.
18. LB broth containing 100 µg/mL ampicillin and 25 µg/mL kanamycin (LB-Amp/Kan); light-sensitive; store at 4°C in the dark for up to 4 wk.
19. Tn5 transposase enzyme (Epicentre Technologies).
20. LB agar plates containing 12.5 µg/mL chloramphenicol (LBA-Cap); light sensitive; store at 4°C in the dark for up to 4 wk.
21. LB agar plates containing 12.5 µg/mL chloramphenicol and 25 µg/mL kanamycin (LBA-Cap/Kan); light-sensitive; store at 4°C in the dark for up to 4 wk.
22. LB broth containing 12.5 µg/mL chloramphenicol and 25 µg/mL kanamycin (LB-Cap/Kan); light-sensitive; store at 4°C in the dark for up to 4 wk.
23. Solution I: 25 mM Tris-HCl, pH 8.0, 10 mM EDTA, 50 mM glucose, 100 mg/mL RNase (DNase free); store at 4°C.
24. Solution II: 0.2 M NaOH, 1% (w/v) sodium dodecyl sulfate (SDS). Should be prepared immediately prior to use. Can be made from stock solutions of 10 M NaOH and 10% (w/v) SDS.
25. Solution III: 3 M potassium acetate, pH 5.5.
26. 70% Ethanol.
27. 10 mM Tris-HCl, pH 7.5.
28. Lipofectamine reagent (Invitrogen).
29. CRIB-1 cells; store in liquid nitrogen for long-term storage.
30. Opti-MEM I Reduced Serum Medium (Invitrogen).
31. *N,N'*-Hexamethylene-*bis*-acetamide (NN-HBA).
32. Dimethyl sulfoxide (DMSO).
33. Absolute ethanol
34. 3 M Sodium acetate, pH 4.6.
35. 75% (v/v) Ethanol.
36. 40% Glycerol (v/v in LB broth).

3. Methods

The method given here describes the construction and characterization of a gene knockout library for a BoHV-1 infectious clone using Tn5 transposon technology. There are five essential steps in this process;

1. Construction of the gene knockout vector.
2. Generation of the knockout library.

3. Confirmation of transgene insertion.
4. Assessment of the effect of gene knockout on virus recovery.
5. Characterization of the location of the transgene insertion in the genome.

3.1. Construction of Tn5 Transfer Vector

To facilitate the generation of a gene knockout library, a Tn5 transfer vector is constructed. The transfer vector has three essential elements; a prokaryotic selectable marker, a eukaryotic reporter gene expression cassette, and, flanking these elements, the hyperactive 19-bp mosaic ends recognized by the Tn5 transposase. The prokaryotic marker is used to select for transposed BAC clones, and the reporter gene is used to confirm successful transfection of the modified BAC if no infectious virus is recovered.

3.1.1. EZ::TN pMOD<MCS>

The EZ::TN pMOD<MCS> plasmid vector (Epicentre Technologies) is a transposon construction vector consisting of a multiple cloning site (MCS) flanked at either end by 19-bp hyperactive EZ transposase recognition sequences (*see* **Note 1**). Following insertion of the desirable elements into the MCS, the completed transposon can be either amplified from plasmid backbone using polymerase chain reaction (PCR) or excised using restriction enzyme digestion (*Pvu*II) depending on the sequences into the MCS.

3.1.2. Cloning of the Eukaryotic Reporter Gene (GFP)

1. To permit insertion of the eukaryotic green fluorescent protein (GFP) expression cassette from pEGFP-N1 (Clontech) into EZ::TN pMOD<MCS>, the MCS from pEGFP-N1 was deleted by digesting the native plasmid with *Eco*47III and *Pin*A1.
2. Following digestion, the 5′ overhanging bases from *Pin*A1 cleavage were filled using T4 DNA polymerase (Promega).
3. The pEGFP-N1 was resolved in a 0.7% (w/v) agarose gel.
4. The DNA was recovered using a QIAquick gel extraction kit (Qiagen) according to the manufacturer's instructions.
5. Approximately 50 ng of the recovered plasmid was religated using T4 DNA ligase (Promega) in a total volume of 10 μL.
6. After incubation at 4°C overnight, 1 μL of the ligation mixture was electroporated into 50 μL *E. coli* strain DH10B.
7. Immediately following electroporation, the DH10B cells were recovered in 950 μL of SOC broth and incubated at 37°C with gentle shaking for 1 h.
8. One hundred microliters of the transformed DH10B cells were plated onto LBA-Kan and incubated at 37°C for 12–16 h.
9. Randomly selected kanamycin-resistant colonies were resuspended in 10 μL LB broth, and 1 μL of this was used as the PCR template; colonies were screened by PCR using primers ECGFPfE 5′-gttcctggaattctgctggccttttgctcacatgt-3′ and ECGFPrE 5′-acgcctgaattcacattgatgagtttggacaaacca-3′.

10. PCR reaction conditions: 1 U Platinum *Taq* DNA polymerase, 20 m*M* Tris-HCl, pH 8.4, 50 m*M* KCl, 30 µ*M* MgCl$_2$, 250 µ*M* of each dATP, dCTP, d,GTP, dTTP, each primer to 20 µ*M,* and 1 µL of template.

11. PCR cycling conditions: (94°C, 5 min) for 1 cycle, (94°C, 30 s; 60°C, 20 s; 72°C, 120 s) for 35 cycles; (72°C, 10 min) for 1 cycle; then hold at 4°C.

12. After completion of the PCR, 5 µL of the reaction was digested with *Eco*RI using standard conditions followed by gel electrophoresis. Plasmids with the MCS deleted were identified as those clones that did not digest with *Eco*RI. MCS-deleted PCR product will be 1670 bp long and does not cut with *Eco*RI, whereas if the MCS is retained, the 1740-bp PCR product is cut into two fragments (1020 and 720 bp) by *Eco*RI.

13. The remainder of the PCR products from clones with the MCS deleted were combined, digested with *Eco*RI, and gel-purified.

14. After the EGFP cassette was recovered it was ligated into pMOD<MCS>, which had been digested with *Eco*RI and dephosphorylated.

15. An aliquot of the ligation mixture was electroporated into DH10Bs.

16. Colonies containing the GFP expression cassette were identified by PCR screening using the pMOD<MCS>-specific primers, MODfwd 5′-attcaggctgcgcaactgt-3′ and MODrev 5′-gtcagtgagcgaggaagcggaag-3′.

17. A PCR-positive colony was selected at random and cultured overnight in 5 mL of LB-Kan.

18. Plasmid DNA was prepared using a High Pure Plasmid Isolation kit (Roche) according to the manufacturer's instructions.

19. Insertion of the GFP cassette was confirmed by *Eco*RI digestion.

20. Expression of GFP was confirmed by transfection of the plasmids into RK13 cells.

21. RK13 cells were seeded at 5×10^5 cells/well in 6-well plates.

22. After seeding at this concentration, wells should then be 80–90% confluent after 18–24 h, which is suitable for transfection.

23. For each transfection reaction 0.6 µg of plasmid DNA was diluted to 100 µL using EC buffer.

24. Enhancer reagent, 3.2 µL, was added to each reaction (followed by mixing and gentle centrifugation to collect the reaction components) and incubated for 5–10 min at room temperature (RT).

25. After this incubation, 10 µL of Effectene was added, and the reaction was mixed by pipeting gently and incubated for 3–5 min at RT.

26. The growth media was removed from the monolayers, and the monolayers were washed once with PBS; then 1.6 mL of growth media was added to each well.

27. Each transfection reaction was diluted with 800 µL of growth media and transferred to each well.

28. The transfections were incubated for 18–24 h at 37°C with 5% CO$_2$.

29. The transfection mixtures were removed from the monolayer and replaced with 3 mL of growth media followed by incubation at 37°C with 5% CO$_2$.

30. GFP expression was typically evident at 48 h post transfection (*see* **Note 2**).

31. A GFP-expressing plasmid, pMOD-GFP, was selected at random for subsequent cloning steps.

3.1.3. Cloning of Bacterial Selectable Marker (Kanamycin Resistance)

1. To permit the selection of BAC clones containing inserted transposons, a kanamycin resistance cassette (Kan^R) was amplified from the EZ::TN"<KAN-1> Transposon (*see* **Note 1**) using the following oligonucleotides, KANfwd 5'-ggaAAGCTTcgttgtgtctcaaaatctctgatg-3' and KANrev 5'-ccAAGCTTcggttgatga-gagctttgttgtaggtg-3'. The resultant amplicon was approx 1150 bp in length.
2. Reaction conditions: as described in **Subheading 3.1.2.**, primers KANfwd and KANrev to 20 µ*M,* and 5 ng EZ::TN <Kan-1> Transposon.
3. Cycling conditions: (94°C, 4 min) for 1 cycle; (94°C, 10 s; 65°C, 15 s; 72°C, 90 s) for 10 cycles; (94°C, 10 s; 72°C, 90 s) for 25 cycles; (72°C, 10 min) for 1 cycle; then hold at 4°C.
4. Following amplification the Kan^R resistance cassette was ligated into pGEM-T (Promega) according to the manufacturer's instructions.
5. Recombinant colonies were then selected on LBA-Amp/Kan (*see* **Note 3**).
6. Colonies resistant to both antibiotics were grown up in overnight cultures, and plasmid DNA was prepared from 3 mL of broth.
7. The presence of the Kan^R resistance cassette was confirmed by *Hind*III digestion and gel electrophoresis analysis.
8. For recloning of the Kan^R resistance cassette into pMOD-GFP, a pGEM-T clone was selected at random, and 2 µg was digested with *Hind*III.
9. Following gel electrophoresis, the Kan^R fragment was recovered from the gel using a Qiagen gel extraction kit according to the manufacturer's instructions.
10. The Kan^R fragment was then ligated into pMOD-GFP that had previously been digested with *Hind*III and dephosphorylated.
11. Following electroporation into DH10B cells, recombinant clones were selected on LBA-Amp/Kan.
12. Insertion of the Kan^R resistance cassette was confirmed by preparing plasmid DNA and *Hind*III digestion and agarose gel electrophoresis.

3.1.4. Amplification of Completed Tn5 Transposon

1. Following confirmation that the required elements were correctly inserted into pMOD, the completed transposon was amplified from pMOD-GFP-Kan^R using MODfwd and MODrev.
2. The PCR reaction was carried out using the following parameters:
   a. Reaction conditions: as described in **Subheading 3.1.2.**, primers MODfwd and MODrev to 20 µ*M,* and 10 ng of pMOD-EGFP-Kan^R.
   b. Cycling conditions: (94°C, 4 min) for 1 cycle; (94°C, 30 s; 55°C, 30 s; 72°C, 3 min) for 35 cycles; (94°C, 10 min) for 1 cycle; then hold at 4°C.
3. After amplification, the PCR transposon was gel-purified and recovered using a QIAquick gel extraction kit (Qiagen) according to the manufacturer's instructions.
4. Following elution, the transposon was quantified by spectroscopic analysis.

Table 1
Composition of the Tn5 Transposition Reactions Used to Generate a Gene Inactivation Library for a *Bovine herpesvirus 1* Infectious Clone

	Reaction			
	A	B	C	D
Transposition buffer 10X	1.0	1.0	1.0	1.0
Infectious clone DNA (240 µg/mL)	0.9	0.9	0.9	0.9
Transposon DNA (40 µg/mL)	0	1	1 (1:10)	1 (1:100)
Sterile milli-Q water	7.1	6.1	6.1	6.1
Transposase (1 U/µL)	1.0	1.0	1.0	1.0
Total reaction	10	10	10	10

3.2. Generation of Gene Insertion Library

3.2.1. Transposition of Infectious Clone

1. Transposition reactions using the BoHV-1 infectious BAC clone and the Tn5 transposon were assembled as described in **Table 1.**
2. All the components were assembled on ice excluding the Tn5 transposase enzyme.
3. The reaction was then transferred to a 37°C water bath for 1–2 h.
4. The reaction was terminated by the addition of 1 µL of stop buffer, mixed, and incubated at 75°C for 10 min.
5. The reaction was stored at –20°C until required for transformation into DB10B cells.

3.2.2. Transformation of Library

1. Two microliters from each transposition reaction were added to 50 µL of electroMAX DH10B cells on ice.
2. The electroporation mix was added to prechilled electroporation cuvets.
3. Cells were pulsed at 1.5 kV, 100 Ω, 50 µF, recovered in 150 µL of SOC broth, and incubated at 37°C with gentle shaking for 1 h.
4. Half of the electroporation mix for reaction A was plated on LBA-Cap; the other half was plated on LBA-Cap/Kan followed by incubation at 37°C for 12–16 h.
5. The total electroporation mixes for reactions B–D were plated onto LBA-Cap/Kan followed by incubation at 37°C for 12–16 h (*see* **Note 4**).
6. The numbers of colonies generated from a typical electroporation experiment of the transposition reactions are shown in **Table 2.**
7. Colonies selected at random for further analyses were transferred to 3 mL of LB-Cap/Kan using sterile toothpicks and incubated with rocking for 12–16 h at 37°C (*see* **Note 5**).
8. The cells from 3 mL of the overnight culture were pelleted at 10,000*g* for 5 min, and the supernatant was discarded.

Table 2
**Total Number of DH10B Colonies Recovered Following Transposition
of the *Bovine herpesvirus 1* Bacterial Artificial Chromosome pBAC-BHV37
With a Tn5 Transposon Containing a Kanamycin Resistance Cassette**

	No. of resistant colonies	
Reaction	LBA-Cap	LBA-Cap/Kan
A	TMC	0
B	ND	300–400
C	ND	120
D	ND	40

LBA-Cap, LB agar plates containing 12.5 µg/mL chloramphenicol; LBA-Cap/Kan, LB agar plates containing 12.5 µg/mL chloramphenicol and 25 µg/mL kanamycin; TMC, too many to count; ND, not done.

9. The pellet was resuspended in 300 µL of solution I; 300 µL of solution II was then added followed by gentle mixing and incubation at RT for 5 min.
10. Three hundred microliters of solution III was added followed by gentle mixing and incubation on ice for 10 min.
11. The precipitate was then pelleted at 15,000g at RT for 10 min.
12. The supernatant was recovered, and 0.7 vol of isopropanol was added.
13. BAC DNA was then pelleted at 15,000g at RT for 10 min.
14. The DNA pellet was washed once with 70% ethanol and air-dried (*see* **Note 6**).
15. The DNA pellet was resuspended in 50 µL of 10 mM Tris-HCl, pH 7.5 (*see* **Note 6**).

3.3. Analysis of Gene Insertion Library

Following generation of the insertion library, three different analyses were carried out to determine whether the transposed clones had retained complexity and infectivity.

3.3.1. Restriction Endonuclease Analysis

1. An aliquot of the resuspended BAC DNA, 5 µL, was digested with *Hind*III at 37°C for 4 h.
2. The digestion fragments were then resolved in a 0.7% agarose gel for 20 h at 2 V/cm.
3. The DNA fragments were then visualized by ethidium bromide staining and UV illumination using standard procedures.
4. The *Hind*III digestion profiles of eight BoHV-1 BAC clones selected at random from the BoHV-1 transposition library are illustrated in **Fig. 1.**

3.3.2. Transfection Analysis of Transposed Clones

The ability of the transposed BoHV-1 BAC clones to regenerate infectious BoHV-1 was tested by transfecting an aliquot of the miniprep DNA into the

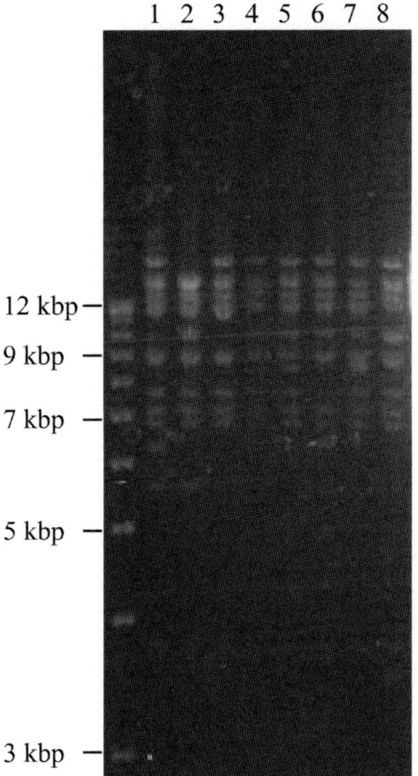

Fig. 1. Restriction endonuclease (*Hin*dIII) analysis of eight transposed bacterial artificial chromosomes containing the *Bovine herpesvirus 1* genome. *Hin*dIII digestion fragments were resolved in 0.7% agarose gel for 20.5 h at 2 V/cm. Lanes 1–8, pBAC-BHV37::Tn(*EGFP-Kan^R*)-1–8. Molecular weight marker, 1-kb DNA ladder (Invitrogen).

BoHV-1–susceptible cell line CRIB-1, a pestivirus-resistant derivative of MDBK cells *(8)* using lipofectAMINE *(see* **Note 7**).

1. CRIB-1 cells were seeded at 5×10^5 cells/well into 6-well plates, followed by incubation at 37°C in 5% CO_2 for 18–24 h. At this time the cells were typically between 80 and 90% confluent.
2. For each BAC clone, 1 μg of DNA was diluted with OPTI-MEM to 100 μL.
3. In addition for each clone 8 μL of lipofectAMINE was diluted with OPTI-MEM to 100 μL.
4. The diluted BAC clone was gently mixed with the diluted lipofectAMINE reagent and incubated at RT for 45 min.
5. Immediately prior to adding the transfection mixture to the monolayers, the growth media was removed from the cell monolayers and rinsed twice with PBS.

6. The monolayers were rinsed once with approx 0.5 mL/well of OPTI-MEM and as much of the wash solution as possible was removed.
7. Each reaction was diluted to 1 mL using OPTI-MEM, and the mixture was added to the washed monolayer.
8. The transfection reactions were incubated for 18–24 h at 37°C with 5% CO_2.
9. The transfection reaction was gently aspirated from the monolayer and replaced with 3 mL of growth media supplemented with 10% donor calf sera and 2 m*M* NN-HBA *(9)* and incubated at 37°C with 5% CO_2.
10. The transfections were monitored for GFP expression, generally visible after 48 h, and for cytopathic effects (CPEs) typical of BoHV-1 infection, usually observed 3–7 d posttransfection (*see* **Note 8**).
11. If no CPE was evident at 7 d post transfection, the supernatant was passaged three times to confirm the noninfectious virus phenotype.
12. Cell supernatants were passaged by preparing a fresh CRIB-1 monolayer as described for the transfection experiments.
13. Following removal of the growth medium (**Subheading 3.3.2., step 1**), the monolayers were washed once with PBS. After washing, a 200-μL aliquot of the transfection supernatant was added to the monolayer followed by incubation at 37°C for 1 h, with gentle rocking every 15 min.
14. After 1 h the liquid was removed, replaced with 2 mL of growth medium, and subsequently incubated at 37°C with 5% CO_2 for 7 d.
15. This procedure was repeated a further two times unless CPEs became evident at an earlier passage.

3.3.3. Sequence Analysis of Transposition Site

1. Sequencing of the transposed BAC requires highly purified DNA. BAC DNA for automated sequence analysis is prepared from a single colony inoculated into 250 mL of LB-Cap/Kan followed by incubation at 37°C overnight with shaking.
2. BAC DNA was recovered using the NucleoBond BAC 100 kit (Macherey-Nagel) according to the manufacturer's instructions. At least 1 μg DNA is used per sequencing reaction.
3. BAC DNA was prepared for sequencing by heating 1 μg DNA at 60°C for 30 min (*see* **Note 9**).
4. To the heat-treated BAC DNA, 16 μL Big-Dye terminator V2 mix, 2 μL DMSO, 20 μ*M* KanME, 5′-CTCCTTCATTACAGAAACGGC-3′ were added, and the volume was adjusted to 40 μL using sterile MQW (*see* **Note 10**).
5. The sequencing cycling conditions used were: (95°C, 5 min) for 1 cycle; (95°C, 30 s; 50°C, 20 s; 60°C, 4 min) for 60 cycles; (60°C, 4 min) for 1 cycle; then held at 4°C.
6. Following completion of cycling, the entire sequencing reaction was added to a microfuge tube containing 100 μL absolute ethanol and 4 μL 3 *M* sodium acetate, pH 4.6, and incubated at RT for 5 min.
7. Sequencing products were pelleted at 10,000*g* for 30 min at RT.
8. The supernatant was aspirated, and the pellet was gently washed with 75% ethanol.

Table 3
Characterization of the Sites of Transposition in Eight Modified *Bovine herpesvirus 1* Infectious Clones

Clone[a]	Infectious[b]	GFP expression	Insertion site	Gene function[c]
1	Yes	Yes	UL43 (virion protein)	Non
2	No	Yes	UL8 (DNA helicase component)	Ess
3	Yes	Yes	UL46 (tegument protein)	Non
4	Yes	Yes	Transposon sequence only	
5	No	Yes	UL52 (DNA helicase)	Ess
6	Yes	Yes	UL2 (uracil glycosylase)	Non
7	Yes	Yes	Repeat region	Ess
10	Yes	No	UL20/UL19 intergenic region	Non

[a] Clone number of pBAC-BHV37::Tn(*EGFP-Kan^R*).

[b] Infectious *Bovine herpesvirus* 1 recovered following transfection in to CRIB-1 cells.

[c] Gene function is determined as essential (Ess) for viral replication when the modified clone is unable to reconstitute infectious virus. Gene function is classified as nonessential (Non) for viral replication following reconstitution of infectious virus.

9. The pellet was air-dried.

10. The sequence fragments were subsequently resolved on an ABI377 Automated DNA Sequencer according to the manufacturer's instruction.

11. The site of insertion was determined by *blast n* searching on Genbank. **Table 3** illustrates some of the results from sequence analyses following transposition of the BoHV-1 infectious clone.

4. Notes

1. Since the construction of the Tn5 transposon for the methods described in this manuscript, Epicentre Technologies have updated their transposition reagents. The plasmid EZ::TN pMOD<MCS> has been replaced by EZ::TN pMOD-2<MCS>. The two plasmids differ only by the insertion of unique sequencing primer sites adjacent to the hyperactive 19-bp mosaic ends, to permit sequencing of the sites of insertion with the transposon derived from EZ::TN pMOD-2<MCS>. In addition to updating the plasmid vector, the EZ::TN <KAN-1> transposon has been updated to the EZ::TN <KAN-2> transposon. The update was required to eliminate sequencing problems caused by lack of primer specificity of the primers supplied by Epicentre Technologies for transposon-mediated sequencing applications; as a result, the updated transposon is 245 bp smaller than the original. The two primers, KANfwd and KANrev, described for amplification of the functional Kan^R cassette for prokaryotic selection, will amplify the cassette from either transposon.

2. Expression of GFP is tested in RK13 cells, as transfection of these cells using Effectene results in very high transfection efficiencies (>90%). However, RK13

cells are not very susceptible to BoHV-1 infection, so MDBK cells and MDBK derivatives are utilized for recovery of infectious virus (*see* **Note 6**). In addition, when one is transfecting mammalian cells with Effectene, it is important to use high-quality DNA. Plasmid DNA using standard techniques is not suitable and needs to be further purified by phenol/chloroform extraction and precipitation. Alternatively, plasmid DNA of higher purity can be prepared for clones of interest using ion exchange-based methods such as the BAC100 kit (Machery-Nagal) or the High Pure Plasmid Isolation kit (Roche). Effectene transfections are also sensitive to the quantity of DNA in the reaction, and as a result no more than 0.6 μg of DNA should be used per reaction.

3. When PCR products are cloned that contain restriction enzyme sites close to the 5′ and 3′ termini, they are routinely cloned using T/A cloning procedures and then re-excised with the relevant enzyme to ensure that the termini of the PCR product are completely digested. In addition, this also gives a reference point to return to if problems are encountered in subsequent manipulation experiments, compared with repeating the PCR, which may generate new populations of products. Furthermore, the recloned products for the EGFP and Kan^R cassettes were assayed for function by demonstrating EGFP expression and Kan^R resistance, respectively, rather than complete sequencing, as this is far less time-consuming. Functional assays are also forgiving for any PCR mutations that do not affect function when such mutations may not be readily apparent from sequence data analysis.

4. Although a single transposition reaction can be utilized to generate the BAC insertion library, the success of an experiment is dependent on the ratio of transposon DNA to BAC clone DNA, which is dependent on the accuracy of DNA quantitation. Therefore, by setting up multiple reactions with differing DNA concentrations, it is possible to allow for any errors in DNA quantitation. It is also recommended to try an initial transposition reaction with a much smaller clone of the target DNA compared with a BAC. Initial experiments using a modified Tn7 transposon system demonstrated a strong bias against random insertion into the BoHV-1 BAC. This type of target bias is easier to identify using a smaller target compared with a BAC.

5. At this stage it is possible to pick a representative library of transposon insertions for later analysis. This is done by adding 100 μL of LB-Cap/Kan to a 96-well tissue culture plate (Nunc) and inoculating each well, which is then inoculated with a single colony. Eight wells on either side of the plate are half-filled with sterile water, and the plate is wrapped in Parafilm to reduce evaporation. The plate is shaken at 37°C overnight. Glycerol stocks are prepared by adding 100 μL of 40% glycerol (v/v in LB broth) to each well and storing the plate at −70°C. Recombinant clones are recovered for analysis by scraping a sterile loop across the surface of the glycerol stock followed by inoculation into LB broth containing Cap/Kan. BAC DNA is then prepared as described in **Subheading 3.3.3.**

6. It is not recommended to dry the BAC DNA pellet under vacuum, as the resultant pellet maybe difficult to resuspend completely. In addition, it is also recommended to allow the air-dried BAC DNA to resuspend overnight at 4°C. The resuspended

BAC DNA can be stored at –20°C or –70°C indefinitely. If the BAC DNA is to be used regularly, short-term storage for up to 1 mo at 4°C is recommended. Repeated freeze/thaw cycles appear to damage the BAC DNA.

7. Routine transfection experiments using MDBK cells and their derivatives such as CRIB-1 have demonstrated low transfection efficiencies. These efficiencies are generally below 10% of all cells within the monolayer. When one is dealing with infectious clones, this low efficiency is not important. The presence of any cells, no matter how few, expressing GFP, even a single cell, demonstrates successful transfection. The expression of GFP also demonstrates that the transfected clone has been successfully transposed. In addition, the low transfection efficiency is unimportant when one is recovering infectious virus, as once virus is reconstituted, it is obviously self-replicating.

8. Transfection of CRIB-1 cells with the lipofectAMINE protocol is much more forgiving of both DNA quality and quantity. BAC DNA prepped using alkaline lysis works well for this method, and whereas 1 µg of DNA is used, as little as 0.5 µg is also effective. If the cells do not recover well from the transfection protocol, it is recommended that the amount of DNA be optimized, starting with 0.5 µg of DNA.

9. Depending on the base composition contained within the BAC, initial heat treatment may not be required prior to sequencing. It has been determined that this is an essential step for successful sequencing of the BoHV-1 BAC clones described in this study. More extensive heat treatment has also been used successfully with an initial denaturation step of 95°C for 30 min prior to cycle sequencing.

10. Prior to using a constructed transposon as a sequencing anchor, the orientation of the inserted elements should be determined. As both the EGFP and the Kan^R cassettes were cloned nondirectionally the transposon was sequenced in pMOD<MCS> using the primers MODfwd and MODrev. After one determines the orientation, sequencing primers can be designed to sequence across the junctions of transposition. This step would not be required if the updated plasmid, EZ::TN pMOD-2<MSC> (*see* **Note 1**), were used, as specific sequencing sites have been incorporated adjacent to the hyperactive mosaic ends as part of updating the vector.

Acknowledgments

This work was supported in part by research grant FLOT.203 from Meat and Livestock Australia.

References

1. Roizman, B. (1996) The function of herpes simplex virus genes: a primer for genetic engineering of novel vectors. *Proc. Natl. Acad. Sci. USA* **93,** 11307–11312.
2. Delecluse, H. J., Hilsendegen, T., Pich, D., Zeidler, R., and Hammerschmidt. W. (1998) Propagation and recovery of intact, infectious Epstein-Barr virus from prokaryotic to human cells. *Proc. Natl. Acad. Sci. USA.* **95,** 8245–8250.
3. Schumacher, D., Tischer, B. K., Fuchs, W., and Osterrieder, N. (2000) Reconstitution of Marek's disease virus serotype 1 (MDV-1) from DNA cloned as a

bacterial artificial chromosome and characterization of a glycoprotein B-negative MDV-1 mutant. *J. Virol.* **74,** 11088–11098.

4. Brune, W., Messerle, M., and Koszinowski, U. H. (2000) Forward with BACs new tools for herpesvirus genomics. *Trends Genet.* **16,** 254–259.

5. Muyrers, J. P. P., Zhang, Y., and Stewart, A. F. (2001) Techniques: recombinogenic engineering—new options for cloning and manipulating DNA. *Trends Biochem. Sci.* **26,** 325–331.

6. Wagner, M., Ruzsics, Z., and Koszinowski, U. H. (2002) Herpesvirus genetics has come of age. *Trends Microbiol.* **10,** 318–324.

7. Mahony, T. J., McCarthy, F. M., Gravel, J. L., West, L., and Young, P. L. (2002) Construction and manipulation of an infectious clone of the *Bovine herpesvirus 1* genome maintained as a bacterial artificial chromosome. *J. Virol.* **76,** 6660–6668.

8. Flores, E. F. and Donis, R. O. (1995) Isolation of a mutant CRIB-1 cell line resistant to bovine viral diarrhea virus infection due to block of viral entry. *Virology* **208,** 565–575.

9. Stavropoulos, T. A. and Strathdee, C. A. (1998) An enhanced packaging system for helper-dependent herpes simplex virus vectors. *J. Virol.* **72,** 7137–7143.

Selective Silencing of Viral Gene E6 and E7 Expression in HPV-Positive Human Cervical Carcinoma Cells Using Small Interfering RNAs

Ming Jiang and Jo Milner

Summary

The newly discovered phenomenon of RNA interference (RNAi) offers the dual facility of selective viral gene silencing coupled with ease of tailoring to meet genetic variation within the viral genome. Such promise identifies RNAi as an exciting new approach to treat virus-induced diseases, including virus-induced cancers. RNAi can be induced using small interfering RNA (siRNA). Synthetic siRNA targets homologous mRNA for degradation, and this process is highly efficient. Using cervical cancer cells as a model, we demonstrate RNAi for viral oncogenes. Cervical cancer is the second most common cancer in women worldwide and is caused by human papillomavirus (HPV). Silencing of HPV E6 and E7 gene expression was achieved using siRNAs to target the respective viral mRNAs. E6 silencing induced accumulation of cellular p53 protein, transactivation of the cell cycle control *p21* gene, and reduced cell growth. By contrast, E7 silencing induced apoptotic cell death. HPV-negative cells appeared to be unaffected by the antiviral siRNAs. Thus siRNA can induce selective silencing of exogenous viral genes in mammalian cells, and the process does not interfere with the recovery of cellular regulatory systems previously inhibited by viral gene expression.

Key Words: RNA interference; siRNA; human cervical cancer; human papillomavirus; HPV E6; HPV E7; apoptosis.

1. Introduction

In recent years, the technique of small RNA interference (siRNA) for selective silencing of mammalian gene expression has rapidly developed *(1–9)*. This technology can be a powerful tool not only for studying the molecular and cellular pathology of virus-induced diseases but also for the development of novel antiviral therapies *(10–25)*. Here we use cervical carcinoma cells positive for

From: *Methods in Molecular Biology, vol. 292: DNA Viruses: Methods and Protocols*
Edited by: P. M. Lieberman © Humana Press Inc., Totowa, NJ

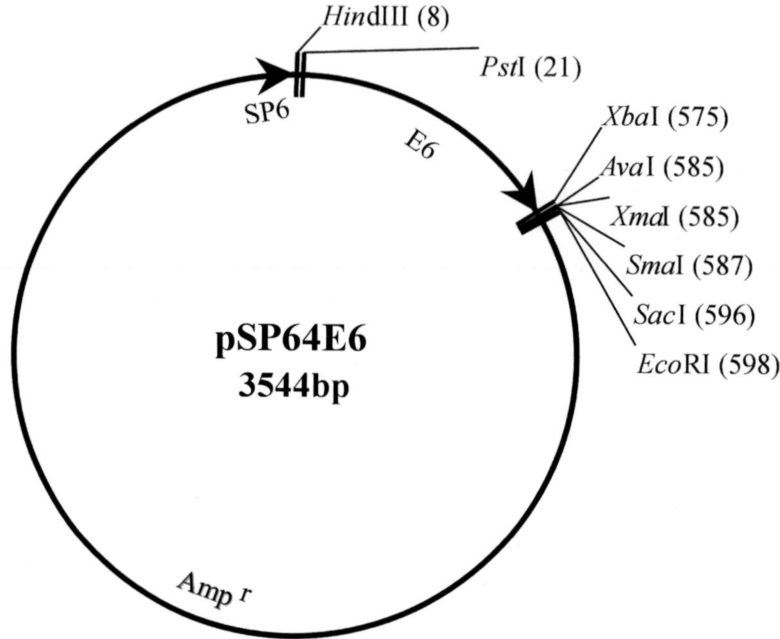

Fig. 1. Restriction map of pSP64E6.

human papillomavirus type 16 (HPV16) to illustrate the method for selective silencing of viral gene E6 and E7 using antiviral siRNA.

2. Materials

2.1. Human Cell Lines and Cell Culture Materials

1. Human cervical carcinoma cell lines CaSKi and SiHa (ATCC).
2. Human normal diploid fibroblasts (NDFs; Coriell Institute for Medical Research).
3. Human colon carcinoma epithelial HCT 116 (ATCC).
4. Cell culture media: RPMI, minimum essential medium (MEM) Dulbecco's modified Eagle's medium (DMEM), and Opti-MEM™ (Invitrogen).
5. Fetal calf serum (FCS; Invitrogen).
6. Oligofectamine™ reagent (Invitrogen).
7. Phosphate-buffered saline (PBS).
8. Fluorescein isothiocyanate (FITC)–dextran 150 (FD-150; Sigma, UK).
9. Centricon-10 (Amicon).

2.2. RNA/DNA-Related Materials

1. Synthetic siRNA oligonucleotides (MWG, Germany).
2. 10X siRNA annealing buffer: 200 mM Tris-HCl, pH 7.5, 100 mM MgCl, 500 mM NaCl; store at –20°C.

3. RNeasy mini kit (Qiagen, UK).
4. Oligotex mRNA mini kit (Qiagen).
5. RNase-free DNase set (Qiagen).
6. Zeta-Probe GT Nylon Membrane (Bio-Rad).
7. DNA oligonucleotide primers (Sigma-Genosys, UK).
8. Human HPV16 E6 cDNA in vector pSP64 (map shown in **Fig. 1**).
9. Reverse-iT™ one-step reverse transcriptase polymerase chain reaction (RT-PCR) kit (ABgene, UK).
10. Reverse Blend transcriptase (ABgene, UK).
11. DNA engine dyad™ Peltier Thermal Cycler (MJ Reseach).
12. DNA engine opticon system (MJ Reseach).
13. Quantitect™ SYBR green PCR mix (Qiagen).
14. Agarose and DNA electrophoresis equipment.
15. Ethidium bromide (EtBr; Sigma).

2.3. Protein-Related Materials

1. Protein sodium dodecyl sulfate-polyacrylamide gel electrophoresis (SDS-PAGE) and gel transfer equipment.
2. Cell lysis buffer A: 150 mM NaCl, 0.5% NP40, 50 mM Tris-HCl, pH 8.0; store at 4°C.
3. 4X Laemli's buffer: 8% SDS, 40% glycerol, 400 mM dithiothreitol (DTT), 240 mM Tris-HCl, pH 6.8, 0.001% bromophenol blue; store at 4°C.
4. Hybond™ ECL™ nitrocellulose membrane (Amersham Pharmacia Biotech, UK).
5. Block reagent (Roche, Germany).
6. TBS: 20 mM Tris-HCl, pH 7.5, 500 mM NaCl.
7. Tween-20.
8. BM chemiluminescence blotting substrate (Roche).

2.4. Antibodies

1. Mouse monoclonal anti-p53 antibody DO-1 (Oncogene).
2. Mouse monoclonal anti-p21 antibody (SX118) (Pharmingen).
3. Mouse monoclonal anti-Rb antibody (G3-245) (Pharmingen).
4. Rabbit polyclonal antiactin antibody (Sigma).
5. Rabbit antimouse immunoglobulins, horseradish peroxidase (HRP)-conjugated (DAKO, Denmark).
6. Goat antirabbit immunoglobulins HRP-conjugated (DAKO).
7. Rabbit antimouse immunoglobulins, FITC-conjugated (DAKO).

2.5. Flow Cytometry

1. Annexin-V-fluos (Roche, Germany).
2. HEPES buffer: 10 mM HEPES/NaOH, pH 7.4, 140 mM NaCl, 5 mM CaCl$_2$.
3. Propidium iodide (PI; Sigma).
4. Flow cytometry FACSCalibur (BD Biosciences).

2.6. Cell Immunostaining Materials

1. Hoechst 33258 (Sigma).
2. Triton X-100.
3. 4% Paraformaldehyde in PBS: 4 g paraformaldehyde in 50 mL dH$_2$O; heat at 60°C with a few drops of 1 *N* NaOH to dissolve, then cool down to room temperature, and add 50 mL 2X PBS. (This is best prepared fresh, but it can be stored at room temperature for up to a month.)

3. Methods

The methods described below comprise (1) design of siRNA, (2) transfection of siRNA, and (3) characterization of transfected cells.

3.1. Design of the E6 and E7 siRNAs

Synthetic 21-nucleotide RNAs targeting viral E6 and E7 mRNA are designed according to the E6 and E7 mRNA sequence (GenBank NC_001526).

1. To design the siRNA oligonucleotides, choose 19-nt sequences from the most highly conserved region at the middle of the gene in this case (*see* **Note 1**); avoid the sequences that contain GC- or AT-rich regions; GC contents should preferably be around 50%.
2. Check the RNA oligonucleotides for the potential to form secondary dimer and/or loop structures, for both strands, using Vector NTI to make sure they have minimal secondary structure. The 21-nt RNA oligonucleotides should be the 19-nt sequence with a 2-nt 3′ overhang of 2′-deoxythymidine (dTdT) (*see* **Note 1**).
3. Finally, do the Genbank blast screening to make sure that the chosen 21-nt sequences (both sense and antisense) only target viral E6 or E7 mRNAs and do not also recognize human mRNA sequences. More criteria may be used when designing siRNA (*see* **Note 1**).
4. The E6 siRNA corresponds to the coding region 142–160 relative to the first nucleotide of the start codon, and E7 siRNA is from 101 to 119 relative to the first nucleotide of the start codon.
5. Design a control siRNA that has no target mRNA sequence in the human HPV-positive cell lines (**Fig. 2**). Additional controls (not shown here) should include positive RNAi controls against an endogenous cellular mRNA such as lamin A/C mRNA *(9)*.

3.2. Preparation of siRNA

1. The 21-nt RNA oligonucleotides are chemically synthesized and purified with high-performance liquid chromatography (HPLC) by MWG (Germany).
2. Synthetic oligonucleotides are then deprotected using the methods and solution supplied by MWG; they are then dissolved in DNase/RNase-free dH$_2$O at 25 µ*M* concentration.
3. The single-strand RNA oligonucleotides are aliquoted and stored at –80°C.

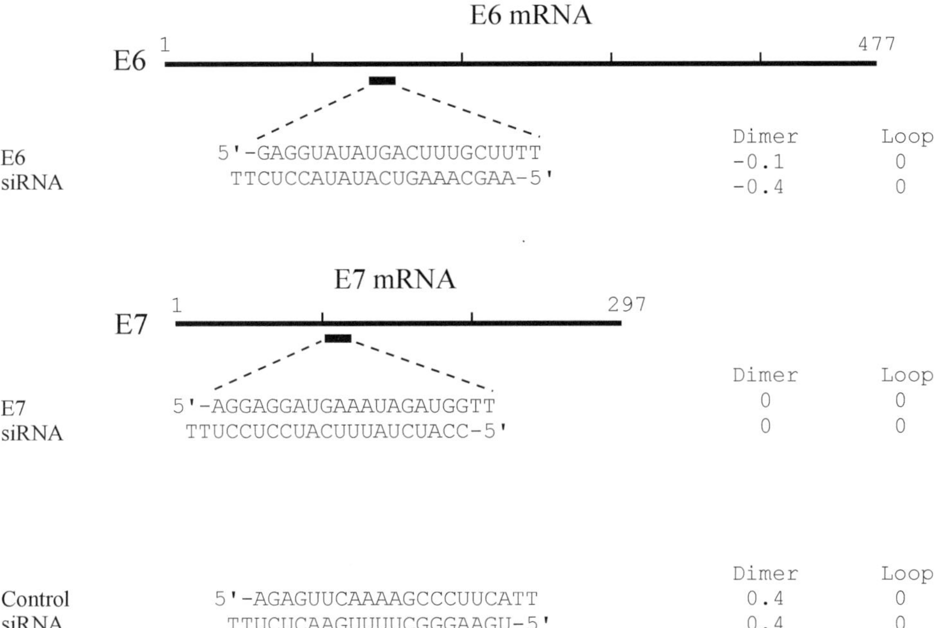

Fig. 2. Sequences of E6, E7, and control siRNAs. (Reproduced with permission from **ref. *11.***)

4. Anneal the siRNAs as follows:

	Volume (µL)	Final conc. (µ*M*)
Sense RNA oligonucleotides (25 µ*M*)	40	20
Antisense RNA oligonucleotides (25 µ*M*)	40	20
10X Annealing buffer	10	1X
ddH$_2$O	10	
Total	100	

for 1 min at 90°C, and then at 37°C for a further hour. The annealed siRNA can be aliquoted and stored in –20°C, but it is preferable to use it fresh.

5. Check annealed siRNA by 15% nondenaturing acrylamide gel electrophoresis (3.75 mL 40% acrylamide; 0.4 mL 2% *bis*; 1 mL 10X TBE; 100 µL 10% ammonium persulfate; 100 µL; 4.7 mL dH$_2$O; and 10 µL TEMED).

6. As shown in **Fig. 1,** the sense and antisense oligonucleotides of E6 siRNA migrate faster and stain with less intensity with EtBr than with annealed E6 siRNA **(Fig. 3).**

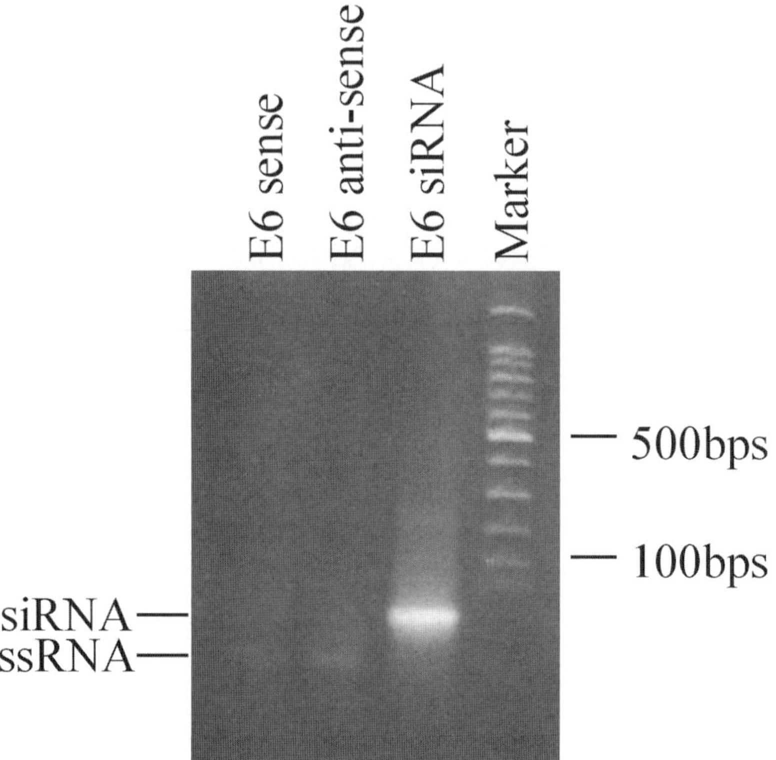

Fig. 3. Migration of single-stranded E6 RNA sense and antisense oligonucleotides and annealed E6 siRNA on 15% nondenaturing acrylamide gel.

3.3. Human Cell Transfection With siRNA

3.3.1. Cell Line Maintenance

1. CaSKi and SiHa epithelial cell lines are derived from human cervical carcinomas and contain the integrated HPV16 genome, about 600 copies (CaSKi) and one or two copies (SiHa). CaSKi cells also contain sequences related to HPV18 (ATCC CRL-1550). Both cell lines contain wild-type p53.
2. The cell lines need to be handled under Biosafety level 2 containment according to the ATCC.
3. Culture CaSKi cells in RPMI plus 10% FCS, 2 m*M* L-glutamine, and 1 m*M* sodium pyruvate.
4. Culture SiHa cells in MEM plus 10% FCS, 1 m*M* sodium pyruvate, and 0.1 m*M* nonessential amino acids.
5. Use human NDFs and human colon carcinoma epithelial HCT116 as HPV-negative cell line controls for this study.
6. Culture NDFs in MEM plus 15% FCS, 1.0 m*M* sodium pyruvate, and 0.1 m*M* nonessential amino acids.

7. Culture HCT116 cells in DMEM with 10% FCS.
8. Maintain all the cell lines with 100 U/mL penicillin and 100 µg/mL streptomycin at 37°C in 5% CO_2 in air.

3.3.2. Cell Transfections

1. Trypsinize CaSKi and SiHa Cells and subculture them into 10-cm^2 6-well plates without antibiotics, 1.5×10^5 cells per well.
2. Transfect the cells after 24 h with siRNA formulated into Oligofectamine Reagent (liposomes) according to the manufacturer's instructions.
3. For each well (about 2.5×10^5 cells), prepare the transfection mix as follows:
   a. Dilute 10 µL 20 µ*M* annealed siRNA with 175 µL Opti-MEM.
   b. Then dilute 3 µL Oligofectamine with 12 µL Opti-MEM, and allow the mixture to sit for 5 min at room temperature.
   c. Mix the diluted siRNA and Oligofectamine together, and allow it to sit for 20 min at room temperature before applying it to the cells as follows.
   d. Wash the cells once with Opti-MEM, and then add 800 µL Opti-MEM to the well followed by 200 µL siRNA mixture.
   e. After 4 h of incubation at 37°C in 5% CO_2, add 0.5 mL normal growth media containing 30% FCS without antibiotics to give the final volume of 1.5 mL media per well.
4. Harvest the cells at different time points for analysis.
5. Note that the transfection could be expanded to a 10-cm dish (50 cm^2) using the same cell/siRNA ratio, but the transfection efficiency may decrease.

3.3.3. Measurement of Transfection Efficiency

1. Transfection efficiency can be measured using FITC–dextran instead of siRNA.
2. Dissolve the FITC–dextran in PBS to 50 µg/µL, and centrifuge at 4000*g* for at least 60 min to get rid of free fluorescence using Centricon-10.
3. Dilute the centrifuged FITC–dextran in PBS at a final concentration of 20 µg/µL, and stored at 4°C.
4. For a 6-well plate (2.5×10^5 cells per well), use 10 µL of FITC–dextran per well instead of 10 µL siRNA to transfect CaSKi and SiHa.
5. To check the transfection efficiency, wash the transfected cells three times gently with PBS, and count the cells with internalized green fluorescence under the microscope, as shown in **Fig. 4.** For all the cell lines used in this study the transfection efficiencies are between 70 and 80%.
6. Confirm the transfection efficiencies using fluorescence-labeled siRNA (*see* **Note 2**).

3.4. Cell Growth Analysis

3.4.1. Cell Growth Curve

Cell growth rates are determined by cell counting. The siRNA-transfected SiHa cell growth curves are shown in **Fig. 5.** Control siRNA does not affect SiHa cell growth, E6 siRNA reduces cell growth, and E7 siRNA results in cell growth decrease and loss of cells at 72 h.

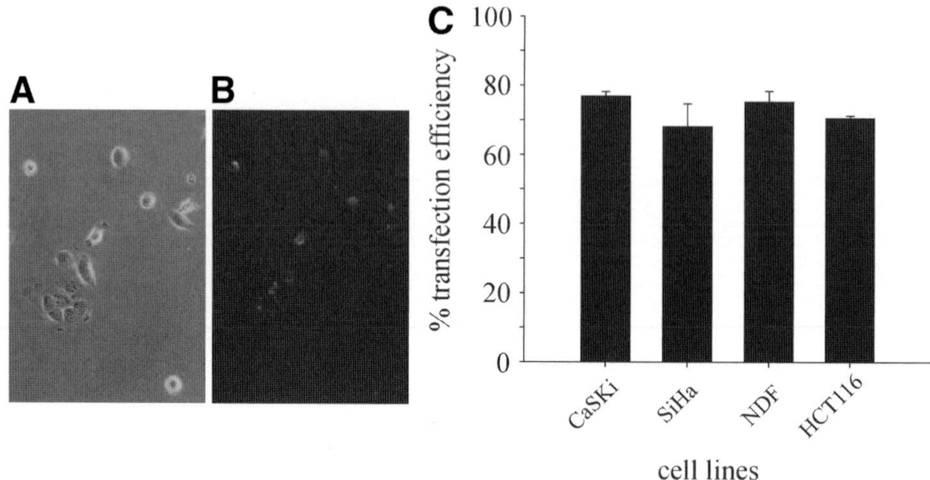

Fig. 4. SiHa cells are transfected with FITC-dextran for 12 h. **(A)** Phase contrast and **(B)** fluorescence images of FITC-dextran–transfected cells at 12 h **(C)** Transfection efficiencies for all the cell lines used in this study. (C, reproduced with permission from **ref. *11*.**)

3.4.2. E6 and E7 siRNA Induce Growth Arrest and Cell Death, Respectively, in SiHa Cells

The control siRNA and E7 antisense oligonucleotides have no effect on SiHa cell growth (**Fig. 6A** and **D,** respectively). The E6 siRNA-transfected cells grow much more slowly, but the cells keep their morphology even after 3 d, whereas the E7 siRNA-treated cells lose their morphology, round up, and undergo apoptosis from 2 d onward. On the other hand, the E7 siRNA-treated HPV-negative cells, NDF and HCT116, appear to be unaffected and continue to grow normally.

3.4.3. Cell Cycle Analysis

1. For cell cycle analysis, harvest 10-cm^2 ($\geq 2.5 \times 10^5$) cells.
2. Wash with 1 mL PBS, and resuspend in 1 mL PBS.
3. Add 4 mL ice-cold absolute ethanol to fix the cells overnight at –20°C. (The samples can be stored at –20°C for up to a month.)
3. Pellet the fixed cells at 500 g for 5 min, and wash once with PBS.
4. Resuspend the cell pellets in 1 mL PBS with 5 µg/mL propidium iodide (PI) + 200 U/mL RNase A.
5. Analyze the samples on a flow cytometer using 488-nm excitation and a filter >560 nm for PI detection.
6. Do the cell cycle analysis using the Cylchred Programme (http://www.uwcm.ac.uk/study/medicine/haematology/cytonetuk/documents/software.htm).
7. Typical cell cycle profiles are shown in **Fig. 7.**

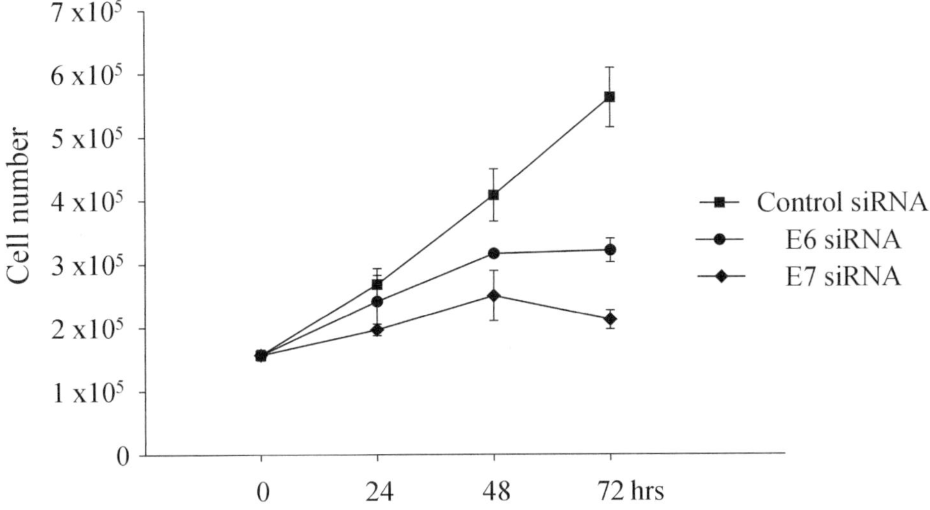

Fig. 5. SiHa cell growth curves after treatment with control siRNA, E6 siRNA, or E7 siRNA as indicated.

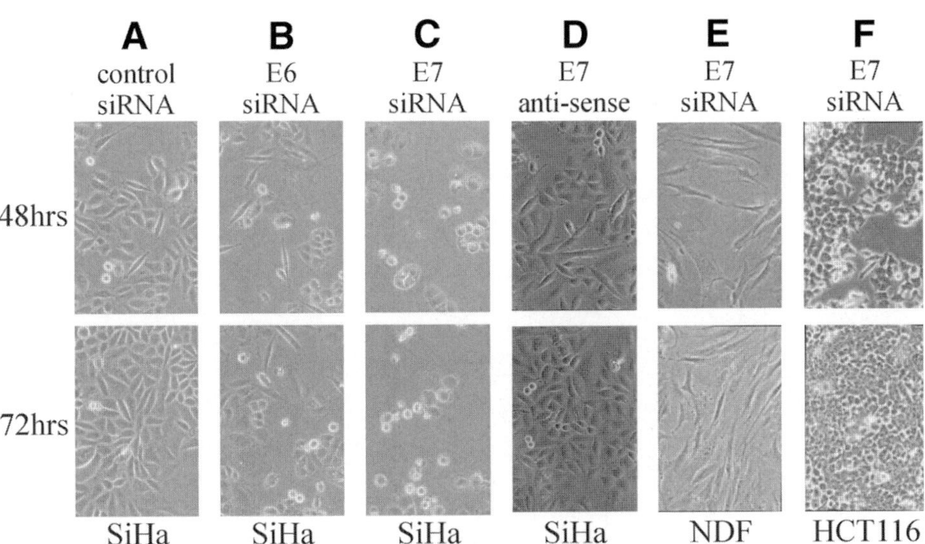

Fig. 6. E7 siRNA induces apoptosis in human cervical carcinoma cells. **(A–D)** Phase contrast images of SiHa cells treated with control siRNA, E6 siRNA, E7 siRNA, and E7 antisense RNA, as indicated. E7 siRNA does not affect the growth of human **(E)** normal diploid fibroblasts (NDF) or **(F)** HCT116 colon carcinoma cells. (Reproduced with permission from **ref. *11*.**)

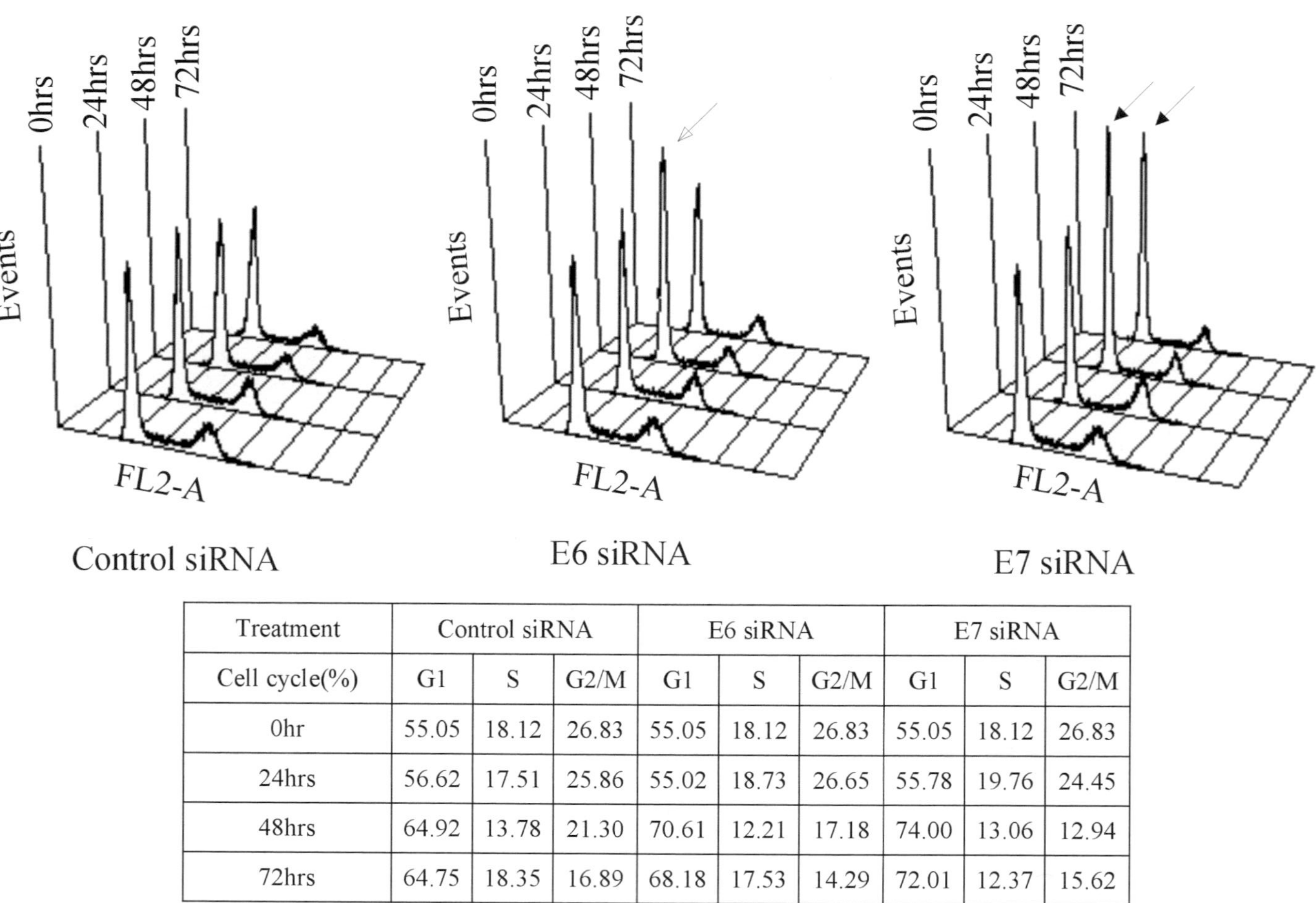

Treatment	Control siRNA			E6 siRNA			E7 siRNA		
Cell cycle(%)	G1	S	G2/M	G1	S	G2/M	G1	S	G2/M
0hr	55.05	18.12	26.83	55.05	18.12	26.83	55.05	18.12	26.83
24hrs	56.62	17.51	25.86	55.02	18.73	26.65	55.78	19.76	24.45
48hrs	64.92	13.78	21.30	70.61	12.21	17.18	74.00	13.06	12.94
72hrs	64.75	18.35	16.89	68.18	17.53	14.29	72.01	12.37	15.62

Fig. 7. Cell cycle analysis of SiHa cells treated with control siRNA, E6 siRNA, or E7 siRNA as indicated at different time points. E6 siRNA treatment induces slight transient G1 increase (open arrow), whereas E7 siRNA induces higher G1 increase (black arrows). Note that the apoptotic cells have been gated out.

3.5. Detection of Apoptotic Cells

Apoptotic cells are identified by annexin-V staining. As seen in **Fig. 6,** the E7 siRNA transfected SiHa cells are rounded up and detached from the plate. Further Annexin-V-Fluos labeling and flow cytometry analysis show that these cells are undergoing apoptosis **(Fig. 8)**.

1. Trypsinize, harvest, and wash 10-cm^2 ($\geq 2.5 \times 10^5$) cells with PBS.
2. Pellet the cells at 200g for 5 min, and resuspend in 100 µL of HEPES buffer + 2 µL Annexin-V-Fluos + 1 µg/mL PI.
3. After 10–15 min of incubation in the dark, analyze the cells on a flow cytometer using 488-nm excitation and a 515-nm bandpass filter for Annexin-V fluorescein detection and a filter > 560 nm for PI detection.
4. Compensation of the instrument is required to exclude overlapping of the two emission spectra. The PI-positive and Annexin-V-negative populations are also gated out.

3.6. Detection of E6 and E7 mRNAs in HPV-Positive Cells

3.6.1. RNA Preparation of Transfected Cells

Total cellular RNA is prepared with an RNeasy kit (Qiagen) according to the manufacturer's instructions. The mRNAs are isolated from total cellular RNA using an Oligotex mRNA mini kit (Qiagen) following the manufacturer's instructions.

3.6.2. Detection of E6 mRNA by Northern Hybridization

1. Preparation of the E6 probe.
   a. Use the HPV16 E6 cDNA fragment from pSP64E6 as the E6 probe.
   b. Purify and label the 0.6-kb *Hind* III–*Xba* I fragment using the random primer labeling method *(26)* with ^{32}P-dCTP.
2. Northern hybridization.
   a. Fractionate the 0.3 µg mRNA of each transfection sample at standard 1.0% agarose–formaldehyde gel electrophoresis.
   b. Transfer the fractionated mRNAs from gel to nylon membrane.
   c. Perform the Northern hybridization as in standard molecular biology methods *(26)*.
   d. As shown in **Fig. 9A,** the E6 mRNA level is much lower in E6 siRNA-treated cells than in control siRNA-treated cells.

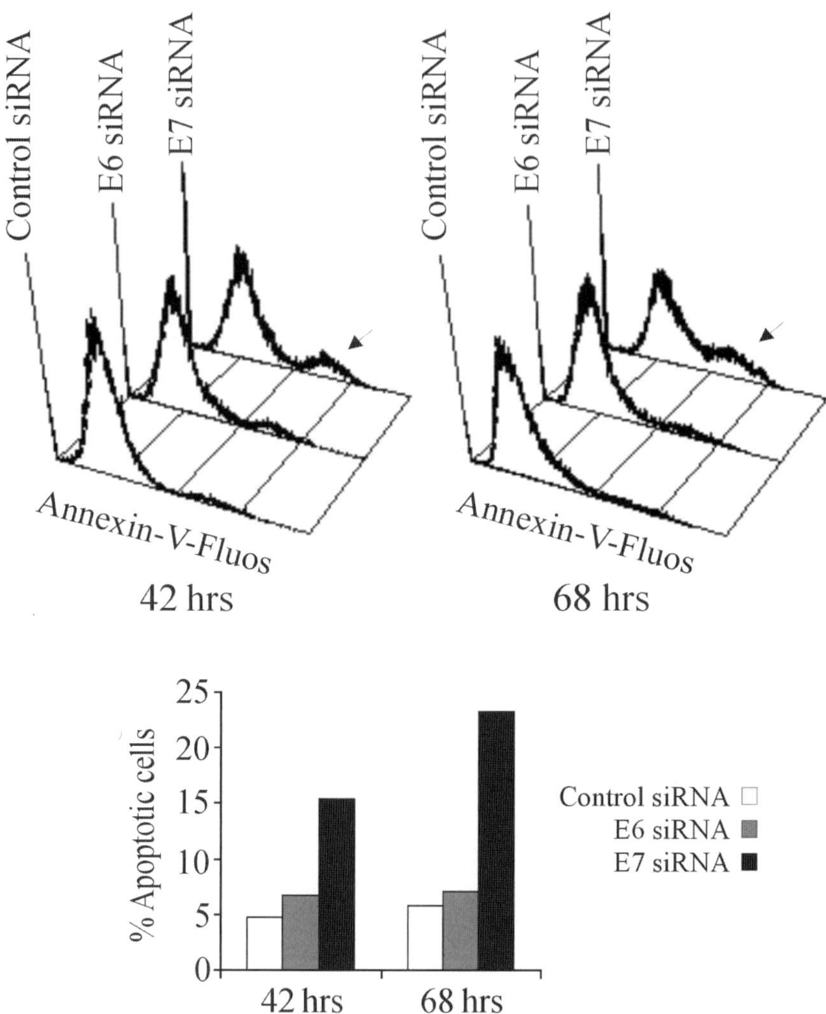

Fig. 8. FACS analysis of apoptotic cells. Control siRNA, E6 siRNA, and E7 siRNA as indicated. Arrows show the population of apoptotic cells. (Reproduced with permission from **ref. *11.***)

3.6.3. Detection of E6, E7, and p53 mRNA by RT-PCR

1. Employ the Reverse-iT one-step kit (ABgene Biotechnologies), and use 0.1 µg total RNA in each reaction. Control RT-PCR without reverse transcriptase is needed to ensure there is no DNA contamination in the total RNA (*see* **Note 3**).
2. For E6 cDNA amplification, the primers are 5′-cggaattcatgcaccaaaagagaactgca-3′ and 5′-cccaagcttacagctgggtttctctacg-3′, and the thermal cycle is 47°C, 30 min;

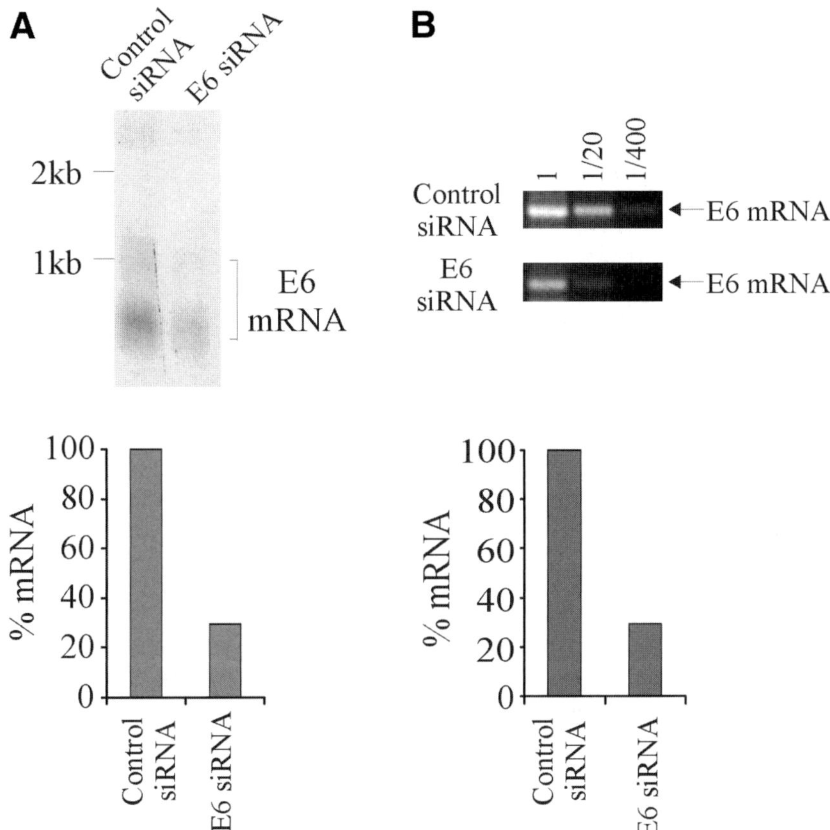

Fig. 9. Quantitation of E6 mRNA level by (**A**) Northern hybridization and (**B**) semi-quantitative RT-PCR. Both methods show equivalent decrease of E6 mRNA level by E6 siRNA in SiHa cells (see histograms). (Reproduced with permission from **ref. *11*.**)

94°C, 2 min; then 35 cycles of 94°C, 45 s; 55°C, 45 s and 72°C 1 min; followed by 72°C for 5 min.

3. For E7 cDNA amplification, the primers 5′-cggaattcatgcatggagatacacctacat-3′ and 5′-cgggaagcttatggtttctgagaacagatgg-3′ are used in the following thermal cycle: 47°C, 30 min; 94°C, 2 min; then 30 cycles of 94°C, 45 s; 58°C, 45 s, and 72°C 1 min; followed by 72°C for 5 min.

4. For p53cDNA, the primers are 5′-atggaggagccgcagtcagat-3′ and 5′-tcagtctgagtcaggc-ccttc-3′, and the thermal cycle is as follows: 47°C, 30 min; 94°C, 2 min; then 30 cycles of 94°C, 45 s; 58°C, 45 s, and 72°C 2 min; followed by 72°C for 5 min.

5. For semiquantitative RT-PCR, 100 ng total RNA is diluted 1:20 and 1:400. Semiquantitative RT-PCR gives similar results as Northern blotting (**Fig. 9B**). Real-time RT-PCR giving the same results as semiquantitative RT-PCR (**Fig. 10**).

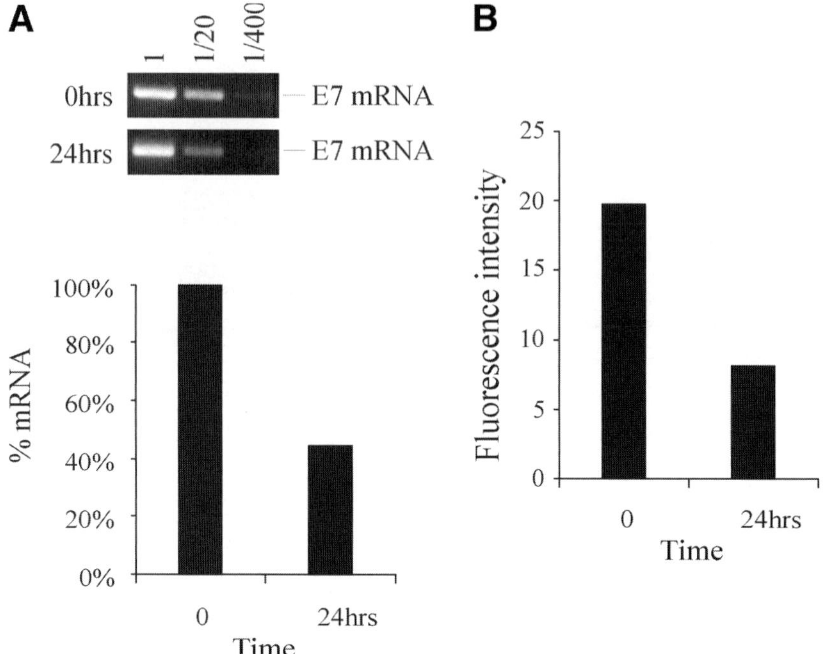

Fig. 10. Quantitation of E7 mRNA level by **(A)** semiquantitative RT-PCR and **(B)** real-time RT-PCR. Both methods show the similar knockdown E7 mRNA level by E7 siRNA in SiHa cells (see histograms).

Real-time RT-PCR was performed using a DNA engine opticon system with Quantitect™ SYBR green PCR mix and Reverse Blend transcriptase. The thermal cycle is as follows: 47°C, 30 min; 94°C, 15 min; then 35 cycles of 94°C, 45 s; 58°C, 45 s, and 72°C, 1 min; followed by 72°C, 10 min.

3.7. Detection of p53, p21, and pRb Expression

3.7.1. Preparation of Cell Lysates

1. Trypsinize the transfected cells and wash them once with PBS.
2. After cell counting, pellet the cells and lyse them in lysis buffer A (~50 µL per 1×10^6 cells) on ice for 30 min.
3. Dilute the samples 1:1 in 4X Laemli's buffer, boil for 5 min, and store at –20°C.

3.7.2. Detection of Protein Expression by Immunoblotting

1. Use the monoclonal antibody DO-1 at a dilution of 1:1000 to detect human p53 protein; use anti-p21 (SX118; diluted 1:500) and anti-pRb (G3-245; diluted 1:500) to detect p21 and pRb, respectively. Detect actin with rabbit polyclonal antibody.

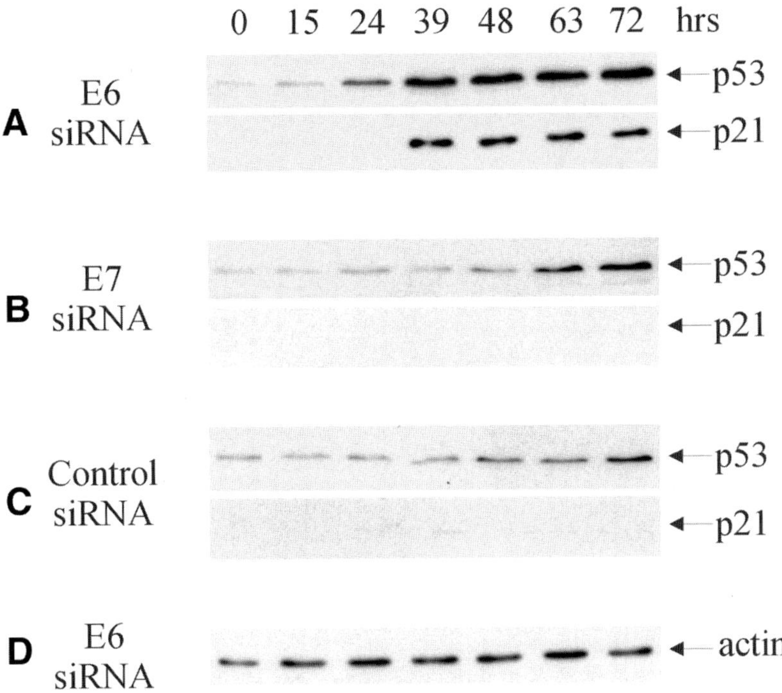

Fig. 11. Treatment with E6 siRNA induces activation of p53 protein. (**A**) SiHa cells treated with E6 siRNA show a marked increase in p53 protein accompanied by p21 expression. (**B**) E7 siRNA and (**C**) control siRNA fail to induce similar effects on p53 and p21 proteins. (**D**) Equivalent sample loading for immunoblots is confirmed by actin levels. (Reproduced with permission from **ref. *11*.**)

Use the HRP-conjugated rabbit antimouse (1:1000) and goat antirabbit (1:2000) antibodies as secondary antibodies, respectively.

2. Use approx 1×10^5 cells per sample for immunoblotting.
3. Resolve proteins by 15% SDS-PAGE and electroblot onto nitrocellulose membrane using standard methods (*27*).
4. Include prestained molecular weight markers as necessary.
5. After transferring, air-dry the membranes, wash them once with 1X TBS, and then block them with 1X TBS plus 1% block reagent for 1 h, followed by incubation in 0.5% block reagent in 1X TBS plus primary antibody for 1 h.
6. Wash the membranes twice using TBS with 0.1% Tween-20 (TBST) for 10 min each time and then twice with TBS with 0.5% block reagent for 10 min each time.
7. After incubating for 1 h with secondary antibody + 0.5% block reagent in TBS, wash the membranes four times with TBST for 15 min each time.
8. Visualization of bound antibodies is enhanced by chemiluminescence.

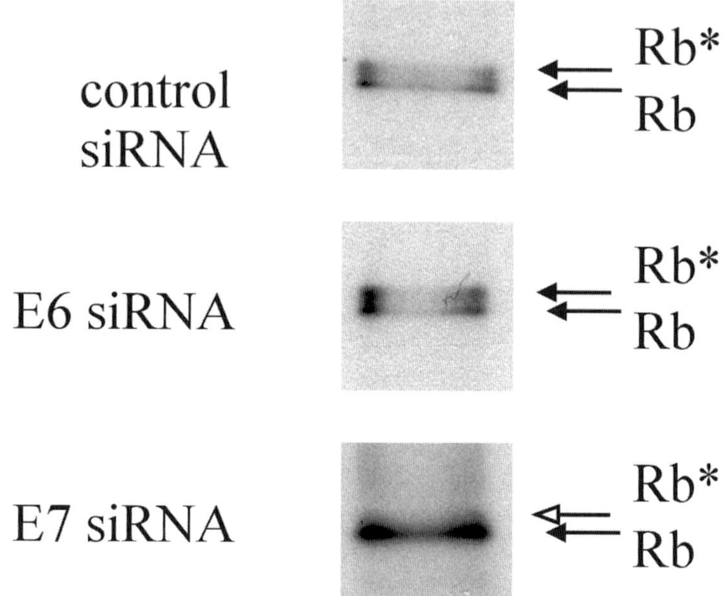

Fig. 12. E7 induces selective loss of hyperphosphorylated cellular pRb. Rb*, hyperphosphorylated pRb; Rb, hypophosphorylated pRb. (Reproduced with permission from **ref. *11*.**)

9. Treatment with E6 siRNA induces accumulation of p53 protein and expression of p21 protein **(Fig. 11A)**, whereas E7 siRNA treatment only slightly increases the p53 level at a very late stage and does not activate p21, a p53 target gene **(Fig. 11B)**. E7 siRNA also induces dephosphorylation of Rb **(Fig. 12)**.

3.7.3. Detection of p53 Expression by Immunostaining

1. Culture the CaSKi (or SiHa) cells on 13-mm cover slips in 24-well plates, and transfect according to **Subheading 3.3.2.**
2. Wash the transfected cells on the cover slips gently with PBS, fix for 10 min in 4% paraformaldehyde at room temperature, and wash with PBS.
3. Permealibize the fixed cells by incubating in 0.2% Triton X-100 in PBS for 15 min, and then wash with PBS five times for 5 min each time.
4. Dilute the anti-p53 antibody DO-1 1:100 in PBS + 3% BSA, apply to the cover slips, and incubate for 30 min.
5. After washing with PBS + 1% Triton X-100 three times for 5 min each times, incubate the cover slips with FITC-conjugated rabbit antimouse antibody (1:40) in PBS + 3% BSA for a further 30 min.
6. Wash the cover slips once with PBS + 1% Triton X-100, then stain the nuclei with Hoechst 33258 (1:2000) in PBS + 1% Triton X-100 for 5 min, and wash a further three times with PBS + 1% Triton X-100 for 5 min each time.

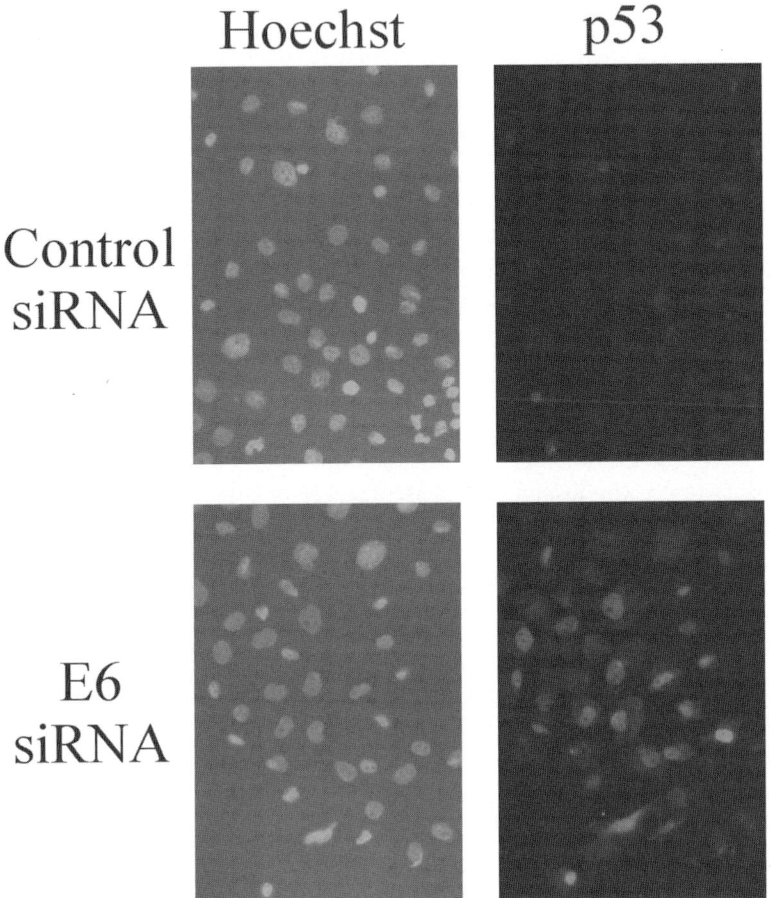

Fig. 13. E6 siRNA induces nuclear accumulation of p53 protein. Cells are fixed at 48 h after transfection and stained with Hoechst (blue) for nuclei and DO-1 (green) for p53. (Reproduced with permission from **ref. *11*.**)

7. Finally, mount the cover slips and check them under the microscope. As shown in **Fig. 13,** the E6 siRNA-treated cells show nuclear accumulation of p53 protein.

4. Notes

1. The design of siRNA is the major point for applying RNA interference in gene silencing and can be very time-consuming and financially costly. Dharmacon Research (USA) and MWG (Germany) have published a guide for choosing and designing siRNAs for individual target mRNAs. The details can be found at http://www.dharmacon.com. Other recommendations include starting 75 bp downstream from the start codon and choosing 21-nt gene sequences that start with an

AA dimer. These do not necessarily increase RNAi efficiency *(28,29)*. The 2-nt 3′ overhang can be 2′-deoxythymidine (dTdT) or ribouridine (UU). Both overhangs work efficiently, but the dTdT may enhance the stability of the siRNA duplex *(1)*. The siRNA does not need to be 2′ACE-protected either. In the present study all the siRNAs were protected by 5′-O-DMT-ON-2′-Fpmp, and the siRNA selection criteria were as follows: choose the conserved sequence from the middle of the mRNA (this is very important because these highly conserved regions are unlikely to have nucleotide variants among individual cell lines, since the siRNA is a highly specific silencing tool); avoid the AT- or GC-rich region if possible; and select a sequence with minimal secondary dimer and loop structures for both strands.

2. The transfection efficiency can also be easily checked by transfecting 3′ or 5′ fluorescence-labeled siRNA (MWG or Dharmacon). Note that the 3′ modified siRNA can lose the silencing function *(30–32)*.

3. Because PCR is a very sensitive detection method, even trace HPV DNA contamination will affect the E6 or E7 mRNA quantitation, and RT-PCR control without reverse transcriptase is an absolute necessity. If there is DNA contamination, the RNA samples can be treated with RNase-free DNase I and then reprecipitated using ethanol.

Acknowledgments

Work in the authors' laboratory is funded by Yorkshire Cancer Research. The HPV-16 E6 cDNA plasmid was kindly provided by Helena Browne, Division of Virology, Department of Pathology, University of Cambridge, UK. We also thank Bert Vogelstein for generously making available the HCT116 cell line.

References

1. Elbashir, S. M., Harborth, J., Lendeckel, W., Yalcin, A., Weber, K., and Tuschl, T. (2001) Duplexes of 21-nucleotide RNAs mediate RNA interference in cultured mammalian cells. *Nature* **411,** 428–429.

2. Harborth, J., Elbashir, S. M., Bechert, K., Tuschl, T., and Weber, K. (2001) Identification of essential genes in cultured mammalian cells using small interfering RNAs. *J. Cell Sci.* **114,** 4557–4565.

3. Elbashir, S. M., Harborth, J., Weber, K., and Tuschl, T. (2002) Analysis of gene function in somatic mammalian cells using small interfering RNAs. *Methods* **26,** 199–213.

4. Sharp, P. A. (2001) RNA interference—2001. *Genes Dev.* **15,** 485–490.

5. Brummelkamp, T. R., Bernards, R., and Agami, R. (2002) A system for stable expression of short interfering RNAs in mammalian cells. *Science* **296,** 550–553.

6. Lewis, D. L., Hagstrom, J. E., Loomis, A. G., Wolff, J. A., and Herweijer, H. (2002) Efficient delivery of siRNA for inhibition of gene expression in postnatal mice. *Nat. Genet.* **32,** 107–108.

7. Martinez, L. A., Naguibneva, I., Lehrmann, H., et al. (2002) Synthetic small inhibiting RNAs: efficient tools to inactivate oncogenic mutations and restore p53 pathways. *Proc. Natl. Acad. Sci. USA* **99,** 14849–14854.

8. Wang, B., Matsuoka, S., Carpenter, P. B., and Elledge, S. J. (2002) 53BP1, a mediator of the DNA damage checkpoint. *Science* **298,** 1435–1438.

9. Jiang, M. and Milner, J. (2003) Bcl-2 constitutively suppresses p53-dependent apoptosis in colorectal cancer cells. *Genes Dev.* **17,** 832–837.

10. Milner, J. (2003) RNA interference for treating cancers caused by viral infection. *Expert Opin. Biol. Ther.* **3,** 459–467.

11. Jiang, M. and Milner, J. (2002) Selective silencing of viral gene expression in HPV-positive human cervical carcinoma cells treated with siRNA, a primer of RNA interference. *Oncogene* **21,** 6041–6048.

12. Thompson, J. D. (2002) Applications of anti-sense and siRNAs during preclinical drug development. *Drug Discov. Today* **7,** 912–917.

13. Shuey, D. J., McCallus, D. E., and Giordano, T. (2002) RNAi: gene-silencing in therapeutic intervention. *Drug Discov. Today* **7,** 1040–1046.

14. Sullenger, B. A. and Gilboa, E. (2002) Emerging clinical applications of RNA. *Nature* **418,** 252–258.

15. Carmichael, G. G. (2002) Silencing viruses with RNA. *Nature* **418,** 379–380.

16. Cullen, B. R. (2002) RNA interference: antiviral defence and genetic tool. *Nat. Immunol.* **3,** 597–599.

17. Lee, N. S., Dohjima, T., Bauer, G., et al. (2002) Expression of small interfering RNAs targeted against HIV-1 rev transcripts in human cells. *Nat. Biotechnol.* **20,** 500–505.

18. Jacque, J. M., Triques, K., and Stevenson, M. (2002) Modulation of HIV-1 replication by RNA interference. *Nature* **418,** 435–438.

19. Novina, C. D., Murray, M. F., Dykxhoorn, D. M., et al. (2002) siRNA-directed inhibition of HIV-1 infection. *Nat. Med.* **8,** 681–686.

20. Capodici, J., Kariko, K., and Weissman, D. (2002) Inhibition of HIV-1 infection by small interfering RNA-mediated RNA interference. *J. Immunol.* **169,** 5196–5201.

21. Gitlin, L., Karelsky, S., and Andino, R. (2002) Short interfering RNA confers intracellular antiviral immunity in human cells. *Nature* **418,** 430–434.

22. Randall, G., Grakoui, A., and Rice, C. M. (2003) Clearance of replicating hepatitis C virus replicon RNAs in cell culture by small interfering RNAs. *Proc. Natl. Acad. Sci. USA* **100,** 235–240.

23. Wilson, J. A., Jayasena, S., Khvorova, A., et al. (2003) RNA interference blocks gene expression and RNA synthesis from hepatitis C replicons propagated in human liver cells. *Proc. Natl. Acad. Sci. USA* **100,** 2783–2788.

24. Hamasaki, K., Nakao, K., Matsumoto, K., Ichikawa, T., Ishikawa, H., and Eguchi, K. (2003) Short interfering RNA-directed inhibition of hepatitis B virus replication. *FEBS Lett.* **543,** 51–54.

25. Ge, Q., McManus, M. T., Nguyen, T., et al. (2003) RNA interference of influenza virus production by directly targeting mRNA for degradation and indirectly inhibiting all viral RNA transcription. *Proc. Natl. Acad. Sci. USA* **100,** 2718–2723.

26. Ausubel, F. M., Brent, R., Kingston, R. E., et al. (eds.) (1995) *Current Protocols in Molecular Biology.* John Wiley & Sons, New York.

27. Harlow, E. and Lane, D. (eds.) (1988) *Antibodies: A Laboratory Manual.* Cold Spring Harbor Laboratory Press, Cold Spring Harbor, NY.

28. Holen, T., Amarzguioui, M., Wiiger, M. T., Babaie, E., and Prydz, H. (2002) Positional effects of short interfering RNAs targeting the human coagulation trigger tissue factor. *Nucleic Acids Res.* **30,** 1757–1766.
29. Kawasaki, H., Suyama, E., Iyo, M., and Taira, K. (2003) siRNAs generated by recombinant human Dicer induce specific and significant but target site-independent gene silencing in human cells. *Nucleic Acids Res.* **31,** 981–987.
30. Song, E., Lee, S. K., Wang, J., et al. (2003) RNA interference targeting Fas protects mice from fulminant hepatitis. *Nat. Med.* **9,** 347–351.
31. Amarzguioui, M., Holen, T., Babaie, E., and Prydz, H. (2003) Tolerance for mutations and chemical modifications in a siRNA. *Nucleic Acids Res.* **31,** 589–595.
32. Hamada, M., Ohtsuka, T., Kawaida, R., et al. (2002) Effects on RNA interference in gene expression (RNAi) in cultured mammalian cells of mismatches and the introduction of chemical modifications at the 3′-ends of siRNAs. *Antisense Nucleic Acid Drug Dev.* **12,** 301–309.

IX

COMPUTATION/SYSTEMS BIOLOGY OF VIRUSES

28

Design of a Herpes Simplex Virus Type 2 Long Oligonucleotide-Based Microarray

Global Analysis of HSV-2 Transcript Abundance During Productive Infection

J. S. Aguilar, Peter Ghazal, and Edward K. Wagner

Summary

The design and construction of a long (75-mer) oligonucleotide-based DNA microarray for herpes simplex virus type 2 transcripts is described. This array is utilized to generate an analysis of HSV-2 transcript abundance as a function of conditions of infection of human cells, and global patterns of HSV-2 transcript abundance are compared with those for HSV-1. General similarities in patterns along with notable differences in specific details are noted. These results reveal a marked conservation in the program of gene activity between phenotypically diverged strains.

Key Words: Herpes simplex virus type 2; human herpesvirus 2; oligonucleotide-based DNA microarrays; kinetics of transcript abundance; DNA replication inhibitor; global profiling of viral gene expression.

1. Introduction

1.1. HSV-2 Compared With HSV-1

Full DNA sequences are available for the two closely related herpes simplex viruses of humans, HSV-1 and HSV-2 (HHV1 and HHV2). HSV-1 is normally associated with primary infection of the lip, with latency being established in the trigeminal nerve ganglia, whereas HSV-2 is associated with genital infections and latency in the sciatic nerve ganglia (*1*). In humans both can be transmitted sexually, although this mode is generally associated with HSV-2. The initial infection normally takes place in epithelial tissue, from which virus can reach the innervating sensory neurons. Here, they can become latent, with their genomes forming an episome in the nuclei of the neurons. In this state, only the

From: *Methods in Molecular Biology, vol. 292: DNA Viruses: Methods and Protocols*
Edited by: P. M. Lieberman © Humana Press Inc., Totowa, NJ

LAT gene is abundantly transcribed, but no viral proteins have been detected. Reactivation can take place sporadically—usually as a consequence of stress—causing a new productive infection at the site of the primary infection.

Although both viruses are medically important, the vast bulk of basic research on the basic virology and molecular biology of HSV including patterns of viral gene expression during replication and latency has been carried out with HSV-1 *(2–4)*. Structurally, herpes virions consist of a double-stranded DNA genome located inside an icosahedral capsid. The capsid is surrounded by a protein-rich tegument, which, in turn, is covered by a membrane envelope. The virion binds and penetrates the cell by means of the interaction between viral membrane–glycoproteins and specific cellular receptors *(5,6)*. Once inside the cell, the capsid with the genome is transported to the nuclear membrane. The capsid delivers the genome inside the nucleus, where transcription and replication takes place.

Transcription takes place in a kinetic cascade. Initially, the immediate early (IE) genes are transcribed. IE transcription can occur in the absence of protein translation. These IE genes regulate the transcription of all viral genes. After IE, the early genes are transcribed. The E genes generally code for proteins needed for DNA replication. Late (L) genes transcription depends on DNA replication. The replication of some L genes can start before DNA replication takes place, but they reach their maximum levels only after DNA replication. These are the leaky late genes. The transcription of another set of late genes—the strict late genes—depends absolutely on the replication of viral DNA.

1.2. Differences Between HSV-2 and HSV-1

Despite their high degree of genomic identity (>80%) *(7,8)*, their significantly different pathology in humans is reflected in differences in the behavior of the virus in animal systems *(9,10)*. An important problem in herpes virology is to understand the molecular basis leading to the different pathologies between HSV-1 and HSV-2. Differences in viral functions as well as in the cellular functions affected by each type of virus could explain these different pathologies. Thus, some viral functions are different in HSV-1 and HSV-2. For instance, whereas in HSV-1 glycoprotein C is the major viral function responsible for virus attachment to cells and glycoprotein B mediates penetration, in HSV-2 glycoprotein B is the major protein involved in both binding and penetration *(11)*. Also, the viral host shutoff activity is much stronger in HSV-2 than in HSV-1 *(12)*. Furthermore, the protein kinase activity associated with ribonucleotide reductase 1 has different ATP requirements in HSV-1 and HSV-2 *(13)*. Cellular functions are also affected differently in HSV-1 and HSV-2 infections. Whereas HSV-1 increases the levels of transcription factors c-Jun *(14)* and NF-κB *(15)*, HSV-2 increases c-fos transcription *(16)*. The regulation of nitric oxide

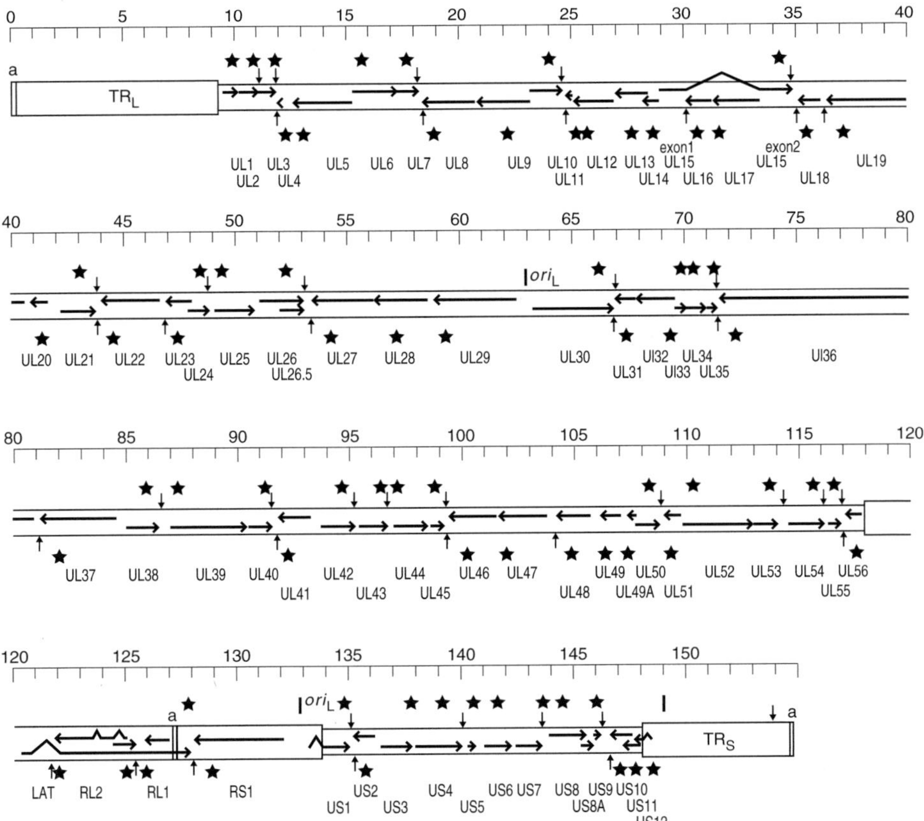

Fig. 1. The HSV-2 transcription map determined from the sequence of the HG52 strain. The major features of the viral genome, the open translational reading frames, polyadenylation signals (vertical arrows), and locations of specific 75-base oligonucleotide probes are shown. (Adapted from **ref. 8**.)

production in epithelial cells is also different, with HSV-2 causing upregulation and HSV-1 downregulation *(17)*. Therefore, to catalog the differences between HSV-1 and HSV-2 and to understand their molecular basis may explain their different pathologies and provide and important knowledge of herpes virology.

1.3. The HSV-2 Sequence-Based Transcription Map

Based on sequence comparisons and specific experimental studies, the general transcription map of HSV-2 (**Fig. 1**) is very similar to that of HSV-1, as are the kinetics of a number of these transcripts (**Table 1**). To date, however, the kinetic characteristics of many HSV-2 genes have not been described, and it is

Table 1
Experimentally Characterized HSV-2 Transcripts

Name	Kinetics	Method Ref.	Protein
RL2	IE	Northern *34*	ICP0
UL3	E	Northern *35*	Nuclear phosphoprotein
UL4	L	Western *36*	Virion(?)
UL7	L	Western *37*	Cleavage/packaging
UL13	L	Western *38*	Protein kinase
UL14	L	Western *39*	Capsid maturation
UL16	L	Western *40*	Capsid maturation
UL17	L	Western *41*	Capsid protein
UL24	L	Western *42*	Capsid protein
UL27	E	Western *43*	Glycoprotein B
UL34	L	Western *44*	Envelope
UL39	E or IE	Cycloheximide reversal *45*	Ribonucleotide Reductase
UL42	E	Western *46*	DNA pol accessory
UL44	E	Western *43*	Glycoprotein C
UL45	E	Western *43*	Tegumen/envelope
UL46	L	Western *46*	Virion tegument phosphoprotein
UL48	L	Western *46*	alpha trans-inducing factor
UL49	E	Western *47*	Tegument (VP22)
UL54	IE	Northern *34*	ICP27
UL55	L	Western *48*	Virion assembly or maturation
UL56	L	Western *49*	Vesicular trafficking
US1	IE	Northern *34*	ICP22
US2	L	Western *50*	Virion protein
US6	E	Western *43*	Glycoprotein D
US12	IE	Northern *34*	ICP47
RS1	IE	Northern *34*	ICP4

IE, immediate early; E, early; L, late.

important to determine the kinetic characteristic of HSV-2 genes and compare them with their counterparts in HSV-1 as a basis for a fuller understanding of the differences between HSV-1 and HSV-2.

Oligonucleotide-based DNA microarrays combine a number of attractive features that make them especially valuable for the analysis of both genetic variation and patterns of gene expression by large DNA-containing viruses *(4,18)*. For this reason, we developed such a microarray system specific for HSV-1 and applied it to a global analysis of HSV-1 transcript abundances during productive infection, and following infection with viruses carrying defined lesions in regulatory genes. This work has been described in **refs.** *4* and *19–21*.

In the present review we describe some of the preliminary results obtained using a oligonucleotide-based array specific for HSV-2. We designed this chip using the same strategy that we have previously used to develop a microarray for HSV-1. Utilizing this newly developed microarray, we have shown that there are some notable differences in the course of transcript accumulation between HSV-1 and HSV-2, even though the general kinetic characteristics are very similar.

2. Materials

1. Oligonucleotides: oligonucleotides (75-mers) were synthesized by Illumina Biologicals (San Diego, CA). Each (25 µmol) was supplied as a lyophilized powder.
2. Chip printing: glass slides (amino-silane-coated) were spotted with oligonucleotides using a GMS417 Arrayer (Genetic Micro Systems), at the facility at the Scottish Genome Centre, Edinburgh, Scotland. Details of printing are described in refs. *18, 19,* and *22.*
3. Chip hybridization, scanning, and quantitation.
   a. Hybridization was carried out under conditions described previously *(21)* in a GeneMachines commercial hybridization chamber.
   b. Hybridization to microarrays was for 16–18 h at 68°C in 15 µL of 5X standard saline citrate (SSC)-0.2% sodium dodecyl sulfate (SDS) under a glass cover slip.
   c. After hybridization, the microarray was washed sequentially in 250 mL of 1X SSC-0.2% SDS, 0.1X SSC-0.2% SDS, and 0.1X SSC for 5 min at room temperature and spun dry in a low-speed centrifuge.
   d. Microarrays were scanned by using a confocal system (Scan Array 4000, General Scanning) and quantitated using Quantarray (General Scanning) or Array Vision (Genicon Science).
4. Virus and cells.
   a. Stocks of HSV-2 (strain HG-52) were prepared in rabbit skin cells.
   b. All procedures have been described in numerous previous publications *(20)*; cells were cultured in Eagle's minimum essential medium (EMEM) containing 5% bovine serum (BSA), 100 µg/mL streptomycin, and 100 U/mL penicillin.
   c. Analysis of gene transcription was carried out in human foreskin fibroblasts (HFFs).
   d. These cells were cultured in EMEM containing 10% fetal calf serum, 100 µg/mL streptomycin, 100 U/mL penicillin, and 100 µ*M* G418.
   e. The cultures were maintained at 37°C in 5% CO_2.
5. Isolation of poly(A)-containing RNA from infected cells.
   a. Culture dishes of confluent HFF cells (~ 10 million cells/plate) were infected at a multiplicity of infection of 5 PFU per cell.

b. The media were removed and the cells washed once with saline.

c. Virus adsorption was carried out for 30 min in 6 mL of phosphate-buffered saline (PBS)-10% glucose and then replaced with fibroblast medium.

d. Infections were continued for different time intervals at 37°C under 5% CO_2.

e. During 100-μg/mL cycloheximide and 300-μg/mL phosphonoacetic acid (PAA) treatments, the drugs were added to the cells 30 min prior to adsorption.

f. The drugs were present during the adsorption step and during the postadsorption period. This postadsorption period was 4 h for cycloheximide and 6 h for PAA.

g. At the desired times after infection, medium was removed, and the cells were washed with saline.

h. Total RNA was extracted with 4 mL of TRIzol (Invitrogen) per dish, following the manufactures procedure.

i. Poly(A)-selected RNA was prepared from total RNA using an Oligotex (Qiagen) kit. The recovery of poly(A)-selected RNA was between 1 and 2% of the total RNA.

6. Synthesis of the fluorescent-labeled cDNA.

a. Fluorescent labeled cDNA was prepared by reverse transcription; we typically used 1 μg of poly(A) selected RNA for each hybridization.

b. The synthesis was carried out utilizing SuperSript Reverse Transcriptase (Invitrogen) and random hexamers as primers.

c. The nucleotide concentrations were 0.5 mM dGTP, dATP, and dTTP, and 0.3 M dCTP and 0.1 mM fluorescent nucleotide Cy3- and Cy5-dCTP (Perkin Elmer).

d. The labeling reaction was continued for 2 h at 42 C.

e. At the end of the reaction, RNA was digested with 0.05 mM NaOH at 70°C for 10 min.

f. After neutralization, unincorporated nucleotides were removed by ultrafiltration and two washes through Microcon YM-30 filters (Amicon).

g. Fluorescent labeled cDNA was denatured for 2 min at 99°C and incubated for 20–30 min.

7. Nick translation.

a. For nick translation we used cloned HSV-2 DNA in pBR322 plasmids (provided by Nigel Fraser).

b. Each labeling reaction utilized 2 μg of plasmid DNA that was nicked by treatment with 2 mU of DNase I (Invitrogen, inactivated by heating at 70°C for 20 min) in 15 μL of reaction mixture at 15°C for 5 min.

c. The reaction mixture was washed by ultrafiltration through YM-30 filters.

d. The nicked DNA was labeled by incubation for 2 h at 15°C with 10 U of *E. coli* DNA pol I (Invitrogen) in 50 μL of reaction mixture containing 20 μM of dATP, dTTP, and dGTP and Cy3- or Cy5-dCTP.

e. The unincorporated nucleotides were removed as for the labeling of cDNA.

f. The nick-translated DNA was heated for 1 min at 65°C and then centrifuged at 14,000g for 2 min to remove solid impurities.

g. The purified labeled DNA was denatured at 99°C for 5 min, and kept cold on ice for another 5 min.

    h. Hybridization, washes, and data analysis were carried out as in the case of cDNA.

8. Data analysis.

    a. Each probe on the chip is printed three times, and hybridization values for that probe on the chip are the median background-subtracted value.

    b. All data for a given experimental condition are based on a minimum of three replicate individual chip hybridizations.

    c. The median values from those experiments were pooled to calculate the final median values displayed in the tables.

    d. To compare data from replicate experiments, the 75th percentile rank for the total viral hybridization was calculated. One experiment was arbitrarily chosen as the reference, and the 75th percentile values of all other determinations were adjusted to this value by appropriate factoring. In this way, chips belonging to each experimental group are scaled accordingly *(2)*.

9. Expression database: the original data are available (accession no. GXE-0000X Username: Reviewer; Password: HSV2; in the MIAME-compliant GTI expression database GPX^db http:/www.gti.ed.ac.uk).

3. Methods

3.1. Design of Oligonucleotides for the HSV-2 Array and Validation by Testing With Nick-Translated DNA

We have fully described the method of design of the 75-base oligomers we have used to construct our HSV-1 chip in several previous reports *(4,19,21)*. Briefly, each HSV gene was surveyed near its 3′-end to find one to three 75-base regions of low internal structure and a G+C content as close to the average value for the genome as possible; this was to ensure efficient labeling of cDNA using oligo-dT primers and to preclude self-annealing. In the case of partially overlapping transcripts sharing the same polyadenylation site, oligonucleotides specific for the unique 5′ transcripts were also chosen. The HSV-1 array contains 67 unique sets of HSV-1 probes, of which 43 represent transcripts from single genes. Another 11 probes detect sets of viral transcripts within a single kinetic class that share polyadenylation signals. Thus, a total of 54 probes detect either individual or overlapping transcript sets whose kinetics can be unambiguously assigned, and an additional 13 probes detect overlapping transcripts of different kinetic classes.

For the HSV-2 chip, a total of 97 probes were synthesized using information obtained for the HG-52 strain in Genbank and printed on chips. These were then tested by hybridization to dye-labeled, nick-translated cloned HSV-2 fragments covering the entire genome (a gift of Nigel Fraser). The 75-mer probes, which showed the highest hybridization values and little or no cross-hybridization with nonhomologous DNA fragments, are listed in **Table 2** both by position on the genome and the G+C content. The HSV-2 chip includes a number

Table 2
Oligonucleotide Probes Specific for HSV-2 Transcripts

Name	Location (5′-3′) HG52	G+C	Hyb signal	HSV-1 equivalent
UL1	9597–9665	56	s	U1
UL2	10861–10935	60	m	—
UL3	11693–11767	56	s	U3
UL4	12022–11948	57	m	U4-5′
UL5	12849–12775	59	s	U4/5
UL6	16698–16772	55	m	U6/7
UL7	18080–18154	55	m	U8/9
UL8	18587–18513	71	m	U8-5′
UL9	23010–22936	61	m	—
UL10	24529–24603	68	m	U10
UL11	25027–24953	57	w	—
UL12	25554–25480	57	m	—
UL13	27470–27396	61	m	—
UL14	28438–28364	63	s	—
UL15	34569–34643	59	s	U15
UL16	31002–30928	60	s	U16/17
UL17	31557–31483	61	s	—
UL18	35963–35889	67	m	—
UL19	38721–38647	61	m	U19-5′
UL20	41451–41377	60	m	—
UL21	43172–43246	73	m	U21
UL22	44241–44167	60	m	U22
UL23	47730–47656	65	m°	U23
UL24	48157–48231	57	m	U24
UL25	49037–49111	61	s	U25
UL26	51710–51784	63	s	—
UL27	53565–53491	61	s	U27-5′
UL28	56541–56467	56	s	U27/8
UL29	59885–59811	59	s	U29
UL30	64574–64648	56	s	U30
UL31	67098–67024	59	m	—
UL32	69149–69075	61	s	—
UL33	69958–70022	65	m	—
UL34	70489–70563	71	s	—
UL35	71281–71355	68	s	U35
UL36	76005–75931	67	s	U36
UL37	81567–81493	63	s	U37
UL38	85711–85715	71	s	U38
UL39	89303–89377	59	m	U39-5′
UL40	91480–91554	61	s	U39/40

Table 2
(Continued)

Name	Location (5′-3′) HG52	G+C	Hyb signal	HSV-1 equivalent
UL41	92074–92000	55	s	U41
UL42	93858–93932	64	s	U42
UL43	96386–96460	63	m	U43
UL44	98347–98421	68	s	U44-5′
UL45	99020–99094	57	w	U44/45
UL46	99447–99373	61	s	U46/47
UL47	101832–101758	65	m	—
UL48	105596–105522	61	m	U48
UL49	106499–106425	63	s	U49
UL49.5	107597–107523	56	s	—
UL50	108738–108812	64	s	U50
UL51	109506–109432	64	m	U51
UL52	112250–112354	59	m	U52-5′
UL53	113964–114038	63	m	U52/53
UL54	116053–116127	49	m	U54[f]
UL55	116692–116766	57	w	U55
UL56	117497–117423	63	m	U56
LAT 5′-end	119544–119618	71	m	RLAT-5′
LAT intron	120761–120835	63	m	RLAT-I
ICP0	124126–124052	57	s	RICP0
ICP34.5	125697–125622	67	m	RICP34.5
ORFOP	125090–125164	61	m	ROP
LAT-3′-end	127878–127940	73	m	RLAT-3′
ICP4	131370–131296	71	m	RICP4
US1/ICP22	134549–134623	64	w	R/S22
US2	135679–135605	60	m	US2
US3	137305–137379	61	s	US3-5′
US4	139752–139826	61	s	US3/4
US5	140507–140581	64	m	US-5′
US6	141347–141421	56	m	—
US7	143348–143423	65	m	US5/6/7
US8	144493–144567	59	m	US8-5′
US8a	145339–145413	57	w	N/A
US9	146062–146136	60	m	US8/9
US10	147063–146989	67	m	US10/11/12
US11	147699–147625	67	w	—
US12	148035–147961	55	m	
—	—	—	—	—

s, strong; m, moderate; w, weak; *see* **Fig. 2.**

of probes for transcripts that were not synthesized in the HSV-1 chip; these include the Us8a gene (not present in HSV-1) and a number of genes expressed as multiply overlapping transcripts such as the UL11–13 and UL31-36 regions. Probes (and genes) that are equivalent in specificity and location to HSV are also indicated in **Table 2.**

An example of the hybridization of nick-translated cloned DNA is given in **Fig. 2**, in which the hybridization of *Hin*dIII fragments B and L to the HSV-2 chip are shown. It can be seen that hybridization values can be generally grouped into strong, moderate, and weak values. It is not known whether this reflects differential hybridization efficiencies of the oligonucleotides or differential nick-translation efficiency of the DNA testers, or both. These values are also included in **Table 2.**

3.2. Grouping HSV-2 Transcripts Into Kinetic Classes

The ability to group HSV-1 transcripts into immediate-early, early, and late classes is readily accomplished by examination of those transcripts abundantly expressed in the absence of *de novo* protein synthesis, such as during a cycloheximide blockage during infection, and those abundantly expressed in the absence of viral DNA replication. This latter condition is best accomplished by analysis of RNA levels expressed following infection with tight mutant of one of the genes essential for viral DNA replication at several times, compared with control levels with a *wt* infection. We have carried out some careful control experiments using the HSV-1 chip to show that generally equivalent results can be attained using a DNA synthesis inhibitor, such as PAA or acyclovir.

In the present study, we isolated RNA from both HSV-1- and HSV-2-infected human fibroblasts 4 h after infection in the presence of 100 μg/mL cycloheximide and then synthesized dye-labeled cDNA. It has long been known that HSV-2 transcripts can be detected with some efficiency using homologous HSV-1 probes in Northern blots *(23,24)*, and we hybridized the dye-labeled cDNA to HSV-1 chips to determine whether there was efficient cross-hybridization under the conditions used. As shown in **Table 3,** the four putative HSV-2 IE transcripts, ICP4, -0, -22, -27, and -47, all hybridized efficiently to HSV-1 chips. The transcript encoding the large subunit of ribonucleotide reductase hybridized to barely significant levels under these conditions, and since, as shown just below, this HSV-2 transcript is abundant in early RNA hybridized to HSV-2 chips, we conclude that (as is the case in HSV-1 infections) it is not an abundant IE transcript in HSV-2 infections.

We next compared the patterns of expression of those HSV-2 transcripts whose probes were essentially equivalent in position and specificity to their HSV-1 homologs under conditions of PAA blockage of viral DNA replication.

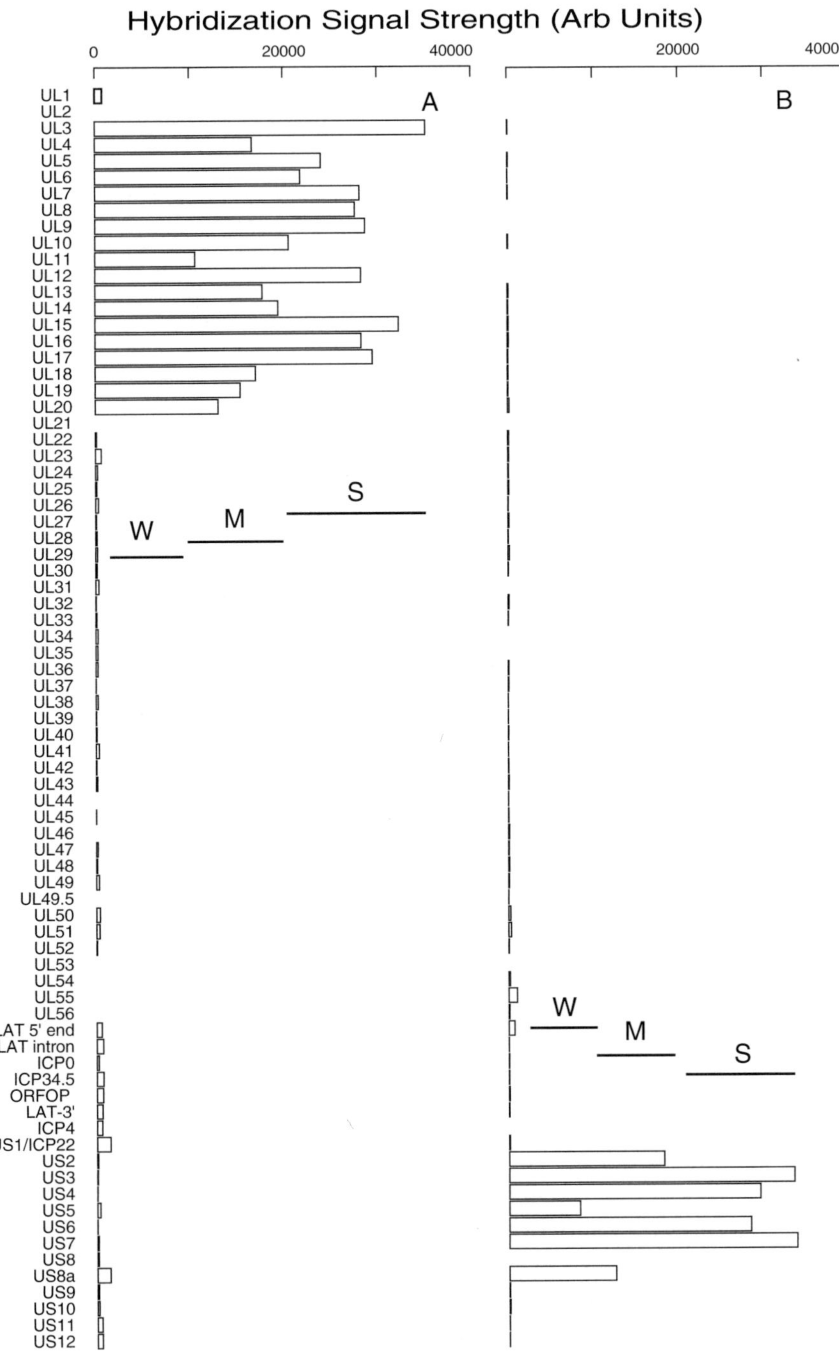

Fig. 2. Representative hybridization of the HSV-2 microarray to nick-translated cloned HSV-2 DNA fragments.

Table 3
Expression of HSV-2 Immediate-Early and Selected Early Transcripts in the Presence of 100 µg/mL Cycloheximide[a]

Gene	HSV-1		HSV-2	
	Median[b]	SD	Median[b]	SD
ICP27	27,100[c]	2100	39,900	11,600
ICP0	60,900	38,300	47,200	4100
ICP4	48,900	11,100	86,800	30,100
ICP27	47,100	19,600	42,800	10,400
ICP47	36,000	25,000	36,200	18,200
UL39-5′	800	400	4400	1400
UL39/40	4500	5000	6300	300
UL23	3000	2100	1100	1000
UL50	2200	2200	2600	2200

[a] Hybridization was done on HSV-1-specific DNA microarrays.
[b] Based on two separate experiments.
[c] Arbitrary units.

For these experiments, RNA was isolated at 6 h after infection in the presence of 200 µg/mL PAA to inhibit DNA replication. Hybridization of HSV-2 RNA to the HSV-2 chip, compared with that seen for HSV-1 RNA to the HSV-1 chip, is shown in **Fig. 3.** Transcripts are grouped by the kinetic class of the HSV-1 representative in **Fig. 3,** and it is evident that there is a general correspondence in the overall patterns of abundance. This confirms the general correspondence of kinetics between HSV-1 and HSV-2 transcripts, but notable exceptions can be seen and are examined in more detail in the next section, **Subheading 3.3.**

3.3. Analysis of Specific Differences Between HSV-1 and HSV-2 Transcript Abundance Patterns

Human fibroblasts were infected with HSV-2 and RNA isolated at 2, 4, 8, and 16 h following infection. Such RNA was then used as a template for synthesis of dye-labeled cDNA, and this was hybridized to HSV-2 chips. Median, normalized hybridization values for the various transcripts arranged according to their position on the HSV-2 genome beginning with the left end of the long unique region are shown in **Table 4.**

There is a general correspondence in kinetic class between HSV-1 and HSV-2, but detailed comparisons of representative IE, early, and late transcripts reveal some notable differences in timing and relative abundance (compare data with those in **ref. 4**). Perhaps the most striking feature of the HSV-2 transcription program is a slower rate of increase of early and late transcript abundance,

Table 4
Differential Expression of HSV-2 Transcripts As a Function of Time After Infection[a]

	Time after infection (h)											
Transcript	2	SD	4	SD	6	SD	6-PAA[b]	SD	8	SD	16	SD
UL1	1200	700	7900	3900	13,300	7500	7000	800	7200	6600	5200	13900
UL2	200	700	500	2200	27,900	21500	14,800	300	2000	23700	8400	14900
UL3	1200	300	4100	1200	10,000	2800	3400	1100	6800	1500	3100	2700
UL4	600	100	1700	900	3700	1500	3200	200	1800	800	900	5800
UL5	1100	500	5300	2100	7300	4200	8800	1400	3000	2800	2700	3000
UL6	900	100	5700	1200	15,300	4200	4500	700	19,200	13200	10,500	5200
UL7	700	200	3400	1600	4100	900	1900	200	4000	1500	2300	10800
UL8	1600	400	5900	3400	14,500	5500	17,000	4100	4200	5100	2700	7100
UL9	300	100	3500	1500	7300	2600	4100	700	10,400	5200	3800	1200
UL10	400	100	2100	900	6900	2000	1000	100	2900	1800	1100	11500
UL11	1000	1000	9000	13400	21,800	9700	16,000	1100	24,900	15900	4200	3500
UL12	2200	1300	31,700	12000	81,500	33900	42,300	15600	57,600	23700	12,000	4900
UL13	2100	600	10,800	1700	15,900	1200	13,000	1500	10,600	1500	6400	2600
UL14	1700	1200	7500	2800	22,700	11300	24,700	2600	7000	6700	6600	3200
UL15	900	200	3900	2300	8200	2900	5100	500	7200	1100	2800	2200
UL16	2100	200	12,100	4200	27,700	6600	16,500	1100	16,400	8500	6200	2800
UL17	1800	300	8000	2100	16,300	3900	9700	2200	10,200	4700	4900	1900
UL18	900	600	22,900	8000	52,800	25700	14,500	1600	50,300	18700	13,700	6400
UL19	1800	200	19,000	6300	38,300	3200	12,200	1800	42,600	13200	15,200	3900
UL20	700	500	5900	3800	8600	2200	8500	1400	8700	2000	4000	1600
UL21	600	300	2500	1200	17,100	10200	2900	500	14,400	7200	4300	2000
UL22	800	100	8000	1300	20,400	3000	8600	600	14,000	5400	2900	2900

(Table continues)

Table 4
Differential Expression of HSV-2 Transcripts As a Function of Time After Infection[a] (Continued)

	Time after infection (h)											
Transcript	2	SD	4	SD	6	SD	6-PAA[b]	SD	8	SD	16	SD
UL23	3100	400	13,000	8900	10,800	6100	19,800	2500	4000	900	1600	6900
UL24	2400	500	8100	3100	22,800	9600	2800	600	14,700	4800	5600	1500
UL25	2000	600	9300	3000	20,100	9700	6200	700	14,300	8600	6500	5800
UL26	1800	400	19,700	7200	37,000	8300	9300	500	31,900	17300	14,400	4800
UL27	2900	1500	26,800	9000	46,900	10200	17,000	3000	44,100	13600	12,500	3100
UL28	1000	300	6000	1600	10,900	6100	6400	1300	11,000	2300	3200	2000
UL29	9200	4700	49,200	14600	64,200	22000	42,800	5100	55,100	5900	14,000	4000
UL30	3000	1100	19,200	5400	40,800	3300	20,300	2400	32,100	10000	9700	6000
UL31	1000	200	9200	5200	16,600	6200	1100	900	20,600	15300	5900	4400
UL32	700	700	8400	1400	8400	3300	4300	1300	3800	1600	3200	1200
UL33	1400	200	23,800	6100	37,600	8500	11,900	600	48,800	11700	9700	4800
UL34	1100	600	13,200	6100	22,500	5000	5100	500	8400	7600	4100	8300
UL35	900	100	4900	300	10,600	5500	2200	300	7700	5800	3500	1300
UL36	500	200	1700	600	8300	4100	1500	300	7100	4600	2900	3000
UL37	2100	300	15,000	2900	24,000	2400	15,000	1400	17,800	3800	8600	1600
UL38	700	100	6600	2800	19,500	5300	4700	1100	21,500	6000	6100	2000
UL39	3800	700	20,900	9200	41,600	10,500	20,300	3000	19,000	9200	7500	2500
UL40	3700	3100	25,500	5500	35,000	6900	22,800	300	36,000	5800	7700	4800
UL41	800	300	5800	2000	14,200	6600	3000	200	6900	1000	2700	1900
UL42	800	100	6400	800	14,200	3700	6900	1000	8700	5700	3400	8500
UL43	500	600	2700	2500	2700	2800	1200	400	2100	1500	1200	700
UL44	200	100	200	800	1200	600	600	100	1600	1400	1000	500
UL45	0	0	0	1100	100	200	0	0	1000	1100	600	14600
UL46	200	100	300	1300	700	900	400	100	2100	1600	1000	600
UL47	1100	400	4100	800	18,500	11500	900	800	23,000	5700	17,600	9500
UL48	2800	1400	41,500	7000	84,500	24100	29,400	3600	64,900	13600	28,600	8900
UL49	1300	400	8200	11700	29,600	3000	600	4600	47,200	18000	22,600	8000

UL49.5	2000	1500	17,400	15000	67,200	26,900	29,500	5300	34,400	20900	20,400	9600
UL50	5300	1300	25,900	8900	36,800	6500	30,100	8000	21,100	12200	11,500	3700
UL51	1400	400	6200	5000	23,900	4300	4200	800	38,700	9300	23,600	11100
UL52	0	0	400	800	400	600	1000	600	800	1600	500	500
UL53	500	100	2600	3400	9600	5900	14,300	2100	2100	1000	1000	4600
UL54	8000	5500	18,800	10100	28,800	2900	24,900	3800	2800	4200	1100	1200
UL55	100	200	1800	1100	1000	1800	2600	400	800	1900	400	600
UL56	200	100	1400	700	3900	1500	2200	300	2800	1600	5100	5400
LAT 5′-end	700	100	400	2200	1900	3900	3100	1200	1900	1900	900	1500
LAT intron	100	200	200	2200	800	800	600	300	1000	1800	600	800
ICP0	12,200	6600	8800	5500	30,600	6600	11,400	1000	24,000	17,900	13,200	8400
ORFOP	100	0	100	2500	0	100	0	0	1100	1400	600	9300
ICP34.5	700	100	3500	8000	6700	3500	6500	1500	5100	2600	2300	2100
LAT-3′-end	700	100	600	1200	5300	2900	400	200	3600	1600	3500	2600
ICP4	500	1600	500	700	2900	1800	2200	2300	2600	3800	1400	800
US1/ICP22	3800	1000	21,800	4200	52,200	9400	27,800	1700	19,900	22,500	6200	6700
US2	1100	400	1900	2600	3600	2200	2000	200	3200	1600	1600	3100
US3	1400	400	12,100	2400	27,000	7100	5900	700	17,800	3900	8600	3600
US4	500	500	4500	3000	5800	2200	7600	500	3300	900	1700	1400
US5	600	200	1400	2500	2500	1400	2200	500	3600	1500	2300	1800
US6	400	100	500	3400	2600	1300	900	100	1900	1300	800	500
US7	1900	600	9400	6400	31,100	9900	10,200	1300	10,100	6800	2300	5300
US8	700	400	6800	3200	12,500	13,500	11,100	1900	6500	4900	3100	2000
US8a	3200	800	30,800	4200	47,900	900	27,300	2100	50,500	4500	18,600	9800
US9	400	100	4600	2000	11,700	3700	4400	700	6300	3800	2200	1400
US10	0	1100	100	3800	1600	2400	2000	500	3000	4300	2200	2700
US11	1600	700	9200	3800	22,100	4700	5600	800	2700	10,600	7400	5800
US12	200	100	200	700	600	700	100	100	1600	1600	800	300

[a] Data are median normalized values of at least three separate experiments. All data were normalized as described, and scans were carried out at the same laser power. Units are arbitrary.

[b] Phosphonoacetic acid (PAA) was present during virus adsorption and during incubation following infection at 200 µg/mL.

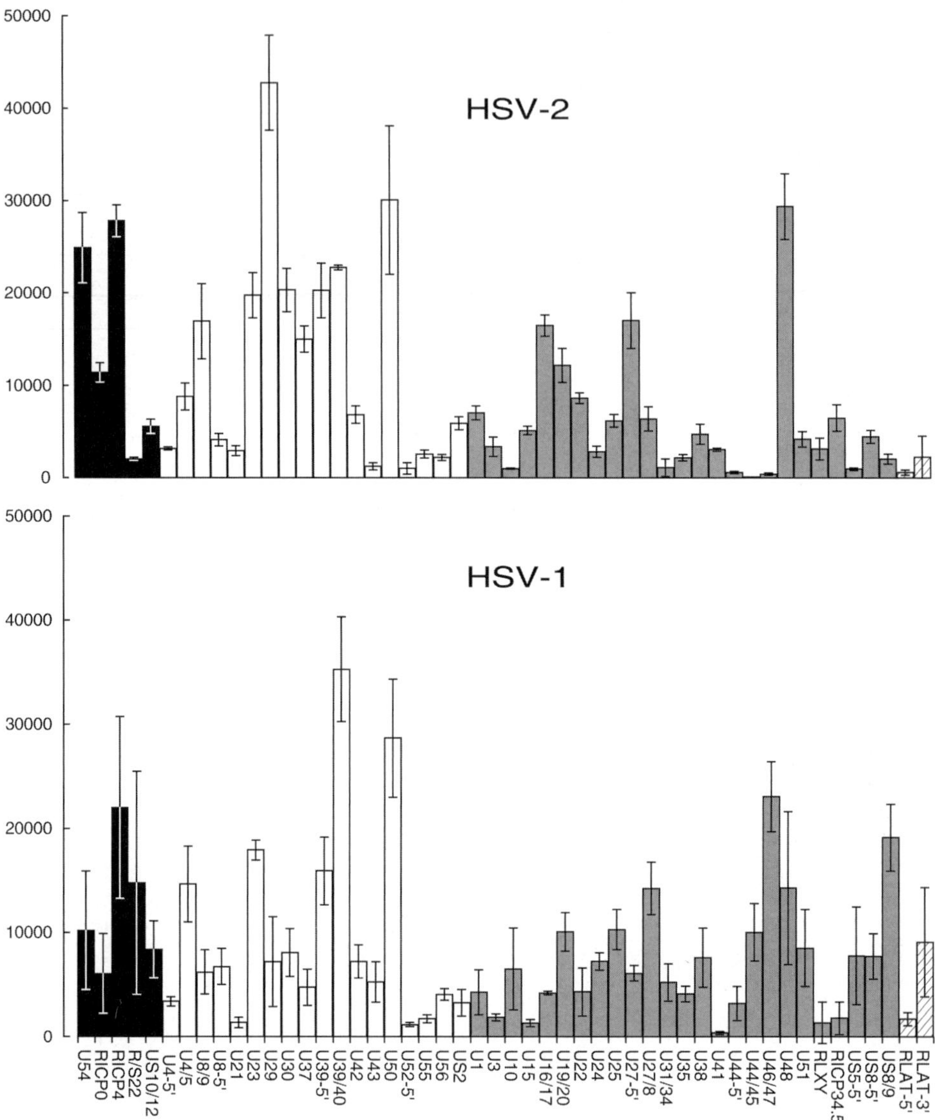

Fig. 3. Comparison of HSV-2 and HSV-1 transcript abundance at 6 h post infection in the presence of 200 µg/mL PAA. As described in the text, transcripts are arranged by the kinetic class of the HSV-1 representative, and only those transcripts with probes that fully correspond between the HSV-1 and HSV-2 chips are shown. The *y* axis is signal strength in arbitrary units.

reflected in a higher relative value of late transcripts seen at the latest times assayed compared with an HSV-1 infection, in which the abundance of the bulk of transcripts has significantly declined. The IE ICP4 transcript follows a generally similar course during both infections, but the decline and reaccumulation of ICP0 is much more pronounced for HSV-2, perhaps as a result of levels of host shutoff activity at the earlier times. The time-course of the ICP27 transcript appears generally similar during the course of infection of the two viruses.

Although comparison of RNA abundance in the presence and absence of viral DNA replication at a single time is only partially useful in a precise determination of kinetics because rates of synthesis and degradation of specific transcripts may vary significantly, it can reveal some further specific differences in the kinetic signatures of homologous transcripts of HSV-2 and HSV-1. The ratios of HSV-2 transcript levels at 6 h in the presence of 200 μg/mL PAA to those without the drug are shown in **Table 5** along with corresponding values for homologous HSV-1 transcripts. The kinetic class of many correlate well with these ratios, late transcripts showing values of 0.7 or less, and early transcripts values of 1 or more.

We have also shown the nucleotide homology between the translational reading frames and the 300 bases upstream of each translation start site in **Table 5.** This includes both the mRNA leader sequence and, based on our extensive analysis of HSV-1 promoters, the promoter/control regions of individual transcripts *(25–31)*. The protein coding homologies are nearly all in the 80% or higher range, whereas leader coding sequences range from values as low as 50% to as high or higher than those seen with the translational reading frames; these values are entirely consistent with an analysis done on a selected region of both viral genomes some time ago *(7)*. The highest values are in those promoter/control regions that overlie the translational reading frames of upstream protein coding regions.

The ratios of 18 transcripts show appreciable differences between the two closely related virus types. Of these, seven proteins found in the virion (UL10, -18, -19, -24, -25, -49, US5, and US10) show less expression under conditions of DNA blockage than their corresponding numbers in HSV-1 infections, whereas UL15, involved in DNA packaging, appears to be less dependent on DNA replication for appreciable expression. The lower values of the IE transcripts ICP0 and ICP4 may reflect the higher level of virion-associated nuclease present at early times in HSV-2 infections. Although it would be expected that many differences in sensitivity to inhibition of viral DNA replication are a result of differences in viral promoter/control sequences between the virus types, the relatively limited extent of such elements within the broad definition of promoter/control regions used here precludes any obvious correlation between sequence homology between such regions and differences in PAA sen-

Table 5
Ratios of HSV-2 Transcript Abundance 6 Hours After Infection in the Presence and Absence of PAA[a]

Transcript	Protein encoded	Homology (%)		6 h PAA/6 h of:	
		CDS	5′-Control[b]	HSV-2[c]	HSV-1[d]
UL1	Glycoprotein L	74.1	57	0.53	0.45
UL2	Uracil-DNA glycosylase	82.7	58.7	0.53	N/P[e]
UL3	Nuclear phosphoprotein	78.8	77.6	0.34	0.23
UL4	Possibly virion protein	78.3	86	0.86	0.85
UL5	DNA helicase–primase component	86.2	86.3	1.21	1.80
UL6	Minor capsid protein	86.9	80.7	0.29	N/P
UL7	Cleavage/packaging	83.7	56.7	0.46	0.89
UL8	DNA helicase–primase component	83.4	50.7	1.17	1.86
UL9	Ori binding protein	88.1	67	0.56	0.80
UL10	**Glycoprotein M**	**83**	**93.3**	**0.14**	**0.66**
UL11	Myristilated tegument protein	81.6	83.3	0.73	0.93
UL12	Dnase	83.6	81.7	0.52	N/P
UL13	Protein kinase, tegument protein	86.8	92.6	0.82	N/P
UL14	Capsid maturation	88.5	77.3	1.09	N/P
UL15	**DNA packaging**	**89.7**	**51.2**	**0.62**	**0.25**
UL16	"Capsid maturation β-capsid"	81.2	75.3	0.60	0.41
UL17		84.6	79	0.60	N/P
UL18	**Capsid protein**	**86.1**	**70.3**	**0.27**	**0.99**
UL19	**Major capsid protein**	**90.1**	**62.3**	**0.32**	**0.78**
UL20	Virion membrane protein	84.8	54	0.99	N/P
UL21	Tegument protein	85.3	65.3	0.17	0.26
UL22	Glycoprotein H	81.3	58.3	0.42	0.65
UL23	Thymidine kinase	81.1	88	1.83	1.48
UL24	**Capsid protein**	**81.5**	**74.3**	**0.12**	**1.36**

UL25	**Virion protein**	**86.8**	**59.5**	**0.31**	**1.03**
UL26	Capsid maturation protease	80.4	69.7	0.25	N/P
UL27	Glycoprotein B	88.3	91	0.36	0.48
UL28	**DNA packaging**	**88.7**	**52.3**	**0.59**	**1.62**
UL29	**ssDNA binding protein**	**89.7**	**57.3**	**0.67**	**1.44**
UL30	**DNA polymerase catalytic subunit**	**89.7**	**59.7**	**0.50**	**1.51**
UL31	**envelopement/egress**	**88.1**	**89.3**	**0.07**	**0.64**
UL32	Unknown	86.8	91.3	0.51	N/P
UL33	DNA packaging	91.4	79.3	0.32	N/P
UL34	Membrane-associated phosphoprotein	79.8	81	0.23	N/P
UL35	Capsid protein	84.4	67.3	0.21	0.30
UL36	Very large tegument protein	83	45.3	0.18	0.21
UL37	Tegument protein	86.3	68	0.63	0.58
UL38	Capsid protein	83.3	66.9	0.24	0.60
UL39	**Ribonucleotide reductase large subunit**	**84.7**	**68.3**	**0.49**	**2.22**
UL40	**Ribonucleotide reductase small subunit**	**88.6**	**82**	**0.65**	**3.14**
UL41	Host shutoff factor	85.2	68	0.21	0.06
UL42	DNA polymerse subunit	80.9	66.7	0.49	0.43
UL43	Probable membrane protein	78.2	49.3	0.44	0.54
UL44	Glycoprotein C	79.8	64	0.50	0.37
UL45	Tegument/envelope protein	79.2	57.7	0.00	N/P
UL46	**Tegument protein**	**82.1**	**86**	**0.57**	**1.65**
UL47	Tegument protein	84.6	60	0.05	N/P
UL48	Vp16	84.2	64	0.35	1.21
UL49	**Tegument protein**	**77.5**	**70**	**0.02**	**1.71**
UL49.5	**Tegument protein**	**73.1**	**80**	**0.44**	**N/P**
UL50	Deoxyuridine triphosphatase	80.2	68.3	0.82	2.62

(Table continues)

Table 5
Ratios of HSV-2 Transcript Abundance 6 Hours After Infection in the Presence and Absence of PAA[a]

Transcript	Protein encoded	Homology (%)		6 h PAA/6 h of:	
		CDS	5′-Control[b]	HSV-2[c]	HSV-1[d]
UL51		80.1	68.3	0.18	0.97
UL52	DNA helicase–primase component	83.9	72.3	2.50	1.45
UL53	Glycoprotein K	84.5	86	1.49	1.02
UL54	ICP27	81.7	70.7	0.86	0.99
UL55	**Virion assemby or maturation**	**83.2**	**68.3**	**2.60**	**0.48**
UL56	**Vesicular trafficking**	**74.5**	**55**	**0.56**	**0.68**
LAT 5′-end	Reactivation	N/A[f]	64.7	0.75	1.19
LAT intron	Stable LAT-intron	N/D[g]	N/D	N/S[h]	1.15
ICP0	ICP0	77.7	70.7	0.97	0.83
ORFOP		N/D	N/D	0.08	N/D
ICP34.5	Neurovirulence factor	70.1	50.7	0.37	0.51
LAT-3′-end	LAT poly(A) site	N/A	N/A	0.76	1.24
ICP4	ICP4	83.6	66.8	0.53	2.56
US1/ICP22	ICP22	73.9	68.1	0.56	0.83

US2	Virion protein	77.3	69.3	0.22	0.40
US3	Protein kinase	80.4	69.3	1.31	2.19
US4	Gycoprotein G	64.7	73	0.88	0.49
US5	**Glycoprotein J**	**67.4**	**58.3**	**0.35**	**2.05**
US6	Glycoprotein D	83.7	67	0.33	N/P
US7	Glycoprotein I	77.7	71.7	0.89	1.40
US8	**Unknown**	**81.1**	**60.7**	**0.57**	**1.67**
US8a	Tegument protein	81.6	N/A[i]	0.38	N/P
US9	Gycoprotein E	82.7	74.3	1.25	2.19
US10	**Virion protein**	**75**	**65.3**	**0.25**	**0.79**
US11	RNA binding protein, nucleolar	74.3	66.2	0.17	N/P
US12	ICP47, inhibits antigen presentation	69.4	69.6	N/S	N/P

PAA, phosphonoacetic acid; CDS, coding sequence.

[a] Bold data are transcripts showing different 6 hr PAA/6 hr untreated ratios in HSV-1 and HSV-2 infections.

[b] Arbitrarily chosen as the 300 bases upstream of the coding start; see text.

[c] See Table 3.

[d] Data from **refs. 4** and **20**.

[e] N/P, no specific probe on the chip.

[f] N/A, no protein encoded.

[g] N/D, not measured.

[h] N/S, signal below background.

[i] N/A, HSV-1 does not express a unique mRNA for this gene.

sitivity. Clearly, further experimental analysis including time variance studies and more careful analysis of individual promoter elements will be required to assess fully the significance of the differences noted. It is to be expected that such differences will have an important role in differences in the patterns of viral pathogenesis noted in the introduction.

3.4. Conclusions

3.4.1. Oligonucleotide-Based DNA Microarrays Provide a Rapid, Global Picture of HSV Transcript Abundance

It is clear from the approaches and data presented in this short review that the construction of an oligonucleotide-based DNA microarray can be readily accomplished for any large DNA virus whose genomic sequence is known. Thus, with a convenient, rapid, and straightforward series of comparative quantitative analyses, the basic similarities as well as potential specific differences in the transcription program of the very closely related HSV-1 and HSV-2 were determined. Clearly, similar approaches can be made with any herpesvirus as well as other viruses of equivalent or greater genomic complexity. The relatively low cost of custom oligonucleotide synthesis and the growing availability of laser scanners and image quantifying software brings the approach into the reach of many reasonably funded, moderate-sized laboratories. Although this technique, as is the case with any single approach, cannot fully establish all areas of significant differences in the transcription patterns of any two large DNA viruses, comparative studies, such as described here, will provide an important basis for more detailed, mechanistic studies. Combined with similar global methods becoming available for the quantitative analysis of levels of viral proteins during the course of infection, a very rich and detailed picture of the overall program of gene expression of any given virus or other pathogen can be obtained.

3.4.2. Areas for Further Experimental Refinement

Although the approaches outlined in this report are useful for the study of viral gene expression in cultured cells, it is clear that significant increases in sensitivity will enhance the method's applicability for the study of viral gene expression in animal models. Toward this end, higher sensitivity detection of hybridized cDNA can be accomplished by using colloidal gold and silver labeling of cDNA and light scattering measurements *(32)*. Currently, in our own laboratory, we have been able to use such methods to allow hybridization and ready analysis of samples of 100–200 ng of poly(A) RNA from infected cells, a 10-fold increase in sensitivity. Such methods, coupled with amplification methods *(33)*, should allow hybridization of RNA from a few tens of cells or less from infected tissue.

In addition to increasing the sensitivity of detection of hybrids, studies with other viral and cellular arrays have determined that oligonucleotides in the size range of 45–65 nt will provide sufficient specificity to allow the unambiguous detection of viral and cellular transcripts on microarrays (unpublished data). Thus, the overall cost of synthesis of probes can be significantly reduced from that used in the present studies. This affords the opportunity to generate chips with greater numbers of gene probes and a corresponding increase in the resolution of the global patterns of transcription seen, with little increase in overall costs.

Finally, the quantitative nature of microarray data can be readily improved. For example, careful analysis of hybridization values using overlapping cloned DNA fragments along with analysis of the uniformity of labeling by nick translation can be used to establish absolute hybridization efficiencies to a much higher degree of accuracy than described in this preliminary study. Such information will allow the conversion of the hybridization values obtained here into much more precise measures of absolute transcript levels. These levels can, in turn, be used to calibrate Northern blot and primer extension data for detailed studies of the effects of specific modifications of control sequences and conditions of infection upon absolute transcript abundance.

Although these and other refinements will serve to continually enrich the resolution and detail of the global patterns of viral (and cellular) transcript abundance obtained with microarrays, the basic approaches outlined in this report will provide the framework for these higher resolution studies.

Acknowledgments

This work was supported by grants CA11861 and CA90287 to E. K. W. and the British Biotechnology Science Research Council and Scottish Higher Education Funding Council to P. G. Marcia Rice, Jim Sunabe, and G. Devi-Rao (UCI) and Douglas Roy and Klemens Vierlinger (GTI, Edinburgh) provided technical assistance.

References

1. Wagner, E. K. (1999) Herpes simplex virus—molecular biology, in *Encyclopedia of Virology* Webster, R. G. and Granoff, A. eds.), Academic Press, London, pp. 686–697.
2. Wagner, E. K. (2003) Web site: http://darwin.bio.uci.edu/~faculty/wagner/index.html.
3. Wagner, E. K. and Bloom, D. C. (1997) The experimental investigation of herpes simplex virus latency. *Clin. Microbiol. Rev.* **10,** 419–443.
4. Wagner, E. K, Garcia, R. J. J, Stingley, S. W., et al. (2002) Practical approaches to long Oligonucleotide-based DNA microarrays: lessons from herpesvirues. *Prog. Nucleic Acids Res. Mol. Biol.* **71,** 445–491.

5. Spear, P. G., Eisenberg, S. P., and Cohen, G. H. (2003) Three classes of cell surface receptors for alphaherpesvirus entry. *Virology* **257,** 1–8.

6. Shukla, D. and Spear, P. G. (2001) Herpesviruses and heparan sulfate: an intimate relationship in aid of viral entry. *J. Clin. Invest.* **108,** 503–510.

7. Draper, K. G, Frink, R. J., Devi-Rao, G. B, Swain, M., Galloway, D., and Wagner, E. K. (1984) Herpes simplex virus types 1 and 2 homology in the region between 0.58 and 0.68 map units. *J. Virol.* **52,** 615–623.

8. Dolan, A., Jamieson, F. E., Cunningham, C., Barnett, B. C., and McGeoch, D. J. (1998) The genome sequence of herpes simplex virus type 2. *J. Virol.* **72,** 2010–2021.

9. Braig, S. and Chanzy, B. (2002) Prise en charge practique de l'herpes genital. *Pathol. Biol.* **50,** 472–476.

10. Whitley, R. J. (1993) Neonatal herpes simplex virus infections. *J. Med. Virol.* **Suppl 1,** 13–21.

11. Herold, B. C, Gerber, S. I., Belval, B. J., Siston, A. M., and Shulman, N. (1996) Differences in the susceptibility of herpes simplex virus types 1 and 2 to modified heparin compounds suggest serotype differences in viral entry. *J. Virol.* **70,** 3461–3469.

12. Strelow, L. I. and Leib, D. A. (1995) Role of the virion host shutoff (vhs) of herpes simplex virus type 1 in latency and pathogenesis. *J. Virol.* **69,** 6779–6786.

13. Peng, T., Hunter, J. R. C., and Nelson, J. W. (1996) The novel protein kinase of the RR1 subunit of herpes simplex virus has autophosphorylation and transphosphorylation activity that differs in its ATP requirements for HSV-1 and HSV-2. *Virology* **216,** 184–196.

14. Jin, P. and Ringertz, N. R. (1990) Cadmium induces transcription of proto-oncogenes c-jun and c-myc in rat L6 myoblasts. *J. Biol. Chem.* **265,** 14061–14064.

15. Feng, C. P., Kulka, M., and Aurelian, L. (1993) NF-kappa B-binding proteins induced by HSV-1 infection of U937 cells are not involved in activation of human immunodeficiency virus. *Virology* **192,** 491–500.

16. Goswami, B. B. (1987) Transcriptional induction of proto-oncogene *fos* by HSV-2. *Biochem. Biophys. Res. Commun.* **143,** 1055–1062.

17. Thakur, A., Athmanathan, S., and Wilcox, M. (2003) The differential regulation of nitric oxide by herpes simplex virus-1 and -2 in a corneal epithelial cell line. *Clin. Exp. Ophthamol.* **28,** 188–190.

18. Chambers, J., Angulo, A., Amarantunga, D., et al. (1999) DNA microarrays of the complex human cytomegalovirus genome: profiling kinetic class with drug sensitivity of viral gene expression. *J. Virol.* **73,** 5757–5766.

19. Stingley, S. W., Ramirez, J. J. G., Aguilar, J. S., et al. (2000) Global analysis of herpes simplex virus type 1 transcription using an oligonucleotide-based DNA microarray. *J. Virol.* **74,** 9916–9927.

20. Aguilar, J. S., Roy, D., Ghazal, P., and Wagner, E. K. (2002) Dimethyl sulfoxide blocks herpes simplex virus-1 productive infection in vitro acting at different stages with positive cooperativity. Application of micro-array analysis. *BMC Infect. Dis* **2,** 9.

21. Yang, W. C., Devi-Rao, G. V., Ghazal, P., Wagner, E. K., and Triezenberg, S. J. (2002) General and specific alterations in programming of global viral gene expres-

sion during infection by VP16 activation-deficient mutants of herpes simplex virus type 1. *J. Virol.* **76,** 12758–12774.

22. http://www.gti.ed.ac.uk/scgti.html.

23. Thompson, R. L. and Wagner, E. K. (1988) Partial rescue of herpes simplex virus neurovirulence with a 3.2 kb cloned DNA fragment. *Virus Genes* **1,** 261–273.

24. Costa, R. H., Draper, K. G., Banks, L., et al. (1983) High-resolution characterization of herpes simplex virus type 1 transcripts encoding alkaline exonuclease and a 50,000-dalton protein tentatively identified as a capsid protein. *J. Virol.* **48,** 591–603.

25. Guzowski, J. F., Singh, J., and Wagner, E. K. (1994) Transcriptional activation of the herpes simplex virus type 1 U_L38 promoter conferred by the *cis*-acting downstream activation sequence is mediated by a cellular transcription factor. *J. Virol.* **68,** 7774–7789.

26. Wagner, E. K., Guzowski, J. F., and Singh, J. (1995) Transcription of the herpes simplex virus genome during productive and latent infection, in *Progress in Nucleic Acid Research and Molecular Biology* (Cohen, W. E. and Moldave, K., eds.), Academic, San Diego, pp. 123–168.

27. Pande, N. T., Petroski, M. D., and Wagner, E. K. (1998) Functional modules important for activated expression of early genes of herpes simplex virus type 1 are clustered upstream of the TATA box. *Virology* **246,** 145–157.

28. Wagner, E. K., Petroski, M. D., Pande, N. T., Lieu, P. T., and Rice, M. K. (1998) Analysis of factors influencing the kinetics of herpes simplex virus transcript expression utilizing recombinant virus. *Methods* **16,** 105–116.

29. Petroski, M. D. and Wagner, E. K. (1998) Purification and characterization of a cellular protein that binds to the downstream activation sequence of the strict late U_L38 promoter of herpes simplex virus type 1. *J. Virol.* **72,** 8181–8190.

30. Lieu, P. T., Pande, N. T., Rice, M. K., and Wagner, E. K. (2000) The exchange of cognate TATA boxes results in a corresponding change in the strength of two HSV-1 early promoters. *Virus Genes* **20,** 5–10.

31. Huang, C.-J. and Wagner, E. K. (1994) The herpes simplex virus type 1 major capsid protein (VP5-UL19) promoter contains two cis-acting elements influencing late expression. *J. Virol.* **68,** 5738–5747.

32. www.invitrogen.com.

33. Wang, J., Hu, L., Hamilton, S. R., Coombes, K. R., and Zhang, W. (2003) RNA amplification strategies for cDNA microarray experiments. *Biotechniques* **34,** 394–400.

34. Easton, A. J. and Clements, J. B. (1980) Temporal regulation of herpes simplex virus type 2 transcription and characterization of virus immediate early mRNA's. *Nucleic Acids Res.* **8,** 2627–2645.

35. Worrad, D. M. and Caradonna, S. (1993) The herpes simplex virus type 2 UL3 open reading frame encodes a nuclear localizing phosphoprotein. *Virology* **195,** 364–376.

36. Yamada, H., Jiang, Y. M., Oshima, S., et al. (1998) Characterization of the UL4 gene product of herpes simplex virus type 2. *Arch. Virol.* **143,** 1199–1207.

37. Nozawa, N., Daikoku, T., Yamauchi, Y., et al. (2002) Identification and characterization of the UL7 gene product of herpes simplex virus type 2. *Virus Genes* **24,** 257–266.

38. Daikoku, T., Shibata, S., Goshima, F., et al. (1997) Purification and characterization of the protein kinase encoded by the UL13 gene of herpes simplex virus type 2. *Virology* **235,** 82–93.

39. Wada, K., Goshima, F., Takakuwa, H., Yamada, H., Daikoku, T., and Nishiyama, Y. (1999) Identification and characterization of the UL14 gene product of herpes simplex virus type 2. *J. Gen. Virol.* **80,** 2423–2431.

40. Oshima, S., Daikoku, T., Shibata, S., Yamada, H., Goshima, F., and Nishiyama, Y. (1998) Characterization of the UL16 gene product of herpes simplex virus type 2. *Arch. Virol.* **143,** 863–880.

41. Goshima, F., Watanabe, D., Takakuwa, H., et al. (2000) Herpes simplex virus UL17 protein is associated with B capsids and colocalizes with ICP35 and VP5 in infected cells. *Arch. Virol.* **145,** 417–426.

42. Zhu, H. Y., Murata, T., Goshima, F., et al. (2001) Identification and characterization of the UL24 gene product of herpes simplex virus type 2. *Virus Genes* **22,** 321–327.

43. Cockrell, A. S. and Muggeridge, M. I. (1998) Herpes simplex virus 2 UL45 is a type II membrane protein. *J. Virol.* **72,** 4430–4433.

44. Shiba, C., Daikoku, T., Goshima, F., et al. (2000) The UL34 gene product of herpes simplex virus type 2 is a tail-anchored type II membrane protein that is significant for virus envelopment. *J. Gen. Virol.* **81,** 2397–2405.

45. Strnad, B. C. and Aurelian, L. (1976) Proteins of herpesvirus type 2. II. Studies demonstrating a correlation between a tumor-associated antigen (AG-4) and a virion protein. *Virology* **73,** 244–258.

46. Kato, K., Daikoku, T., Goshima, F., Kume, H., Yamaki, K., and Nishiyama, Y. (2000) Synthesis, subcellular localization and VP16 interaction of the herpes simplex virus type 2 UL46 gene product. *Arch. Virol.* **145,** 2149–2162.

47. Geiss, B. J., Tavis, J. E., Metzger, L. M., Leib, D. A., and Morrison, L. A. (2001) Temporal regulation of herpes simplex virus type 2 VP22 expression and phosphorylation. *J. Virol.* **75,** 10721–10729.

48. Yamada, H., Jiang, Y. M., Oshima, S., et al. (1998) Characterization of the UL55 gene product of herpes simplex virus type 2. *J. Gen. Virol.* **79,** 1989–1995.

49. Koshizuka, T., Goshima, F., Takakuwa, H., et al. (2002) Identification and characterization of the UL56 gene product of herpes simplex virus type 2. *J. Virol.* **76,** 6718–6728.

50. Jiang, Y. M., Yamada, H., Goshima, F., et al. (1998) Characterization of the herpes simplex virus type 2 (HSV-2) US2 gene product and a US2-deficient HSV-2 mutant. *J. Gen. Virol.* **79,** 2777–2784.

29

Real-Time Quantitative PCR Analysis of Viral Transcription

James Papin, Wolfgang Vahrson, Rebecca Hines-Boykin, and
Dirk P. Dittmer

Summary

Whole-genome profiling using DNA arrays has led to tremendous advances in our understanding of cell biology. It has had similar success when applied to large viral genomes, such as the herpesviruses. Unfortunately, most DNA arrays still require specialized and expensive resources and, generally, large amounts of input RNA. An alternative approach is to query entire viral genomes using real-time quantitative PCR. We have designed such PCR-based arrays for every open reading frame of human herpesvirus 8 and describe here the general design criteria, validation procedures, and detailed application to quantify viral mRNAs. This should provide a useful resource either for whole-genome arrays or just to measure transcription of any one particular mRNA of interest. Because these arrays are RT-PCR-based, they are inherently more sensitive and robust than current hybridization-based approaches and are ideally suited to query viral gene expression in models of pathogenesis.

Key Words: Real-time quantitative PCR; TaqMan; herpesvirus; microarray.

1. Introduction

Polymerase chain reaction (PCR) *(1)* has allowed many scientific fields, including virology, to develop assays for the detection of their template of interest. PCR has risen as the gold standard for detection of the presence of a pathogen in many instances in which cell culture or serological assays were once considered unsurpassed. However, post-PCR handling steps required to evaluate the product are a cumbersome part of PCR assays. The ability to track the amplification and quality of the product without post-PCR steps was first seen with the description of quantitative assays using replicable hybridization probes *(2)*. This technique has since become the foundation from which real-

From: *Methods in Molecular Biology, vol. 292: DNA Viruses: Methods and Protocols*
Edited by: P. M. Lieberman © Humana Press Inc., Totowa, NJ

time quantitative PCR has been developed *(3)*. Real-time quantitative (QPCR) PCR measures the amount of PCR product at each cycle of the reaction either by binding of a fluorescent, double-strand-specific dye (SYBRgreen™) or by hybridization to a third sequence-specific, dual-labeled fluorogenic oligonucleotide (molecular Beacon, TaqMan™). Since the introduction of real-time QPCR, many applications have arisen using this technology. The kinetics and chemistries of real-time QPCR are covered in detail by Mackay et al. *(4)*.

Coupling reverse transcription (RT) to PCR yields the most sensitive method yet evolved to detect the presence of specific mRNAs. Unlike hybridization-based methods, RT-PCR can distinguish between various spliced mRNAs, through exon-specific primers. RNAse protection can also be used to distinguish between differently spliced messages, but this method is more difficult to adapt to high-throughput application. One of the many applications of real-time QPCR is its use in transcriptional profiling of DNA viruses. Because viruses encode on the order of 2–200 different mRNAs, many of which are coregulated, a limited number of PCR reactions can be used to query the entire viral transcriptome. This is in contrast to bacterial or mammalian genomes, which typically produce 1000–60,000 different mRNAs. We initially developed a real-time QPCR array to study the transcriptional profile of Kaposi's sarcoma-associated herpesvirus (KSHV) in culture and in different clinical samples *(5,6)*. By designing and evaluating primers specific for every predicted open reading frame (ORF) in the KSHV genome, real-time QPCR arrived at essentially the same result as array studies for this virus, yet, at low throughput, no special training beyond good laboratory practices was required. In contrast to hybridization, we found real-time QPCR very forgiving with regard to sample quality, reagents, and handling (*see* **Note 1**).

This chapter covers the necessary guidelines and protocols for the development of a real-time QPCR assay for single genes or viral arrays. The guidelines listed in this chapter were adopted from the development of the KSHV genome-wide array *(5,6)*.

2. Materials

1. TRI-Reagent™ (Sigma, St. Louis, MO).
2. 1.5-mL Phase lock tubes (Eppendorf, Brinkman Instruments, NY).
3. 96-Well real-time PCR plates, skirted.
4. Tissue homogenizer.
5. Oligonucleotide primers/probes (MWG, NC).
6. SYBR Green Enzyme Mix (Applied Biosystems, CA).
7. TaqMan Enzyme Mix (Applied Biosystems).
8. Real-Time PCR thermocycler/equipment. (e.g., ABI Prizm 7700, ABI Prizm 5700).
9. 10X Reverse transcriptase buffer (Applied Biosystems).
10. 25 mM MgCl$_2$ (Applied Biosystems).
11. 10 mM dNTPs (Applied Biosystems).
12. Reverse transcriptase (SuperScript II, Invitrogen, CA).

13. Random hexamer RT primer (Applied Biosystems).
14. Oligotex dT beads (Qiagen).
15. 70% Isopropanol in diethyl pyrocarbonate (DEPC) water.
16. DEPC water.
17. *Primer3* (version 3.0.9) software *(7)*.
18. *EMBOSS* (version 2.7.1) software *(10)*.
19. Ruby (version 1.6.7) software *(22)*.
20. PrimeTime (Vahrson and Dittmer, in preparation).
21. Excel (Microsoft, Redmond, WA).

3. Methods

The methods described below explain: (1) primer/probe design for QPCR, (2) mRNA isolation from tissues/cells, (3) reverse transcription of mRNA into DNA, (4) the setup of SYBR green-based QPCR, (5) the setup of probe-based QPCR, and (6) the setup of multiplex QPCR.

3.1. Primer Design

Primer/probe design is one of the most important aspects in achieving a successful QPCR assay. The following guidelines have been included to help attain the best primers possible for the assay. There are many computer programs and web-based applications available to assist in the design of primers and probes; for the purposes of this chapter, some have been listed in **Subheading 3.1.3.** and **3.1.4. Figure 1** exemplifies three primers and a possible nomenclature. Note that the primers are located toward the 3′-end of the ORF.

3.1.1. Primer Guidelines

1. The melting temperature (T_m) of the primers should be in the range of 59±2°C (*see* **Note 2**).
2. The maximal difference between two primers within the same primer pair should be ≤2°C.
3. Total guanidine (G) and cytosine (C) content within any given primer should be between 20 and 80%.
4. There should not be any GC clamp designed into any of the primers.
5. Primer length should fall into the range of 9–40 nucleotides.
6. Hairpins with a stem length of four residues or more should not exist in the primer sequence.
7. Fewer than four repeated G residues should be present within a primer.
8. The resulting amplicon should be at least 50 nt in length, but typically no larger than 100 nt.

3.1.2. Probe Guidelines

Real-time QPCR products can be detected either by an intercalating dye or annealing of a third, specific probe. The guidelines for designing TaqMan

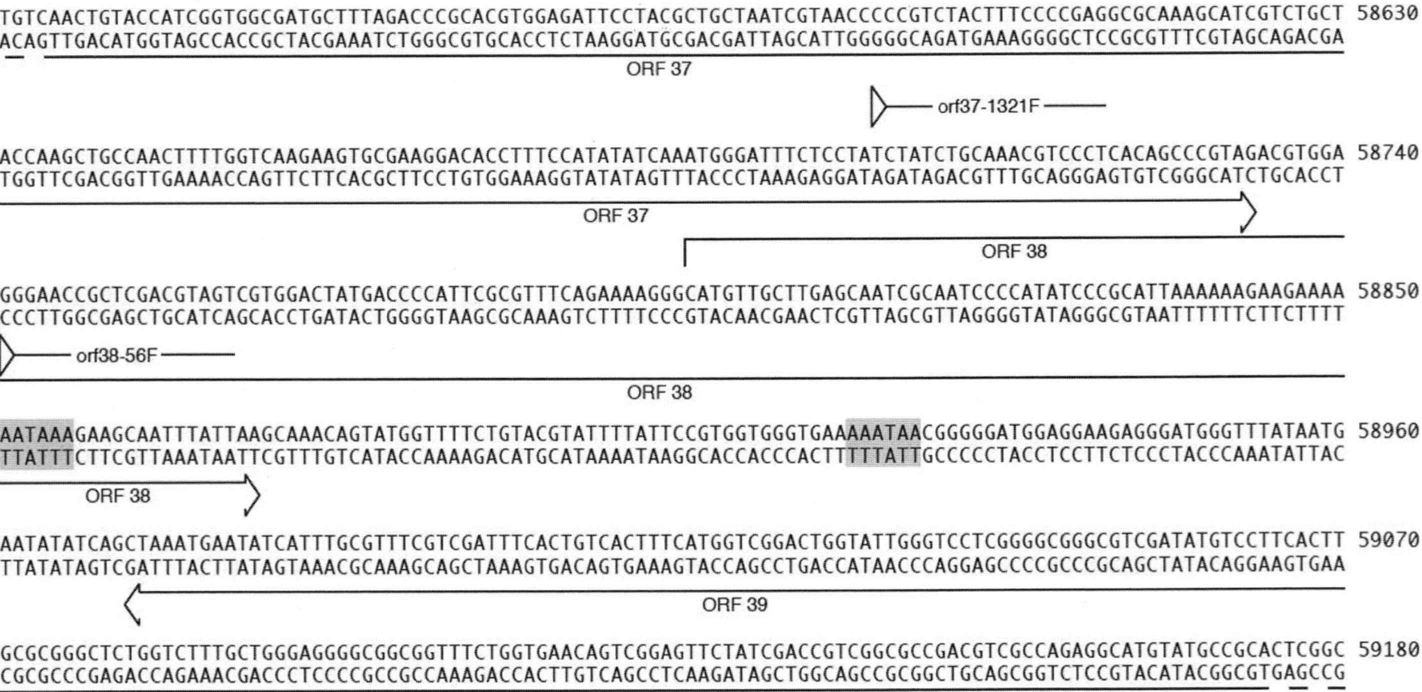

Fig. 1. Illustration of primer position within an open reading frame. The diagram demonstrates open reading frames within a sequence and the primer placement within those open reading frames. Open reading frames are labeled ORF 37–39 and designated by the large arrows. Examples of primers are shown by the small arrows followed by the name of the primer. (Primers are named after the open reading frame and position they represent.) Shaded area signifies predicted mRNA termination signal).

probes follow those used for the construction of primers, with the following two exceptions:

1. The T_m of the probe should be 10°C higher, compared with the T_m of the corresponding primer pair.
2. There should not be a G residue at the 5′-end of the probe.

3.1.3. Primer Express®

The Primer Express® software v2.0 (Applied Biosystems, cat. no. 4330710) is available for Windows NT and Windows 2000. Prior versions (up to v1.5) were also available for Macintosh OS9. Hopefully, the program will eventually be ported to Windows XP, although no release date has been announced. This program does a good job at designing TaqMan-based primer and probe sets (as well as a host of other primers). It is easy to use and is well documented in the manual. We typically use the settings for TaqMan probes and primers to design SYBR-based primers (the other primers are identical).

There are three disadvantages to Primer Express: (1) the program must be purchased, although usually one copy is included with purchase of an ABI machine; (2) it has limitation in handling large batches of sequences or large genomes; at least version 1.5 was not able to design primers for one entire herpesvirus genome (120,000 bp) at a time; (3) at least version 1.5 scanned the sequence from the 5′-end, while we experienced better results when selecting TaqMan sets near the 3′-end of the open-reading-frame (ORF). This location allows detection in instances of lower quality RNA or of low-processivity reverse transcription.

Other commercial primer design programs are also available and should be suitable for design using the guidelines outlined above.

3.1.4. EMBOSS *and* Primer3

EMBOSS (European Molecular Biology Open Software Suite) (10) is a comprehensive collection of free open-source programs for sequence analysis. It represents a freely available and more robust alternative to proprietary programs such as PrimerExpress (Applied Biosystems) and others.

Eprimer3, a program for searching PCR primers, is based on the *Primer3* program *(7)* from the Whitehead Institute/MIT Center for Genome Research. It allows one to search a DNA sequence for both PCR primers and oligonucleotide beacons. More than 60 parameters can be specified to adapt the program for various purposes. They include constraints on physicochemical properties of the primers, probes, and product, like T_M, GC content and size; constraints on sequence properties like the amount of self-complementarity and 3′-overlapping bases; positional constraints within the template sequence;

avoidance of sequences specified in a misspriming library; and many more (*see* **Note 3**).

3.1.4.1. EXTRACTING ORFs FROM DATABASE ENTRIES

For many applications it is useful to restrict the target of the primer search to coding regions within a larger sequence. Using the UNIX *grep* command, you can inspect the annotations of GenBank or EMBL database entries for coding regions: *grep 'CDS' sequence-file.*

The EMBOSS *extractfeat* program lets you extract the respective sequences as individual sequences: extractfeat-type cds *sequence-file.*

As an alternative to extracting the coding sequences, one can use their positions as constraints for the primer search, as demonstrated below in **Subheading 3.1.4.2.**

3.1.4.2. GENERATING PRIMER PAIRS FOR REAL-TIME PCR

Eprimer3 takes a vast number of parameters influencing the way primers are selected (*see* **Note 4**). When searching for primers suitable for real-time PCR, the most important parameters are the ones controlling T_M, sequence of the primers, and size of the product. Here is a sample invocation of *eprimer3* (explanations below):

```
eprimer3
    -otm 59.0 -mintm 57.0 -maxtm 61.0 -maxdifftm 2.0
    -mingc 20.0 -maxgc 80.0 -maxpolyx 4 -selfany 4
    -productosize 500 -productsizerange 200–800
    -includedregion 1736,5692 sequence-file
```

In the first line of the example, we specify the T_M for the primers: The optimal T_M (-otm) would be 59.0°C with a tolerance of ±2°C (-mintm, -maxtm) and the additional constraint that the difference in T_M between the two primers must not exceed 2°C (-maxdifftm). On the next line parameters constraining the primer sequence are specified: the GC content must be 20–80% (-mingc, -maxgc), there must be no runs of identical nucleotides longer than four (-maxpolyx), and the maximal alignment score when testing for self-complementarity and for matches between forward and reverse primers must not be more than 4, which corresponds to an overlap of four nucleotides (-selfany). In the third line the desired optimal size of the product is given as 500 (-productosize)–300 (-productsizerange). Finally, in the last line, the portion of the sequence in which *eprimer3* searches for primers is restricted to a region between and including positions 1736 and 5692.

3.2. mRNA Isolation

The method described in this section outlines the purification of total RNA from either tissues (**Subheading 3.2.1.**) or cells (**Subheading 3.2.2.**) using

TRI-reagent (Sigma-Aldrich). Other companies offer similar chemicals that will yield similar results. A subsequent step using dT beads is then used for the selection of mRNA from the total RNA pool (3.2.3).

3.2.1. RNA Isolation From Tissues

1. Transfer 750 µL of TRI-reagent into the tube containing the tissue sample, and place the samples on ice.
2. Before one uses the homogenizer, clean it with TRI-reagent and ethanol in the following manner: TRI-reagent, 70% Ethanol, and then clean TRI-reagent.
3. Samples can then be homogenized one at a time and returned to ice. However, the homogenizer should be cleaned between every sample by the same method described in **step 2.**
4. Incubate the samples on ice for 5 min following homogenization.
5. Add 150 µL of chloroform to each sample, and mix well. This can be done by shaking the tubes, or by briefly vortexing the tubes.
6. After mixing, centrifuge the sample at full speed in a bench-top centrifuge for 15 min at 4°C.
7. Once the tube is removed from the centrifuge, three phases should be visible within the tube. *The upper phase contains the RNA and should be removed and placed into a new phase-lock tube* (Eppendorf, Brinkman Instruments, Westbury, NY). The middle phase contains the DNA from the sample, and the lower phase contains protein. The remaining two phases in the tube should be disposed of properly once the upper phase of interest is removed, as they are organic waste and must be handled as such.
8. Add 250 µL of phenol/chloroform/isoamyl alcohol to the sample in the phase-lock tube, and vortex the sample for 1 minute.
9. Incubate the samples on ice for 5 min.
10. Centrifuge the sample again at full speed for 10 min at 4°C.
11. After centrifugation, transfer 175 µL of the clear upper phase to a new tube, and mix with an equal volume (175 µL) of isopropanol.
12. Mix the sample thoroughly, and incubate at –80°C overnight.
13. Remove RNA in isopropanol from the freezer, and allow the sample to thaw.
14. Once the sample is thawed, centrifuge at full speed for 20 min at 4°C.
15. Aspirate the supernatant being cautious not to disturb the RNA pellet.
16. Add 1 mL of 70% ethanol (in DEPC water). *Do not* dislodge or attempt to redissolve the pellet.
17. Carefully aspirate the supernatant, and air-dry the tube for 10 min.
18. Resuspend the RNA pellet in 250 µL of DEPC-treated water. You can now proceed to the mRNA enrichment step **(Subheading 3.2.3.)**, or the total RNA pool can be frozen at –80°C for future use.

3.2.2. RNA Isolation From Cells

The process for extracting total RNA from cultured cells is identical to that for extracting total RNA from tissues **(Subheading 3.2.1.)**, with the exception of using a tissue homogenizer. When one is isolating total RNA from cells, a tissue homogenizer is not needed; simply resuspending the cell pellet in TRI-reagent is sufficient to lyse the cells. The following protocol describes the isolation of total RNA from a cell pellet; cells should be pelleted for this procedure by centrifugation at 200*g* for 5 min at 4°C.

1. Transfer 750 µL of TRI-reagent into the tube containing the tissue sample, and place the samples on ice. If the samples are not already in an 1.5-mL centrifuge tube, they should be transferred once they are resuspended into TRI-reagent.
2. Vortex the samples for 30 s, and place them on ice for 5 min.
3. Add 150 µL of chloroform to each sample, and mix well. This can be done by shaking the tubes, or by briefly vortexing the tubes.
4. After mixing, centrifuge the sample at full speed in a bench-top centrifuge for 15 min at 4°C.
5. Once the tube is removed from the centrifuge, three phases should be visible within the tube. *The upper phase contains the RNA and should be removed and placed into a new phase-lock tube* (Eppendorf, Brinkman Instruments). The middle phase contains the DNA from the sample, and the lower phase contains protein. The remaining two phases in the tube should be disposed of properly once the upper phase of interest is removed, as they are organic waste and must be handled as such.
6. Add 250 µL of phenol/chloroform/isoamyl alcohol to the sample in the phased lock tube, and vortex the sample for 1 min.
7. Incubate the samples on ice for 5 min.
8. Centrifuge the sample again at full speed for 10 min at 4°C.
9. After centrifugation, 175 µL of the clear upper phase should be transferred to a new tube and mixed with an equal volume (175 µL) of isopropanol.
10. Mix the sample thoroughly, and incubate at –80°C overnight.
11. Remove the RNA in isopropanol from freezer, and allow the sample to thaw.
12. Once the sample is thawed, centrifuge at full speed for 20 min at 4°C.
13. Aspirate the supernatant, being cautious not to disturb the RNA pellet.
14. Add 1 mL of 70% ethanol (in DEPC water). *Do not* dislodge or attempt to redissolve the pellet.
15. Carefully aspirate the supernatant and air-dry the tube for 10 min.
16. Resuspend the RNA pellet in 250 µL of DEPC-treated water. You can now proceed to the mRNA enrichment step (**Subheading 3.2.3.**), or the total RNA pool can be frozen at –80°C for future use.

3.2.3. Enrichment of mRNA

The Oligotex mRNA purification system (Qiagen) exploits the observation that cellular mRNAs contain a polyadenylated (poly[A]) tail of 20–250 adenosine residues *(11)*. Since mRNAs are the only cellular RNAs that contain a poly(A) tail, this feature can be taken advantage of to purify and enrich mRNA exclusively from a total RNA pool. The Qiagen Oligotex system uses a dT oligomer coupled to a solid phase matrix to bind the poly(A) tail of mRNA while the remaining RNA, which does not contain a poly(A) tail, is washed away. Hybridization of the poly(A) tail to the dT oligomer is dependent on high-salt conditions, so the complex can be easily disrupted by lowering the ionic strength. The protocol for purification of mRNA using the Oligotex system is covered in detail in the supplementary material of the kit and therefore is not included in this chapter.

3.3. Reverse Transcription

Reverse transcription takes advantage of reverse transcriptase, an enzyme found in retroviruses, to synthesize a strand of DNA that is complementary (cDNA) to the sequence of the RNA used in the reaction. This cDNA can then be used as a template in a PCR reaction. The following steps in this process involve the creation of cDNA from the previously isolated RNA/mRNA. Subheadings in this section describe the setup of the reverse transcription reaction (**Subheading 3.3.1.**), the cycling conditions necessary for reverse transcription (**Subheading 3.3.2.**), and the process to ready the sample for use as a template in a PCR reaction, including the digestion of the remaining RNA.

3.3.1. Setup of the Reverse Transcription Reaction

To begin setting up the reverse transcription reaction, a master mix should be created containing all the necessary reagents excluding the sample. Listed below are the reagents and the volume necessary for a 1X reaction. The total volume of master mix created should be sufficient for $n + 1$ reactions, where n equals the number of reactions you wish to carry out. As an example, the volume necessary for a 11X reaction master mix is also listed.

Reagents	1X	11X (10 + 1)
	(µL)	(µL)
10X RT-buffer (ABI)	2	22
25 mM MgCl$_2$ (ABI)	4.4	48.4
10 mM dNTPs (ABI)	4	44
Reverse transcriptase (Invitrogen)	1	11
Random hexamer	1	11
RNase inhibitor (ABI)	0.4	4.4
Total	12.8	140.8

Once the master mix is created, aliquot 12.8 µL of the master mix into a 0.2-mL thin-walled PCR tube for each reaction. Subsequently add 7.2 µL of RNA for each sample to a tube of aliquoted master mix. The total volume of the reaction is 20 µL (12.8 µL of master mix and 7.2 µL of RNA), and the amount of input RNA can be anywhere in the range of 3 to 4000 ng. The reaction is now ready for cycling; continue to **Subheading 3.3.2.**

3.3.2. Cycling Conditions for RT

The creation of cDNA using RT is a simple one-cycle, three-step reaction in the thermocycler. The times and temperatures are as follows: 42°C, 45 min; 52°C, 30 min; 70°C, 10 min.

After the cycling, continue to **Subheading 3.3.3.** or the reactions can be stored at 4°C until continuing with the sample preparation.

3.3.3. RNA Digestion and Sample Preparation

Following the RT reaction, the sample is prepared for PCR amplification by digestion of the remaining RNA and by increasing the volume of the sample. The RNA digestion is necessary to remove remaining RNA that might interfere with the subsequent PCR reaction. To perform the RNA digestion, simply add 1 U of RNase H to the sample, and incubate at 37°C for 30 min. This is sufficient to remove all the remaining RNA from any RNA/DNA hybrids. Since the total volume of the sample is only approx 20 µL, the volume should be increased using DEPC water to yield enough sample for multiple PCR reactions. The reaction should be diluted to a total of at least 200 µL, but it can be brought up to as much as 600 µL.

3.4. SYBR Green-Based QPCR

The setup and cycling of a QPCR reaction using SYBR Green as a detection system is covered in **Subheadings 3.4.1.** and **3.4.2.** This includes the preparation of the master mix, the concentration of reagents within the reaction, and the cycling conditions for QPCR (*see* **Note 5**).

One of the crucial aspects of PCR is to guard against contamination. Ideally all steps of the setup reaction are conducted in separate rooms: (1) a so-called white room to assemble the primers and reagents; (2) a so-called gray room to prepare the RNA and add the sample to the PCR; and (3) a post-PCR array black room, which is normal laboratory space. All surfaces should be washed with 10% bleach weekly, and if possible overhead ceiling UV lights should be installed.

3.4.1. Setup of the Reaction

1. The first step in setting up the reaction is creating the primer mix. The primer mix consists of both the forward and reverse primers at a concentration of 1 pmol/µL. However, individual primers are stored at 100 pmol/µL at –80°C and combined and diluted to yield enough forward and reverse primer mix for 100 reactions. In the case of single-primer real-time QPCR, we did not find it necessary to test a range of primer concentrations and to optimize them individually.
2. This primer mix is then combined with the SYBR Green 2X PCR mix (Applied Biosystems) to create the master mix. The volume of each mix that is added to cre-

96-well SYBR-Green QPCR Set-up

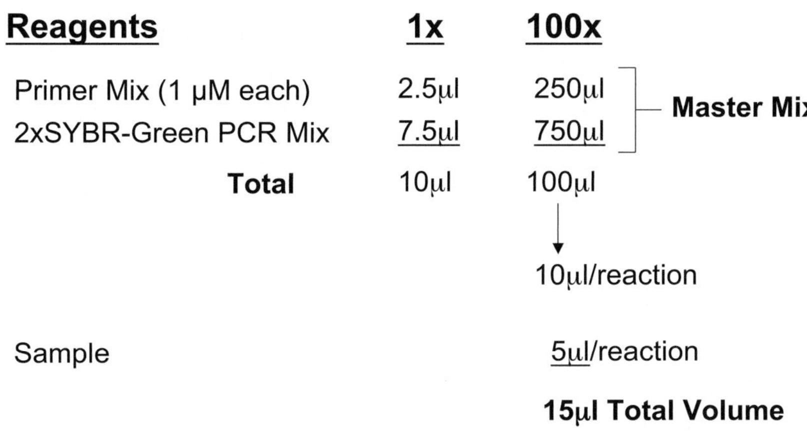

Fig. 2. Diagram of SYBR-Green QPCPR setup. The diagram lists the reagents and volumes necessary for the setup of a 96-well reaction real-time QPCR. The master mix consists of primer mix and 2X SYBR-Green PCR mix. 1X, volumes necessary for one reaction; 100X, L volumes necessary for 96 reactions. Once the master mix is combined to a total volume of 100 µL, 10 µL is aliquoted per well (the volume for a 1X reaction), and 5 µL of sample is added later for a total volume of 15 µL.

ate the master mix depends on the final reaction volume you are trying to achieve and the number of reactions for which the master mix is being prepared. For the purposes of this chapter, a final volume of 15 µL will be used. This volume was chosen as we have previously demonstrated its efficacy *(5,6)*. Depending on the individual equipment, smaller volumes may be possible, but we found that without automation the pipeting error becomes substantial. In comparison with the 50 µL volume originally recommended by many manufacturers, a smaller volume lowers the cost of QPCR per reaction by 70%.

3. To create the master mix for a 15 µL final volume, 2.5 µL of the primer mix (166 n*M* final concentration) should be added to 7.5 µL of 2X SYBR Green PCR Mix (Applied Biosystems) for a 1X reaction. The amount of master mix created should be equal to *n* + 1, where *n* represents the number of reactions for which the master mix is being created.
4. Once the master mix is created, 10 µL of master mix should be aliquoted per reaction into a 96-well skirted PCR plate.
5. The plate is then moved into the next room and 5 µL of the sample is added to bring the final reaction volume to 15 µL.
6. **Figure 2** shows the contents and volumes necessary for the setup of 96-well QPCR reactions.

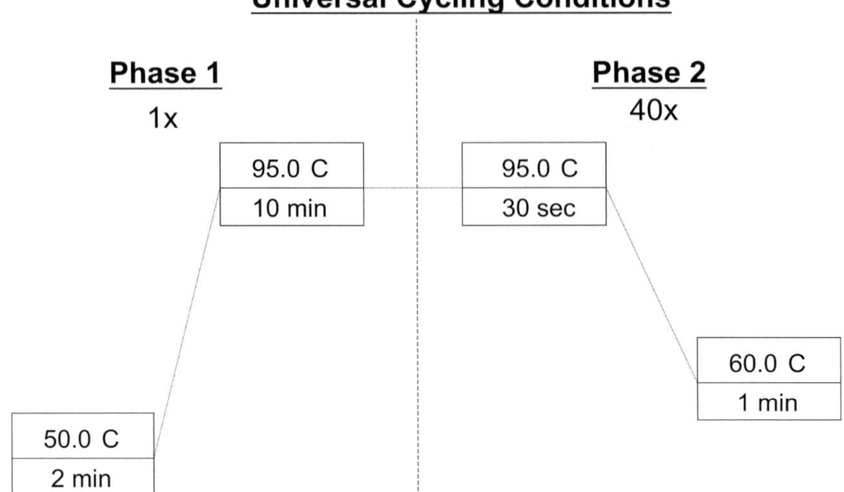

Fig. 3. Universal cycling conditions: representation of the cycling conditions necessary to conduct real-time QPCR. The cycling conditions are separated into two phases (1 and 2). The phases are divided by the dashed line, and the steps are represented by the boxes listing the temperature and the time of the step. The numbers listed below the phase (e.g., 40X in phase 2) shows the number of cycles for which the phase is run.

3.4.2. Cycling Conditions

Listed here are the universal cycling conditions for real-time QPCR (*see* **Note 6**). The cycling conditions consist of two phases (**Fig. 3**). Phase one contains two steps. The first step, 2 min at 50.0°C, is an equilibration step to allow for the action of UNGase. The second step, 10 min at 95.0°C, is used to activate the polymerase within the PCR mix (hot-start PCR). Phase two is the cycling or amplification phase of the reaction. During this phase the first step is a denaturing phase, 30 s at 95.0°C, and the second step is the annealing and elongation phase, 1 min at 60.0°C. This second phase is run for 40 cycles (*see* **Note 7**).

3.5. Probe-Based QPCR

The setup and cycling of a QPCR reaction using a fluorogenic probe (*see* **Note 8**) as a detection system is covered in **Subheadings 3.5.1.** and **3.5.2.**

3.5.1. Setup of the Reaction

Setting up a reaction for probe-based QPCR is similar to the setup of SYBR Green-based QPCR as covered in **Subheading 3.4.1.**

1. First, a primer/probe mix is created. This mix, as the name suggests, contains the primer set and the probe that will be used for detection. The volume of the primer/probe mix that will be added to a 1X reaction is 2.5 µL. As with SYBR Green-based QPCR, the final concentration of the primer in the reaction should be 166 n*M,* so the concentration of the probe in the primer probe mix should be 1 pmol/µL. The probe should be at a final concentration of 166 n*M* in the reaction.

2. Once the primer/probe mix is created, a master mix can be made. A 1X master mix contains 2.5 µL of the primer probe mix and 7.5 µL of the 2X TaqMan PCR mix (Applied Biosystems). The volume of master mix required depends on the number of reactions. Create a volume sufficient for *n* + 1 reactions, where *n* equals the number of reactions you wish to run.

3. Once it is created, the master mix should be aliquoted into the reaction plate at a volume of 10 µL/reaction.

4. The sample can then be added to each reaction at a volume of 5 µL/reaction to yield a total reaction volume of 15 µL.

3.5.2. Cycling Conditions

The universal cycling conditions for SYBR Green-based QPCR are also applicable to primer/probe-based QPCR and are listed in **Subheading 3.4.2.** The conditions are also shown in **Fig. 3** (*see* **Note 9**).

3.6. Multiplex (Multiple Probe) QPCR

The setup and cycling of a QPCR reaction using multiple probes within a single reaction are covered in **subheadings 3.6.1.** and **3.6.2.**

3.6.1. Setup of Multiplex QPCR

The most complex aspect of setting up multiplex QPCR is the creation of the primer/probe mix. This mix should contain all primers and probes that are to be used in the reaction. The concentration of all primers within the mix should be at 1 pmol/µL to yield a final concentration of 166 n*M* in the reaction. The concentration of all the probes within the mix should be 1 µ*M,* so as to yield a final concentration of 166 n*M* within the reaction. It is important that all primers and probes for the reaction be diluted together within the same tube so that only 2.5 µL of the primer/probe mix is needed to yield the proper concentrations of all primers and probes in the final reaction.

One of the complications of multiplex PCR lies in the fact that the more abundant message may plateau before the less abundant mRNA and consume all reagents. If the relative abundance of both targets in the reaction is known, the primer concentration for the more abundant mRNA should be rate-limiting (typically 1/5 to 1/10 of the less abundant mRNA) and has to be determined empirically. For the detection of viral mRNAs, we found that cellular house-keeping mRNAs are 10–1000-fold more abundant, and we typically use those as internal standards, rather than trying to measure two different viral transcripts in the same reaction.

The second step is to create a master mix containing 2.5 µL of the primer/probe mix and 7.5 µL of the TaqMan PCR mix. Once the primer/probe mix is created, the setup of multiplex QPCR is identical to primer/probe-based PCR. Determine the number of reactions needed, and create a sufficient amount of master mix for *n* + 1 reactions, where *n* equals the number of reactions needed. The master mix is then aliquoted into the reaction plate at a volume of 10 µL/reaction. The sample can then subsequently be added to the reaction at a volume of 5 µL/reaction to yield a total reaction volume of 15 µL.

3.6.2. Cycling Conditions

The universal cycling conditions for SYBR Green-based QPCR are also applicable to multiplex QPCR and are listed in **Subheading 3.4.2.** The conditions are also shown in **Fig. 1.**

3.7. Real-Time (Multiple Primer) QPCR Arrays

Once one has become comfortable with setting up real-time QPCR for one mRNA of interest, there is no *a priori* reason not to set up a real-time QPCR for multiple mRNAs. We previously developed a set of 96 primers that query every single ORF of the KSHV genome *(5,12)*.

3.7.1. Design of Real-Time QPCR Arrays

Real-time PCR primers are designed using the same criteria and software as before, either by hand, extracting one ORF at a time from Genbank, or by feeding the entire genomic sequence into the PrimeTime program. The design criteria stay the same. When predicted ORFs overlap, primers are selected outside the region of overlap. Unless a complete transcript map for the virus is known, however, one cannot exclude the possibility that some primers are located in regions in which 3′-UTR or 5′-UTR segments of two different genes overlap.

3.7.2. Setup of the Reaction

Here it is essential to conduct everything in a 96-well format **(Fig. 4)**. It is important that forward primers be synthesized on a separate plate, then the reverse primers. We use a MerMade 96-well synthesizer (BioAutomation, Plano, TX) and store the primers in dH$_2$0 at –80°C at a concentration of 100 pmol/µL. Primer length and purity are verified using a P/ACE MDQ Capillary Electrophoresis System (Beckman Coulter, Fullerton, CA). This service is available in most institutional core facilities or from any number of commercial oligo provider companies (e.g., MWG). It is important to minimize freeze–thaw cycles of the master primers. Hence, we transfer 50 µL forward and 50 µL reverse

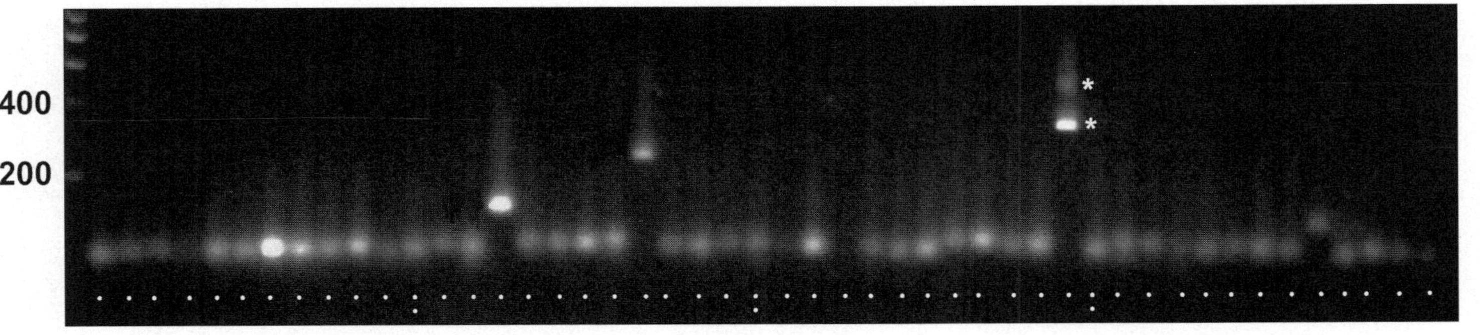

Fig. 4. Ethidium bromide-stained 2% agarose gel of the PCR products for each KSHV ORF after 40 cycles. The template was reverse-transcribed poly(A) mRNA from BCBL-1 cells 48 h after 12-0-tetradecanoyl-phorbol-13-acetate (TPA) induction. Molecular weight markers are shown on the left. Most amplicons are of the same size; exceptions are either housekeeping genes or primers against splice variants (*).

Fig. 5. The CAS-1200 pipeting robot (Corbett Research, Australia). The CAS-1200 is designed specifically for pipeting PCR reactions. The operating system for the robot is straightforward, easy to use, and runs in Windows XP. PCR reactions can be set up in 0.2-mL thin-walled PCR tubes with a 96-well or 384-well format.

primer into a new 96-well plate once a month, which yields enough primer for 100×96 arrays. The reaction is set up essentially as listed in **Subheading 3.4.,** except that a multipipetor can be used to add primer mix to the reaction plate and later the sample to each well. Mastermix is added to each well using a repeat pipetor. Alternatively, a robot **(Fig. 5)** can be used in this step.

3.7.3. Cycling Conditions

The universal cycling conditions for SYBR Green-based QPCR are also applicable to multiplex QPCR and are listed in **Subheading 3.4.2.** The conditions are also shown in **Fig. 1.**

3.8. Analysis of Real-Time QPCR and Real-Time QPCR Arrays

3.8.1. Theoretical Considerations

Prior to reaching saturation (owing to exhaustion of primers and nucleotides, loss of polymerase activity, and so on), PCR amplification proceeds exponentially and can be described by $N_i = N_0 \times (1 + k)^i$, where N_0 represents the number of molecules in the original sample and N_i the number of mRNA molecules at cycle i (i = 1–40). During the exponential phase, the amplification efficiency k ($0 \leq k \leq 1$) of a given primer pair is constant. Before real-time PCR, it was not easy to identify the exponential phase of the reaction. Either the same reaction was run for different cycle numbers (20, 22, 24, and so on) and the product quantified by gel electrophoresis using the same amount of sample in each case, or different dilutions of sample were used in multiple PCR reactions for the same cycle number. During real-time QPCR the amount of product at each cycle is quantified *(3)*. Fluorescence intensity, *Rn*, has a logarithmic dependence on fluorophor (the PCR product) concentration, yielding $Rn = \log(N_i) = \log[N_0 \times (1 + k)^i]$. Real-time quantitative PCR compares two samples with target concentrations N_a and N_b by recording the cycle numbers (C_T) for *a* and *b* at which the amplification product yields enough fluorescence to cross an operator-determined threshold *T* (set at five times the SD of the nontemplate control [NTC]). Consequently, $Rn_a = Rn_b$ and $\log[N_a \times (1 + k)^a] = \log[N_b \times (1 + k)^b]$ or $\log(N_a) - \log(N_b) = \log_{(1+k)}b - \log_{(1+k)}a = \log_{(1+k)}b - a$ (for i = 0, $N_{i=0} = N_0 \times (1 + k)^0$, i.e., $N_{i=0} = N_0$). Ideally, $k = 1$ and $(1 + k) = 2$, i.e., at each cycle two reactions products are produced per target molecule. This leads to $N_i = N_0 \times (1 + 1)^i = N_0 \times 2^i$. Assuming $\log = \log_2$, $N_a/N_b = 2^{b-a}$, where N_a/N_b represents the fold difference in mRNA levels of two samples with $C_T = a$ and $C_T = b$.

Hence, it is possible to extract the relative ratio of abundance in two samples based on this calculation. Interestingly, hybridization-based DNA arrays have similar characteristics, since the color intensity ratio in a fluorescent Cy3/Cy5 DNA array exhibits a logarithmic dependence on the amount of hybridized probe *(13)*. Analogous to the amplification efficiency k for PCR, a hybridization-efficiency K_0 applies to DNA arrays, which is a function of the length and base composition of the particular cDNA fragment at a given hybridization temperature.

3.8.2. Absolute Quantification for One Primer Pair on Multiple Samples

To quantify the abundance of a single mRNA and/or viral species in diagnostic applications, a standard curve is generated that plots the C_T number in relation to the copy number per unit, for instance, copy number per 10^6 cells or per 1 µg DNA *(14)*. Actual values are interpolated by linear regression analysis **(Fig. 6)**. The slope of the dilution curve defines the amplification efficiency k. A decision is made based on the interpolated copy numbers, and the signifi-

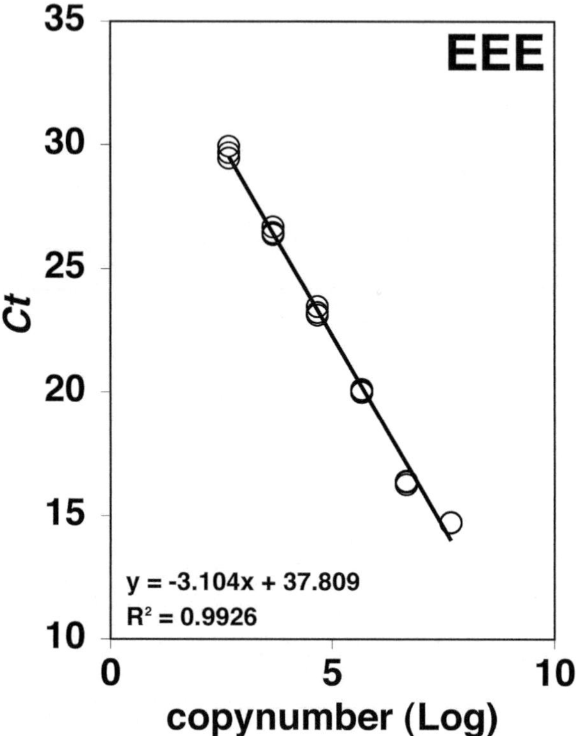

Fig. 6. Linear regression of a real-time QPCR primer pair for eastern equine encephalitis virus (EEE). Plotted on the *X*-axis is the log of the copy number against the C_T achieved for each dilution on the *Y*-axis. The slope of the line is listed in the form of $y = mx + b$, and R^2 = the regression coefficient. Each dilution was amplified in triplicate.

cance of the observation is established by multiple measurements per sample. Mean, standard deviation (SD), and/or confidence intervals (CVs) can be calculated from the interpolated copy number per sample, and goodness of fit of the standard curve can be gauged by its regression coefficient R^2 (reviewed in **ref. *15***). Calculations are performed using Excel or more advanced statistical software such as SPSS (SPSS Science, Chicago, IL). A standard curve will also reveal the linear range of real-time QPCR for a particular primer pair, such as in viral load assays. This type of validation is ideally suited for the quantification of multiple samples with a single primer pair. However, it is very cumbersome, and in order to maintain perfect accuracy, a standard curve has to be included with each amplification group (96-well plate) and for each primer pair (*see* **Note 10**).

3.8.3. Absolute Quantification of Two or More Different Primer Pairs

Standard curves are generated for each primer pairs as in **Subheading 3.8.2.,** and actual copy numbers are interpolated for each target. Copy numbers can then be compared for each target over multiple samples using conventional statistics as outlined in **Subheading 3.8.2.** Furthermore, copy numbers for the two (or more) different targets can be compared with each other. For example, the relative mRNA levels for two different mRNAs in the same tissue can be recorded. Calculations are performed using Excel or more advanced statistical software such as SPSS.

3.8.4. Relative Quantification for One Primer Pair

Often transcriptional profiling is concerned only with relative differences between two samples, *a* and *b*, which are expressed in unit less fold change. Hence raw C_T numbers can be used directly. Relative quantification eliminates the intermediate use of a standard curve and allows for the direct comparison of the fold differences between two target populations. This only requires the data for each sample, not a standard curve *(16–19)*. By applying rank-based statistics (Wilcoxon's sum rank test) or a simple *t*-test, we can determine, for instance, whether one of the tissues or treatment yields to a relative (and statistically significant) change in mRNA levels between different samples.

3.8.5. Relative Quantification for Multiple Primer Pairs

The unmanipulated C_T data for multiple primer pairs and multiple samples can also be used to extrapolate the relative expression pattern for many genes. To do so, we need to apply hierarchical clustering, as previously described *(20)*. Importantly, the same clustering algorithms that are in use for hybridization array analysis can be used to analyze real-time QPCR arrays (**Fig. 7**). Instead of feeding in the individual spot intensities as recorded in hybridization arrays *(21)* as a gene by experiment table into the program, individual C_T values in the format of a PCR primer by experiment table are used as input.

In order for relative quantification to be valid between different primer pairs, three constraints are placed on the amplification efficiency for each primer pair *k:*

1. The amplification efficiency k or $E = (1 + k)$ must not change with increasing cycle number. This assumption is valid only during exponential amplification, when primers, polymerase activity, or nucleotides do not limit the reaction. Setting the threshold appropriately guarantees that this key assumption is not violated.
2. The amplification efficiency k is constant over a wide range of concentrations (typical five orders of magnitude for real-time QPCR, which determines the linear range of the assay) but may not be accurate for comparing very low or high levels

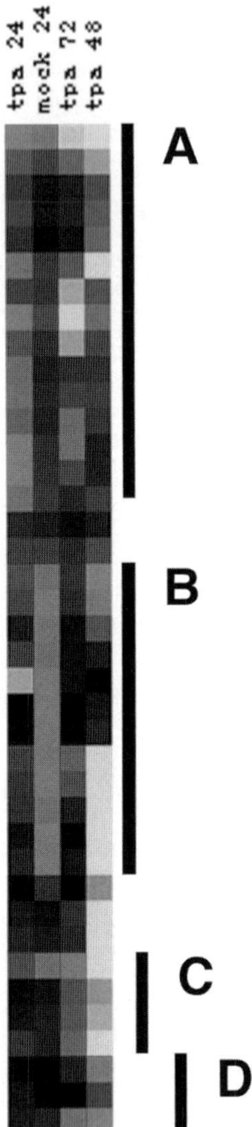

Fig. 7. Representation of hierarchical cluster analysis. Shades of gray indicate transcription level with lighter shades representing increased transcription. Groups of genes clustering together are shown by the thick black lines next to the clustogram, and each group is labeled with a letter. A total of four clusters are found in this figure (A, B, C, and D).

of target DNA. In contrast to conventional end-point or gel-based PCR methods, real-time QPCR instruments and fluorescent chemistry record the entire amount of product at each cycle and thus allow for the direct visual observation of constraints (1) and (2) for each data point.

3. The amplification efficiency k determines the spread, i.e., into how many fold target-level-difference a given C_T difference translates. Under ideal amplification conditions, exactly two molecules are produced per parent at each cycle. This assumption leads to the widely used shortcut to convert C_T differences into fold differences: fold difference $(a - b) = 2 \exp(C_{Ta} - C_{Tb})$.

Is this a reasonable supposition? **Figure 8A** visualizes the effect of changes in k by plotting relative fold difference for various amplification efficiencies $E = k + 1$. Assuming ideal amplification ($k = 1$, i.e., $E = 2$), a C_T difference of five cycles between two samples C_{Ti} and C_{Tj} translates into a 32-fold difference in input levels. However, if the amplification reaction proceed with 20% less efficiency than ideal ($k = 0.8$, i.e., $E = 1.8$), $C_{Ti} - C_{Tj} = 5$ represents only a 19-fold difference. If the PCR efficiency drops below $k = 0.6$, even a 10-cycle difference in C_T does not yield an appreciable fold difference. Since most PCR reactions do not proceed under ideal conditions, assuming $k = 1$ ($E = k + 1 = 2$) almost always overestimates the true difference in target levels. This explains some of the outrageous discrepancies in fold induction/suppression, observed when DNA hybridization array data were verified by real-time QPCR.

Multiple-primer real-time QPCR arrays compound this problem, since the aim is to compare many different primer pairs with each other. This is strictly possible only under than additional constraint, that the amplification efficiencies E_a and E_b for any two primer pairs a and b in the array do not differ from each other. It makes a comparison between different primer pairs (measuring the transcription profile of different mRNAs) impossible, without first determining the standard curves for each primer pair j, $j = 1 \ldots m$, and then comparing fold differences obtained after absolute quantification. Surprisingly, however, most primer pairs have very similar amplification efficiencies. We typically calculate the amplification efficiency by dilution once for each primer pair and exclude primers that fail to amplify with $E < 1.8$ (**Fig. 8B**). According to the considerations in **Fig. 8A,** the maximal error introduced by different amplification efficiencies in this case is twofold, or one C_T unit. This is less than the experimental error in most cases. More elaborate schemes have been and are still being developed to compare multiple primers *(16,17)*. For transcription profiling, however, in which only the change between any two samples is recorded (e.g., increase over time), relative clustering of the C_T values will easily discern different response classes (*see* **Note 11**).

Finally, it is important to realize that by and large the error for each primer pair is dependent on the experimental error and handling error only, but not on the amplification efficiency or the amount of input sample (**Fig. 8C**).

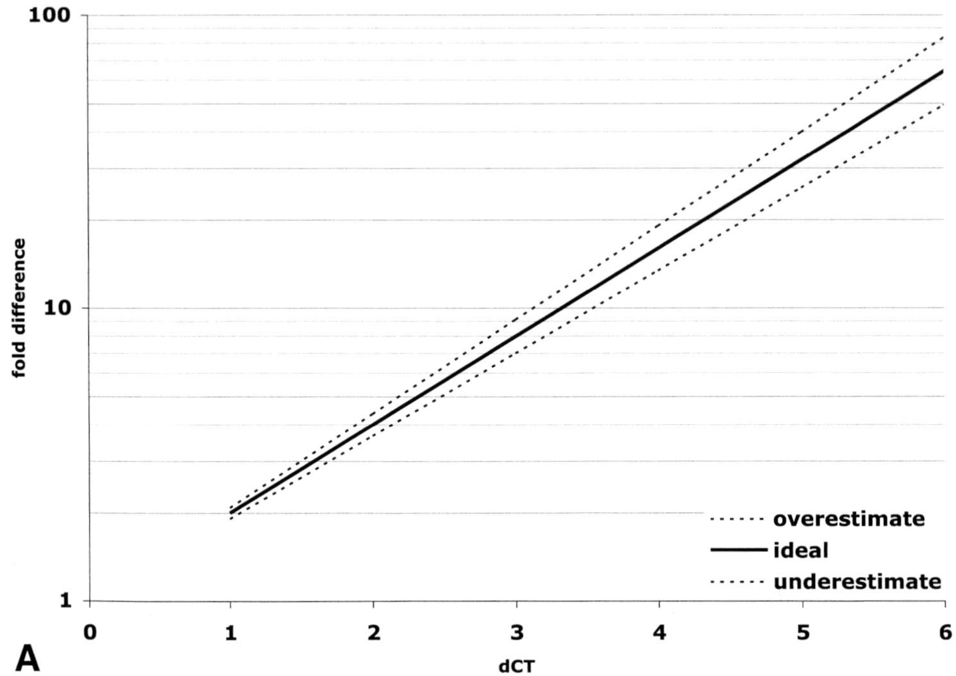
100
10
1
fold difference
overestimate
ideal
underestimate
0
1
2
3
4
5
6
dCT
A

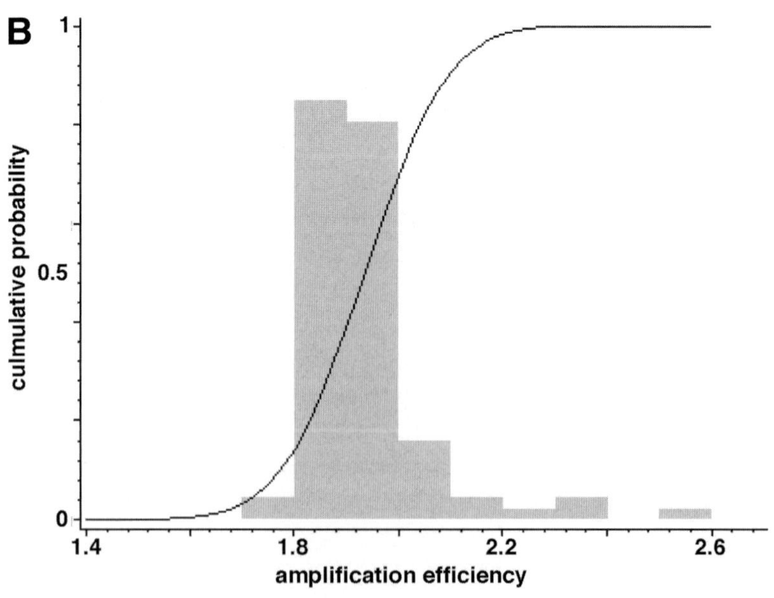
B
1
0.5
0
culmulative probability
1.4
1.8
2.2
2.6
amplification efficiency

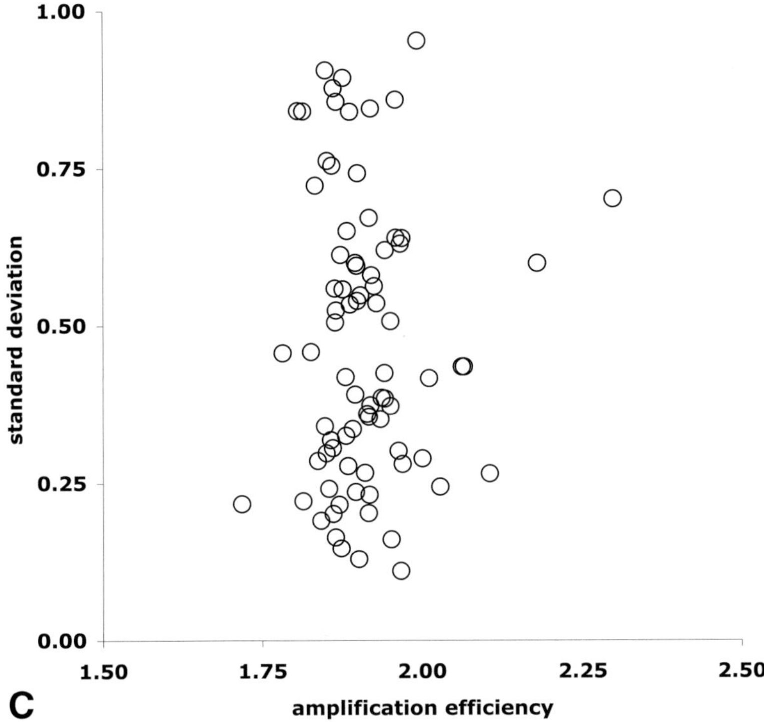

Fig. 8. Amplification efficiency of real-time QPCR. **(A)** Visualization of the effect of changes in amplification efficiency when comparing fold differences between primers. Theoretical normalized C_T values are plotted on the *x*-axis against relative fold difference of the *y*-axis. Ideal amplification is represented by the solid black line; over- and underestimates are shown with dotted black lines. **(B)** Amplification efficiencies for 91 primers. The gray histogram shows the number of primers for a given amplification efficiency. Overlaid is the culmulative probability. **(C)** Error is independent of amplification efficiency. This is shown by plotting amplification efficiency (*x*-axis) against the standard deviation (*y*-axis) for each primer in the KSHV 96-primer array.

3.8.6. Normalization

For real-time QPCR, two types of normalization can be applied: type I normalization relative to a reference sample t_0 or median for each gene yielded dCT, and type II normalization relative to the reference gene, e.g., GAPDH, yielded DCT. The latter eliminates differences caused by variation in the overall input cDNA concentration. Using experimental samples (e.g., response to a particular drug in culture), one should set up the experiment and normalize the

Table 1
Normalization Possibilities for Two Genes (*A* and *B*)

Gene	Time				Median	Mean	SD
	0	1	2	3			
Raw data							
A	10	12	14	16	13	13	2.58
B	25	25	25	25	25	25	0.00
GAPDH	20	21	20	19	20	20	0.82
Normalization							
Type I (*T0*)							
A	0	2	4	6	3	3	2.58
B	0	0	0	0	0	0	0.00
GAPDH	0	1	0	−1	0	0	0.82
Type II							
A	−10	−9	−6	−3	−7.5	−7	3.16
B	5	4	5	6	5	5	0.82
GAPDH	0	0	0	0	0	0	0.00
Type II followed by type I							
A	0	1	4	7	2.5	3	3.16
B	0	−1	0	1	0	0	0.82
GAPDH	0	0	0	0	0	0	0.00

input material (e.g., same number of cells) such that the variation in the reference gene is $\leq 1 \times C_T$ unit.

During type I normalization, only C_T values of a single primer pair are compared with each other. Hence amplification efficiency differences between primer pairs do not enter the calculation. In contrast, type II normalization compares two different primers pairs, such as for gene *A* and gene GAPDH, with associated, possibly different, amplification efficiencies k_A and k_{GAPDH}. After both normalizations were applied successively, we obtained dDCT. (unfortunately, the current literature uses DCT for type II and DDCT to denote the outcome after both normalizations (ABI user bulletin P/N4303859), which masks the different properties of the two operations). For relative analysis, clustering is performed in log-space (C_T values) rather than interpolated mRNA levels, and only a linear term is subtracted during normalization, which does not impact on the rank order between samples (*see* **Note 12**).

A simple time-course experiment can exemplify the analysis (**Table 1**). It shows imaginary C_T values at different times after treatment for a set of three mRNAs named *A, B,* and GAPDH. At this point, the biological interpretation is clear: gene *A* mRNA levels decrease over time (as evidenced by increasing C_T values), whereas gene *B* and gene GAPDH mRNA levels remain constant. This is often the only conclusion that an investigator needs to draw from a particular inquiry: which mRNAs are induced and which are suppressed relative to each other at any given time (or tumor sample or treatment modality).

In its easiest incarnation, we analyze the data as follows (*see* **Table 1**):

1. Transfer all C_T values into Excel.
2. Set up an $m \times n$ chart: the columns for condition, here $t0$–$t3$, and the rows for each primer pair, here gene *A*, gene *B*, and gene GAPDH (*see* the raw data group in **Table 1**).
3. Calculate the mean, median, and SD.
4. Apply type I normalization by subtracting the $C_T(t0)$ for each gene from all subsequent samples (*see* the type I normalization group in **Table 1**) to yield dCT. Instead of any particular sample, either the mean or median may be used as well. Note that the SD did not change.
5. Sort the data (in Excel) according to the SD. This will identify genes that are up- and downregulated over time, and the response will be monotonous. Since GAPDH has an SD of 0.82 (or ~1 C_T U), any gene that exhibits an SD great than 2X the SD of GAPDH is thought to respond to treatment.
6. Import the Excel spreadsheet into cluster from Eisen et al. *(20)* and http://genome-www5.stanford.edu/MicroArray/SMD/restech.html, cluster the array, and generate an image with Treeview. Note that the red–green scale in the cluster program operates on a range of –1 to +6. Graduations in color correspond to change with treatment. An mRNA that changes from 200 to 500 copies will cluster next to an mRNA that changes from 20,000 to 50,000 copies.
7. Type I normalization highlights the change with treatment and eliminates any differences in the basal levels (all genes are 0 at $t0$).
8. Generate a standard curve (simply cloning the PCR product will yield a target), and perform absolute quantification.

To accommodate samples of different experiments or different amounts of cDNA pools, we add type II normalization (*see* **Table 1**):

1. Transfer all C_T values into Excel.
2. Set up an $m \times n$ chart: the columns for condition, here $t0$ $t3$, and the rows for each primer pair, here gene *A*, gene *B*, and gene GAPDH (*see* the raw data group in **Table 1**).
3. Calculate the mean, median, and SD.
4. Apply type II normalization by subtracting the C_T for GAPDH from each gene for all samples (*see* the type II normalization group in **Table 1**) to yield DCT. Instead of any particular sample, either the mean or median may be used as well.
5. Sort the data (in Excel) according to the SD. This will identify genes that are up- and downregulated over time, and the response will be monotonous. Since GAPDH has an SD of 0.82 (or ~1 C_T U), any gene that exhibits an SD great than 2X the SD of GAPDH is thought to respond to treatment. Note that the SD of GAPDH is now zero, and the SD for all other genes has changed and reflects the combined SD for GAPDH and the gene of interest.

6. In addition, we can obtain an impression of the relative levels: assuming ideal amplification $E = 2$, at $t3$ gene A has a DCT of 7 [= 6 − (−1)] or is expressed at approx 2^{-7} or 0.8% the level of GAPDH.

7. Import the Excel spreadsheet into cluster from Eisen et al. *(20)* and (http://genome-www5.stanford.edu/MicroArray/SMD/restech.html, cluster the array, and generate an image with Treeview. Note that the red–green scale in the cluster program operates on a range of −10 to +6. Graduations in color correspond to the overall level of mRNA rather than the change with treatment.

8. Generate a standard curve (simply cloning the PCR product will yield a target), and perform absolute quantification.

Finally, we can combine both normalization procedures as follows (**Table 1**):

1. Transfer all C_T values into Excel.
2. Set up an $m \times n$ chart: the columns for condition, here $t0$ $t3$, and the rows for each primer pair, here gene A, gene B, and gene GAPDH (*see* the raw data group in **Table 1**).
3. Calculate the mean, median, and SD.
4. Apply type II normalization and type I normalization to each data point (*see* the type II followed by type I normalization group in **Table 1**) to yield dDCT. Instead of any particular sample, either the mean or median may be used as well.
5. Sort the data (in Excel) according to the SD. This will identify genes that are up- and downregulated over time, and the response will be monotonous. Since GAPDH has an SD of 0.82 (or ~1 C_T U), any gene that exhibits an SD great than 2X the SD of GAPDH is thought to respond to treatment.
6. We still obtain an impression of the relative levels for all data points except at $t0$; assuming ideal amplification $E = 2$, at $t3$ gene A has a DCT of 7 (= 7 − 0) or is expressed at approx 2^{-7} or 0.8% the level of GAPDH.
7. Import the Excel spreadsheet into cluster from Eisen et al. *(20)* and (http://genome-www5.stanford.edu/MicroArray/SMD/restech.html, cluster the array, and generate an image with Treeview.
8. Generate a standard curve (simply cloning the PCR product will yield a target), and perform absolute quantification.

4. Notes

1. The following web sites offer free programs and further discussion:
   Cluster analysis: shareware and publications
   http://genome-www5.stanford.edu/MicroArray/SMD/restech.html.
   http://www.biochem.ucl.ac.uk/bsm/virus_database/vgbg.html.
   http://lymphochip.nih.gov/index2.html.
   http://www.gene-regulation.com/pub/databases.html.
   http://srs.ebi.ac.uk/srsbin/cgi-bin/wgetz?-page+srsq2+-noSession.
   http://llmpp.nih.gov/.
   http://nciarray.nci.nih.gov/.
   http://srs.embl-heidelberg.de:8000/srs5/.

http://www.embl-heidelberg.de/ chenna/clustal/darwin/.
Real-time QPCR: shareware and resources
http://www.wzw.tum.de/gene-quantification/.
http://medgen31.rug.ac.be/primerdatabase/links_menu.php.
Commercial sites
http://www.optimaldesign.com/index.html.
http://www.silicongenetics.com/cgi/SiG.cgi/Products/GeneSpring/index.smf.
http://www.affymetrix.com/community/index.affx.
http://www.panomics.com/.
http://www.gene-regulation.com/pub/databases.html.

2. Although the stipulations for primer design do not state that the primer T_m should be 60°C, it is best if the primers are as close to this temperature as possible. The reason for this strict adherence is because 60°C is the temperature used in the universal cycling conditions. If the primers are designed to work at this temperature, then the cycling conditions do not need to be altered for the primers to amplify. It is better to design your primers around the conditions than to alter the conditions to the primer pair, especially in the comparison of multiple primer pairs, as occurs when normalizing to cellular housekeeping genes.

3. Whole Genome Primer Design using PrimeTime *PrimeTime* (Vahrson and Dittmer) is a software program for large-scale primer design. It automates the procedures of designing PCR primers, making it feasible to generate hundreds of primers for a whole viral genome in a few minutes. It is written in the Ruby programming language *(22)* and is built on top of *EMBOSS (10)* and *Primer3 (7)*. It extends the capabilities of the underlying programs by allowing one to position primers relative to the start or end of an ORF, by a fail-soft option that retries a primer search with slightly varied parameters when the original request failed, and by employing basic quality-control procedures to ensure the uniqueness of a primer combination in the genome. *PrimeTime* is invoked with the name of the file containing the sequence as parameter. The file must be in EMBL database format *(10)*: *PrimeTime* writes the results of the primer search into file *sequence-file.primetime*. For each primer pair, it reports the positions, sequences, GC ontents, T_M, and lengths for both primers, as well as the length of the amplification product. The file is in a tab-delimited format that can easily be read into a spreadsheet program like Excel.

4. A general problem in genome-wide primer design is the presence of repeats in the template sequence, which may lead to ambiguous amplification products. *PrimeTime* checks the resulting primers in a way that is independent of the original primer search, as performed by *Primer3*. It identifies and flags problematic regions, which then need to be inspected manually.

5. The use of SYBR Green as a detection method does not allow for the recognition of a specific amplicon, as with a labeled probe. This is because SYBR Green binds to all dsDNA in the reaction including primer dimers. This problem is overcome by two solutions. The first is that the concentration of the primer is very low (166 n*M*), and therefore the primer will preferably bind to the sample, reducing the risk of primer dimer formation. The second comes with the application of melting curve

analysis. By measuring the fluorescence of the sample over a range of temperatures (e.g., 60°C–92°C) after completion of the reaction, one can determine the melting point of the amplicon by the fluorescence emitted. Only a single, sharp peak should be in evidence. The T_m of the amplicon can then be compared with that of the positive control to determine that the correct sequence has been amplified.

6. It is important to choose the proper real-time QPCR equipment. Different molecular beacons (e.g., FAM and JOE) as well as SYBR Green emit fluorescence at different wavelengths, and not all thermocyclers are compatible with all chemistries. Some models can only read a within a certain wavelength, only allowing the detection of a single chemistry, and therefore are not compatible for multiplex QPCR. Other machines, such as the ABI Prizm 7700, can read all available chemistries and would therefore be a better choice for running multiplex over a single detection machine.

7. All samples need to be analyzed in triplicate. In our hands the combined pipeting and instrument error was less than 6%. The ABI7900HT can distinguish twofold differences (between 5000 and 10,000 copies) with a 97.7% confidence level. Most biologically relevant changes in viral transcription exhibit a much higher level of variation.

8. When using TaqMan probes to identify amplified products, we have not experienced any problems with nonspecific signals *(14)*. However, should such problems be encountered, a number of recent developments that have increased the specificity of real-time quantitative PCR can be used. These are (1) substitution of the conventional, fluorescent quencher TAMRA with a nonfluorescent (dark) quencher, (2) incorporation of a minor-groove binder (MGB) *(27,28)* or (3) incorporation of 5-propyne-2′-deoxyuridine into the probe *(29,30)*. These should solve any and all specificity problems that might arise. It is also a good idea to monitor the web site 7700taqman@listserv.acns.nwu.edu to keep abreast of the latest improvements in quantitative real-time PCR technology.

9. What are the sensitivity and specificity of PCR and RT-PCR, respectively? We previously used 2.5 µg total DNA (corresponding to ~5 × 10^5 cells) and were able to detect 1000 copies of KSHV in the sample. Although this sensitivity proved sufficient for our studies in animals, clearly better sensitivity can be achieved, for instance, by extending the PCR to 50 cycles, and DNA isolation can be improved using a QiAmp DNA isolation kit instead of the traditional proteinase K digestion. This sensitivity equals published reports that demonstrate a linear range of TaqMan-based quantification of 10^2–10^6 copies of KSHV per 10^6 peripheral blood mononuclear cells with a CV of 10% *(24,25)*. In the case of influenzavirus, real-time quantitative PCR was shown to be as sensitive as nested PCR, with less nonspecific amplification *(26)*.

10. Normalization is a recurring problem in comparative mRNA analysis. We now include five TaqMan amplicons (gapdh, actin, actin-2, c-myc, and hprt) that are specific for human housekeeping genes and use iterative geometric averaging *(18)* to determine the most appropriate control for a given data series. Adding a synthetic mRNA of known copy number (coding for the bacterial gene for neomycinR

and β-galactosidase), prior to reverse transcription, may be used to control for enzyme efficiency of both the reverse transcriptase and the *Taq* polymerase.

11. In conventional PCR, different primers perform with different amplification efficiencies and require different annealing temperatures. In contrast, real-time QPCR primers are designed to fit very narrow performance criteria (ABI Bulletin #P/N4303859). We found no need to compute initial mRNA levels or to use an external standard curve for purely comparative analysis, since the average amplification efficiency E for each primer pair in our published viral real-time QPCR array was 1.94±0.12 (n = 91), and the associated standard error across these primers directed against the same target (purified viral linear genomic DNA) was 0.06-fold *(6)*. By contrast, cellular mRNA levels typically change several fold in response to specific stimuli. We estimate that for any target in the array, the biological variation associated with clinical specimens is well above the experimental error for this technology. By including a defined copy number of an exogenous, synthetic RNA prior to reverse transcription, another layer of standardization may be added. This yields a truly quantitative assay with a linear range of six orders of magnitude.

12. The primary achievement of real-time QPCR is that for the first time PCR delivers reliable quantitative information without the need for dilution series, internal competitors, and so on. The quantitative information can be extracted because the PCR reaction is monitored in real time, i.e., the reaction product is quantified at every cycle, and only data points during exponential amplification are used to compute the target concentration *(3)*. In adapting real-time QPCR to comparative transcription profiling *(5,6)*, we realized that we could use the real-time QPCR output (the so-called C_T value) for all primers in the array directly in existing cluster analysis programs such as those developed by Eisen et al. *(20)*. In fact, the initial step in hybridization-based analysis is to compute the logarithm of the signal intensity, in order to improve statistical performance *(21)*, whereas the real-time QPCR output (CT) already represents a logarithmic measure of the target concentration and can therefore be used directly for robust analysis.

References

1. Mullis, K. B. and Faloona, F. A. (1987) Specific synthesis of DNA in vitro via a polymerase-catalyzed chain reaction. *Methods Enzymol.* **155,** 335–350.
2. Lomeli, H., Tygai, S., Pritchard, C. G., Lizardi, P. M., and Kramer, F. R. (1989) Quantitative assays based on the use of replicable hybridization probes. *Clin. Chem.* **35,** 1826–1831.
3. Heid, C. A., Stevens, J., Livak, K. J., and Williams, P. M. (1996) Real time quantitative PCR. *Genome Res.* **6,** 986–994.
4. Mackay, I. M., Arden, K. E., and Nitsche, A. (2002) Real-time PCR in virology. *Nucleic Acids Res.* **30,** 1292–1305.
5. Fakhari, F. D. and Dittmer, D. P. (2002) Charting latency transcripts in Kaposi's sarcoma-associated herpesvirus by whole-genome real-time quantitative PCR. *J. Virol.* **76,** 6213–6223.

6. Dittmer, D. P. (2003) Transcription of Kaposi's sarcoma-associated herpesvirus in Kaposi's sarcoma lesions. *Cancer Res.* **63,** 2010–2015.

7. Rozen, S. Primer3 Software Distribution. Retrieved on May 18, 2004. http://frodo.wi.mit.edu/primer3/primer_code.html. Boston, MA.

8. Santos, R. A., Hatfield, C. C., Cole, N. L., et al. (2000) Varicella-zoster virus gE escape mutant VZV-MSP exhibits an accelerated cell-to-cell spread phenotype in both infected cell cultures and SCID-hu mice [In Process Citation]. *Virology* **275,** 306–317.

9. Lin, S. F., Robinson, D. R., Oh, J., et al. (2002) Identification of the bZIP and Rta homologues in the genome of rhesus monkey rhadinovirus. *Virology* **298,** 181–188.

10. Rice, P., Longden, I., and Bleasby, A. (2000) *EMBOSS: the European Molecular Biology Open Software Suite. Trends Genet* **16,** 276–277.

11. Bentley, D. (2002) The mRNA assembly line: transcription and processing machines in the same factory. *Curr. Opin. Cell Biol.* **14,** 336–342.

12. Dittmer, D. P. (2003) Transcription profile of Kaposi's sarcoma-associated herpesvirus in primary Kaposi's sarcoma lesions as determined by real-time PCR arrays. *Cancer Res.* **63,** 2010–2015.

13. Cantor, C. R. and Schimmel, P. R. (1980) *Biophysical Chemistry: The Behavior of Biological Macromolecules.* Freeman, New York.

14. Dittmer, D., Stoddart, C., Renne, R., et al. (1999) Experimental transmission of Kaposi's sarcoma-associated herpesvirus (KSHV/HHV-8) to SCID-hu Thy/Liv mice. *J. Exp. Med.* **190,** 1857–1868.

15. Glantz, S. and Slinker, B. (2000) *Primer of Applied Regression & Analysis of Variance.* McGraw-Hill, New York.

16. Pfaffl, M. W. (2001) A new mathematical model for relative quantification in real-time RT-PCR. *Nucleic Acids Res.* **29,** E45–55.

17. Pfaffl, M. W., Horgan, G. W., and Dempfle, L. (2002) Relative expression software tool (REST) for group-wise comparison and statistical analysis of relative expression results in real-time PCR. *Nucleic Acids Res.* **30,** e36.

18. Vandesompele, J., De Preter, K., Pattyn, F., et al. (2002) Accurate normalization of real-time quantitative RT-PCR data by geometric averaging of multiple internal control genes. *Genome Biol.* **3,** RESEARCH0034.

19. Lossos, I. S., Czerwinski, D. K., Wechser, M. A., and Levy, R. (2003) Optimization of quantitative real-time RT-PCR parameters for the study of lymphoid malignancies. *Leukemia* **17,** 789–795.

20. Eisen, M. B., Spellman, P. T., Brown, P. O., and Botstein, D. (1998) Cluster analysis and display of genome-wide expression patterns. *Proc. Natl. Acad. Sci. USA* **95,** 14863–14868.

21. Bowtell, D. and Sambrook, J. (2003) *DNA Microarrays: A Molecular Cloning Manual.* Cold Spring Harbor Laboratory Press, Cold Spring Harbor, NY, p. 712.

22. Thomas, D. and Hunt, A. (2001) *Programming Ruby.* Addison-Wesley, Boston.

23. Hwang, S., Lee, D., Gwack, Y., Min, H., Choe, J. (2003) Kaposi's sarcoma-associated herpesvirus K8 protein interacts with hSNF5. *J. Gen. Virol.* **84,** 665–676.

24. Lallemand, F., Desire, N., Rozen Daum, W., et al. (2000) Quantitative analysis of human herpesvirus 8 viral load using a real-time PCR assay. *J. Clin. Microbiol.* **38,** 1404–1408.
25. White, I. E. and Campbell, T. B. (2000) Quantitation of cell-free and cell-associated Kaposi's sarcoma-associated herpesvirus DNA by real-time PCR. *J. Clin. Microbiol.* **38,** 1992–1995.
26. Schweiger, B., Zadow, I., Heckler, R., et al. (2000) Application of a fluorogenic PCR assay for typing and subtyping of influenza viruses in respiratory samples. *J. Clin. Microbiol.* **38,** 1552–1558.
27. Kutyavin, I. V., Afonia, I. A., Mills, A., et al. (2000) 3′-Minor groove binder-DNA probes increase sequence specificity at PCR extension temperatures. *Nucleic Acids Res.* **28,**. 655–661.
28. Afonina, I., Zivarts, M., Kutyavin, I. V., et al. (1997) Efficient priming of PCR with short oligonucleotides conjugated to a minor groove binder. *Nucleic Acids Res.* **25,** 2657–2660.
29. Buhr, C. A., Wagner, R. W., Grant, D., and Froeheen, B. C. (1996) Oligodeoxynucleotides containing C-7 propyne analogs of 7-deaza-2′-deoxyguanosine and 7-deaza-2′-deoxyadenosine. *Nucleic Acids Res.* **24,** 2974–2980.
30. Wagner, R. W., Matteucci, M. D., Lewis, J. G., et al. (1993) Antisense gene inhibition by oligonucleotides containing C-5 propyne pyrimidines. *Science* **260,** 1510–1513.

30

Rapid Screening of Chemical Inhibitors That Block Processive DNA Synthesis of Herpesviruses

Potential Application to High-Throughput Screening

Robert P. Ricciardi, Kai Lin, Xulin Chen, Dorjbal Dorjsuren, Robert Shoemaker, and Shizuko Sei

Summary

Processivity factors associate with DNA polymerases, enabling them to incorporate thousands of nucleotides without dissociating from the template. The processivity factors encoded by each of the herpesviruses are ideal targets for specifically blocking viral replication, particularly since they have unique primary amino acid sequences. Here we provide details of a rapid mechanistic plate assay and its potential application to high-throughput screening of libraries of tens of thousands of chemical compounds to identify inhibitors of processive DNA synthesis. Methods of validation testing are presented.

Key Words: Processivity factor; DNA polymerase; PF-8; Pol-8; KSHV; herpesviruses; inhibitor screening; mechanistic plate assay; high-throughput screening; chemical combinatorial library.

1. Introduction

To copy their genomes successfully, DNA polymerases need to be processive, i.e., they need to be able to incorporate thousands of nucleotides without dissociating from the template. To accomplish this task, most DNA polymerases, including those from bacteria, yeast, mammals, and viruses, are dependent on an accessory protein, often referred to as a processivity factor. Two notable processivity factors are eukaryotic proliferating cell nuclear antigen and the *E. coli* β-subunit, which encircle the DNA in the form of sliding clamps that tether their polymerases to the template, enabling them to synthesize extended strands (reviewed in **ref. 1**). Interestingly, the well-studied bacteriophage T7 DNA polymerase recruits *E. coli* thioredoxin to function as its processivity fac-

From: *Methods in Molecular Biology, vol. 292: DNA Viruses: Methods and Protocols*
Edited by: P. M. Lieberman © Humana Press Inc., Totowa, NJ

tor. Thioredoxin does not actually form a circular sliding clamp, but, through its association with T7 DNA polymerase, essentially converts the polymerase into a clamp-like structure to prevent it from dissociating from DNA *(2)*. Thus, although processivity factors perform a common function, they have varied sequences and can be varied in structure.

Many viruses encode processivity factors, making them ideal targets for blocking viral replication. For example, there are eight known human herpesviruses, human herpes simplex viruses 1 and 2 (HSV-1 and HSV-2); varicella-zoster virus (VZV), Epstein-Barr virus (EBV), cytomegalovirus (CMV), human herpesviruses 6 and 7 (HHV-6 and HHV-7), and human herpesvirus 8 or Kaposi's sarcoma-associated herpesvirus (KSHV or HHV-8); they all contribute to a range of human diseases and malignancies. The especially attractive feature of these herpesvirus processivity factors is the apparent specificity they have for their cognate viral DNA polymerases *(3)*. This is consistent with the primary amino sequences of the herpesvirus processivity factors being largely unique *(4)*. In addition, more recent evidence indicates that the functional subunit composition of the individual herpesvirus processivity factors can be different, i.e., the processivity factor of HSV-1 appears to function as a monomer *(5,6)* whereas KSHV *(7)* and CMV *(7,8)* function as dimers.

Here we describe a rapid mechanistic plate assay *(9,10)* and its potential application to high-throughput screening of a chemical library of tens of thousands of chemical compounds to identify inhibitors of processive DNA synthesis. We use the DNA polymerase (Pol-8) and processivity factor (PF-8) of KSHV as an illustrative example of how this assay can be applied to identification of DNA synthesis inhibitors of any herpesvirus.

2. Materials

1. TNT® T7 Quick Coupled Transcription/Translation System (Promega).
2. pTM1-Pol-8 and pTM1-PF-8 expression plasmids *(3)*.
3. 20-mer Oligonucleotide primer 5′ GCCAATGAATGACCGCTGAC-3′.
4. 5′ Biotinylated 100-mer oligonucleotide template: 5′ biotin-GCACTTATTG CATTCGCTAG TCCACCTTGG ATCTCAGGCT ATTCGTAGCG AGCTACGCGT ACGTTAGCTT CGGTCATCCC GTCAGCGGTC ATTCATTGGC-3′.
5. Streptavidin-coated, transparent, nuclease-free plates (Roche).
6. DIG-dUTP (digoxigenin-11-2′-deoxy-uridine-5′-triphosphate), alkali-stable (Roche).
7. DIG Detection ELISA (ABTS, Roche) that contains antidigoxigenin/peroxidase, Fab fragments conjugated with peroxidase (POD). POD reacts with the substrate ABTS (2,2′-azino-*bis* [3-ethylbenzthiazoline-6-sulfonic acid]) (*see* **Note 1**).

2.1. Solutions

1. Phosphate-buffered saline (PBS), 10X stock solution. Per liter: 80 g NaCl, 2 g KCl, 11.5 g $Na_2HPO_4 \cdot 7H_2O$, and 2 g KH_2PO_4.

Table 1
Preparation of Premix Solution for DNA Synthesis[a]

Component	Final conc.	Stock solution	Amount of stock solution (µL) per well
$(NH_4)_2SO_4$	100 mM	1 M	5.0
Tris-HCl, pH 7.5	20 mM	200 mM	5.0
$MgCl_2$	3 mM	100 mM	1.5
EDTA	0.1 mM	5 mM	1.0
DTT	0.5 mM	10 mM	2.5
Glycerol	4%	50%	4.0
BSA	40 µg/mL	10 mg/mL	0.2
dATP	50 µM	5 mM	0.5
dGTP	50 µM	5 mM	0.5
dCTP	50 µM	5 mM	0.5
DIG-dUTP	10 µM	1 mM	0.5
Pol-8 (in vitro)	—	—	2.0
PF-8 (in vitro)	—	—	2.0

BSA, bovine serum albumin; DIG, digoxigenin; DTT, dithiothreitol.
[a] *See* **Notes 2–5.**

2. PBS working solution, pH 7.3: 137 mM NaCl, 2.7 mM KCl, 4.3 mM Na_2HPO_4, and 1.4 mM KH_2PO_4.
3. Wash buffer: PBS and 0.1% Tween-20.
4. Blocking stock solution (10X): Dissolve blocking reagent (Roche) in maleic acid buffer by constant stirring on a heating block (65°C) or heat in a microwave, autoclave, and store at 4°C. The solution remains opaque.
5. PBS/1% blocking solution: Dilute the blocking stock solution (10X) 1:10 in PBS to 1%, and store at 4°C.
6. Maleic acid buffer: dissolve 10.6 g maleic acid (0.1 M) and 8.76 g NaCl (0.15 M) in 900 mL dH_2O, and adjust pH to 7.5 with 10 N NaOH. Add ddH_2O to 1 L.
7. Preparation of premix solution for DNA synthesis (*see* **Table 1**).

3. Methods

3.1. The Rapid Plate Assay Overview

1. Basically, a template with biotin attached to its 5′-end and a primer annealed to its 3′-end is bound to the streptavidin-coated wells of 96-well plates.
2. The premix solution, containing polymerase (Pol-8) and processivity factor (PF-8) and dNTPs with DIG-dUTP (substituted for dTTP), is then added to the wells of the plates.
3. This is followed by the addition of the chemical test compounds.
4. The plates are incubated to enable the DNA synthesis reaction to proceed.

5. The reaction is stopped, and the incorporation of dNTPs into synthesized DNA is detected by an ELISA reaction that employs anti-DIG antibody conjugated to peroxidase.
6. Criteria for high-throughput screening of thousands of chemical compounds are presented.
7. In addition, an example of confirmation testing to validate inhibitors of processive DNA synthesis is described.

3.2. Annealing Primer to Template

1. Mix 250 pmol (1.525 μg) of primer and 250 pmol (8.25 μg) of biotinylated template in 0.25 mL PBS (pH 7.3), and anneal by heating to 90°C for 5 minutes followed by cooling to room temperature.
2. Dilute the annealed primer/template (P/T) to 0.1 p*M* with cold PBS, and store at –20°C.

3.3. Binding Primed Template to 96-Well Plates

1. Add 0.2 pmol of P/T (1.22 ng of primer and 6.6 ng of template) in 100 μL to each well of the streptavidin-coated plates.
2. Cover all the wells with foil (which comes with the plates) and incubate for 90 min at 37°C or overnight at 4°C.

3.4. DNA Synthesis Reaction

1. Remove the foil, and completely remove the P/T binding solution from the 96-well plates (*see* **Note 6**).
2. Directly add 25.2 μL of the premix solution to each well, and then add the chemical test compound or H_2O (24.8 μL) to a final reaction volume of 50 μL (*see* **Note 7**).
3. Incubate at 37°C for 90 min.
4. For a negative control, use a separate premix solution in which Pol-8 or PF-8 or both Pol-8 and PF-8 are absent.
4. Add 1 μL of 0.5 *M* EDTA to stop the reaction.

3.5. First Plate Wash

1. Remove the reaction mixture, and wash wells of plates six times with 200, 225, 250, 275, 300, and 325 μL of wash buffer.
2. In each washing step, the wash buffer should remain in the wells for 30 s.

3.6. Binding of Anti-DIG-POD

1. Add 100 μL anti-DIG-POD working solution (final concentration of 200 mU/mL in PBS/1% blocking reagent) and incubate at 37°C for 1 h.

3.7. Second Plate Wash

1. Remove the anti-DIG-POD working solution. Wash six times with 200, 225, 250, 275, 300, and 325 μL of wash buffer.
2. For each wash step, the wash buffer should remain in the wells for 30 s.

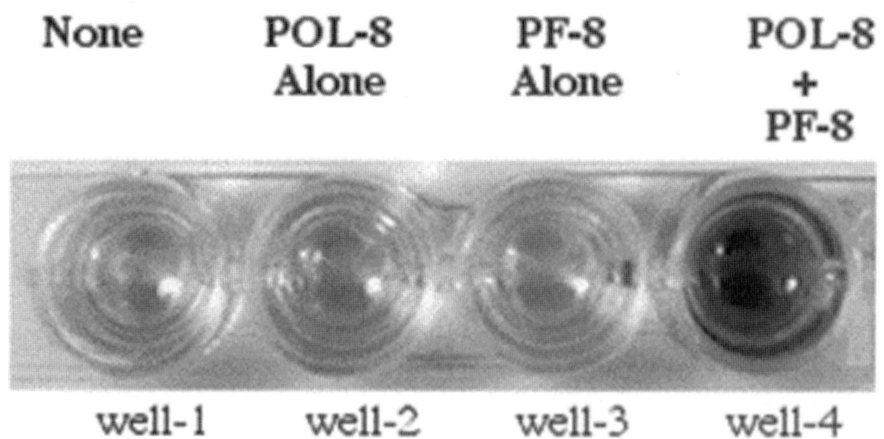

Fig. 1. The rapid mechanistic plate assay detects processive DNA synthesis directed by Pol-8 and PF-8. A premix solution, containing polymerase (Pol-8), processivity factor (PF-8), and dNTPs with DIG-dUTP (substituted for dTTP), was added to the streptavidin-coated plates in which a 100-nt biotinylated template annealed to a 20-nt primer was bound. After the reaction was stopped, the incorporation of DIG-dUTP into DNA was quantified by ELISA using peroxidase-conjugated anti-DIG antibody. A color reaction was generated only when both Pol-8 and PF-8 were present (well 4) but not with the buffer control (well 1), or Pol-8 or PF-8 alone (wells 2 or 3).

3.8. POD Color Detection as a Measure of DNA Synthesis

1. Add the ABTS substrate for POD to the wells.
2. The plates can be either agitated on a rotating plate shaker or stood on the lab bench at room temperature.
3. Color development generally occurs within 5–30 min to produce an OD 405 nm of 0.4–1.0.
4. **Figure 1** indicates no DNA synthesis with negative controls (buffer; Pol-8 alone, PF-8 alone) compared with significant DNA synthesis observed with the positive control (Pol-8 plus PF-8)

3.9. Quantification of DNA Synthesis by ELISA

1. Absorbance that is proportional to the incorporated DIG-dUTP in DNA is determined for each well of the plate using an ELISA reader at 405 nm.

3.10. General Strategies for Design and Development of High-Throughput Screening

The rapid mechanistic assay described here is applicable for screening hundreds to several thousands of compounds both manually or with instrumentation. However, the screening of chemical combinatorial libraries, often consisting of 1–2 million compounds and/or tens of thousands of natural extracts, entails the adaptation of methodologies to a high-throughput mode, often referred to as high-throughput screening (HTS). To give some idea of the magnitude of HTS, a recent general survey of industry leaders provided a minimum estimate of 80,000 compounds screened weekly to qualify as HTS (**11**).

HTS assays must possess robust, sample-handling capacity to match vast collections of chemical libraries as well as adequate sensitivity and reliable accuracy to discriminate active compounds (referred to as hits) with reasonable certainty. Although the rate of sample throughput can be greatly increased by automated liquid handling and signal detection systems in most HTS assays, maintaining the assay quality with the greatest sensitivity and smallest possible variability remains the most challenging aspect of the design and development of HTS assays, as methodology- and instrumentation-driven variations in measurement parameters are inherent to biological or biochemical reactions of the adopted assays. To evaluate the quality of established HTS assays, various statistical methods are utilized. Zhang et al. *(12)* recently introduced a new statistical parameter concept, a screening window coefficient termed the Z-factor. The Z-factor defines the relative separation of two groups of signal datasets, actives and inactives, within a given dynamic range. The Z-factor is formulated such that in theoretically perfect assays with zero variation and infinite dynamic range, the value reaches a maximum of one. When Z-factors range from 0.5 to 1.0, the separation is considered significant enough to discriminate true hits with a high degree of confidence. Using the-Z factor, one can uniformly assess and compare the quality of various assays, in which observed signal measurements are interpreted either as active or inactives hits with certain degrees of probability. This innovative concept has helped tremendously to standardize the optimization process for many HTS assays.

3.11. Modification of the Rapid Mechanistic Plate Assay for HTS

1. The original rapid mechanistic plate assay for KSHV Pol-8/PF-8 DNA synthesis assay *(9)* described above (**Subheadings 3.1.–3.9.**) can be modified for HTS. The assay is modified so that known herpesvirus polymerase inhibitors cause at least a 50% reduction in DNA synthesis at micromolar concentrations. Lowering the compound test dose to the micromolar range is one of our critical goals for optimization, to avoid selecting promiscuous inhibitors of KSHV Pol-8/PF-8, as it has previously been cautioned that higher doses of test compounds commonly result in many nonspecific screening hits from molecular target-based in vitro HTS assays *(13)*.

2. Cidofovir diphosphate, a phosphorylated form of cidofovir, has been selected as a reference inhibitory compound, since cidofovir has been shown to suppress KSHV DNA production in cell-based assays *(14–17)* (cidofovir was a kind gift from Dr. Michael Hitchcock of Gilead Sciences, Foster City, CA; its diphosphate form was synthesized by TriLink BioTechnologies, San Diego, CA).

3. The KSHV Pol-8/PF-8 HTS assay thus employs a reaction buffer containing 50 mM (NH4)$_2$SO$_4$, 20 mM Tris-HCl, pH 7.5, 3 mM MgCl$_2$, 0.1 mM EDTA, 0.5 mM DTT, 2% glycerol, 40 µg/mL BSA, in addition to 0.625 µM dNTPs, 0.125 µM DIG-dUTP, and KSHV Pol-8 and PF-8. Z-factors of the modified KSHV Pol-8/Pf-8 assay generally ranged from 0.5 to 0.8.

3.12. Application of the Modified Rapid Mechanistic Plate for Pilot HTS

1. We first examined the suitability of the modified KSHV Pol-8/PF-8 DNA synthesis assay for HTS by screening of the NCI Training Set, a collection of 230 well-characterized compounds *(18)*. Reproducibility of the optimized assay was validated by replicate testing of the Training Set compounds with r^2 greater than 0.9 at 20 µM drug concentration *(19)*.
2. Subsequently, we have begun testing the NCI Diversity Set, which comprises approx 2000 synthetic compounds derived from nearly 140,000 compounds available in sufficient quantity from the NCI DTP Repository *(18)*. Testing of the Diversity Set will be useful not only for HTS assay development and optimization, but also for identification of potential lead structures. We have already identified several hit compounds.
3. The inhibitory activity of the hit compounds is verified by secondary experiments that are designed to formulate mean inhibitory concentration (IC$_{50}$) values if applicable. Compounds deemed truly active are subjected to validation testing as described below in **Subheading 3.13.**
4. However, before we can embark on large-scale HTS campaigns for KSHV Pol-8/PF-8 inhibitors, we must also establish a reliable source of homogeneous Pol-8 and PF-8 production. To this end, we have constructed recombinant baculovirus vectors and successfully expressed and purified functionally active recombinant KSHV Pol-8 and PF-8 from the viral vector-infected Sf9 insect cells *(20)*.
5. Functional integrities of the purified rPol-8 and rPF-8 were verified by in vitro DNA synthesis activity, which was effectively blocked by cidofovir diphosphate in a dose–response manner *(20)*. We have confirmed that the specificity of the rPol-8/rPF-8–based DNA synthesis assay is virtually comparable to that of the in vitro translated protein-based assay.

3.13. Validation Testing

It is essential to establish that the hits from the rapid mechanistic plate assay block DNA synthesis directed by Pol-8 and PF-8 specifically, e.g., by inhibiting the catalytic activity of Pol-8 or the interactions among Pol-8/PF-8, Pol-8/DNA, or PF-8/DNA. However, many compounds may inhibit DNA synthesis nonspecifically, e.g., by intercalating DNA. To confirm the specificity of the inhibition, and further, to distinguish whether the inhibition affects the catalytic Pol-8 vs the processive Pol-8/PF-8 interaction, requires greater detailed analyses. One informative test is the in vitro M13 DNA synthesis gel assay *(21)*, which reveals the sizes of the DNA products resulting from processive DNA synthesis on agarose gels. The details of the M13 gel assay performed with herpesvirus polymerases and their cognate processivity factors are presented here, followed by an illustration of the assay using candidate inhibitors from the rapid mechanistic plate assay.

1. Preparation of the M13 primer-template.
   a. Anneal the M13 Universal sequencing primer to M13mp18 (+) single-stranded DNA (Pharmacia) in the presence of 100 mM NaCl, 1 mM EDTA, and 50 mM Tris-HCl, pH 7.6.
   b. Remove excess primer by filtration through a Centricon-100 spin filter (Amicon).
2. Performing the M13 DNA synthesis assay.
   a. Using a 1.5-mL centrifuge tube for each reaction, add 2 µL of each protein (Pol-8 and/or PF-8), X µL of compound and $(21 - X)$ µL of H_2O to 25 µL of 2X buffer to make a 50 µL final reaction volume that contains 100 mM $(NH_4)2SO_4$, 20 mM Tris-HCl, pH 7.5, 3 mM $MgCl_2$, 0.1 mM EDTA, 0.5 mM dithiothreitol, 4% glycerol, 40 µg/mL bovine serum albumin, 60 µM each of dATP, dGTP, and dTTP, 10 µM [α-^{32}P]dCTP (3000 Ci/mmol; NEN), and 50 fmol of M13 primed template.
   b. Incubate the reaction at 37°C for 60 min, and then stop it by adding 50 µL stop solution containing 1% SDS, 10 mM EDTA, 10 mM Tris-HCl, pH 8, and 200 µg/mL proteinase K followed by incubation at 37°C for 60 min.
3. Analyzing labeled M13-derived DNA products by gel electrophoresis.
   a. Extract the DNA products by phenol/chloroform followed by ethanol precipitation in the presence of 1 M ammonium acetate.
   b. Resuspend the precipitated DNA in 20 µL gel loading buffer (50 mM NaOH, 2.5 mM EDTA, 25% glycerol, and 0.025% bromocresol green) and then fractionate on a 1.3% alkaline agarose gel.
   c. Dry the gel and subject it to autoradiography or analysis by a PhosphorImager (Molecular Dynamics).

3.13.1. Example of Validation Testing

1. Twenty-eight putative inhibitors (hit compounds) were obtained by performing a rapid mechanistic plate assay screen of 2000 compounds from the NCI repository.
2. The specificity of these putative inhibitors, which were chosen on the basis of DIG-dUTP incorporation, were validated by the M13 gel assay, which allows newly synthesized DNA strands to be visualized as labeled products on a alkaline agarose gel. A subset of these compounds tested in the M13 assay is represented in **Fig. 2.**
3. Several of these compounds (represented by compound 130813) completely inhibited DNA synthesis directed by Pol-8/PF-8 in the M13 gel assay, whereas other compounds (represented by compound 86372) were partially effective (*see* upper panel in **Fig. 2**). Notably, one compound (i.e., compound 147744) that proved to be inhibitory in the rapid mechanistic plate assay failed to inhibit Pol-8/PF-8 in the M13 gel assay (*see* upper panel in **Fig. 2**).

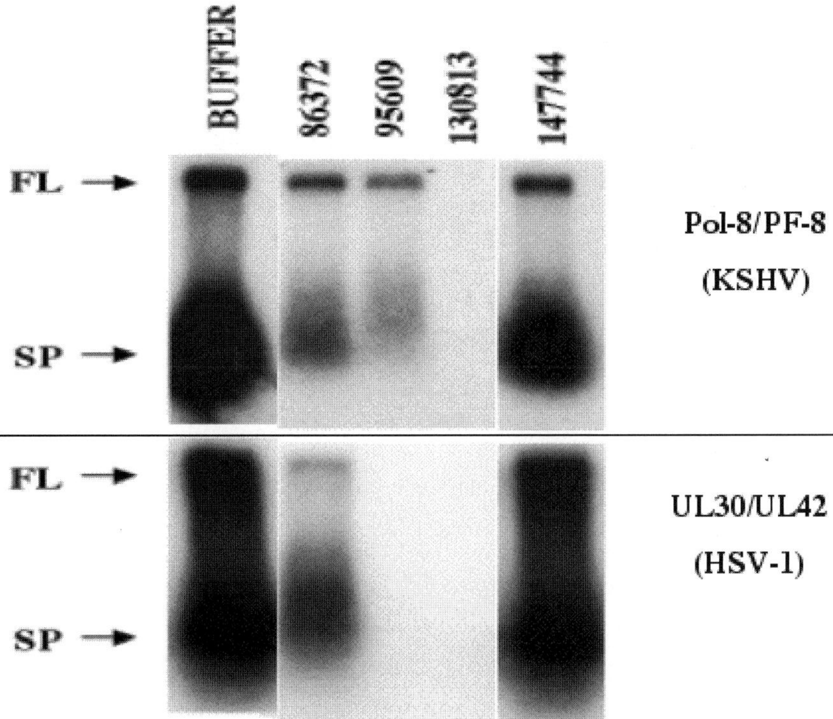

Fig. 2. Use of the M13 gel assay to validate putative DNA synthesis inhibitors identified by the rapid mechanistic plate assay. M13 single-stranded DNA with an annealed oligonucleotide primer was incubated with the premix solution containing in vitro translated Pol-8 and PF-8 and [^{32}P]dCTP for radiolabeling. The synthesized DNA products were fractionated on a 1.3% alkaline agarose gel and then analyzed by autoradiography. The control reaction (buffer, no compound) generated DNA strands corresponding to the full-length (FL) template (7249 nt) as well as characteristic shorter products (SP) that were several hundred nts in length. Numbers above each lane designate a particular test compound from the NCI repository. The effect of each compound on DNA synthesis by the Kaposi's sarcoma-associated virus (KSHV) proteins Pol-8 and PF-8 (upper panel) is directly compared with DNA synthesis by the herpes simplex virus (HSV)-1 proteins UL30 and UL42 (lower panel).

4. To establish their validity and specificity further, this same panel of compounds was tested in the M13 gel assay for inhibition of DNA synthesis directed by the HSV-1 DNA polymerase (UL30) and processivity factor (UL42). The effect of most of the compounds on DNA synthesis by HSV-1 is similar to that observed with KSHV, with the major exception being compound 95609, which caused complete inhibition HSV-1 but only partially inhibited KSHV (*see* **Fig. 2**).

3.13.2. Interpretation of Validation Testing

In summary, the M13 gel assay serves to rigorously substantiate the compounds classified as inhibitory by the rapid mechanistic plate assay. Because the advantage of the rapid mechanistic plate assay is to examine thousands of compounds conveniently and relatively inexpensively, it is not surprising to find occasionally that a putative inhibitory compound actually proves not to be inhibitory using the more "visual" M13 gel assay (e.g., compound 147744; *see* **Fig. 2**) Importantly, unlike the mechanistic plate assay that employs a template of 100 nt, the M13 gel assay demands robust processivity by employing a full-length (FL) template of 7249 nt. It is noted that a large proportion of short products (SP) of several hundred nts are typically generated in this assay (*see* **Fig. 2**). The major question is whether inhibitors observed by both the rapid mechanistic plate assay and M13 gel assays are specific to the proteins in question. This demands testing the same compounds on other DNA polymerase/processivity complexes, e.g., HSV-1. Inhibitors that completely block more than one polymerase/processivity complexes may be acting nonspecifically, e.g., by intercalating DNA or by inhibiting the conserved catalytic site present in many polymerases. Of interest are inhibitors that completely inhibit one polymerase/processivity complex, but not another. For example, the complete inhibition of DNA synthesis directed by the HSV-1 proteins with compound 95609, but partial inhibition of the KSHV proteins, suggests that this inhibitor has greater specificity for HSV-1 than KSHV. Further requisite testing for specificity involves determining whether this inhibitor blocks DNA synthesis directed by other herpesvirus and other cellular DNA polymerases and processivity factors. Since the molecular structures of the NCI inhibitors have been characterized, new chemical compounds may be designed to obtain even greater specificity.

4. Notes

1. The ABTS substrate kit contains all the reagents necessary to prepare a working solution of 2,2′-azino-*bis* (3-ethylbenzthiazoline-6-sulfonic acid) (ABTS). ABTS will produce a water-soluble, green-colored product upon reaction with horseradish peroxidase. The substrate is light-sensitive and must be kept in the dark both as a stock solution and as a working solution.
2. The total volume of a DNA synthesis reaction per well is 50 µL. This includes 25.2 µL of premix solution and 24.8 µL of the test compound with water to 50 µL final volume.
3. For Anti-DIG-POD stock and working solutions and for ABTS POD substrates, refer to the instruction manual of the DIG Detection ELISA (ABTS) kit (Roche).
4. Use separate premix solutions as negative controls in which Pol-8, PF-8, or both are absent and replaced with water.

5. The PF-8 and Pol-8 proteins can be either in vitro translated or purified from *E. coli,* as in the case of PF-8 *(7)* or from baculovirus, as in the case of Pol-8 *(20)*.
6. To ensure a clean background, it is important to remove completely all the P/T binding solution.
7. The reaction may need to be mixed to distribute the test compound.

Acknowledgments

This work was supported by the National Cancer Institute, National Institutes of Health, under research grant CA80602 to R.P.R. and federal funds under contract number NO1-CO-12400. The contents of this publication do not necessarily reflect the views or policies of the Department of Health and Human Services, nor does mention of trade names, commercial products, or organizations imply endorsement by the U.S. Government. The authors wish to thank M. Ciustea and Y. Chen for careful reading of the manuscript.

References

1. Vivona, J. B. and Kelman, Z. (2003) The diverse spectrum of sliding clamp interacting proteins. [Review] *FEBS Lett.* **546,** 167–172.
2. Doublie, S., Tabor, S., Long, A. M., Richardson, C. C., and Ellenberger, T. (1998) Crystal structure of a bacteriophage T7 DNA replication complex at 2.2 A resolution. *Nature* **391,** 251–258.
3. Lin, K., Dai, C. Y., and Ricciardi, R. P. (1998) Cloning and functional analysis of Kaposi's sarcoma-associated herpesvirus DNA polymerase and its processivity factor. *J. Virol.* **72,** 6228–6232.
4. Takeda, K., Haque, M., Nagoshi, E., et al. (2000) Characterization of human herpesvirus 7 U27 gene product and identification of its nuclear localization signal. *Virology* **272,** 394–401.
5. Zuccola, H. J., Filman, D. J., Coen, D. M., and Hogle, J. M. (2000) The crystal structure of an unusual processivity factor, herpes simplex virus UL42, bound to the C terminus of its cognate polymerase. *Mol. Cell.* **5,** 267–278.
6. Crute, J. J. and Lehman, I. R. (1989) Herpes simplex-1 DNA polymerase. Identification of an intrinsic 5′-3′ exonuclease with ribonuclease H activity. *J. Biol. Chem.* **264,** 19266–19270.
7. Chen, X., Lin, K., and Ricciardi, R. P. (2004) Human Kaposi's sarcoma herpesvirus processivity factor-8 functions as a dimer in DNA synthesis. *J. Biol. Chem.* (in press).
8. Appleton, B. A., Loregian, A., Filman, D. J., Coen, D. M., and Hogle, J. M. (2004) The cytomegalovirus DNA polymerase subunit UL44 forms a C clamp-shaped dimer. *Mol. Cell.* **15,** 233–244.
9. Lin, K. and Ricciardi, R. P. (2000) A rapid plate assay for the screening of inhibitors against herpesvirus DNA polymerases and processivity factors. *J. Virol. Methods* **88,** 219–225.
10. Ricciardi, R. P. and Lin, K. (2001) Methods for identifying antiviral agents against human herpesviruses. United States Patent No. 6,204,028.

11. Razvi, E. (2003) Emerging themes and quantitative metrics in high-throughput screening space. *Drug Market Dev.* **14.**
12. Zhang, J. H., Chung, T. D., and Oldenburg, K. R. (1999) A simple statistical parameter for use in evaluation and validation of high throughput screening assays. *J. Biomol. Screen.* **4,** 67–73.
13. McGovern, S. L., Caselli, E., Grigorieff, N., and Shoichet, B. K. (2002) A common mechanism underlying promiscuous inhibitors from virtual and high-throughput screening. *J. Med. Chem.* **45,** 1712–1722.
14. Medveczky, M. M., Horvath, E., Lund, T., and Medveczky, P. G. (1997) In vitro antiviral drug sensitivity of the Kaposi's sarcoma-associated herpesvirus. *AIDS* **11,** 1327–1332.
15. Neyts, J. and De Clercq, E. (1997) Antiviral drug susceptibility of human herpesvirus 8. *Antimicrob. Agents Chemother.* **41,** 2754–2756.
16. Kedes, D. H. and Ganem, D. (1997) Sensitivity of Kaposi's sarcoma-associated herpesvirus replication to antiviral drugs. Implications for potential therapy. *J. Clin. Invest.* **99,** 2082–2086.
17. De Clercq, E., Naesens, L., De Bolle, L., Schols, D., Zhang, Y., and Neyts, J. (2001) Antiviral agents active against human herpesviruses HHV-6, HHV-7 and HHV-8. [Review] *Rev. Med. Virol.* **11,** 381–395.
18. Shoemaker, R. H., Scudiero, D. A., Melillo, G., et al. (2002) Application of high-throughput, molecular-targeted screening to anticancer drug discovery. [Review] *Curr. Top. Med. Chem.* **2,** 229–246.
19. Dorjsuren, D., Brennan, A., Chen, X., et al. (2003) Development of a high-throughput screening assay for identification of novel inhibitors of KSHV polymerase and processivity factor, in 8th International Conference on Malignancies in AIDS and Other Immunodeficiencies: Basic, Epidemiologic and Clinical Research, Bethesda, MD.
20. Dorjsuren, D., Badralmaa, Y., Mikovits, J, Li, A., Fisher, R., Ricciardi, R., Shoemaker, R., and Sei, S. (2003) Expression and purification of recombinant Kaposi's sarcoma- associated herpesvirus DNA polymerase using a Baculovirus vector system. *Protein Expr Purif.* **29,** 42–50.
21. Digard, P., Chow, C. S., Pirrit, L., and Coen, D. M. (1993) Functional analysis of the herpes simplex virus UL42 protein. *J. Virol.* **67,** 1159–1168.

Index

W

X

Y